Molecular Sieve Zeolites-II

The Second International Conference co-sponsored by the Division of Colloid and Surface Chemistry, the Division of Petroleum Chemistry, and the Division of Physical Chemistry of the American Chemical Society and Worcester Polytechnic Institute at Worcester Polytechnic Institute, Worcester, Mass., Sept. 8–11, 1970.

Edith M. Flanigen

Leonard B. Sand

Conference Co-chairmen

ADVANCES IN CHEMISTRY SERIES **102**

AMERICAN CHEMICAL SOCIETY

WASHINGTON, D. C. 1971

Coden: ADCSHA

Library of Congress Catalog Card 77–156974

ISBN 8412–0115–3

PRINTED IN THE UNITED STATES OF AMERICA

Advances in Chemistry Series

Robert F. Gould, *Editor*

FOREWORD

ADVANCES IN CHEMISTRY SERIES was founded in 1949 by the American Chemical Society as an outlet for symposia and collections of data in special areas of topical interest that could not be accommodated in the Society's journals. It provides a medium for symposia that would otherwise be fragmented, their papers distributed among several journals or not published at all. Papers are refereed critically according to ACS editorial standards and receive the careful attention and processing characteristic of ACS publications. Papers published in ADVANCES IN CHEMISTRY SERIES are original contributions not published elsewhere in whole or major part and include reports of research as well as reviews since symposia may embrace both types of presentation.

CONTENTS

CATALYSIS

CONCLUSION

PREFACE

During his attendance at a Gordon Conference in the USA in 1962, Professor R. M. Barrer of Imperial College, London, expressed the opinion that periodic molecular sieve conferences would be worthwhile. On the premise that it was appropriate to hold the first conference in London to honor Professor Barrer's contributions, a group of British scientists was encouraged to initiate its organization. As a result, the first International Conference on Molecular Sieves was held successfully in London, April 4–6, 1967, under the chairmanship of Professor Barrer and under the sponsorship of the Society of Chemical Industry. The conference was attended by some 200 scientists, and 34 papers were presented. The proceedings, "Molecular Sieves," is published by the Society of Chemical Industry, 14 Belgrave Square, London, S.W. 1, 1968.

In 1968, Professor Barrer and one of us (LBS) initiated the continuation of the conferences on a triennial basis by inviting the second conference to be held at Worcester Polytechnic Institute, with the American Chemical Society invited to be the sponsoring society. It was believed appropriate to hold the second conference in the United States to recognize the pioneering commercialization of molecular sieve zeolites by Union Carbide Corporation and the petroleum refining industry. The organizing committee for the second conference included the co-chairman representing the American Chemical Society (EMF), Dr. W. L. Kranich, WPI, secretary-treasurer, Dr. P. K. Maher, Davison Chemical Division, W. R. Grace & Co., industrial liaison, and Prof. Barrer, honorary member. Dr. D. W. Breck participated helpfully in several discussions. The Petroleum Research Fund contributed to the expenses of invited distinguished speakers, and contributions for support of the conference were received from BP (North America) Ltd.–British Petroleum Co., Ltd., Esso Research & Engineering Co., W. R. Grace & Co.–Davison Division, Gulf Research & Development Co., Mobil Research & Development Corp., Nalco Chemical Co., Norton Co., Shell Development Co., Sun Oil Co., Texaco, Inc., Union Carbide Corp., Union Oil Co. of California, and Universal Oil Products Co. A number of special services were donated by the Union Carbide Corp.

Approximately 300 scientists from 18 countries participated in the conference. The meeting provided an opportunity for presentation of new scientific information and discussion and exchange of ideas both formally and informally, in an extremely pleasant and congenial environment. The success and enjoyment of the conference resulted partly from the careful planning and programming of the organizing committee, but especially from the untiring work of the local committees headed by: Dr. Y. H. Ma, Prof. W. B. Bridgman, Mr. W. Trask, Mr. G. Fuller, Mrs. A. H. Weiss, Mr. A. Begin, Mr. C. A. Keisling, Dr. I. Zwiebel, Dr. A. H. Weiss, Mrs. D. C. French, Mr. G. Gage, and Dr. R. E. Wagner.

On the final day of the conference, a permanent International Zeolite Conference Committee was elected, consisting of Miss Edith M. Flanigen, Chairman, Prof. R. M. Barrer, Dr. H. B. Habgood, Prof. A. V. Kiselev, Dr. P. K. Maher, Prof. W. M. Meier, Dr. C. Naccache, Prof. J. V. Smith, Prof. J. B. Uytterhoeven, and Dr. P. B. Venuto. It was voted to accept the invitation of Prof. Walter M. Meier to host the third conference in Zurich, Switzerland, in September 1973, under the chairmanship of Prof. Meier. Unfortunately, he could not attend the second conference because en route his plane was hijacked to the Jordanian desert where he was held hostage.

The Conference Committee would like to commend Robert F. Gould, Editor, Mrs. Colleen Stamm, Assistant Editor, and the staff of the ADVANCES IN CHEMISTRY SERIES for their efficient handling of an unexpectedly large number of manuscripts and their outstanding job in the publication of the Preprints and these two final volumes. The cooperation and contributions of the authors, discussants, session chairmen, panel chairmen and panelists, and other participants are also gratefully acknowledged.

The Proceedings are divided into two volumes following the session divisions at the Conference. Volume I contains the Introductory Lecture and 40 papers with discussion, presented under Synthesis, Structure and Mineralogy, and Modification and General Properties. Volume II includes 35 papers with discussion in Sorption and Catalysis, and the Concluding Remarks.

It appears from the contributions and participation in the first and second conferences that they serve a need in the rapidly growing field of molecular sieve science and technology, and they have been established accordingly on a triennial basis. May we take this opportunity to thank the many others who contributed to the success of the conferences and extend best wishes to those organizing the next. We can assure them that the pleasure of meeting and working with the many people involved in the conference far outweighs any burden of work in its organization.

EDITH M. FLANIGEN
LEONARD B. SAND

Tarrytown, New York
Worcester, Massachusetts
March 1971

42

Intracrystalline Diffusion

R. M. BARRER

Physical Chemistry Laboratories, Chemistry Department, Imperial College, London, S.W. 7, England

The rate of intracrystalline migration of molecules is significant for molecule sieving and for molecular sieve catalysis. In view of this, an account has been given of various aspects of diffusion in porous aluminosilicates. Three interrelated diffusion coefficients can be measured: differential, integral, and tracer coefficients. Ways of measuring these and their concentration dependence are considered, using single crystals and powders. Examples include various penetrants, such as water, small nonpolar molecules, and complex molecules. Diffusion anisotropy and effects of intracrystalline channel dimensions, exchange ions of different size and valence, temperature, concentration, and the presence of foreign molecules have also been discussed.

Heterogeneous catalysts are usually high-area porous materials which may be amorphous or crystalline. An important aspect of all such materials is the rapidity with which reactant molecules reach active sites and products leave these sites. Apart from flow in gas or liquid phase, there may be surface migration into and from micropores, whether in amorphous catalysts or in crystalline ones, such as the zeolites. It is still an open question how important such migration processes are as rate-controlling steps. However, it seems likely that active sites deep in a porous crystal will be less important than sites near the surface because many more unit diffusion steps will be needed to transport molecules to and from deeply buried sites. As corollaries, one would expect that only a limited volume fraction of a crystal of a zeolite such as sieve Y is catalytically effective, and that for best performance crystals in the catalyst support should be well exposed and as small as possible, in order to provide the largest surface-to-volume ratio.

Studies of intracrystalline diffusion can delineate factors limiting transport within crystals and hence will aid understanding of molecule sieving and site accessibility in catalysis. One may hope to learn from basic diffusion studies how diffusion coefficients within zeolites are related to:

1. Intracrystalline channel geometry and dimensions.
2. Shape, size, and polarity of penetrant molecules.
3. Cation dispositions, size, charge, and numbers.
4. Lattice defects, such as stacking faults.
5. Presence of "impurity" molecules in channels.
6. Structural changes brought about by penetrants.
7. Structural damage associated with physical and chemical treatments.
8. Concentration of penetrant within the crystals.

Despite the obvious importance of diffusion studies, information on these topics is still limited. The position reflects experimental difficulties and, in part, a failure to realize how useful a probe diffusion may be. It is hoped that the importance of diffusion will become apparent from the present review.

Diffusion Coefficients

In the two-component system gas + zeolite, in parallel with a general two-component system, one may consider 5 diffusion coefficients: D_{AB} (interdiffusion coefficient of sorbate and zeolite); D_A, D_B (intrinsic diffusion coefficients, respectively, of sorbate and zeolite); and D_A^*, D_B^* (tracer diffusion coefficients, respectively, of sorbate and zeolite). However, because D_B and D_B^* are effectively zero, as for a nonswelling crystal,

$$D_{AB} = D_A$$

It is sometimes assumed that

$$D_A = D_A^* \frac{d\ln a_A}{d\ln c_A} \tag{1}$$

where $a_A = p_A$ is the activity of sorbate, equal to the local equilibrium pressure of sorbate, where the concentration of A is c_A.

However, this relation is not correct. The true relation (*1*) is

$$D_A = D_A^* \frac{d\ln a_A}{d\ln c_A} \left(\frac{1}{1 - \dfrac{c_A L_{AA}^*}{c_A^* L_{AA}}} \right) \tag{2}$$

Here $c_A{}^*$ is the concentration of isotopically labelled species at a point where the concentration of unlabelled species is c_A. L_{AA} and $L_{AA}{}^*$ are the straight and cross phenomenological coefficients of the irreversible thermodynamic formulation of diffusion. The original relation, Equation 1, assumes a zero cross coefficient, which in dense intracrystalline fluids certainly is not likely to be true.

The only information for sorbate–zeolite systems from which Equation 2 can be discussed was obtained by Barrer and Fender (*7*). They studied intrinsic and tracer diffusion of water in chabazite, heulandite, and gmelinite, using H_2O and D_2O. The activity correction:

$$\frac{d\ln a_A}{d\ln c_A} = \frac{d\ln p_A}{d\ln c_A}$$

can be derived from their water sorption isotherms. The results of an analysis of their data are presented in Table I. The ratio $c_A L_{AA}{}^*/c_A{}^* L_{AA}$ can have considerable values, although the correction term $d\ln a_A/d\ln c_A$ is dominant. The results in the table refer to zeolite crystals nearly saturated with water, and should be extended, for various sorbates, over a greater range of sorbate concentrations.

The significance of cross coefficients, $L_{AA}{}^*$, has been demonstrated also for intrinsic and tracer diffusion of SO_2 in surface flow through carbon compacts (*1*). Cross coefficients must arise from direct interactions between A and A^*, and should be significant whenever sorbate is sufficiently concentrated for such encounters to become frequent. In the

Table I. Relation Between D_A and $D_A{}^*$ for Water in Several Zeolites

Zeolite	T,° *C*	$D_A{}^* \times 10^8$, *Cm² Sec⁻¹*	$D_A \times 10^6$, *Cm² Sec⁻¹*	$\frac{\partial \ln a_A}{\partial \ln c_A}$	$\frac{c_A L_{AA}{}^*}{c_A{}^* L_{AA}}$
Chabazite B	75	$46._2$	10.6_5	23.0	0.09_1
	65	$31._8$	7.6	24.0	0.15
	55	$21._4$	5.4_5	25.5	0.17
	45	$14._1$	3.8_0	27.0	0.21
	35	9.0	2.5_0	28.0	0.27
Heulandite	75	9.8	3.0_0	30.0	0.39
	65	6.1	1.9_5	32.0	0.44
	55	3.7_0	1.26	34.0	0.48
	45	2.1_9	0.78	35.5	0.51
	35	1.24	0.46_5	37.0	0.56
Gmelinite	55	7.3_3	1.9_5	26.5	0.15
	45	5.0	1.40	28.0	0.06_5
	35	3.3_1	0.97	29.5	0.030

dilute Henry's law sorption range, $d\ln a_A/d\ln c_A = 1$. Since encounters of A and A^* will then be much less frequent, the cross coefficient should be minimal and

$$D_{AB} = D_A \rightarrow D_A^*$$

The diffusion coefficients D_A and D_A^* referred to above are differential values for particular concentrations c_A, or else determined only over a narrow concentration range. Over a wide concentration range—*e.g.*, in many sorption rate experiments—a more accessible quantity is the integral value, $\widetilde{D}_A$, where

$$\widetilde{D}_A = \frac{1}{c_A}\int_0^{c_A} D(c_A)dc_A$$

or, for interval experiments,

$$\widetilde{D}_A = \frac{1}{(c_A - c'_A)}\int_{c'_A}^{c_A} D(c_A)dc_A$$

Measurement of Molecule Diffusion in Zeolite Crystals

Single Crystals. Naturally-occurring zeolites are sometimes found as large single crystals. Tiselius (*34, 35*) used this feature to study diffusion in zeolites. Diffusion of water in heulandite crystals was followed by an

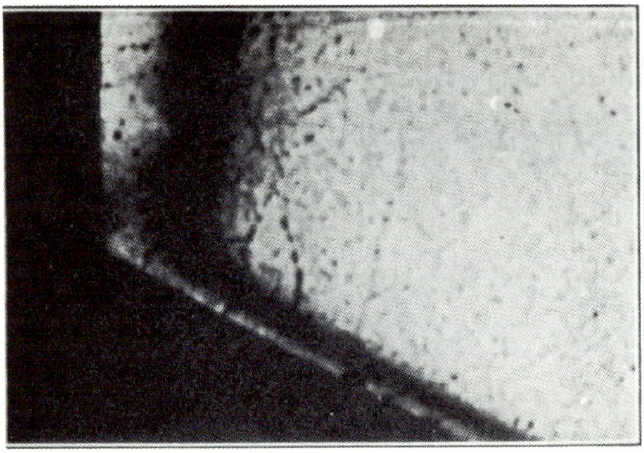

Zeitschrift fuer Physikalische Chemie

Figure 1. A stage of diffusion in a heulandite single crystal viewed through crossed Nicols (34). *The diferent distances moved by the band in 2 directions show the diffusion anisotropy. The direction of more rapid penetration is normal to 201, and that of less rapid pentration is normal to 001*

optical procedure in which birefringence of a crystal plate was observed under crossed Nicols. Birefringence was a function *inter alia* of the concentration of sorbate. Accordingly, under crossed Nicols dark bands progress into the crystal as diffusion proceeds (Figure 1). At time t, if the band has progressed through distance x, then

$$x^2 = 2\widetilde{D}_A t$$

The figure shows strong anisotropy in the diffusion of zeolitic water, the dark band having moved further from the 201 than from the 001 face. $\widetilde{D}_A$ is given in Table II; the activation energies, E, in the Arrhenius equation $D_A = D_0 \exp - E/RT$ are 5400 and 9140 cal/mole, respectively, for diffusion normal to 201 and 001.

Table II. Diffusion Anisotropy in Heulandite

T, °C	$\widetilde{D}_A \times 10^7 \perp 201$, $Cm^2\ Sec^{-1}$	$\widetilde{D}_A \times 10^7 \perp 001$, $Cm^2\ Sec^{-1}$
20	2.7	0.23
33.8	4.1	0.45
46.1	4.8	0.66
60.0	7.6	1.45
75.0	11.1	2.8

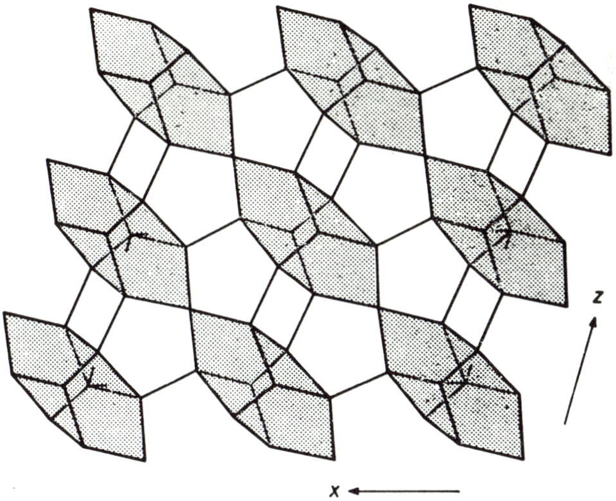

W. Meier, "Molecular Sieves," Society of the Chemical Industry

Figure 2. Sheets of linked tetrahedra present in heulandite (25). Diffusion across these sheets does not occur. Between the sheets, different intersecting channels run parallel to a *and* c, *respectively, forming two-dimensional networks*

Heulandite consists of sheets (Figure 2) linked to other like sheets and enclosing between each pair a two-dimensional channel system (25). Normal to the sheets, Tiselius found the diffusion coefficient to be negligible, as would be expected from the compact nature of the sheets. Between the sheets are channels circumscribed by 10-rings and 8-rings:

(a) Parallel to *a*, elliptical 8-rings of free diameter 2.4 and 6.1A.

(b) Parallel to *c*, 3.2 and 7.8A (10-rings) and 3.8 and 4.5A (8-rings). The anisotropy for different directions parallel to the sheets is in accord with the structure, which was unknown at the time of Tiselius' measurements.

Tiselius, again using suitably cut plates of heulandite, established actual concentration gradients from birefringence studies as a function of distance x from the surface. If the concentration at $t = 0$ is c_0 throughout the crystal, and at time t is c_x at distance x; while at $x = 0$, $c = c_\infty$ for all $t > 0$, then

$$D_{c=c_x} = \frac{1}{2t} \cdot \frac{dx}{dc} \int_{c=c_x}^{c=c_\infty} x\,dc$$

The concentration-distance curve at time t serves to give dx/dc and

$$\int_{c_x}^{c_\infty} x\,dc$$

Hence, $D_{c=c_x}$ is obtained. This method is associated with Boltzmann (*14*) and was developed by Matano (*24*). Some results are given in Table III. The saturation water content, c_∞, was 19.67% by weight of water; the values of c_0 are given in the table. D_A should be, and was, within error, independent of c_0. The variation of D_A with concentration

Table III. $D_A \perp$ 201 Plane in Heulandite at 20°C, Cm^2 Sec^{-1}

c, *Wt %*	$c_0 = 8.3\%$, $D_A \times 10^7$	$c_0 = 13.2\%$, $D_A \times 10^7$	$c_0 = 16.2\%$, $D_A \times 10^7$
10	0.04	—	—
11	0.2	—	—
12	0.7	—	—
13	1.3	—	—
14	2.0	2.1	—
15	2.7	2.6	—
16	3.0	3.5	—
17	4.0	4.2	4.0
18	4.0	4.1	4.2
19	3.3	3.5	4.1

may be associated with change in the binding energy of water as c increases and with small lattice changes associated with rising water content.

Powders. Synthetic zeolites and naturally occurring zeolites of sedimentary origin normally occur as powders. For them different methods must be employed, which fall into 2 subdivisions:

(a) Sorption or desorption is measured, keeping the sorbate pressure constant.

(b) Sorption or desorption occurs in a constant-volume variable-pressure vessel.

CONSTANT-PRESSURE SORPTION OR DESORPTION. The boundary conditions for any crystallite are

$c = c_\infty$ just within the surface for all $t > 0$.
$c = c_0$ throughout the crystallite at $t = 0$.

while for flow

$$\frac{\partial c}{\partial t} = \text{div}\,(D_A \text{ grad } c).$$

When D_A is independent of c and the crystals are all of the same shape and size, e.g., spheres of radius r_0

$$\frac{Q_t - Q_0}{Q_\infty - Q_0} = 1 - \frac{6}{\pi^2}\sum_{n=1}^{\infty}\frac{1}{n^2}\exp - \frac{D_A\, n^2\, \pi^2 t}{r_0^2}$$

For larger times, all but one exponential term become negligible, and the equation reduces to

$$\ln\frac{Q_\infty - Q_t}{Q_\infty - Q_0} = \ln\frac{6}{\pi^2} - \frac{D_A\, \pi^2 t}{r_0^2}$$

so that a plot of the left side against t gives a straight line of slope $\frac{D_A \pi^2}{r_0^2}$. Once D_A is known, this can be substituted in the full equation, which then can be tested for all t.

For small t, simpler forms can be obtained:

(a) $$\frac{Q_t - Q_0}{Q_\infty - Q_0} \doteq 6\left(\frac{D_A t}{\pi r_0^2}\right)^{\frac{1}{2}} - \frac{3D_A t}{r_0^2}$$

or

(b) $$\frac{Q_t - Q_0}{Q_\infty - Q_0} \doteq 6\left(\frac{D_A t}{\pi r_0^2}\right)^{\frac{1}{2}}$$

Plots of the left side against $\sqrt{t}$ give D_A/r_0^2, and so D_A.

For rectangular parallelepiped crystallites of edges a, b, and c, the limiting expressions are

$$\frac{Q_\infty - Q_t}{Q_\infty - Q_0} = \frac{512}{\pi^2} \exp - \frac{D_A \pi^2 t}{4}\left(\frac{1}{a^2} + \frac{1}{b^2} + \frac{1}{c^2}\right)$$

$$\frac{Q_t - Q_0}{Q_\infty - Q_0} = 2\left(\frac{D_A t}{\pi}\right)^{\frac{1}{2}} \left\{\frac{1}{a} + \frac{1}{b} + \frac{1}{c}\right\}$$

again giving the $\sqrt{t}$ law.

The % deviation from the $\sqrt{t}$ law is shown for particles of several shapes as functions of $\frac{Q_\infty - Q_t}{Q_\infty - Q_0} = \frac{M_t}{M_\infty}$ in Figure 3 (7).

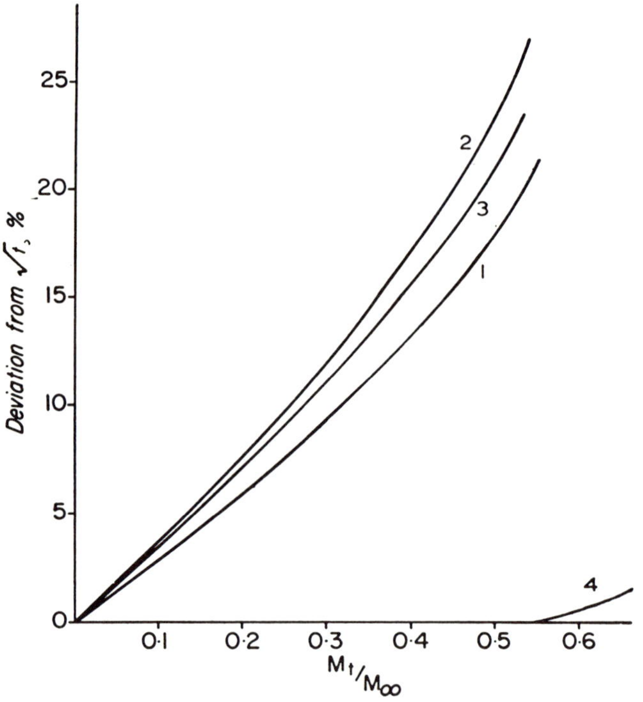

Figure 3. The % deviations from the $\sqrt{t}$ *law as functions of* $\frac{Q_\infty - Q_t}{Q_\infty - Q_0} = \frac{M_t}{M_\infty}$ *for particles of different forms* (7)

Curve 1: Sphere
Curve 2: Cube
Curve 3: Rectangular parallelepiped (2:2:1)
Curve 4: Plane sheet

The correspondence with the $\sqrt{t}$ law is illustrated in Figure 4 (7) for the intrinsic diffusion of water in chabazite, gmelinite, and heulandite.

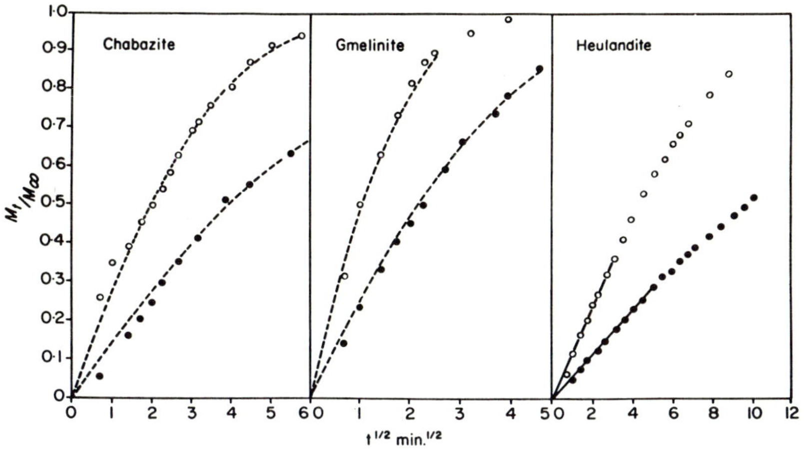

Figure 4. Applicability of the $\sqrt{t}$ law for the intrinsic diffusion of water in chabazite, gmelinite, and heulandite (7)

Calculated curves: – – –
Chabazite: ○, *75.4°C;* ●, *30.8°C*
Gmelinite: ○, *62.5°C;* ●, *31.7°C*
Heulandite: ○, *77.8°C;* ●, *37.4°C*

CONSTANT-VOLUME VARIABLE-PRESSURE SORPTION OR DESORPTION. At time $t = 0$, for sorption, a dose of gaseous sorbate is introduced into the sorption volume and left to distribute itself between gas phase and crystals. Exact analytical solutions are available, provided sorption follows Henry's law (2, 27). Thus, for spheres of radius r_0

$$\frac{Q_t - Q_0}{Q_\infty - Q_0} = 1 - 6K(K+1)\sum_{\alpha_n} \frac{\exp - \alpha_n^2\tau}{9(K+1) + \alpha_n^2 r_0^2 K^2}$$

where $\tau = \dfrac{D_A t}{r_0^2}$ and α_n is the nth positive root of the relation

$$\operatorname{Tan} \alpha r_0 = \frac{3r_0\alpha}{3 + Kr_0^2\alpha^2}$$

and where

$$K = \frac{V_g}{kV_s} = \frac{(Q_\infty)_g}{Q_\infty} = \frac{(Q_0)_g - (Q_\infty - Q_0)}{Q_\infty}$$

In this last expression, k is the Henry's law constant, and K denotes the equilibrium ratio of gas in the gas phase and in crystals. V_g and V_s are volumes of gas phase and crystals, respectively. $(Q_\infty)_g$ and $(Q_0)_g$ are amounts of gas finally and initially in the gas phase.

For small enough times, the $\sqrt{t}$ law again results in the form (6)

$$\frac{Q_t - Q_0}{Q_\infty - Q_0} = \frac{6}{r_0}\frac{(K+1)}{K}\sqrt{\frac{D_A t}{\pi}} = \frac{6}{r_0}\left(\frac{(Q_0)_g + Q_0}{(Q_0)_g - (Q_\infty - Q_0)}\right)\sqrt{\frac{D_A t}{\pi}}$$

The validity of the $\sqrt{t}$ law is shown under constant-volume variable pressure conditions when Henry's law is approximately valid in Figure 5.

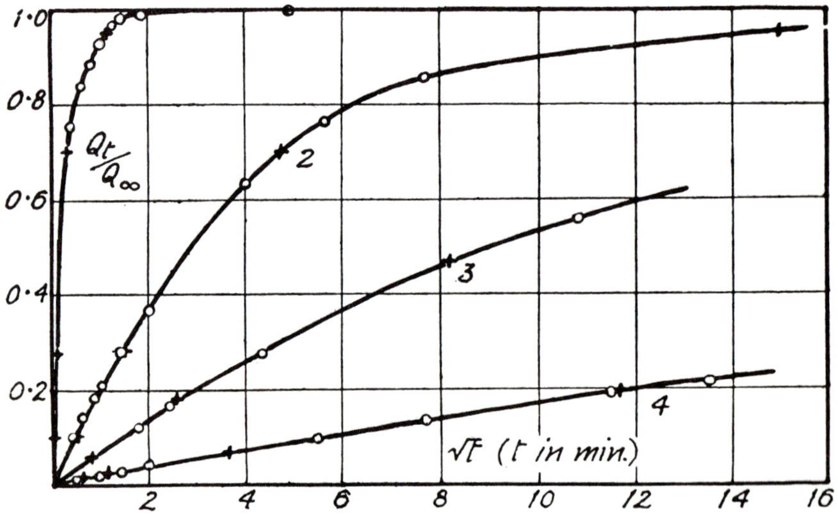

Transactions of the Faraday Society

Figure 5. Validity of the $\sqrt{t}$ law in a constant-volume variable-pressure system where the sorption isotherms approach Henry's law (2)

Curve 1: Ne in Li-mordenite at −185°C
Curve 2: Ne in Ca-mordenite at −185°C
Curve 3: Kr in Ba-mordenite at 24°C
Curve 4: Kr in Levynite at 0°C
○*: Experimental points*
×*: Calculated points, using the full solution of the diffusion equation*

The $\sqrt{t}$ Law and Particle Size and Shape Distributions in Powders. A special importance of the $\sqrt{t}$ laws is that, provided time is small enough for each particle to act as a semi-infinite medium, the laws hold whatever the size and shape distribution, with $1/r_0$ replaced by $A/3V$. Thus,

$$\frac{Q_t - Q_0}{Q_\infty - Q_0} = \frac{M_t}{M_\infty} = \frac{2A}{V}\sqrt{\frac{D_A t}{\pi}} \qquad \text{(constant pressure)}$$

$$\frac{Q_t - Q_0}{Q_\infty - Q_0} = \frac{M_t}{M_\infty} = \frac{2A}{V} \cdot \frac{(1 + K)}{K} \sqrt{\frac{D_A t}{\pi}} \quad \text{(constant volume, Henry's law)}$$

Provided total surface areas, *A*, are measured, D_A can be found. *A* can be determined in several ways:

1. By projected area (true area = 3 × projected area).
2. By flow [Kozeny-Carman (*16*) procedure].

These give geometrical or smoothed areas, not taking surface roughness into account.

3. By adsorption, using either
   (*a*) water-filled crystals, so that N_2 or Kr sorption is limited to external areas, or
   (*b*) spherical sorbate molecules too large to enter the crystals—*e.g.*, iso-C_4H_{10} or neo-C_5H_{12} for chabazite, sieve A, or any zeolite less open than these.

The adsorption area includes all fine-scale roughness and hence in general will exceed the smoothed areas found by the other methods. The two areas will lead thus to different values of D_A, and it is not clear which is best. For longer times, though possibly still in the $\sqrt{t}$ range, effects of surface roughness upon the diffusion will have died out. In the first moments they will not. It is useful to determine both areas, and so to give upper and lower limits to D_A. This has not always been done, and in comparing values of D_A found by different authors, the method of determining *A* should be taken into account.

As expected from the diffusion theory, the larger the particles, the slower the sorption; *A* becomes smaller for constant *V*. This is illustrated in Figure 6 (*8*) for sorption of propane at 200°C in chabazite.

Concentration-Dependent D_A in Powders. It has been assumed in the powder techniques discussed so far that D_A does not depend upon intracrystalline concentration. For the dilute range of sorptions—*e.g.*, where Henry's law is valid—this is correct. However, as already considered (Table III), over extended ranges in concentration Tiselius's results for water in a single crystal of heulandite showed strong concentration dependence. For powders, 2 methods may be mentioned.

INTERVAL METHOD. The quantity Q_0 of sorbate initially in the crystal can be varied systematically by steps from zero to near saturation. At each value of Q_0 a small extra amount, ∂Q, is then sorbed, and the value of D_A obtained can be regarded as constant over each interval $Q_0 + \partial Q$. Thus D_A is found as a function of Q_0. An interval method was used for several diffusing species in chabazite by Barrer and Brook (*4*), in which ∂Q was often considerable, compared with Q_0. Thus, an integral value of D_A was obtained over the interval Q_0 to Q_∞. Some results are given in Table IV*a,* in which the areas *A* were those determined by a flow method. $\widetilde{D}_A$ appears to decrease rather strongly with increasing average concentra-

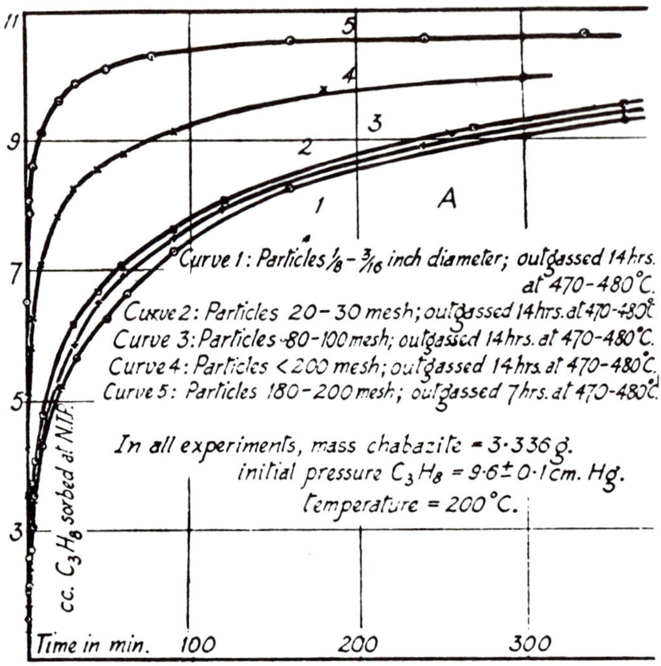

Transactions of the Faraday Society

Figure 6. Influence of particle size and outgassing time upon the sorption of propane into chabazite at 200°C (8)

tion. Similar results were found with CH_2Cl_2 and $(CH_3)_2NH$. However, smaller values of ∂Q would be preferable, and the experiments should be extended to various zeolites for a greater range of sorbates. Results obtained by Schirmer *et al.* (32) for n-$C_{14}H_{20}$ in Zeosorb 5A give D_A increasing with concentration at 205°C and decreasing at 307°C.

METHOD OF BARRER AND BROOK (4). Even when D_A depends upon concentration, a $\sqrt{t}$ law may be observed experimentally:

$$Q_t = Q_0 + Q_a + Ak\sqrt{t}$$

where k is a rate coefficient, and Q_a is the amount (if any) sorbed upon external surfaces; k depends upon $D_A(c)$, Q_0, and Q_∞ but not on t. Since also, from Fick's law

$$Q_t - (Q_0 + Q_a) = -A\int_0^t \left(D_A \frac{\partial c}{\partial x}\right)_{x=0} dt = Ak\sqrt{t}$$

it follows without approximation that

$$k = -2\left(D_A \frac{\partial c}{\partial x}\right)_{x=0} t^{\frac{1}{2}}$$

and the validity of the $\sqrt{t}$ law means that the flux through the ingoing surface is proportional to $t^{-1/2}$.

Keeping Q_∞ constant, k is evaluated for various values of Q_0. The curve of k *vs.* Q_0 must cut the axis of Q_0 at the point $Q_0 = Q_\infty$, because then $\left(\frac{\partial c}{\partial x}\right)_{x=0}$ is zero and so is k. Thus, the experimental curve of k *vs.* Q_0 can be extrapolated to the point $Q_0 = Q_\infty$.

As the interval $Q_\infty - Q_0$ is decreased progressively in successive runs, the system approaches more and more nearly the $\sqrt{t}$ law for constant D_A. Thus, the solution more and more nearly approaches a standard solution for constant D_A, *i.e.*,

$$1 - erf\ 2 \frac{x}{\sqrt{D_A t}} = \frac{c - c_0}{c_\infty - c_0}$$

from which

$$-\left(D_A \cdot \frac{\partial c}{\partial x}\right)_{x=0} t^{\frac{1}{2}} = (c_\infty - c_0)\sqrt{\frac{D_A}{\pi}}$$

As c_0 approaches c_∞, the curve of k *vs.* c_0 approaches asymptotically a tangent of slope $-2(D_A/\pi)^{1/2}$, passing through $c = c_\infty$.

Table IVa. Integral Values of D_A in Chabazite over Intervals Q_0 to Q_∞, $Cm^2\ Sec^{-1}$

Sorbate	T, °C	Q_0, *Cm³ at STP/G*	Q_∞, *Cm³ at STP/G*	$\frac{1}{2}(Q_0 + Q_\infty)$ *Cm³ at STP/G*	$\widetilde{D}_A$
C_3H_8	150	0.00	11.4	5.7	17.6×10^{-13}
		1.60	13.0	7.3	8.3
		4.00	15.4	9.7	3.6
		8.60	20.0	14.3	2.0
		15.10	26.5	20.8	1.5
	200	0.00	7.5	3.75	26.1×10^{-13}
		2.80	10.0	6.4	9.4
		5.25	12.45	8.85	5.1
		10.20	17.40	13.7	3.4
n-C_4H_{10}	150	0.00	14.2	7.1	9.5×10^{-13}
		1.91	16.1	9.0	2.8
		5.28	19.48	12.39	1.59
		9.50	22.8	16.15	1.42
	200	0.00	13.0	6.5	23.6×10^{-13}
		1.60	14.6	8.1	8.51
		2.20	15.2	8.7	6.41
		8.00	19.2	13.6	2.89

Figure 7 shows the way in which k approaches the asymptote when D_A decreases with c (curve 1) and when D_A increases with c (curve 2). Negative values of k refer to desorption ($c_0 > c_\infty$). Some values of D_A obtained in chabazite using this method are given in Table IV*b*, using flow areas for *A*.

Other studies have been made of concentration dependence of D_A, for example in Linde Sieve A by Habgood (*22*) and in chabazite by Brandt and Rudloff (*15*), but the results show little regularity, and much more systematic work is required under constant outgassing conditions to establish what patterns of concentration dependence of D_A emerge, for both polar and nonpolar sorbates.

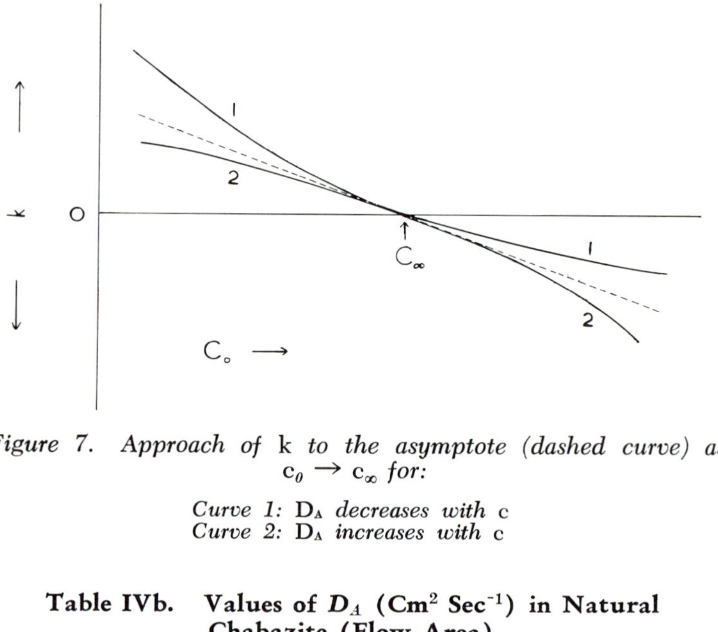

Figure 7. Approach of k *to the asymptote (dashed curve) as* $c_0 \rightarrow c_\infty$ *for:*

Curve 1: D_A *decreases with* c
Curve 2: D_A *increases with* c

Table IVb. Values of D_A ($Cm^2\ Sec^{-1}$) in Natural Chabazite (Flow Area)

Sorbate	T, °*C*	c_∞, *Cm^3 at STP/G*	D_A *at* c_∞
CH_2Cl_2	0	70	$2.1_7 \times 10^{-16}$
	50	47	4.9×10^{-15}
$(CH_3)_2NH$	0	87.0	$1.9_0 \times 10^{-19}$
	50	78.5	$1.3_5 \times 10^{-17}$
	100	70.0	$2.5_1 \times 10^{-15}$
n-C_4H_{10}	200	13.5	$2.1_0 \times 10^{-12}$
		17.1	$1.8_5 \times 10^{-13}$
		19.1	$1.7_0 \times 10^{-13}$

One limited investigation has been made of the concentration dependence of tracer diffusion coefficients, D_A^*, of water at values of θ quite close to saturation (*7*). In the range $0.91 < \theta < 0.98$, D_A^* tends to increase slightly with θ (Table V). Again, however, more information is needed for a greater concentration range and for other sorbates.

Diffusion of Zeolitic Water

Measurements have been made of tracer or intrinsic diffusion coefficients of water in a number of zeolites. Most refer to crystals nearly saturated with zeolitic water, while smoothed rather than adsorption areas were used in calculating the coefficients. From the temperature dependence of D_A or D_A^*, the activation energy, E, may be found using the Arrhenius equations $D_A = D_0 \exp - E/RT$ or $D_A^* = D_0^* \exp - E/RT$. Some results for tracer diffusion are summarized in Table VI. (*18*). These shed considerable light upon certain aspects of intracrystalline diffusion for small polar molecules:

(a) The van der Waals diameter of the water molecule is $\sim$ 2.8A. In many of the zeolites, therefore, there should be no steric restriction to the passage of this small molecule through the windows. Analcite and natrolite are exceptions; here the large energy barriers, E, for diffusion (which exceed those for tracer diffusion in ice) may be partially caused by the squeezing of the water through the narrow windows in the anionic frameworks.

(b) The energy barriers for migration where no restrictions arise due to the free dimensions of the windows are always larger than in liquid water, although less than in ice.

It thus seems clear that tracer diffusion in zeolites involves the "sticking" of water to certain preferential positions—*e.g.*, in clusters around cations—and to variations of this "stickiness," largely of electrostatic origin, along the diffusion channels. There is an approximate in-

Table V. Concentration Dependence of D_A^* for Water near Saturation

Zeolite	T, °*C*	θ	$D_A^* \times 10^7$, *Cm*^{2} *Sec*$^{-1}$
Chabazite B	75	0.97_6	4.8_5
	75	0.93_9	4.1_7
	75	0.91_3	4.3_9
Gmelinite	76.9	0.97_8	1.6_1
	76.9	0.94_4	1.5_4
	76.9	0.91_6	1.4_8
Heulandite	76.1	0.98_4	1.1_4
	76.1	0.95_2	0.97
	76.1	0.91_8	0.83

Table VI. Tracer Diffusion Coefficients,

Zeolite	*Free Dimensions of Windows, A*	D_A* *at* T, °*C*
Analcite (*18*)	2.2 – 2.4	$1.9_7 \times 10^{-13}$ (46°)
Natrolite (*21*)	2.6 – 3.9	—
Heulandite (*7*)	2.4 – 6.1 3.2 – 7.8 3.8 – 4.5	$2.0_7 \times 10^{-8}$ (45°)
Chabazite (*7*)	3.7 – 4.2	$1.2_6 \times 10^{-7}$ (45°)
Gmelinite (*7*)	6.9 3.4 – 4.1	5.8×10^{-8} (45°)
Sieve Na-X (*26*)	~7.4	$2.1_1 \times 10^{-5}$ (40°)
Sieve Ca-X (*26*)	~7.4	$2.4_1 \times 10^{-5}$ (40°)
Sieve Ca-Y (*26*)	~7.4	—
Ice (*23,29*)	—	1×10^{-10} (−2°)
Liquid water (*26,37*)	—	$3.8_7 \times 10^{-5}$ (45°)

verse correlation between E and the intracrystalline porosity which suggests that zeolitic water is most liquid-like in the most open structures, where the largest clusters of water molecules occur, and where these clusters join to give fluid filaments permeating the whole structure (zeolites X and Y). This is supported by x-ray evidence (*33*). In the least open zeolites, such as analcite or natrolite, water occupies particular lattice sites. As porosity increases and water clusters grow in size, the water positions become less definite. In zeolites X and Y, water positions cannot be located.

E for tracer diffusion of monovalent cations in chabazite (*3*) does not seem to be very different from E for water in the same zeolite:

Na^+	$E =$	6.5_5 kcal per gram ion
K^+		7.0_1
Rb^+		6.7_5
Cs^+		7.5_5
Ca^{2+}		13.8
Sr^{2+}		14.6
Ba^{2+}		8.8

Thus, the monovalent cations show "stickiness" to preferred sites comparable with that shown by water, which seems reasonable if cations and water migrate as far as possible in association. The divalent ions, however, especially Ca^{2+} and Sr^{2+}, show much enhanced "stickiness," so that the argument of ion–water association in diffusion must not be taken too far.

***D_A**, for Water in Zeolites, Cm^2 Sec^{-1} (*18*)**

D_0	E, *Kcal per Mole*	*Method*
$1.5_2 \times 10^{-1}$	17.0 ± 0.3	Tracer (HTO)
—	15.0	NMR
7.6×10^{-1}	11.0 ± 0.3	Tracer (D_2O)
1.2×10^{-1}	8.7 ± 0.3	Tracer (D_2O)
2.0×10^{-2}	8.1 ± 0.3	Tracer (D_2O)
—	6.9	NMR
—	6.8	NMR
—	5.6	NMR
—	13.5 ± 1.1	Tracer (H_2O^{18})
		Dielectric relaxation
5.6×10^{-2}	4.6	Tracer (H_2O^{18}) and NMR

Table VII. Characteristics of Porous Crystals

Crystal	*Density, G/Cm³*	*Aperture, A*	*Estimated Saturation Sorptions, Cm³ at STP/G, Anhydrous*
α-, β- Cristobalite	2.32	2.2 – 2.6	187 (He, Ne)
α-, β- Tridymite	2.28	2.2 – 2.6	187 (He, Ne)
Silica glass	~2.2	variable	—
Analcite	2.09	2.2 – 2.4	110 (H_2O)
Sodalite hydrate	2.01	2.3 – 2.6	210 (H_2O)
Erionite	~1.68	3.6 – 4.8	260 (H_2O)
Chabazite	~1.6	3.7 – 4.2	366 (H_2O)
Linde A	1.51	4.3	357 (H_2O)
Faujasite	1.43	~7.4	445 (H_2O)

Diffusion of Small Nonpolar Molecules

The "stickiness" shown by water is not found in small molecules without permanent electric moments. Their mobility within crystals is governed much more by the openness of the frameworks, modified by the siting of the cations, which may partially obscure the window apertures.

The intracrystalline porosity may be available in the form of large cavities, linked by large or small shared windows (zeolites X, A, or chabazite), or this porosity may be subdivided into many small cavities joined by small windows (*e.g.*, in the nonzeolites, cristobalite and tridymite, or in analcite). Table VII illustrates densities, window apertures, and saturation capacities for porous crystals, showing a range in all these properties. Saturation capacities for cristobalite and tridymite are estimated, assuming one guest molecule in each interstice. Those for water are based upon experimental water contents of the crystals.

The energy barriers for He and Ne diffusion in silica glass, tridymite, and cristobalite (*12*) increase rapidly with increasing density (Figure 8). There is also a clear correlation between energy barriers for diffusion and molecular dimensions of the diffusant. This is seen in Figure 9 for α-tridymite (1), silica glass (2), K-A (3), K-mordenite (4), levynite (5), and basic sodalite (6). The larger the dimensions of the diffusant, the smaller D_A becomes, as shown in Table VIII for K-A and K-mordenite. The change can be by many orders of magnitude. The length as well as the cross sectional diameter of the diffusant can play a part, as is seen by comparing D_A for N_2 and Ne in K-A.

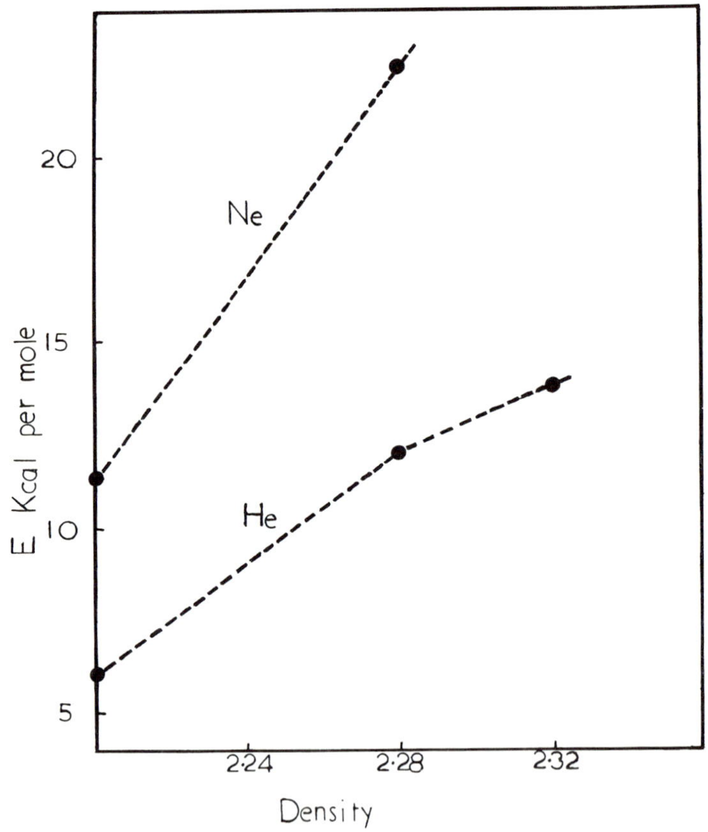

Figure 8. Variation of the energy of activation for diffusion of Ne and He with density of the diffusion media. In order of increasing density, the media are silica glass, tridymite, and cristobalite

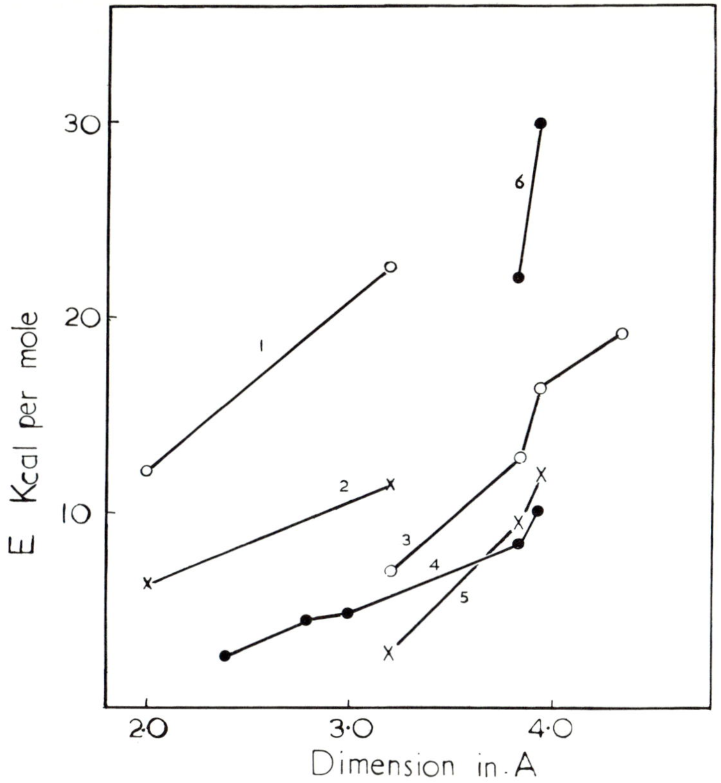

Figure 9. Dependence of energy of activation for diffusion in several media upon the critical van der Waals dimension of the diffusing molecules

1: α-Tridymite
2: Silica glass
3: K-A (Linde Sieve 3A)
4: K-mordenite
5: Levynite
6: Basic sodalite
Reading from left to right, the diffusing molecules are: He, H_2, O_2, N_2, Ne, Ar, Kr, and Xe

The free dimensions of the windows in zeolite A are 4.3A and those in mordenite are 6.7 × 7.0A. Any of the sorbates of Table VIII could migrate therefore into the crystals with minimal values of *E*, were it not for an influence of the K^+ ions (diameter 2.66A). These ions must obstruct the diffusion paths partially in each case. Indeed, ion exchange is one of the most effective ways of changing the molecular sieve characteristics of zeolites. Decationation, as in H-mordenite or H-clinoptilolite, produces maximum clearance of diffusion paths from metallic ions, so

Table VIII. Relationship of Dimensions to Diffusion Coefficients

Gas	*Van der Waals Dimensions, A*	D_A	D_0	*E, Kcal per Mole*
	D_A *in K – A at 20°C, Cm² Sec⁻¹(36)*			
H_2	2.4 × 3.1	2.0×10^{-13}	4.5×10^{-6}	9.9
N_2	3.0 × 4.1	9.8×10^{-18}	1.0×10^{-5}	16.2
Ne	3.2	2.9×10^{-13}	4.5×10^{-8}	7.0
Ar	3.8_3	1.6×10^{-17}	3.7×10^{-8}	12.6
Kr	3.9_4	4.3×10^{-19}	6.7×10^{-7}	16.4
Xe	4.3_4	6.7×10^{-21}	1.2×10^{-6}	19.2
	D_A *in K – mordenite at –78°C(4)*			
H_2	2.4 × 3.1	2.7×10^{-13}	—	2.5
O_2	2.8 × 3.9	2.0×10^{-15}	—	4.4
N_2	3.0 × 4.1	9.2×10^{-16}	—	4.8
Ar	3.8_3	2.4×10^{-16}	—	8.4
Kr	3.9_4	1.8×10^{-18}	—	10.0

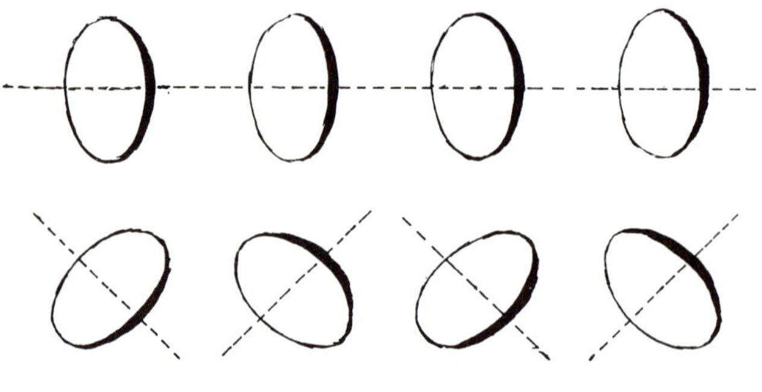

Figure 10. Windows of equal dimensions, all parallel or oriented at an angle

that the full dimensions of the intracrystalline channels are those which govern entry to the crystal and diffusion within it (*6, 10, 30*).

Diffusion of Complex Molecules

From the industrial viewpoint, particular interest attaches to the migration of large molecules in zeolites, especially of paraffins. In Ca-chabazite and Ca-A, diffusion is regulated by apertures through which *n*-paraffin chains can pass, with something of a squeeze, but not iso-paraffins, aromatics, or naphthenes. There are 2 factors which may reduce ease of passage into chabazite as compared with Ca-A. Firstly, the apertures regulating entry are slightly smaller (3.7 × 4.2A in chabazite, compared with 4.3A in zeolite A); secondly, the orientation of successive apertures with respect to each other and their distance apart is less favorable in chabazite than in A. Schematically, this can be illustrated by the aperture sequences in Figure 10. The flexibility of the diffusant chain could be of greater importance when the *n*-paraffin migrates in chabazite than in Ca-A. In chabazite, as shown in Figure 11, when the *n*-paraffin chains grow longer (C_3 to C_7) the friction between the diffusant and its environment increases rapidly, and sorption rates decline (*8*). A similar behavior is expected for any long-chain paraffin–zeolite system. If the chain is long enough, its "head" may be passing one window while its

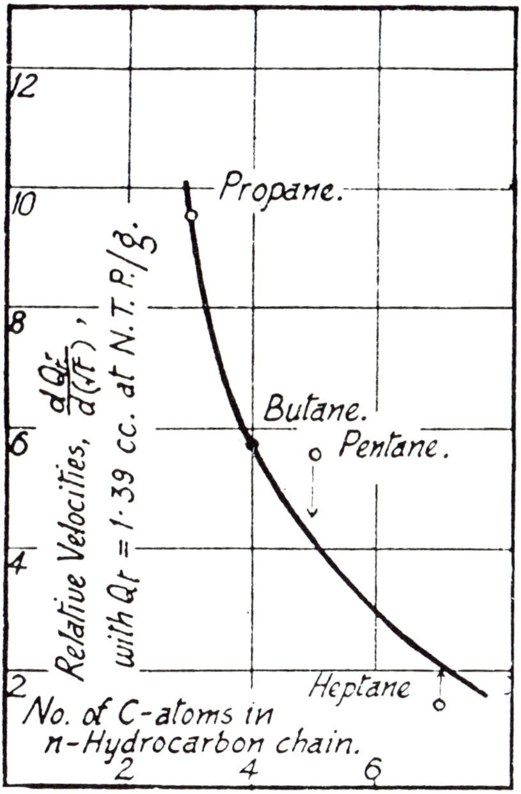

Transactions of the Faraday Society

Figure 11. Initial slopes of sorption curves (Q_t vs. $\sqrt{t}$) *for several* n-*paraffins of increasing chain lengths in chabazite at 212°C (8)*

"tail" is still passing through a preceding window. Activation energies for D_A have been reported in cal per mole as

3100 for C_3C_8 in chabazite (*4*)
7300 for n-C_4H_{10} in chabazite (*4*)
16,100 for n-$C_{14}H_{30}$ in zeosorb 5A (*32*)

The effect of chain length upon E can be inferred from these figures.

A study of C_1 to C_4 paraffins in synthetic mordenite by Satterfield and Frabetti, Jr. (*31*) has shown that diffusion coefficients for desorption are many times smaller than those for sorption, and the coefficients are reduced substantially by grinding the crystals. This investigation has raised problems which may have some general importance. For example, dry grinding may cause substantial local temperature rises associated with local stresses, and may cause at least some surface breakdown of the crystal structure.

Satterfield and Frabetti's work has been supplemented by that of Eberly (*19*), who studied diffusion of n-C_5 to n-C_9 paraffins in erionite and 5A molecular sieve, in the range 93° to 207°C. In erionite they found desorption rates to be smaller than sorption rates. This behavior is found in polymer-penetrant systems when differential diffusion coefficients increase with increasing concentration (*20*), although the differences in apparent integral diffusion coefficients for sorption and desorption seem large for this explanation, being about ten- and seventy-fold for n-heptane and n-octane, respectively. As with chabazite (*8*), the rates of uptake or desorption depended upon the chain length of the diffusant and were influenced by particle size. In Linde 5A, no significant decrease in $\widetilde{D}_A$ was observed with increasing molecular weight, but apparent desorption values were considerably less than those for adsorption.

Apart from concentration dependance, sorption is strongly exothermic and desorption endothermic, so that sorption heats and desorption cools the crystals locally. Since $D_A = D_0 \exp - E/RT$, D_A for sorption could exceed D_A for desorption.

Theory of Concentration-Dependance of D_A

Evidence has been presented of values of D_A or $\widetilde{D}_A$ which increase or decrease with concentration. It is of particular interest to consider the theoretical basis of this behavior in sorbed fluids (*9*). If the driving force per molecule is proportional to $-d\mu_A/dx$ for flow in the x-direction, then the flux, J, through unit area normal to x, is for species A

$$J_A = - B_A c_A \cdot \frac{d\mu_A}{dx}$$

Here μ_A is the chemical potential, and $B_A c_A = L_A$ is the phenomenological coefficient of irreversible thermodynamics. B_A is the intrinsic mobility of A. This expression can be rewritten as

$$J_A = -RT \cdot B_A \frac{d\ln a_A}{d\ln c_A} \cdot \frac{dc_A}{dx}$$

so that

$$D_A = RT \cdot B_A \frac{d\ln a_A}{d\ln c_A} = RT \cdot B_A \frac{d\ln a_A}{d\ln \theta}$$

Values of $d\ln a_A/d\ln\theta$ are given in Table IX for some isotherm equations. β is a coefficient which may be positive or negative; $\alpha = 2a/bkT$, where a and b are the interaction coefficient and covolume in van der Waals equation of state, and $A_1, A_1' \ldots; A_2, A_2' \ldots$ are virial coefficients in the equation of state

$$\frac{\pi}{c_A RT} = 1 + A_1 c_A + A_2 c_A^2 + \ldots = 1 + A_1' \theta + A_2' \theta^2 + \ldots$$

Table IX. Values of $\frac{d\ln a_A}{d\ln \theta}$ for Several Isotherms

Model	*Isotherm*	$\frac{d\ln a_A}{d\ln \theta}$
Localized ideal	$K = \frac{\theta}{p(1-\theta)}$	$\frac{1}{1-\theta}$
Localized with interaction	$K = \frac{\theta}{p(1-\theta)} \exp \beta\theta$	$\frac{1}{1-\theta} + \beta\theta$
Volmer equation of state	$K = \frac{\theta}{p(1-\theta)} \exp \frac{\theta}{1-\theta}$	$\frac{1}{(1-\theta)^2}$
van der Waals equation of state	$K = \frac{\theta}{p(1-\theta)} \exp \left(\frac{\theta}{1-\theta} - \alpha\theta\right)$	$\frac{1}{(1-\theta)^2} - \alpha\theta$
Virial equation of state	$K' = \frac{c_A}{p} \exp (2A_1 c_A + {}^3/_2 A_2 c_A^2 + \ldots)$	
	or	
	$K = \frac{\theta}{p} \exp (2A_1'\theta + {}^3/_2 A_2 \theta^2 + \ldots)$	$1 + 2A_1'\theta + 3A_2'\theta$

In this equation, $\pi/c_A RT$ is the osmotic pressure ratio for the sorbate in the sorbate–zeolite "solution."

Barrer and Jost (9) considered the mobility coefficient, B_A, having in mind localized sorption. For a single jump which could occur with equal probability in any one of p directions,

$${}_1B_A = \frac{\nu}{p} d^2 (1 - \theta) \exp - E_1/RT$$

where ν is the vibration frequency, E_1 the energy of activation, and d the distance jumped per unit diffusion process. If the activation energy persisted long enough for an n-fold jump in a straight line, then

$${}_nB_A = \frac{\nu}{p} (nd)^2 (1 - \theta)^n \exp - \frac{E_n}{RT}$$

and so, allowing for all jumps ($n = 1,2, \ldots$),

$$D_A = \frac{\nu d^2}{p} RT \frac{d\ln a_A}{d\ln\theta} \sum_{n=1}^{\infty} n^2 (1 - \theta)^n \exp - \frac{E_n}{RT}$$

Unless E_n increases strongly with n, multiple jumps for small θ could contribute appreciably to D_A, as seen by considering the relative size of the terms $n^2 (1 - \theta)^n$ for different values of n. However, because the energy of activation is rapidly dissipated by vibrational collisions, E_n for multiple jumps must be relatively large, and in practical cases such jumps probably do not contribute much to D_A.

This argument, though strictly appropriate only for localized systems, may be extended also to the last 3 isotherms of Table IX, in the hope in these cases of visualizing qualitatively or semiquantitatively the behavior of D_A. If we consider separately the terms where $n = 1$ or 2, Figure 12 shows the concentration dependances, respectively, of $(1 - \theta)\, d\ln a_A/d\ln\theta$ and $(1 - \theta)^2\, d\ln a_A/d\ln\theta$. When $n = 1$, for Langmuir's isotherm D_A is is independent of θ, but localized sorption with interaction gives maxima or minima for positive or negative values of β. The less likely case where $n = 2$ gives a linear decrease in D_A with θ for Langmuir's and constant D_A for Volmer's isotherm.

Concentration dependances for $n = 1$ or 2 are given for isotherms based on equations of state or sorbed fluid in Figure 12*b*. Where $n = 1$, for larger values of θ the activity correction is dominant for both Volmer's and van der Waal's isotherms, and D_A rises rapidly with increasing concentration. On the other hand, for the virial isotherm with $n = 1$ and the present choice of coefficients A', the reverse is true. The curvature of the isotherm requires that A_1' be positive; values of A_2' are usually smaller than A_1' and negative.

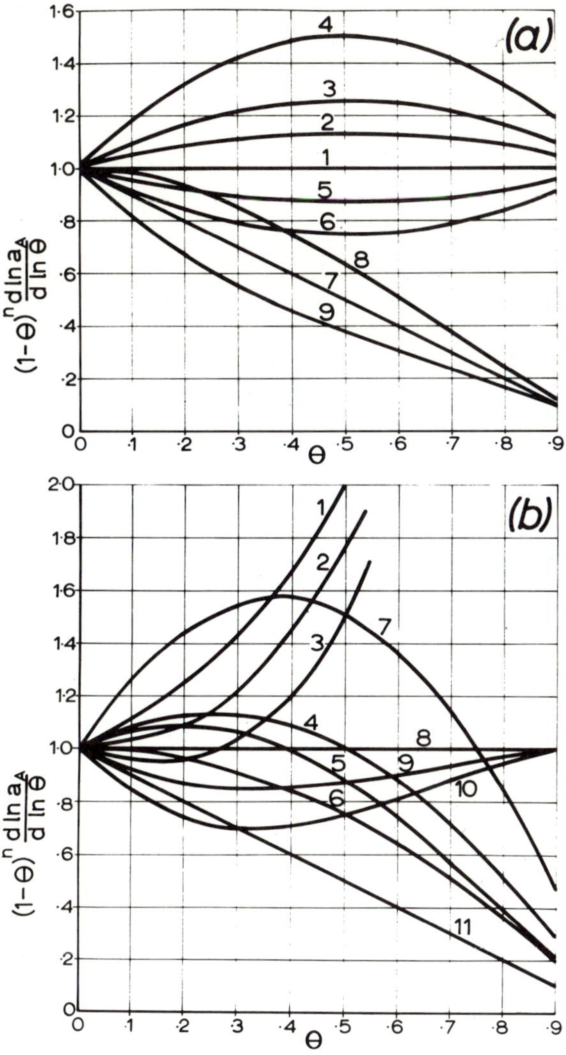

Figure 12. Concentration dependences $(1 - \theta)^n \times \frac{d \ln a_A}{d \ln \theta}$ *for* n = *1 and 2*

(a) Localized sorption with interaction. For curves 1, 2, 3, 4, 5, and 6, n = *1 and* $\beta = 0$ *(ideal Langmuir case), 0.5, 1.0, 2.0, −0.5, and −1.0, respectively. For curves 7, 8, and 9,* n = *2 and* $\beta = 0, +1,$ *and −1, respectively.*

(b) Isotherms based on equations of state. For curves 1, 2, and 3, n = *1 and equations are respectively that of Volmer, and of van der Waals with* $\alpha = 1$ *and* $\alpha = 2$*. For curves 4, 5, 6, and 7,* n = *1 for the virial isotherm with* $A'_1 = 1.0$, $A'_2 = 0$; $A'_1 = 1.0$, $A'_2 = -0.33$; $A'_1 = 0.50$, $A'_2 = 0$; *and* $A'_1 = 2.0$, $A'_2 = 0$*, respectively. Curves 8, 9, and 10 are for* n = *2 and isotherm equations of Volmer, and of van der Waals with* $\alpha = 1$ *and 2, respectively. Finally, curve 11 is for Volmer's isotherm with* n = *3.*

It emerges from this treatment, therefore, that for zeolitic fluids the patterns of behavior of D_A could be complex. D_A might remain constant, increase, decrease, or show maxima or minima as c_A increases, each type of behavior being influenced strongly by the isotherm equation followed and quantitatively determined by this, together with the concentration dependance of B_A. In various zeolite systems, examples of accord with specific isotherm equations have been observed, although agreement seldom covers the full range $0 < \theta < 1$. In general, isotherm equations for localized models tend to be more successful in describing sorption equilibria than the equations of Volmer or van der Waals. The virial isotherm, on the other hand, can be made as successful as desired. The physical meaning of the virial coefficients (which depend on temperature, sorbate, and sorbent, but not on amount sorbed) is not clear yet and the isotherm in this sense is empirical.

Moderated Diffusion

The diffusion coefficients of nonpolar molecules in zeolites can be changed greatly by the addition of controlled amounts of small polar molecules. These are sorbed very strongly and are immobile at the temperature of the subsequent runs with the nonpolar sorbates. Moderated diffusion was studied first in 1954 (*11*) for H_2, O_2, N_2, Ar, and C_2H_6 in crystals of mordenite and chabazite. The moderators were H_2O, NH_3, and CH_3NH_2. These measurements were extended subsequently to O_2, N_2, and Ne diffusing in Na-, (Ca,Na)-, and (K,Na)-A moderated with controlled amounts of NH_3 (*28*).

The influence of moderators upon diffusion coefficients is shown in Figure 13 for mordenite moderated with ammonia for a number of diffusing species. D_A may change by orders of magnitude for relatively small amounts of modifier. This can result in cut-offs in the amounts sorbed, on the time scale of normal experiments, which occur at different uptakes of NH_3 for molecules of different dimensions. Different moderator molecules can, molecule for molecule, have different effects upon the diffusion coefficients. These effects tend to follow the sequence of the molecular volumes of the moderator. As the amount of moderator increases, the energies of activation for diffusion increase (Table X), to parallel the decrease in D_A.

In this rather complex situation, the number and valence of the ions also plays an important part, as shown by Rees and Berry (*28*) (Figure 14). They used ammonia as moderator and progressively exchanged sodium by calcium. The diffusant was O_2 at 90.2°K (filled circles) and at 77.4°K (open circles). For small Ca contents, the changes in D_A are produced readily and are extremely large; but when the Ca^{2+} content

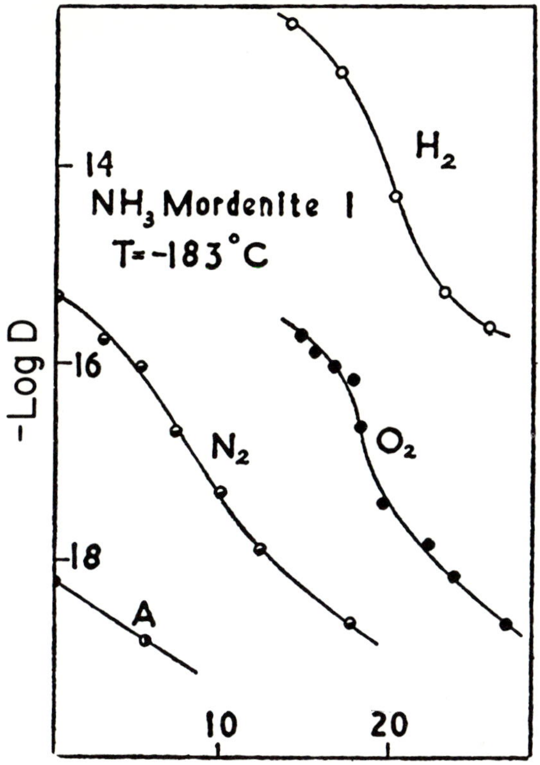

Figure 13. Effect of NH_3 in Na–mordenite as moderator of the diffusion coefficients of H_2, O_2, N_2, and Ar. D *is in* $cm^2\ sec^{-1}$

Table X. *E* for Diffusion in NH_3–Moderated Chabazite

Diffusant	*Amount NH_3, Cm^3 at STP/G*	E, *Kcal/Mole*
C_2H_6	90.2	12.0
	95.9	12.5
	102.0	12.9
	106.5	13.3
	115.0	14.2
Ar	85.4	5.6
	86.4	5.8
	87.4	6.0
	88.2	6.4
H_2	90.7	5.4
	92.5	5.8
	95.3	5.9

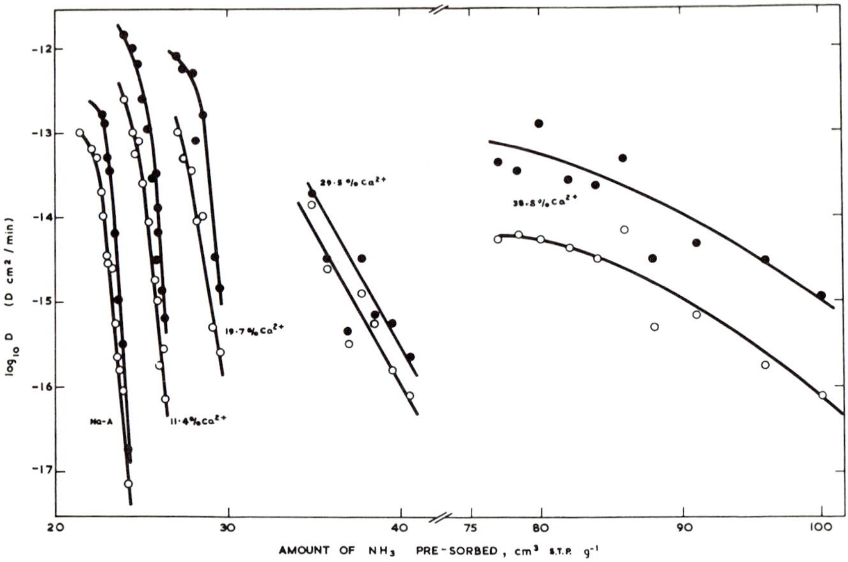

Figure 14. Dependence of log_{10}D for oxygen on the amount of presorbed NH_3 in pure Na-A (Linde sieve 4A) and various Ca^{2+} ion-exchanged samples (20). D *is in* cm^2 min^{-1}

● *90.2°K*
○ *77.4°K*

has reached 38.8% of the exchange capacity, D_A changes very much more slowly for much larger amounts of moderator. This behavior suggests association between some NH_3 molecules and Na^+ ions near the 8-ring windows. Ca^{2+} ions, on the other hand, as would be inferred from the known molecular sieve behavior of unmoderated Ca-A, do not block 8-ring windows. In part, this effect may rest upon their smaller numbers (1 Ca^{2+} replaces 2 Na^+) as well as upon different siting.

Theory of Hindered Diffusion

A theoretical treatment of modified diffusion has been developed (*11*) according to the following model. Molecules entering the zeolite crystal migrate along channels often little wider than themselves. They may encounter periodic energy barriers, the height, E_1, of which varies according to the relative dimensions of channels and molecules. Introduction of immobile modifier molecules or impurities at sites distributed along the channels creates higher energy barriers, E_2, at these sites. The model is at its best for zeolites with parallel nonintersecting main channels (mordenite, gmelinite, cancrinite hydrate, offretite, zeolite I, zeolite Ω). Evidence recently has been obtained that in the crystallization of cancrinite hydrate, discrete silicate anions may be incorporated (*5*) which greatly hinder passage along the main channels (*13*); also, stacking faults may create periodic high-energy barriers in gmelinite, cancrinite hydrate, and offretite. In these contexts, the treatment seems particularly relevant.

The distance between adjacent sorption sites along the channels is d, and N_s is the number of sites between successive energy barriers of height E_2. The number of crossings per second from left to right across a low barrier (height E_1) is $k_1P_1(1 - P_2)$, where P_1 and P_2 are the respective probabilities that sites to the left and right of the barrier are occupied, and where $k_1 \sim \nu_1 \exp - E_1/RT$, ν_1 being a vibration frequency of the sorbed molecule. The number of crossings in the reverse direction is $k_1P_2(1 - P_1)$, and so the net flow from left to right is $k_1(P_1 - P_2)$. The net flow past the rth high-energy barrier is likewise $k_2(P_L - P_R)_r$ with $k_2 = \nu_2 \exp - E_2/RT$, and where P_L and P_R are the respective probabilities of occupancy of the sites immediately to left and right of the barrier. In the steady state, on equating the 2 flows and assuming $\nu_1 \sim \nu_2$, one has

$$\frac{k_1}{k_2} = \exp\frac{(E_2 - E_1)}{RT} = \frac{P_L - P_R}{P_1 - P_2}$$

and

$$\frac{\text{Conc. drop across high barrier}}{\text{Conc. drop between successive high barriers}} = \frac{1}{N_s}\exp\frac{(E_2 - E_1)}{RT}$$

For many values of N_s, E_2, and E_1, this ratio will be $>> 1$. When this condition is fulfilled in the steady state, we may expect in both steady and transient states where concentration gradients occur that P will be nearly independent of position in the region between a given adjacent pair of high-energy barriers.

One may consider next the transient state filling of sites between the rth and $(r + 1)$th high barriers. If dN_r/dt denotes the net flux of molecules into this region, then

$$\frac{dN_r}{dt} = k_2\,(P_L - P_R)_r - k_2\,(P_L - P_R)_{r+1}$$

If, however, $1/N_s \exp \frac{E_2 - E_1}{RT} >> 1$, we have $(P_R)_r \sim (P_L)_{r+1}$ and so

$$\frac{dN_r}{dt} = k_2 \{ (P_L)_r + (P_R)_{r+1} \} = \frac{k_2}{N_s} \cdot \{ N_{r-1} + N_{r+1} \}$$

where N_{r-1} and N_{r+1} are, respectively, the numbers of molecules in the region preceding the rth and following the $(r + 1)$th high barriers. The above relation is one of a series of difference equations which may be set up for successive regions. The solution is (*17*)

$$\frac{N_r}{N_0} = \frac{n - r + 1}{n + 1} - \sum_{s=1}^{n} \frac{\operatorname{Sin}\left(\frac{rs\pi}{n+1}\right) \operatorname{Sin}\left(\frac{s\pi}{n+1}\right)}{(n + 1)\left(n - \operatorname{Cos}\left(\frac{s\pi}{n+1}\right)\right)} \cdot$$

$$\exp - \frac{tk_2}{N_s}\left[1 - \operatorname{Cos}\left(\frac{s\pi}{n+1}\right)\right]$$

where N_0 denotes the number of molecules which would be sorbed at equilibrium on N_s sites at the pressure obtaining at the surfaces of the crystals. n is the total number of high-energy barriers and s is an integer which takes all values between 1 and n.

For a semi-infinite solid where n is very large (or for small enough tk_2/N_s, where n is not large), the above equation can be transformed to

$$\frac{N_r}{rN_0} = \int_0^{\frac{tk_2}{N_s}} \frac{I_r(Z) \exp(-Z)\, dZ}{Z} = I$$

where $I_r(Z)$ is the Bessel function

$$\sum_{m=0}^{\infty} \frac{\left(\frac{1}{2} Z\right)^{r+2m}}{m!\Gamma(r + m + 1)}$$

These integrals have been evaluated for values of $Z_1 = tk_2/N_s$ of 1, 2, 3, and 4 (*11*). The figures in the first column of Table III of Ref. *11* are the values of r, and when $r > 12$ the integral I can be evaluated from the approximate expression

$$I = \frac{Z_1^r (r + Z_1)}{2^r r^2 r!} \exp\left(\frac{Z_1^2}{4r} - Z_1\right)$$

with an error not exceeding 2%.

The total number of molecules which has passed the first high-energy barrier is then

$$N_t = \sum_{r=1}^{n} \cdot N_r$$

and the fractional approach to equilibrium is taken as

$$\frac{Q_t}{Q_\infty} = \frac{1}{n} \cdot \sum_{r=1}^{n} \frac{N_r}{N_0}$$

This assumes the first high barrier is at the surface. If it is not, then a correction may be added corresponding with equilibrium uptake in the N_s sites preceding the first high barrier. This correction, which is negligible when n is large, except as $t \rightarrow 0$, gives

$$\frac{Q_t}{Q_\infty} = \frac{1}{n}\left(\sum_{r=1}^{n} \frac{N_r}{N_0} + 1\right)$$

For the case where $n = 9$, Figure 15*a* shows N_1/N_0, N_2/N_0 . . . N_9/N_0 and also Q_t/Q_∞ as functions of tk_2/N_s, while in Figure 15*b*

$$\sum_{r=1}^{9} \frac{N_r}{N_0}$$

is plotted against tk_2/N_s. Evidently the $\sqrt{t}$ diffusion law is valid. Also, the curve for $n = 9$ coincides with that for $n = \infty$, at least up to $tk_2/N_s = 4$.

The problem thus far has been considered as one of flow down a single typical channel. The total flow, expressed as number of molecules which has passed through unit area of crystal surface normal to the channels, is

$$N_2 = N_c \Sigma N_r$$

where N_c is the number of channels passing through this unit area. One is particularly interested in the ratios N_1/N_2, where N_1 is the total flow in absence of high-energy barriers. In absence of such barriers, the $\sqrt{t}$ diffusion law gives

$$N_1 = \frac{2\, N_c N_0}{N_s d}\left(\frac{D_1 t}{\pi}\right)^{\frac{1}{2}}$$

since

$$\frac{N_c N_0}{N_s d}$$

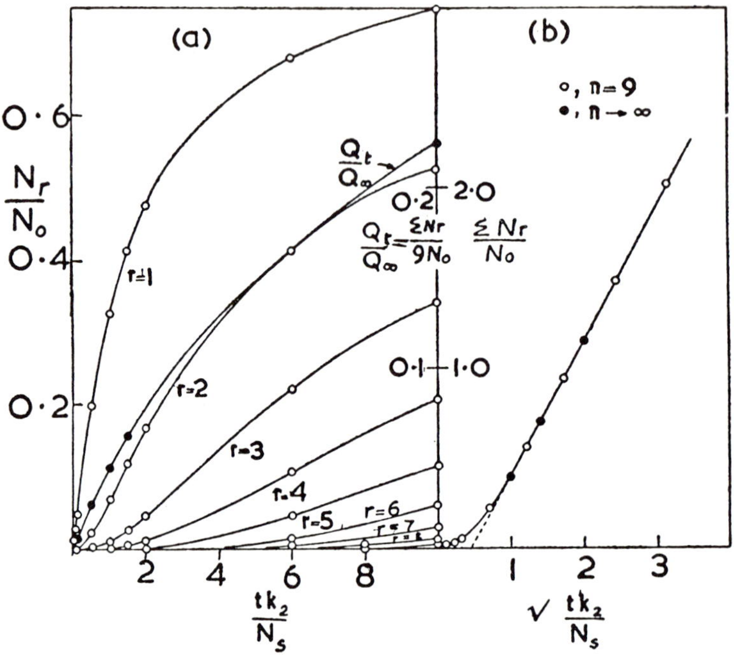

Figure 15. (a) $\frac{N_r}{N_0}$ *plotted against* $\frac{tk_2}{N_s}$ *with* n = 9 *for different values for* r. $\frac{Q_t}{Q_\infty} = \frac{\Sigma N_r}{9N_0}$ *is shown on the same diagram*
(b) *The* $\sqrt{t}$ *diffusion law for hindered diffusion with* n = 9 *and* n → ∞

is the concentration in molecules per cm³ at the ingoing surface. Thus

$$\frac{N_1}{N_2} = \frac{2}{N_s d} \cdot \left(\frac{D_1 t}{\pi}\right)^{\frac{1}{2}} \left\{ \sum_{r=1}^{\infty} \left(\frac{N_r}{N_0}\right) \right\}^{-1}$$

Some values of N_1/N_2 are given in Table XI for N_s = 50 and 5, at −183°C, taking $\nu = 10^{12}$ sec^{-1} and $(E_2 - E_1)$ = 1800, 2700, and 4500 cal per mole. Also, $D_1 = d^2\ \nu \exp - E_1/Rt$. The results may be summarized as follows:

(a) N_1/N_2 has large values which increase rapidly as $(E_2 - E_1)$ increases. Since this difference does not exceed 4500 cal per mole in the calculations and since some impurity ions—*e.g.*, silicate anions in basic cancrinite—or stacking faults can create very high barriers for many diffusants, either may be expected virtually to inhibit diffusion except at high pressures and temperatures.

(b) N_1/N_2 increases with increasing density of high-energy barriers—*i.e.*, decreasing N_s—but less than proportionally to their number. For N_s = 50 and 5, the proportionality approximates to the square root

Table XI. N_1/N_2 for Possible Energy Barriers E_1 and E_2, and Values of N_s at −183°C

N_s	*Cal per Mole* E_1	E_2	$\frac{N_1}{N_2}$ *for* $\frac{tk_2}{N_s} = 1$	$\frac{N_1}{N_2}$ *for* $\frac{tk_2}{N_s} = 4$
50	1800	3600	59.2	41.2
5	1800	3600	187.4	130.7
50	1800	4500	721	503
5	1800	4500	2.28×10^3	1.59×10^3
50	1800	6300	1.06×10^5	7.40×10^4
5	1800	6300	3.36×10^5	2.34×10^5
50	2700	5400	721	503
5	2700	5400	2.28×10^3	1.59×10^3
50	2700	6300	8.77×10^3	6.12×10^3
5	2700	6300	2.77×10^4	1.94×10^4
50	2700	7200	1.06×10^5	7.40×10^4
5	2700	7200	3.36×10^5	2.34×10^4

of the number of high-energy barriers. These are both high barrier densities. For lower densities, N_1/N_2 may decrease less rapidly than this (Ref. *11,* Figure 5*c*).

(c) As time (or tk_2/N_s) increases, the ratio N_1/N_2 declines, but rather slowly.

The very great effect of periodic high-energy barriers upon diffusion phenomena seems able to explain the "cut-off" phenomena reported in the previous section. Diffusion can become so slow for a particular (larger) sorbate molecule that equilibrium is not established and little sorption occurs on the time scale of the experiment; while for another (smaller) molecule, with a smaller value of $(E_2 - E_1)$, the rate of uptake is still large enough for substantial sorption to take place. Further addition of modifier—*i.e.*, more high-energy barriers—will eventually cause a "cut-off" in sorption even of this smaller molecule. Hindered diffusion may occur also in zeolite-based catalysts, where products clog the crystals for inflow of reactants to active sites.

Conclusion

The present survey has outlined some properties of porous crystals as diffusion media. It has shown some aspects which have been studied insufficiently, such as concentration dependence of D_A, which may be very important for catalysis at intracrystalline sites. Many more accurate determinations of differential diffusion coefficients are required to give added understanding of the role of concentration, chain length (of paraffins), polarity, exchange ions, channel geometry, and chemical damage. Nevertheless, the progress made is encouraging and points the way to further advance.

Literature Cited

(1) Ash, R., Barrer, R. M., *Surface Sci.* **1967,** 8, 461.
(2) Barrer, R. M., *Trans. Faraday Soc.* **1949,** 45, 358.
(3) Barrer, R. M., Bartholomew, R. F., Rees, L. V. C., *J. Phys. Chem. Solids* **1963,** 24, 51.
(4) Barrer, R. M., Brook, D. W., *Trans. Faraday Soc.* **1953,** 49, 1049.
(5) Barrer, R. M., Cole, J., Villiger, H., *J. Chem. Soc.*, in press.
(6) Barrer, R. M., Coughlan, B., "Molecular Sieves," p. 141, Soc. Chem. Ind., London, 1968.
(7) Barrer, R. M., Fender, B. E. F., *J. Phys. Chem. Solids* **1961,** 21, 12.
(8) Barrer, R. M., Ibbitson, D. A., *Trans. Faraday Soc.* **1944,** 40, 207.
(9) Barrer, R. M., Jost, W., *Trans. Faraday Soc.* **1949,** 45, 928.
(10) Barrer, R. M., Makki, M. B., *Can. J. Chem.* **1964,** 42, 1481.
(11) Barrer, R. M., Rees, L. V. C., *Trans. Faraday Soc.* **1954,** 50, 989.
(12) Barrer, R. M., Vaughan, D. E. W., *Trans. Faraday Soc.* **1967,** 63, 2275.
(13) Barrer, R. M., Vaughan, D. E. W., in preparation.
(14) Boltzmann, L., *Ann. Phys. Lpz.* **1894,** 53, 959.
(15) Brandt, W. W., Rudloff, W., *J. Phys. Chem.* **1967,** 71, 3948.
(16) Carman, P., "The Flow of Gases through Porous Media," Ch. 4, Butterworths, London, 1956.
(17) Carslaw, H. S., Jaeger, J. C., "Conduction of Heat in Solids," p. 332, Oxford Univ. Press, 1947.
(18) Dyer, A., Molyneux, A., *J. Inorg. Nucl. Chem.* **1968,** 30, 829.
(19) Eberly, P. E., Jr., *Ind. Eng. Chem. Prod. Res. Develop.* **1969,** 8, 140.
(20) cf. Fujita, H., in "Diffusion in Polymers," J. Crank and G. S. Park, Ed., Ch. 3, p. 77, Academic Press, 1968.
(21) Gabuda, S. P., *Dokl. Akad. Nauk., SSSR* **1962,** 146, 840.
(22) Habgood, H. W., *Can. J. Chem.* **1958,** 36, 1384.
(23) Kuhn, W., Thurkauf, M., *Helv. Chim. Acta* **1958,** 41, 938.
(24) Matano, C., *Japan. J. Phys.* **1930-33,** 8, 107.
(25) Meier, W., "Molecular Sieves," p. 10, Soc. Chem. Ind., London, 1968.
(26) Parravano, C., Baldeschweiler, J. D., Boudart, M., *Science* **1967,** 155, 1535.
(27) Paterson, S., *Proc. Phys. Soc.* **1947,** 59, 50.
(28) Rees, L. V. C., Berry, T., "Molecular Sieves," p. 149, Soc. Chem. Ind., London, 1968.
(29) Riehl, N., Dengel, O., Colloq. Phys. Ice Crystals, Zurich, 1962; Onsager, L., Runnels, L. K., *Proc. Natl. Acad. Sci.* **1965,** 50, 208.
(30) Sand, L. B., "Molecular Sieves," p. 71, Soc. Chem. Ind., London, 1968.
(31) Satterfield, C. N., Frabetti, A. J., *Am. Inst. Chem. Engrs.* **1967,** 13, 731.
(32) Schirmer, W., Fiedrich, G., Grossman, A., Stach, H., "Molecular Sieves," p. 276, Soc. Chem. Ind., London, 1968.
(33) Smith, J. V., "Molecular Sieves," p. 28, Soc. Chem. Ind., London, 1968.
(34) Tiselius, A., *Z. Phys. Chem.* **1934,** A, 169, 425.
(35) Tiselius, A., *Z. Phys. Chem.* **1935,** A, 174, 401.
(36) Walker, P. L., Jr., Austin, L. G., Nandi, S. P., in "Chemistry and Physics of Carbon," Vol. 2, pp. 257-371, M. Dekker, New York, 1966.
(37) Wang, J. H., Robinson, C. V., Edelman, I. S., *J. Am. Chem. Soc.* **1953,** 75, 466.

RECEIVED November 13, 1969.

Discussion

J. Turkevich (Princeton University, Princeton, N. J. 08540): What practical advice could you give us for the preparation of the zeolite before adsorption studies, and how long must one wait for equilibrium to be established?

R. M. Barrer: Outgassing times depend upon the openness of the crystals, the cations present, and the size of the crystals. As a practical working rule for water removal, a final temperature of not less than 360°C maintained under the best vacuum for not less than 24 hours is desirable.

Equilibration of adsorbate with a zeolite may be extremely rapid for small nonpolar molecules in open zeolites (a few minutes). Small polar molecules which stick where they hit (*e.g.*, water) may be so strongly sorbed that in a bed of powder the redistribution is very slow (equilibrium pressure very small). It helps to raise the temperature, hold the system at the high temperature in the presence of the sorbate, and then slowly cool to the desired low temperature. Molecules whose physical dimensions are as large or a little larger than the "windows" in the crystals may be sorbed very slowly indeed (by activated diffusion). In extreme cases, weeks may be needed.

J. Turkevich: We have been presented a series of correlations on variation with temperature of adsorption isotherms for single adsorbate on a given adsorbant. Until the discovery and development of crystalline zeolite adsorbants, charcoal, silica gel, and alumina were not well characterized as to pore distribution and homogeneity of surfaces. Now we have well defined adsorbants. Are they so well defined that we can characterize their adsorbant power by an isotherm, for any adsorbate, by one or two numbers?

R. M. Barrer: Because a given zeolite may vary in (1) Si:Al ratio and hence cation density; (2) type and number of exchange ions; (3) amounts of hydrolytic replacement of metal cations through washing; (4) number of defects such as stacking faults; and (5) extent of chemical damage—*e.g.*, through grinding—the position, although obviously better than for an amorphous material, is also not as simple as one could wish. For example, zeolite X, even if homoionic, may still show pronounced variability as a sorbent, especially for polar molecules, as a result of (1), (3), and (5).

D. M. Ruthven and **K. F. Loughlin** (University of New Brunswick, Fredericton, N. B., Canada): Barrer mentions the effect of crystal size on diffusivity and refers, in his paper, to the work of Satterfield and Frabetti, who observed that grinding of zeolite crystals caused a substantial decrease in diffusivity. This effect was attributed to partial blocking of micropores and damage to the zeolite lattice. On this basis, a substantial difference between the adsorptive properties of pelleted and unpelleted zeolites may be generally expected. We have studied the diffusion of the light paraffins in 5-A zeolite and have observed no such difference. This led us to seek alternative possible explanations for the pronounced effect of crystal size on diffusivity which was observed, not only by Satterfield and Frabetti, but also by Barrer and Ibbotson (quoted in the present paper). If the larger crystals contained significant structural defects, the effective crystal diameter might well be smaller than the over-all crystal diameter. This would lead to an erroneously large value of diffusivity for the larger crystals. Defects in the smaller crystals would be expected to be of less significance. Reduction of crystal size may therefore produce a significant decrease in the apparent diffusivity as a result of diminution of defects. We would welcome your comments as to the significance of the effect of defects in this respect.

R. M. Barrer: Factors such as you suggest may well be further complications, although I can offer no evidence for or against the suggestion.

43

Vapor Adsorption on Zeolites Considered as Crystalline Specific Adsorbents

A. V. KISELEV

Department of Chemistry, M.V. Lomonosov State University of Moscow and Institute of Physical Chemistry, USSR Academy of Sciences, Moscow

Adsorption on zeolites is discussed within the framework of a general theory of molecular adsorption on nonporous and porous adsorbents, considering: (1) selection of equations relating in a general way the thermodynamic characteristics of a system taking into account the adsorbate–adsorbent and adsorbate–adsorbate interactions; (2) the use of these equations in deriving from the experimental data physicochemical constants characterizing these interactions; (3) investigation of the dependence of these physicochemical constants on the structure of the zeolite and the adsorbate molecule; (4) calculation of the potential energy of the molecule–zeolite and the molecule–molecule interactions for different points within the cavity, and the possible use of spectroscopic data in conjunction with the adsorption measurements; (5) molecular–statistical calculations of the equilibrium between the zeolite and particular molecules.

The molecular field within the zeolite cavities, especially near the cavity walls, is heterogeneous, but the distribution of the field potential, which is determined by the crystalline structure of the zeolite, is regular throughout the sample (all cages are almost identical). Owing to the regular distribution of the field potential, the main features of adsorption on porous zeolites are similar to those on nonporous crystals. For example, for both nonporous adsorbents and porous zeolites, the heat of adsorption often increases with the adsorption level because of adsorbate–adsorbate interaction. Sometimes, owing to the sequence of increase and decrease of this interaction, the heat of adsorption curve acquires a wave-shaped pattern (Figure 1) and the adsorption isotherms pass through inflection points (Figure 2). This similarity shows that it

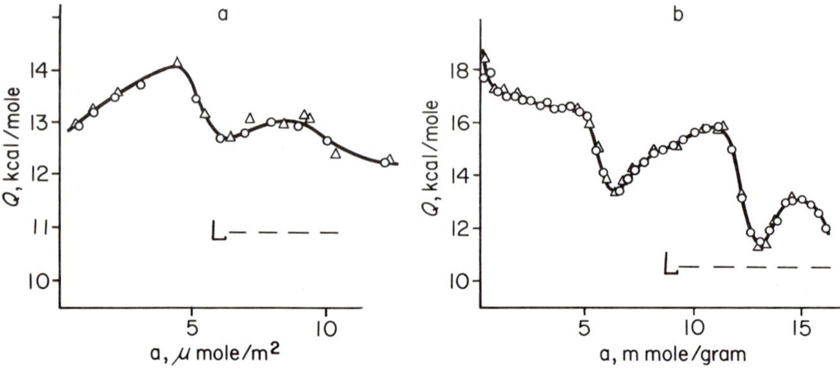

Figure 1. Heat of adsorption, Q, at different values of adsorption level a
a. Isobutyl alcohol on graphitized carbon black (6)
b. Water (4) *on zeolite KX*
L = heat of condensation of the vapor

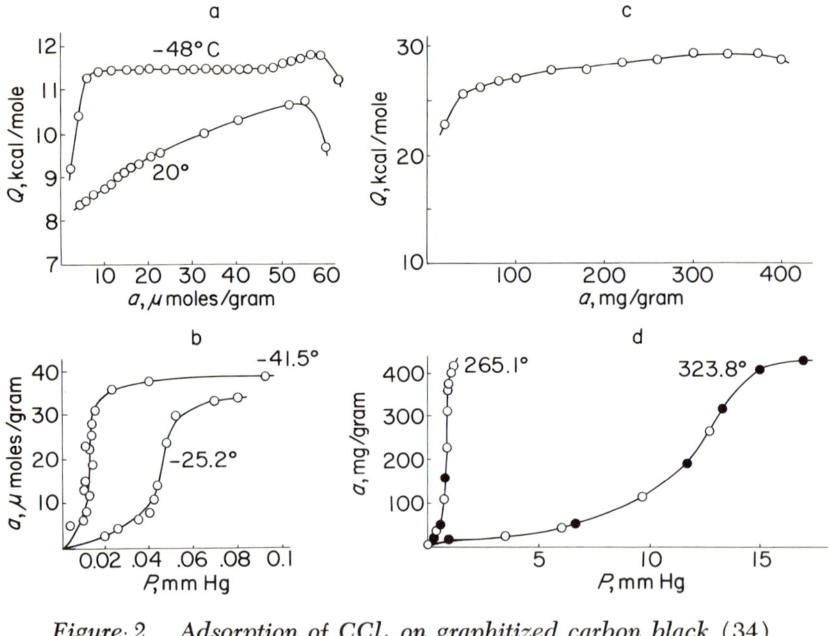

Figure 2. Adsorption of CCl_4 on graphitized carbon black (34)
a. Heats of adsorption (15)
b. Adsorption isotherms (34)
Adsorption of phosphorus vapor on zeolite NaX (13)
c. Heats of adsorption
d. Adsorption isotherms

is possible to apply the molecular theory of adsorption (*28, 35, 38*) to porous zeolites. Such an approach makes it unnecessary to assign to the adsorbate inside the micropores the properties of continuous compressed liquid, which usually is done in the case of adsorption by microporous adsorbents.

The development of the molecular theory of adsorption should consist of the following aspects:

(1) Equations should be selected or a new more general equation developed which have a molecular–statistical basis and describe the relationships between the amount of adsorption, the heat of adsorption, the heat capacity of the system, the pressure, and the temperature. Various constants in these equations must have a clear physical meaning. These equations must describe the exact shape of the isotherms (Figure 2) and must reduce to the Henry equation at low coverage.

(2) The developed equations should be used for analyzing experimental thermodynamic data to obtain physicochemical constants—*e.g.*, Henry constants—whose magnitude is independent of the fitting procedure.

(3) The relationships should be established between these constants and such parameters as the type of the zeolite lattice, the type and concentration of cations, the degree of decationization, and the structure of the adsorbate molecule. As a result, semiempirical relationships may be obtained which could be used for the practical calculation of the adsorption equilibrium.

(4) The molecular field distribution within the channels must be investigated, taking into consideration the structure of the zeolite, and the calculation of the potential energy of interaction between the zeolite and particular molecules must be made. These investigations would be assisted greatly by spectroscopic studies which would make it possible to establish the nature of the zeolite surface, the presence and the nature of structural defects, and the state of the adsorbed molecules.

(5) The statistical calculation of thermodynamic constants for the zeolite–adsorbate system should be made. A comparison of these constants with the experimental values obtained in Stages 2 and 3 would allow the introduction of some correction factors for the constants describing the potential functions of the interaction. As a result, a more accurate general scheme of molecular–statistical calculation of the molecule–zeolite system would be obtained.

In the present review, an attempt is made to analyze the experimental data and to relate the various aspects in the development of the molecular theory of adsorption as applied to zeolites.

Selection of the Adsorption Isotherm Equation

The adsorption isotherms with (Figure 2) and without (*11, 30*) points of inflection can be described by the same equation. This equation necessarily takes into account the adsorbate–adsorbent and adsorbate–

adsorbate interaction. It is based on 2 simplified models. At sufficiently high temperatures, use is made of the model for nonlocalized adsorption together with the equations of state of the adsorbate. The equations of the van der Waals and virial types are related to the isotherms of adsorption of Hill–deBoer (*16*) and Wilkins (*1, 11, 28, 40*), respectively. Another model involving localized adsorption followed by association of adsorbed molecules leads to isotherm equations containing a term which includes the interaction energy of neighboring molecules and their coordination number (*9*) or the association constant (*6, 30*). Both models satisfactorily describe adsorption isotherms with or without an inflection point.

From the point of view of the more general statistical theory of adsorption, if the energy distribution on the surface is known with sufficient accuracy, there is no necessity to use any special model—*e.g.*, localized adsorption.

On the basis of this general theory (*35, 38*), the initial region of the isotherm is described by an exponential series with virial coefficients:

$$a \simeq G = K_1p + K_2p^2 + \ldots \quad (1)$$

or by the reverse series:

$$p = K_1'a + K_2'a^2 + \ldots \quad (1a)$$

Here, G is the adsorption per gram of zeolite (at low vapor pressure, p, the value of G is close to the adsorption level, a); K_1 and K_1' are Henry constants characterizing the adsorbate–adsorbent interaction; K_2 and K_2' are constants characterizing the pair-wise adsorbate–adsorbate interaction which is influenced by the adsorbent field.

Equation 1a is a particular case (at low p and a) of the Wilkins equation:

$$p = a \exp (C_1 + C_2a + C_3a^2 + \ldots) = a \exp (\sum_{i=1}^{i} C_ia^{i-1}) \quad (2)$$

This series usually converges rapidly. Thus, the isotherms of adsorption on the nonporous adsorbent (*5*), as well as in the cavities of zeolite, can be described satisfactorily—*i.e.*, up to $\sim$ 75% saturation of zeolite—by using the first 3 or 4 C_i coefficients (Figure 3a). Equation 2 clearly describes the isotherms, either with or without inflection, and also those including a phase transition (Figure 3b).

Equation 2 describes not only the nonspecific adsorption by zeolites of molecules of group A (*28, 29*) but also the adsorption of quadrupole-type molecules such as CO_2, with respect to which zeolites are more heterogeneous (*1, 11*). This heterogeneity, like the adsorbate–adsorbate interaction, is reflected in the sign and magnitude of C_i. Thus, Equation 2

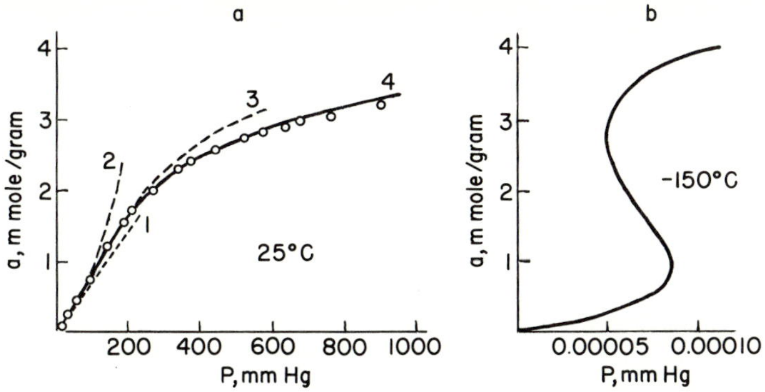

Figure 3. Adsorption isotherms of ethane on zeolite LiX

a. Experimental points (18) *at 25°C are shown with the curves calculated by Equation 2;* p = a *exp* (*4.8600 − 0.22166*a + *0.10249*a^2 + *0.009465*a^3) (*A. G. Bezus*). *Curves 1–4 were calculated using 1, 2, 3, and 4 terms in the exponent, respectively.*

b. At −150°C; calculated according to Equation 4; constants are indicated in the caption for Figure 6.

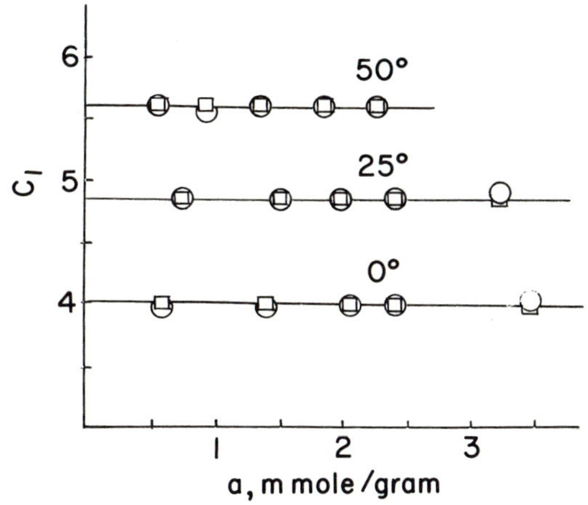

Figure 4. Values of constant C_1 *determined by A. G. Bezus from Equation 2 at* i = *3* (○) *and* i = *4* (□) *for different ranges of* a *in experimental adsorption isotherms of ethane on zeolite LiX* (14)

may be used with advantage in describing the vapor adsorption by zeolites, though other approximate equations (involving the different methods of taking into account the adsorbate–adsorbate interaction) give satisfactory results for certain systems (*7, 9, 30*). Until now, Equation 2 has been applied to simple cases where the adsorption isotherm contains no more than one inflection point.

Treatment of Experimental Data Using Virial Equations

Various complex curves can be satisfactorily approximated by equations of the 1a and 2 type. However, the second stage in the development of the molecular theory of adsorption concerns the determination of K_1 and K_2 or C_1 and C_2, which must depend only on the properties of the system (the structure of the adsorbent and adsorbate molecule) and be independent of the fitting procedure. With the aid of Equation 2, therefore, one can obtain values for C_1 and C_2 which are practically independent of the number of terms within the series and of the interval of experimental values of a (beginning at a low coverage). Figure 4 shows an example illustrating the determination of C_1.

The dependence of C_1 on temperature is shown in Figure 5. The approximate form of these dependences is (*1, 6, 28*):

$$C_1 \simeq B_i - Q_i/RT \tag{3}$$

The linear dependence of C_1 ($=\ln K_1'$) on $1/T$ implies that the heat of adsorption, Q_1, at zero coverage in the particular interval of T is independent of temperature. If the same assumption is made for Q_2, Q_3, etc., the dependence of a on p and T becomes (*1, 6, 28*):

$$p = a \exp \left(\sum_{i=1}^{i} B_i a^{i-1}\right) \exp \left(\sum_{i=1}^{i} - Q_i/RT\right) \tag{4}$$

Figure 6 gives examples of isotherms that have inflection points and also of those that do not. In these cases, use was made of 4 terms in Equation 4. It can be seen from Figure 6 that Equation 2 satisfactorily describes adsorption on microporous zeolites, as in the case of adsorption on nonporous graphitized carbon black (*6*).

It is very important to obtain accurately the first 2 constants (C_1 and C_2 or K_1 and K_2) from the experimental data in order to compare them with those calculated from the molecular–statistical theory. Here, C_1 and C_2 constants were determined from the experimental isotherm by the least squares method with the aid of a computer. However, for practical purposes, it is necessary to describe the isotherms to a higher adsorption level. From Figures 3 and 6, it can be seen that in order to reproduce the isotherm up to ~75% saturation of the adsorbent and to calculate a at different values of p and T, it is sufficient to determine only 2 more constants, C_3 and C_4.

Constants B_i and Q_i in Equation 4 can be determined either from adsorption measurements of a at various p and T or from measurements

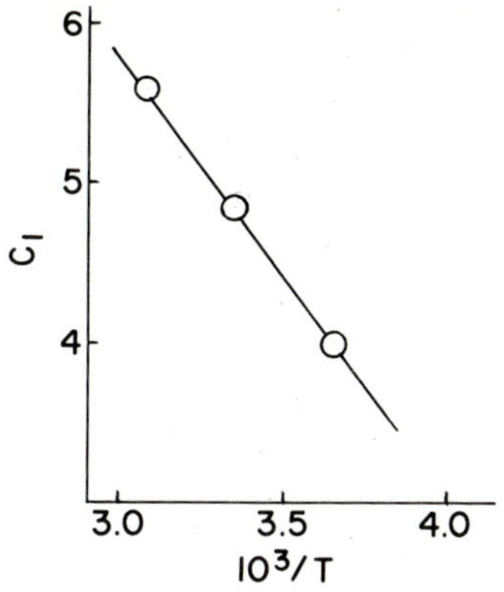

Figure 5. Dependence of constant C_1 *(see Figure 4) on* $1/T$

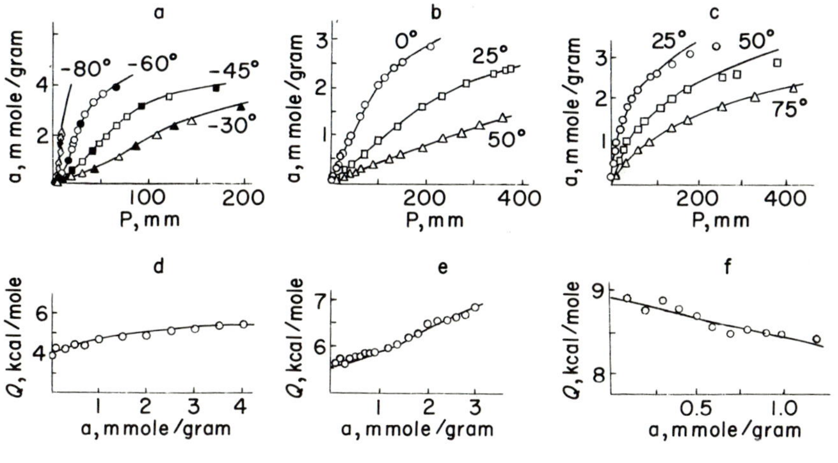

Figure 6. Adsorption of vapors on zeolite LiX; curves are calculated by Equations 4 and 5 (A. G. Bezus). For C_2H_6*, Equation 4 is* $p = a\ exp\ (14.20 - 2785/T) + (0.1364 - 106.8/T)a + (0.4029 - 89.57/T)a^2 - (0.0500 - 17.73/T)a^3$

a, b, c. Dependence of adsorption level a *for Xe,* C_2H_6*, and* C_2H_4*, respectively, on* p *and* T

d, e, f. Curves are calculated by Equations 4 and 5 (A. G. Bezus)

of one adsorption isotherm plus calorimetric measurements (*1, 28*) of the heat of adsorption for different *a* (Figure 6):

$$Q = \sum_{i=1}^{i} Q_i a^{i-1} \qquad (5)$$

These methods for describing the dependences of *a vs. p* and *T*, and *Q vs. a* have the following advantages. First, Equation 2 can be used to describe adsorption isotherms either with or without inflection points (Figure 6). Second, at least the first 2 pairs of constants, C_1, C_2 and Q_1, Q_2, are not empirical and are meaningful physicochemical constants for a given adsorbate–zeolite system. In order to obtain stable values for C_1 and C_2 (*see* Figure 4), measurements of *a* at different *p* and *T* must be precise and should be obtained within narrow intervals of *a*, beginning at low *a*. Precision in determining C_i decreases rapidly with the increase in *i*. However, since the theoretical evaluation of C_3 and C_4 is difficult, the determination of C_3 and C_4 from the experimental isotherm may be empirical, and no advantage is gained by obtaining these values to a great accuracy.

Dependence of Equilibrium Constants and Heat of Adsorption on the Structure of the Zeolite and the Adsorbate Molecule

Sufficiently precise values for C_1, C_2 and Q_1,Q_2 which can be used for comparison with statistical calculations so far have been obtained only for the adsorption of gases. Figure 7 shows the dependences of Q_1 and C_1 on the radius, *r*, for zeolite X containing alkali cations. The composition of the zeolite is described by Avgul (*4*). For nonspecific adsorption of ethane, Q_1 gradually increases with the increase in *r*. Accordingly, C_1 (= $-\ln K_1$) decreases. In this case, Q_2 increases and C_2 decreases. For specific adsorption of ethylene, the heat of adsorption, Q_1, on chang-

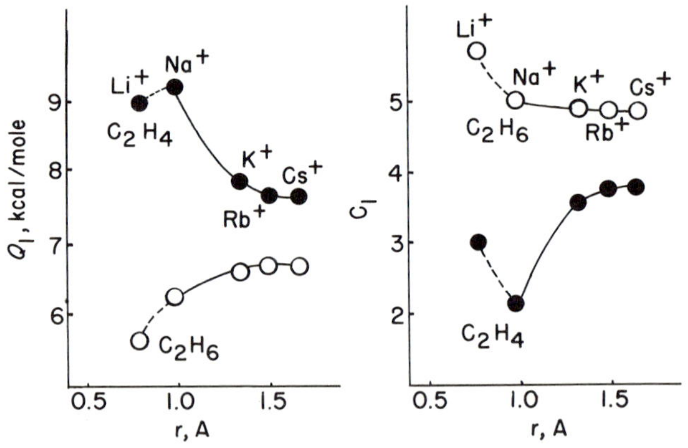

Figure 7. Adsorption of ethane (open circles) and ethylene (solid circles) on zeolites X; dependence of Q_1 *and* C_1 *on the ionic radii,* r*, of the exchange cations (calculated by A. G. Bezus and Pham Quang Du)*

ing from NaX to CsX zeolite, gradually decreases while C_1 correspondingly increases; the isotherms of these systems do not exhibit inflection points (Figure 6). A small increase in Q_1 for ethylene on going from LiX to NaX zeolite and the corresponding decrease in C_1 may be caused by a lower (by 7%) total concentration of the Li^+ plus Na^+ cations in this particular sample of LiX zeolite (*2, 12, 27*).

For sufficiently strong adsorbate–adsorbate interaction and with a corresponding decrease in temperature, Equation 4 leads to an isotherm having an inflection point both in the case of adsorption by nonporous adsorbent—*e.g.*, on graphitized thermal carbon black (*6*)—and of adsorption inside the zeolite channels. Apparently, these properties are possessed by the system phosphorus—zeolite NaX (*13*), Figure 2. In this case, a rapid increase in the heat of adsorption should be expected in the region of low coverage, as was observed for nonporous adsorbent (*15*) (compare Figures 2a and 2c).

A decrease in the adsorption level and in the heat of specific adsorption of CO_2 has been reported for zeolites with a decreasing concentration of exchange cations (*2, 12, 27*). There is a decrease in the heat of specific adsorption of benzene on changing from NaX zeolite to HY zeolite (*26*). In the X-type zeolite, the effect of substitution of Na^+ by doubly charged Ca^{2+} on the heat of molecular adsorption also has been investigated (*26, 32*). In the walls of large cavities there appear not only centers occupied by Ca^{2+} ions but also vacant centers free of any cations. A comparison of the dependences of the heat of adsorption for benzene for zeolites having different concentration of Ca^{2+} indicates that the higher heat of adsorption is related to the S_{II}-centers occupied by Ca^{2+} ions, whereas on centers free of cations the heat of adsorption of benzene lies below that of the cationized centers. Further work should be carried out on deriving quantitative relationships for Q_1, C_1 and Q_2, C_2 as the function of the Al:Si ratio, types and concentration of exchange cations, and degree of decationization. The examples given above indicate that these problems can be solved.

Specific interaction causes greater localization of the adsorbate molecules at small *a*. In the case of strongly adsorbing molecules containing such functional groups as HO—, H_2N—, etc., cations in zeolite can acquire a certain mobility as *a* increases. The relationships between *Q* and *a* in these cases become very complex (*see* Figure 1). Water molecules are localized first by the cations, then they possibly form a kind of water cage by mutual hydrogen bonds inside the zeolite cavity, and finally fill up the central part of this zeolite and water cage. This process depends on the type of cation (*4*). Two steps in the plot of *Q vs. a* for ammonia adsorption on zeolite CaA were found in Ref. *36*. The coordination of water and ammonia around Cu^{2+} ions at the saturation of zeolite similar to that in solution complexes were found by ESR and spectroscopic methods (*33*). It is inferred thus that at high adsorption level of small specifically adsorbed molecules, theory may acquire some features similar to the theory of bulk solutions.

Thus, the determination of Q_1 and C_1 values on the basis of measurements carried out at ordinary temperatures in these cases becomes difficult. In such cases, however, one can approximately estimate the additional contribution of the energy of specific interaction arising from π-bonds or the functional groups at a somewhat higher *a*—*e.g.*, when values of *a* correspond to 1 molecule per cavity. $\Delta Q_{specific}$ is determined by subtracting the heat of adsorption of the reference molecule (incapable to specific interaction) from the heat of adsorption of the specifically adsorbed molecule (*10, 28–30*). Both molecules must have similar geometry, close values of polarizability and of heat of adsorption on nonspecific adsorbent (*6*).

Figure 8 shows the relationships between the heat of adsorption and the polarizability of molecules for a homologous series of *n*-alkanes, ethers, and *n*-alcohols. $\Delta Q_{specific}$ of the functional groups —O— and

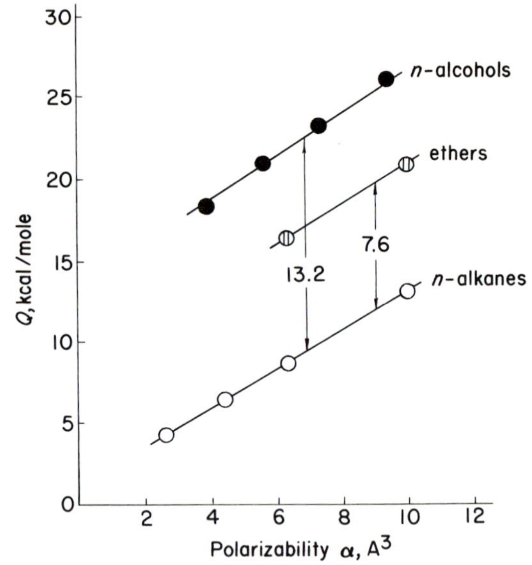

Figure 8. Dependence of the heat of adsorption, Q, for zeolite NaX (at a = 1 molecule per cavity) on the polarizability of the molecule α for n-alkanes (30), ethers (4), and n-alcohols (4)

HO— for the series of compounds examined is practically independent of the length of the hydrocarbon chain in these molecules.

The data summarized in Table I illustrate how values for Q and $\Delta Q_{\text{specific}}$ depend on the structure of the adsorbate molecules. In the series of group-B molecules, the maximum energy of specific interaction is observed for nitromethane and acetonitrile, both of which have large dipole moments. Even stronger specific interaction is observed for molecules of the D-group, which can form an additional hydrogen bond with the negatively charged oxygen atoms of the zeolite structure. For the investigated compounds, the relative contribution of the specific interaction energy, $\Delta Q_{\text{specific}}/Q$, decreases with the increase in the length of the hydrocarbon chain of the molecule. The larger value of the heat of adsorption for tetrahydrofuran compared with that of furan (*23*) is attributed mainly to the difference in the electronic structure of these geometrically similar molecules. The conjugation of the electrons of the oxygen atom with π-bonds in furan causes a more uniform distribution of the electron density in this molecule. This decreases the dipole moment and $\Delta Q_{\text{specific}}$. The heat of adsorption of furan on the zeolite is thus close to that of benzene.

The magnitude of $\Delta Q_{\text{specific}}$ thus obtained may be of practical value for estimating the heat of adsorption for hydrocarbon derivatives containing functional groups, taking into account the molecule geometry. The heat of adsorption of a complex molecule, capable of specific interaction, can be approximately expressed as the sum of the nonspecific heat of adsorption, Q_A, of the suitable reference molecule of group A and of the contribution of the specific interaction energy for the corresponding functional group, $\Delta Q_{\text{specific}}$, taken from Table I. According to another method (*see* Table III in Ref. *4*), the heat of adsorption of a complex molecule can be expressed as the sum of the increments of the total heat of adsorption for each segment of the molecule. For these calculations at different a, it is necessary to study the dependences of $\Delta Q_{\text{specific}}$ and of increments of Q on the adsorption level.

Potential Energy of Interaction of Molecules with Zeolite

The potential energy of adsorption for various cationized X- and A-type zeolites has been calculated by various authors (*9, 12, 30*). Recently (*17, 18*), the potential energy of adsorption was calculated in a large cavity of the zeolite type A. As the Si:Al ratio is 1:1, it was assumed that the excess of negative charge of the $AlO_{4/2}$ tetrahedron is distributed uniformly over all the oxygen atoms. Thus, a charge $z = -0.25e$ (here, e = elementary charge) was assigned to each oxygen atom. On the basis of the assumption of the additive property of pair-wise interactions, the

Table I. Heat of Adsorption, Q, and Contribution of the Specific Interaction Energy, $\Delta Q_{specific}$, for NaX Zeolite,[a] Kcal/Mole

Properties Which Determine the Specific Nature of Adsorption	*Adsorbate*	Q	$\Delta Q_{specific}$	$\frac{\Delta Q_{specific}}{Q}$
	Molecules of Group B			
π-bonds, quadrupole moment	C_2H_4	8.9	3.0	0.34
	C_6H_6	18.0	4.5	0.25
	N_2	5.0	2.3	0.46
	CO_2	10.0	6.0	0.55
Dipole moment, ability to form hydrogen bonds through oxygen and nitrogen atoms only	$(CH_3)_2O$	16.4	7.6	0.46
	$(C_2H_5)_2O$	21.0	7.6	0.36
	CH_3NO_2	19.9	10.0	0.50
	CH_3CN	10.0	11.0	0.58
	Molecules of Group D			
Dipole moment, ability to form hydrogen bonds through oxygen or nitrogen atoms, as well as through OH or NH hydrogen atoms	H_2O	18.5	(15.5)	(0.8)
	CH_3OH	18.4	13.2	0.72
	C_2H_5OH	20.9	13.2	0.63
	C_3H_7OH	23.2	13.2	0.57
	n-C_4H_9OH	26.0	13.2	0.51
	NH_3	16.0	(12.5)	(0.8)
	CH_3NH_2	18.0	11.5	0.64

[a] Value of a = 1 molecule per cavity, experimental data (*2, 3, 4, 7, 23, 30*).

total potential energy, V, of N adsorbed spherical molecules in the field of a large cavity of zeolite A has the following form:

$$V(\vec{r}_1, \vec{r}_2, \ldots, \vec{r}_N) = \sum_{i=1}^{N} \Phi(r_i) + \sum_{1 \leq i < j \leq N} \psi(|\vec{r}_i - \vec{r}_j|) \quad (6)$$

Here, $\Phi(\vec{r}_i)$ is the potential energy of ith molecule located at $\vec{r}_i$ inside the zeolite cavity interacting with the zeolite lattice; $\psi(|\vec{r}_i - \vec{r}_j|)$ is the potential energy of interaction of the ith and jth molecules located, respectively, at the points $\vec{r}_i$ and $\vec{r}_j$. In the calculations of $\Phi(r_i) = \sum_k \phi(|\vec{r}_i - \vec{r}_k|)$, where ϕ is the interaction energy of ith molecule with kth center of zeolite, account was taken only of the cations Li^+, Na^+, and Ca^{2+} in the various positions and of the oxygen atoms bearing negative charge $z_o = -0.25e$. It was assumed that the contribution arising from the charges distributed within $AlO_{4/2}$ and $SiO_{4/2}$ tetrahedra can be expressed approximately by a charge of $z_o = -0.25e$ located at the centers of the oxygen atoms. The polarizability of oxygen in this state was estimated by Brauer *et al.* (*18*). On the other hand, high charges on Al and Si were assigned in Ref. *22*.

The energy $\Phi(\vec{r}_i)$ was expressed as in earlier studies (*18, 30*) as the sum of the dispersion, induction, and repulsion interactions. It was assumed that:

$$\Phi_{\text{disp.}}(\vec{r}_i) = \sum_k C_k \, |\vec{r}_i - \vec{r}_k|^{-6} \quad (7)$$

$$\Phi_{\text{induc.}}(\vec{r}_i) = -\frac{1}{2}\alpha_m \, [E(\vec{r}_i)]^2 \quad (8)$$

$$\Phi_{\text{repul.}}(\vec{r}_i) = \sum_k B_k \, |\vec{r}_i - \vec{r}_k|^{-12} \quad (9)$$

Constant C_k for the dispersion attraction was determined from the Slater-Kirkwood equation:

$$C_{SK} = -\frac{3eh}{4\pi(m)^{1/2}} \cdot \frac{\alpha_m \alpha_k}{(\alpha_m/\bar{n}_m)^{1/2} + (\alpha_k/\bar{n}_k)^{1/2}} \quad (10)$$

where α_m and α_k are the polarizabilities; $\bar{n}_m, \bar{n}_k$ are the effective numbers of electrons for the adsorbate molecule and exchange cation or oxygen atom in zeolite, respectively; h is the Planck's constant and m the electron mass. The electrostatic field strength vector at point $\vec{r}_i$ was calculated from the equation:

$$E(\vec{r}_i) = -\,\text{grad}\left(\sum_k \frac{e_k}{|\vec{r}_i - \vec{r}_k|}\right) \quad (11)$$

where e_k is the cation charge or charge of the oxygen atom in the zeolite. Constants B_k were determined by minimizing $\Phi(\vec{r}_i)$ at the equilibrium distance between the molecule and the corresponding kth center in the walls of the cavity. The equilibrium distance was calculated from the van der Waal's and ionic radii of molecules and kth center.

Owing to the high symmetry of the lattice for the LiA and NaA zeolites, an analysis was made only of the section comprising 1/48 of the total cavity volume (Figure 9). The volume of the corresponding section for CaA zeolite was 1/24 of the total volume of the cavity. The potential energy was calculated for the inner cavity of the selected section for 16 different directions in the case of LiA and NaA zeolites and for 31 directions in the case of the CaA zeolite. For each of these directions, the $\Phi(\vec{r}_i)$ was calculated for 40 different positions of the molecule. The lattice sums

$$\sum_k |\vec{r}_i - \vec{r}_k|^{-6}$$

and

$$\sum_k |\vec{r}_i - \vec{r}_k|^{-12}$$

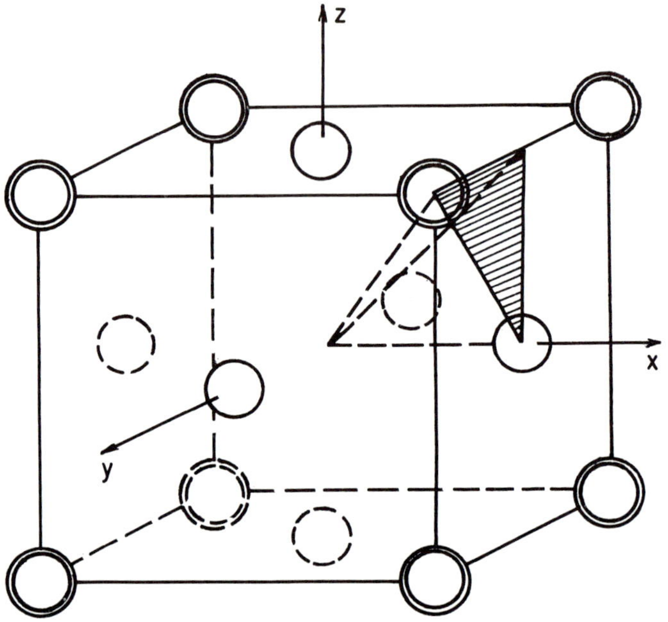

Figure 9. Elementary cell of zeolite A containing 8 univalent cations, ⊙, inside six-membered rings and 6 cations, ○, in eight-membered rings; a section is shown in which there are 16 selected directions

and the field strength were calculated in order to determine the contributions arising from $\Phi_{disp.}$, $\Phi_{induc.}$, and $\Phi_{repul.}$. The summation was carried out for all the exchange cations and charged oxygen atoms, $O^{-0.25e}$, which were located inside 8 neighboring cube-octahedrons, beyond whose limits summation was replaced by integration. The result was close to that obtained by the summation in 64 neighboring cube-octahedrons. The calculation for different distributions of cations gave similar results.

The strength of the electrostatic field inside the cavity of the LiA, NaA, and especially CaA zeolites varies greatly. For different directions, the electrostatic field strength vector is often reversed in sign.

Figure 10 shows examples illustrating the calculated potential curves, which describe the interaction between argon and the NaA and CaA zeolites. The solid lines indicate potential curves for directions corresponding to the lowest and highest depth of the potential well. Potential curves were obtained also for the remaining 14 directions in the section shown in Figure 10 for the zeolite NaA and for 29 directions in the corresponding section for the zeolite CaA. Figure 10 shows only the position of the minimum in these curves. Data in this figure show that the molecular field in the cavities in NaA zeolite are, with respect to adsorption of argon, relatively not too heterogeneous along the various directions; the depth of the potential well along directions 4 and 7 differ only by ~30%. The molecular field in the cavities of zeolite CaA is much more heterogeneous; the potential well depth has a three-fold difference along directions 15 and 19.

The calculation of the potential energy of interaction of a long chain of *n*-alkanes with the zeolite CaA has been described (*24*). These results agree with the experimental values for the heat of adsorption.

More detailed information on the structure of zeolites and the interaction mechanism will make possible a more precise calculation of the potential interaction energy. In the case of zeolite A, more complex molecules should be included in the study, both with respect to their geometry and to the nature of specific interaction for nonzero levels of adsorption. This latter point is especially important because in many cases experimental data for the initial region of the isotherm are unreliable. The study also should include zeolites having different structures, first X and Y zeolites, since many different molecules can penetrate into the cavities of these zeolites. The results obtained in such a study will make it possible to introduce into the statistical calculation the necessary corrections of the constants for the potential interaction functions (*6, 31*). In the case of adsorption of saturated hydrocarbons, this will

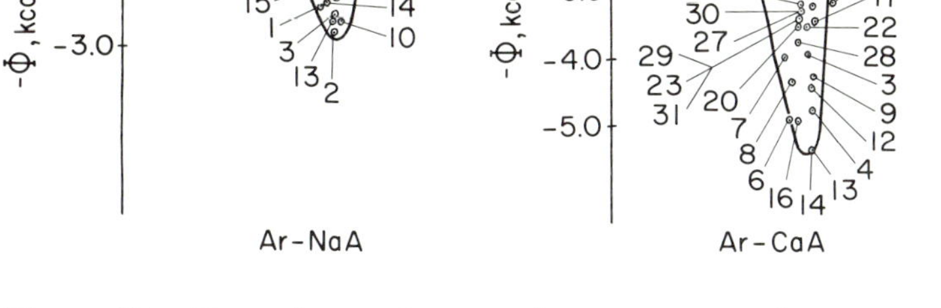

Figure 10. Dependence of potential energy, Φ, on the position of the center of molecule i *along the respective directions inside selected section for NaA;* see *Figure 9*

Same numbers as in (17)

Solid lines = potential curves with the minimum and maximum potential well depth

Points = position of minimum of the remaining curves

give the basic ion–atom potential functions. For zeolites, these potential functions

$$\phi_{\mathrm{Cat}\ldots\mathrm{C}} \quad \phi_{\mathrm{Cat}\ldots\mathrm{H}} \quad \phi_{\mathrm{O}\ldots\mathrm{C}} \quad \text{and} \quad \phi_{\mathrm{O}\ldots\mathrm{H}}$$

depend on the type of the lattice, the Si:Al ratio, and the direction inside the cavity (here, subscript Cat is the exchange cation in the zeolite having charge $+z'e$; subscript O is the oxygen atom of the zeolite having charge $-z''e$, which in simplest case is equal to $-0.25e$). Further and more detailed studies should be carried out of interaction with zeolites of various molecules having different types of π-bonds, quadrupole, and dipole moments.

Of great aid to the study of adsorption on zeolites will be spectroscopic methods used in conjunction with quantitative mass spectroscopic analysis of hydrogen concentration in the zeolite by deuterium exchange with D_2O (*21*).

Up to now, infrared spectroscopy has been used mainly to determine the types of hydroxyl groups and the acidity of zeolites (*39*). The frequencies of the vertical and horizontal vibrations (with respect to the cavity wall) of H_2O molecules adsorbed in zeolite A were determined by measurements in the far infrared ($\sim$220 and $\sim$75 cm^{-1}) (*37*). These values are in agreement with a simple theoretical model. A number of ultraviolet and ESR studies are reviewed (*33*). The difference has been established between the specific molecular interaction of aromatic molecules on zeolites cationized with alkali cations and the more complex interactions involving charge transfer in CaX and decationized X and Y zeolites. These more complex interactions with CaX zeolites containing protonized vacancies and with decationized zeolites are similar. These phenomena are related to the interactions of molecules with acidic centers in zeolites which are stronger, as compared with the molecular adsorption.

In further work, adsorption and spectroscopic studies should be combined. This will make it possible to compare with advantage the thermodynamic and spectroscopic data at the same adsorption level. An example of such a comparison is shown in Ref. *25*.

Molecular–Statistical Calculation of Adsorption Equilibria of Vapors on Zeolites

For NaA zeolite, an approximate calculation of the configurational integrals was made in Ref. *19* on the basis of potential curves for 3 symmetry axes of the second, third, and fourth order. All together, 26 directions were taken into account. In a subsequent work, 482 directions have been considered (*17*). It was assumed that during the adsorption the zeolite itself remains unchanged, and that the cavities in zeolite A together with the adsorbed molecules could be considered as quasi-independent subsystems (*8*). Calculations were made of the first 2 constants, K_1 and K_2, in the virial Equation 1. The equation for the Henry constants has the following form:

$$K_1 = N_c Z_1 / N_A kT \tag{12}$$

Here, $N_c = 3.6 \times 10^{20}$ gram^{-1} is the number of cavities per gram of zeolite A, which has been assumed to be the same for all cationized forms included in this study; N_A is the Avogadro number; k is the Boltzmann constant, and

$$Z_1 = \int_{v_z} \exp\left[-\Phi(\vec{r_i})/kT\right] d\vec{r_i} \tag{13}$$

where $v_z = 800$ A^3 is the volume of a large cavity. By calculating K_1 and its first and second derivatives with respect to T, the specific retention volume at zero coverage ($V_m = K_1RT$) (*29*) can be determined, as well as the isosteric heat of adsorption, the change in the standard chemical potential and standard entropy during adsorption, and the corresponding change in the differential heat capacity of the adsorbate. This had been done for the adsorption on graphite (*6, 31*). A detailed account of these calculations is given in Ref. *17*.

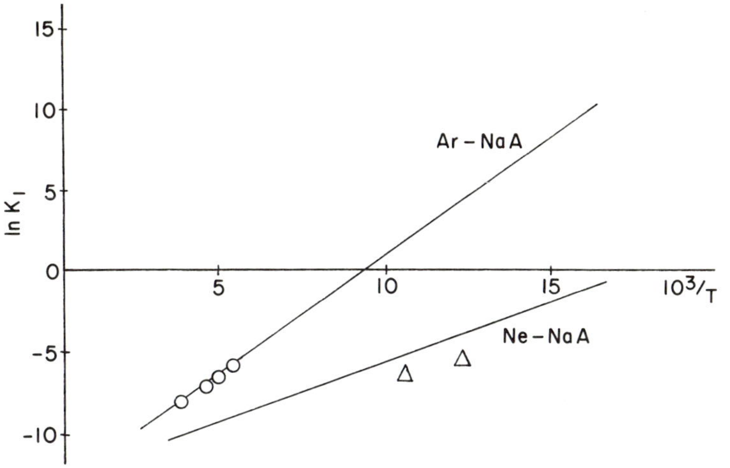

Figure 11. Dependence of the values of the Henry constant, lnK₁, on 1/T for the adsorption of Ne and A on zeolite NaA; calculated curves and experimental points are shown

Figure 11 shows a comparison of the calculated and the experimental dependencies of $\ln K_1$ *vs.* $1/T$ for the adsorption of argon and neon on zeolite NaA. Table II gives a comparison of the theoretical and experimental values for the heat of adsorption, Q_1 (where $a \rightarrow O$), for the systems Ne–NaA, A–NaA, A–CaA, and CH_4–CaA.

The calculation of the next member of virial expansion (K_2 in Equation 1), which would include pair-wise mutual interactions of the adsorbate molecules inside the zeolite cavity, is much more complex. An estimation of K_2 and Q_2 values for systems A–NaA and CH_4–CaA is made in Ref. *20*. Further studies should try to obtain more precise measurements of the thermodynamic and spectroscopic characteristics of adsorption on zeolites. Such measurements are necessary both for introducing corrections to the potential interaction functions and for developing methods for approximate calculations, especially for complex molecules.

Table II. Calculated and Experimental Values for the Heat of Adsorption, Q_1, on Various Zeolites

Zeolite	*Adsorbate*	*Temperature, °K*	Q_1, *Kcal/Mole* *Calcd.*	*Exptl.*
NaA	Ne	80	1.3	1.1
	A	280	2.9	2.8
CaA	A	315	5.3	4.6
	CH_4	290	5.2	3.8; 3.9; 5.6

Conclusion

The examples cited in this review show that all stages of the development of the molecular theory of adsorption mentioned above have been studied to a limited extent for the adsorbate–zeolite system. The examples also show, however, that in order to determine experimentally the stable constants and to establish their dependence on the details of the zeolite structure and the structure of the adsorbate molecule, more systematic investigations will be needed.

Literature Cited

(1) Avgul, N. N., Aristov, B. G., Berezin, G. I., Kiselev, A. V., Kurdyukova, L. Ya., Sinitzyn, V. A., "Molecular Sieves," p. 250, Society of the Chemical Industry, London, 1968.

(2) Avgul, N. N., Aristov, B. G., Kiselev, A. V., Kurdyukova, L. Ya., Frolova, N. V., *Zh. Fiz. Khim.* **1968**, 42, 2682.

(3) Avgul, N. N., Bezus, A. G., Dzigit, O. M., Kiselev, A. V., Kurdyukova, L. Ya., Melnikova, T. A., Mikos, K. N., Muttik, G. G., Seldaček, Z., "Molecular Sieves," p. 264, Society of the Chemical Industry, London, 1968.

(4) Avgul, N. N., Bezus, A. G., Dzhigit, O. M., ADVAN. CHEM. SER. **1971**, 102, 184.
(5) Avgul, N. N., Guzenberg, A. S., Kiselev, A. V., Kurdyukova, L. Ya., Ryabkin, A. M., *Zh. Fiz. Khim.*, in press.
(6) Avgul, N. N., Kiselev, A. V., "Chemistry and Physics of Carbon," Vol. 6, p. 1, P. L. Walker, Ed., Dekker, New York, 1970.
(7) Aristov, B. G., Bosaček, V., Kiselev, A. V., *Trans. Faraday Soc.* **1967**, 63, 2057.
(8) Bakaev, V. A., *Dokl. Akad. Nauk SSSR* **1966**, 167, 369.
(9) Barrer, R. M., "Non-Stoichiometric Compounds," p. 309, L. Mandelcorn, Ed., Academic Press, New York, 1963.
(10) Barrer, R. M., *J. Colloid Interface Sci.* **1966**, 21, 415.
(11) Barrer, R. M., Coughlan, B., "Molecular Sieves," p. 233, Society of the Chemical Industry, London, 1968.
(12) *Ibid.*, p. 241.
(13) Barrer, R. M., Whiteman, J. L., *J. Chem. Soc.* **A1967**, 13.
(14) Bezus, A. G., Kiselev, A. V., Sedlaček, Z., *Zh. Fiz. Khim.* **1969**, 43, 1223.
(15) Berezin, G. I., Kiselev, A. V., Serdobov, M. V., Sagatelyan, R. T., *Zh. Fiz. Khim.* **1969**, 43, 224.
(16) de Boer, J. H., "The Dynamic Character of Adsorption," Clarendon Press, Oxford, 1953.
(17) Brauer, P., Dissertation, Moscow University, Moscow, 1969.
(18) Brauer, P., Kiselev, A. V., Lesnik, E. A., Lopatkin, A. A., *Zh. Fiz. Khim.* **1968**, 42, 2556.
(19) *Ibid.*, **1969**, 43, 1519.
(20) Brauer, P., Lopatkin, A. A., Stepanez, G. F., ADVAN. CHEM. SER. **1971**, 102, 97.
(21) Davydov, V. Ya., Kiselev, A. V., Zhuravlev, L. T., *Trans. Faraday Soc.* **1964**, 60, 2254.
(22) Dempsy, E., "Molecular Sieves," p. 293, Society of the Chemical Industry, London, 1968.
(23) Dzhigit, O. M., Zakharova, E. E., Kiselev, A. V., Mikos, K. N., Muttik, G. G., Rakhmanova, T. A., *Zh. Fiz. Khim.* **1969**, 43, 968.
(24) Fiedler, K., Spangenberg, H., Schirmer, W., *Monatsber. Deut. Akad. Wiss. Berlin* **1967**, 9, 516.
(25) Geodakyan, K. T., Kiselev, A. V., Lygin, V. I., *Zh. Fiz. Khim.* **1969**, 43, 201.
(26) Khudiev, A. T., Klyachko-Gurvich, A. L., Brueva, T. P., Isakov, Ya. I., Rubinstein, A. M., *Izv. Akad. Nauk SSSR, Ser. Khim.* **1968**, 717.
(27) Khvoshchev, S. S., Zhdanov, S. P., Belotzerkovsky, G. M., Redin, V. I., *Zh. Fiz. Khim.* **1968**, 42, 171.
(28) Kiselev, A. V., *Zh. Fiz. Khim.* **1967**, 41, 2470.
(29) Kiselev, A. V., Yashin, Ya. I., "Gas-Adsorption Chromatography," Plenum, New York, 1969.
(30) Kiselev, A. V., Lopatkin, A. A., "Molecular Sieves," p. 252, Society of the Chemical Industry, London, 1968.
(31) Kiselev, A. V., Poshkus, D. P., Afreimovich, A. Ya., *Zh. Fiz. Khim.* **1968**, 42, 2546.
(32) Klyachko-Gurvich, A. L., Khudiev, A. T., Isakov, Ya. I., Rubinstein, A. M., *Izv. Akad. Nauk SSSR, Ser. Khim.* **1967**, 1355.
(33) Lygin, V. I., ADVAN. CHEM. SER. **1971**, 102, 86.
(34) Pierce, C. P., *J. Phys. Chem.* **1968**, 72, 1955.
(35) Poshkus, D. P., "Intermolecular Forces," *Discussions Faraday Soc.* **1965**, 40, 195.
(36) Schirmer, W., Grossmann, A., Fiedler, K., Bulow, M., Sichart, K. H., *Chem. Tech.*, in press.
(37) Spangenberg, H. J., Moller, K., Kunath, D., *Spectrochim. Acta*, in press.

(38) Steele, W. A., "The Solid-Gas Interface," Vol. 1, p. 307, Dekker, New York, 1967.
(39) Uytterhoeven, J. B., Schoonheydt, R., Liengme, B. V., Hall, W. K., *J. Catalysis* **1969**, 13, 425.
(40) Wilkins, F. J., *Proc. Roy. Soc.* **1948**, A164, 496.

RECEIVED February 11, 1970.

Addendum

It is well known that any theory of molecular adsorption which has a molecular statistical basis must lead to the Henry equation in the limiting case of small concentrations. The Polanyi theory and its modifications do not satisfy this condition. In contrast to this, the theories of adsorption which take into consideration the adsorbate–adsorbent and adsorbate–adsorbate interactions on the molecular level give the limiting Henry equation. Therefore, we can use these theories to calculate the Henry constant, the constant of adsorbate–adsorbent equilibrium. I have shown

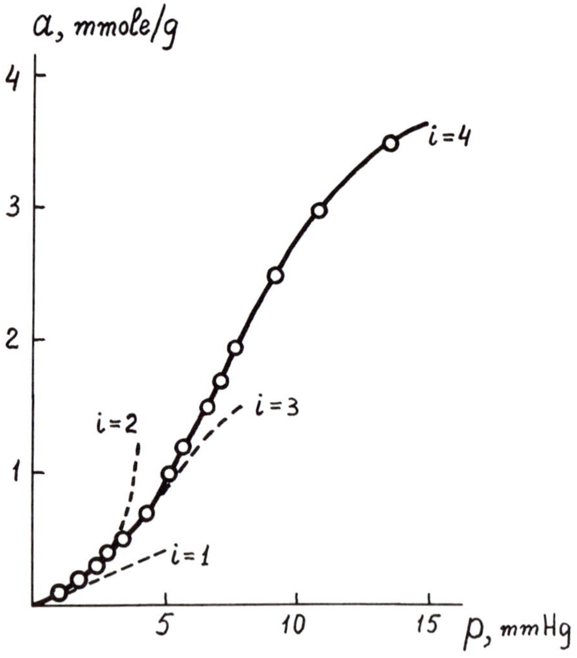

Transactions of the Faraday Society

Figure 1. Isotherm of xenon adsorption on LiNaX-1 zeolite

in my paper examples of the calculation of the Henry constant for ethane and ethylene adsorption on type X zeolites with different alkali cations (pp. 40–44) using the exponential virial equations of adsorption equilibria (Equations 2 and 4 on pp. 40 and 42). The equations describe quite satisfactorily the nonspecific and specific molecular adsorption over a wide range of temperature, both on homogeneous and heterogeneous nonporous as well as microporous crystalline adsorbents. To study the influence of the degree of the adsorbent field heterogeneity, it is possible to use the method suggested by S. Ross and T. P. Olivier ("On Physical Adsorption," Interscience, New York, 1964). This method permits us to evaluate γ, which is the characteristic of the degree of surface heterogeneity, and the value of the Henry constant K_1 [equal to exp $(-C_1)$ in Equation 2, p. 40] for the most homogeneous part of the surface. In the case of homogeneous adsorption field distribution, this method must give a very high γ value and a K_1 value which is close to that obtained from experiment (by means of virial Equation 2, p. 40).

Figure 1 shows the representation of the experimental isotherm (B. G. Aristov, V. Bosacek, A. V. Kiselev, *Trans. Faraday Soc.* **1967,** 63, 2057) of xenon adsorption on partly decationized zeolite LiX-1 (the composition of this zeolite is given on p. 185) with the aid of the virial equation in the exponential form with a different number of coefficients in the series: $i = 1$ (Henry constant), $i = 2$ (second virial coefficient of adsorbate in the adsorbent molecular field), $i = 3$, and $i = 4$ (coefficients determined at fixed values of the first and the second coefficients which are found by the method indicated for the adsorption of ethane, *see* Figure 4 on p. 41). In this case, the isotherm has an inflection point. The figure shows the role of each of these four constants in the description of this isotherm (as was also shown on Figure 3a, p. 41, for the adsorption of ethane on the same zeolite sample). The first two of these constants—Henry constant (the "first" virial constant) and second virial coefficient of adsorbate–adsorbate interaction in the field of the adsorbent —have definite physical meanings.

Table I shows that in the case of adsorption on zeolites, it is possible to determine the Henry constant by different methods (some of these

Table I. Henry Constant, K_1, for Xenon Adsorption on Zeolite LiNaX-1 at −45°C (Mmole/G. Torr)

Hill–De Boer's Method[a]	*Kiselev's Method*[a]	*From Virial Equation*[a]	*Ross and Olivier's Method*
0.021	0.018	0.019	0.019

[a] *See* references on these methods in the review: N.N. Avgul and A. V. Kiselev, "Chemistry and Physics of Carbon," P. L. Walker, Ed., vol. 6, p. 1, Dekker, New York, 1970.

Table II. Henry Constants, K_1, and Zero Coverage

Zeolite and Adsorbate	*Temperature, °C*	K_1, *Mmole/G. Torr* — *Ross and Olivier's Method*	*From Virial Equation*
		$\gamma = 1000$	
LiNaX-1, Xe	−80	0.107	0.080
	−60	0.036	0.033
	−45	0.019	0.019
	−30	0.011	0.012
		$\gamma = 70$	
NaX, Xe	−60	0.060	0.077
	−45	0.028	0.034
	−30	0.016	0.018
		$\gamma = 7$	
LiNaX-1, CO_2	25	0.056	0.353
	45	0.025	0.126
	70	0.008	0.042

determinations were made by B. G. Aristov, V. Bosacek, A. V. Kiselev, *Trans. Faraday Soc.* **1967**, 63, 2057) which take into consideration the adsorbate–adsorbate interaction. The results are in satisfactory agreement, so we can consider this value of K_1 as a physico-chemical constant for the zeolite of indicated type and composition.

Table II shows that the adsorption field in the cavities of zeolite LiX-1 (the degree of decationization is 13%) with respect to nonspecific xenon adsorption is very homogeneous (the distribution constant, γ, calculated by Ross and Olivier's method is 1000). In this case, the simple Ross and Olivier's method gives the same results as the virial equation. We have nearly the same in the case of xenon adsorption on zeolite NaX, for which the γ value is still high enough ($\gamma = 70$). But in the case of specific adsorption of CO_2 on the same zeolite (N. N. Avgul, B. G. Aristov, A. V. Kiselev, L. Ya. Kurdyukova, N. V. Frolova, *Zh. Fiz. Khim.* **1968,** 42, 2682), γ is equal only to 7 and the zeolite surface is inhomogeneous for CO_2 molecules. In this case of specific adsorption, Ross and Olivier's method gives too small values for the Henry constant in comparison with the virial equation. The latter is the best way to represent the experimental data in the initial and moderate filling region for any kind of molecular adsorption on nonporous crystals (N. N. Avgul and A. V. Kiselev, "Chemistry and Physics of Carbon," P. L. Walker, Ed., vol. 6, p. 1, Dekker, New York, 1970) and microporous crystals (A. G. Bezus, A. V. Kiselev, Z. Sedlachek, and Pham Quang Du, *Trans. Faraday Soc.*, to be published; A. G. Bezus, A. V. Kiselev, and Pham Quang Du, *J. Colloid Interface Sci.*, to be published).

Heats of Adsorption, Q_1

	Q_1, *Kcal/Mole*	
Ross and Olivier's Method	*From Virial Equation*	*Experimental (Extrapolated)*
4.1	3.8	3.9
4.6	4.7	4.7
9.2	9.8	9.8

Figure 2 shows the C_1 and Q_1 values for the adsorption of ethane and ethylene on zeolite LiX-2. This sample has only a small degree of decationization (the composition of the samples used is indicated on p. 185). In the case of this sample, LiX-2 (open and filled circles on Figure 2), the dependences of C_1 and Q_1 on cation radius are quite smooth (compare Figure 7, page 44, for highly decationized LiX-1). The influence of decationization is, of course, more pronounced in the case of specific adsorption of ethylene (filled circles). The detailed results of this work were included in a paper by A. G. Bezus, A. V. Kiselev,

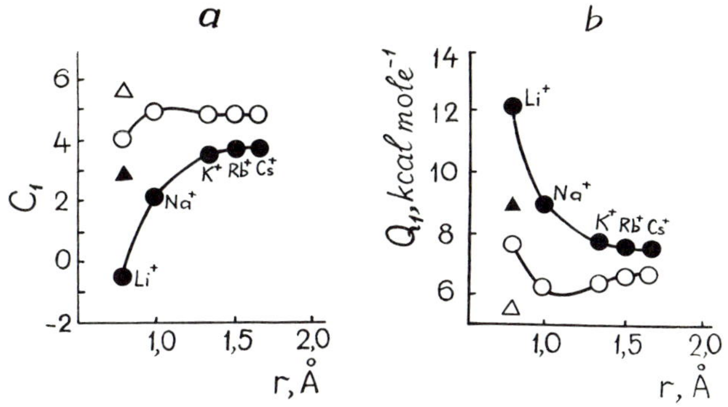

Figure 2. Dependence of (a) constant C_1 and (b) zero coverage heat of adsorption, Q_1, on exchange cation radius in type X zeolite for the adsorption of ethane (open circles) and ethylene (filled circles). The C_1 and Q_1 values for LiNaX-1 zeolite are indicated as triangles.

Table III. Adsorption of Neopentane on Site S_{II} of NaX Zeolite

Model of Zeolite Assumed	*Orientation of Molecule*	*Calculated Potential Energy*	*Experimental Heat of Adsorption at Zero Coverage Q_1, Kcal/Mole*
I	1	8.0	
	2	11.0	
		average 9.5	
			9.5
II	1	9.4	
	2	10.6	
		average 10.0	

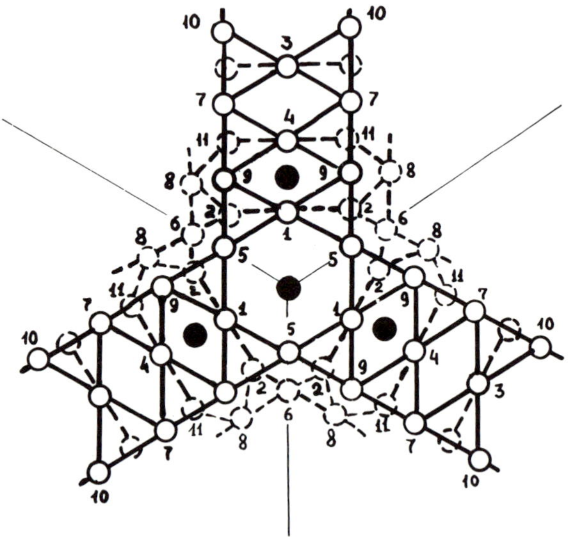

Figure 3. Force centers assumed in Model I for the calculation of the potential energy of interaction of the neopentane molecule with NaX zeolite; site S_{II} with the charged oxygen atoms (open circles) and cations (filled circles) are indicated

Z. Sedlacek, and Pham Quang Du, *Trans. Faraday Soc.*, accepted for publication.

Table II on p. 54 gives some results of our molecular statistical calculations of the heat of nonspecific adsorption of simple molecules in the cavities of type A zeolites. Molecular statistical calculation of thermodynamic characteristics of adsorption in the cavities of type X zeolites are more difficult because of the more complicated structure of the lattice.

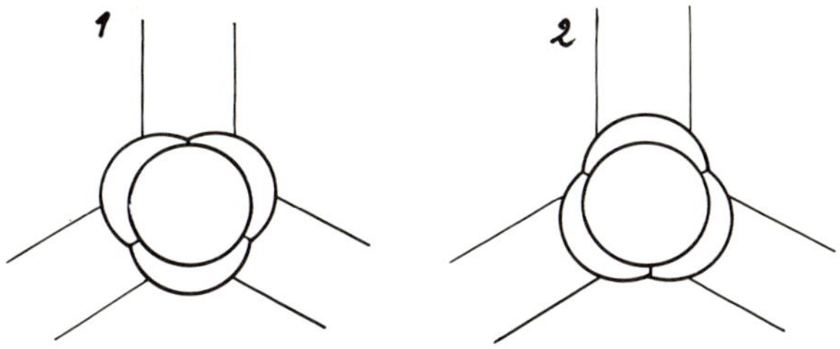

Figure 4. Two orientations of a neopentane molecule with respect to site S_{II} of type X zeolite which are assumed in the calculation of potential energy of adsorption

Table III shows the comparison of the calculated potential energy of neopentane on NaX zeolite with the experimental value of the heat of adsorption (O. M. Dzhigit, A. V. Kiselev, L. G. Ryabukhina, *Zh. Fiz. Khim.* **1970**, 44, 1790). Two models, I and II (Figure 3), of the distribution of the force centers in zeolite NaX and two orientations, (1) and (2), of the neopentane molecule at site S_{II} (Figure 4) were used. The calculations show that for approximate evaluation of potential energy, it is only necessary to take into consideration about 50 nearest charged oxygen atoms and 4 cations. In model II, the influence of other cations (which are situated in sites S_{II} of the supercage) were taken into consideration. The other cations, which are distributed more randomly, do not influence seriously the electrostatic field in the space occupied by the neopentane molecule near one of the S_{II} sites (calculations by L. G. Ryabukhina and A. A. Lopatkin).

Discussion

K. Klier (Lehigh University, Bethlehem, Pa. 18015): Referring to Figure 3,b of this paper, the isotherm at −150°C describes a liquefaction phenomenon with a saturation value which does not correspond to normal liquid density. It rather seems as if this isotherm described two-dimensional condensation. Considering that there are all indications that the two-dimensional critical temperature is much lower than the three-dimensional critical temperature (Devonshire, A. F., *Proc. Roy. Soc. (London)* **1937**, A163) no matter how strong the adsorption bond is, the theory should also show that we are getting three-dimensional condensation before the two-dimensional isotherm assumes a shape as in Figure 3,b.

A. V. Kiselev: Klier supposes that a kind of three-dimensional condensation could start before the two-dimensional type of condensation of ethane shown on Figure 3,b, but the equilibrium pressure of adsorbed ethane in zeolite at −150°C is much smaller (0.0007 torr on Figure 3,b) than the saturated vapor pressure of three-dimensional liquid at the same temperature (p_s = 4.0 torr). Figure 3,b is only a qualitative indication of the possibility of the description of a condensation-type process in the adsorbate in the zeolite because the temperature range (down to −150°C) is too large for the assumption that the heat of adsorption is independent of temperature. This question could be cleared up by direct measurement of the heat capacity of the adsorbate–zeolite system.

R. M. Barrer (Imperial College, London): Type V isotherms similar to those shown in your Figure 2 have also been found by us (Barrer and Kanellopoulos, *J. Chem. Soc.*) when equimolecular mixtures of NH_3 and HCl are sorbed between 200° and 300°C in zeolites. The sorptions of NH_3 alone and of HCl alone in the same temperature ranges are relatively small, but together they interact strongly and are copiously sorbed. Thus, interaction may be strong between unlike molecules and is responsible for the peculiar isotherm shape observed in mixed sorption.

A. V. Kiselev: Barrer indicated the interesting case when HCl and NH_3 themselves are adsorbed relatively weakly in zeolite at 200°–300°C but their mixture is adsorbed very strongly. The corresponding isotherms have approximately vertical sections. There could be two possible explanations for this phenomenon. (1) Qualitatively, it is not important what kind of adsorbate–adsorbate interaction we have: molecular nonspecific (Xe, C_2H_6, etc.), molecular specific (hydrogen bond between adsorbate molecules), or chemical (as in the case of NH_3 and HCl). The energy of the NH_3 + HCl interaction must be larger than the energies of NH_3...NH_3 and HCl...HCl interactions. As a result, the admixture of NH_3 to HCl or vice versa must lead to a rise in the adsorbate–adsorbate interaction energy in this system. (2) The adsorbate–adsorbent interaction for NH_4Cl is weaker than for HCl and NH_3 but adsorbate–adsorbate interaction for NH_4Cl is greater than HCl...HCl and NH_3...NH_3 interactions at these temperatures.

M. Dubinin (Academy of Sciences of the USSR, Moscow, USSR): The survey paper considers the fundamental problems of adsorption of vapors on zeolites within the framework of a general theory of molecular adsorption on nonporous and porous adsorbents. Among other things, it quotes examples of description of adsorption equilibria on zeolites over temperature ranges not exceeding 50° for initial and intermediate regions of isotherms by an exponential series with virial coefficients taking into account both adsorbent–adsorbate and adsorbate–adsorbate interactions. It is assumed that these equations can be used for practical calculations of adsorption equilibria.

Basically, the molecular–statistical approach to the consideration of adsorption equilibria is rational without any doubt. However, its possibilities for such structurally complex microporous systems as zeolites should not be overestimated. At best, it may indicate approximately, for instance, the virial form of the dependence of the equilibrium pressure on the adsorption value of the type:

$$p = K_1'a + K_2'a^2 + K_3'a^3 + \ldots\ldots \qquad (1)$$

If the dependence of the parameter K'_1 on temperature is theoretically substantiated, the temperature dependence of K'_2, to say nothing of the subsequent parameters, is based on empirical assumptions. It seems to me that the Wilkins type equation

$$p = a \exp\,(C_1 + C_2a + C_3a^2 + \ldots) \qquad (2)$$

is most likely purely empirical, in any case at not too small fillings.

The survey paper gives specific examples of such equations for certain systems studied. For one of them, namely for adsorption of ethane on zeolite LiX over the temperature range from 25° to 75°C and the filling range up to 0.67, it supplies all coefficients of the type of Equation 2:

$$p = a \exp\,[(14.20 - 2785/T) + (0.1364 - 106.8/T)a + (0.4029 - 89.57/T)a^2 + (0.0500 - 17.73/T)a^3] \qquad (3)$$

In order to describe adsorption equilibria for ethane on zeolite LiX from three given adsorption isotherms for temperatures of 25°, 50°, and 75°C, eight constants have to be determined, judging by Equation 3. Obviously, additional experimental data are required, such as isotherms or differential heats of adsorption. Thus, to describe adsorption equilibria by the method proposed, one should first use, in order to determine the equation constants, such extensive experimental information as will supply answers to all questions of interest to the practical worker.

The advantage of Equation 3 is that it provides an accurate description of adsorption equilibria over the ranges of temperatures and pressures used for the determination of the constants. This may be of interest, for example, in obtaining expressions in an analytical form for differential heats and entropies of adsorption most closely corresponding to experimental data.

The coefficient of the first term of the virial series in Equation 1 is the inverse Henry constant for the temperature considered. From the graph of the isotherm in Figure 3 of the survey paper, it follows that the linearity of the initial section of the adsorption isotherm of ethane on zeolite LiX in accordance with the calculated Henry constant is observed for adsorption values not exceeding 0.7 mmole/g. According to K. N.

Mikos (5), dehydrated zeolite LiX contains 3.99×10^{20} large voids per gram, corresponding to 0.68 mmole/g. Thus, for ethane, the upper boundary of the Henry region corresponds, on the average, to a maximum of one molecule per large void of the zeolite. In the case under consideration, each molecule adsorbed within the Henry region occupies a space in a separate large void of the zeolite, which leads to the constancy of the adsorption energy and to the absence of interaction among the molecules adsorbed. Judging by the graph of Figure 6 of the survey, the differential heats of adsorption of ethane do not vary significantly in this range of fillings.

The increase in the differential heats of adsorption with filling is usually attributed to the manifestation of interaction, namely of attraction between the molecules adsorbed. This effect is chiefly expressed by the second term of the virial equation. Such examples are illustrated by the curves of the graph in Figure 6 of the survey paper for adsorption of

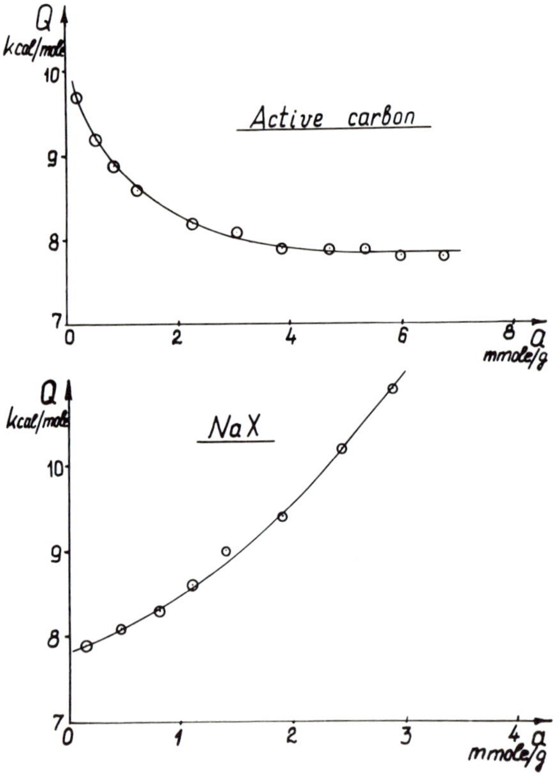

Figure 1. Dependence of differential heats of adsorption of propane, Q, on adsorption values, a, *for active carbon (above) and zeolite NaX (below)*

xenon and ethane on zeolite LiX. However, for carbonaceous microporous adsorbents, for which micropore sizes determined by the small-angle x-ray scattering method are very close to those of large voids of the X-type zeolite, the differential heats of adsorption decrease with filling in the region of small and medium fillings. Other examples are the curves of differential heats of adsorption of propane on active carbon and zeolite NaX which were studied in our laboratory (2) and are given in the graph of Figure 1. Since propane molecules are apolar and dispersion interactions between them are practically independent of the orientation of the molecules, the effect of attraction between the molecules adsorbed should manifest itself regardless of the nature of the adsorbent. In actual fact, one can observe a qualitative difference in the curves of differential heats of adsorption which calls for thorough discussion.

I believe that at this juncture elegant (as regards the general scheme) molecular–statistical calculations as applied to real adsorption systems are based on such numerous simplifications and approximations in calculation, in estimating adsorption energies among other things, that their results are rather of a qualitative nature. Their quantitative significance should not be overestimated. The multi-constant virial equations following from these concepts in empirical determination of constants naturally describe adsorption equilibria with a high degree of accuracy.

The concepts of adsorption in micropores which are being developed by us and are of general importance for microporous adsorbents of various natures are aimed at other targets. Their common goal is approximate description of adsorption equilibria and calculation of differential heats and entropies of adsorption over sufficiently wide ranges of temperature and pressure on the basis of minimum experimental information; for instance, from a single adsorption isotherm for the mean temperature. They are also aimed at substantiating an adsorption equation with a minimum number of experimentally determined constants; for instance, two. The form of this equation, which expresses the dependence of adsorption on pressure and temperature, should permit its application in developing the fundamentals of the theory of kinetics and dynamics of adsorption.

We consider a divergence of $\pm 10\%$ between calculated and experimental adsorption values acceptable for practical purposes. On this basis, criteria are established which determine the limits of applicability of equations.

The concept of volume filling of micropores under review is based on a radical difference of adsorption in micropores as a limiting case as compared with the opposite limiting case of adsorption on the surface of nonporous adsorbents of identical chemical nature. The concept of adsorption in micropores is of a thermodynamical nature, and its main initial principle, which has the meaning of a rational approximation, con-

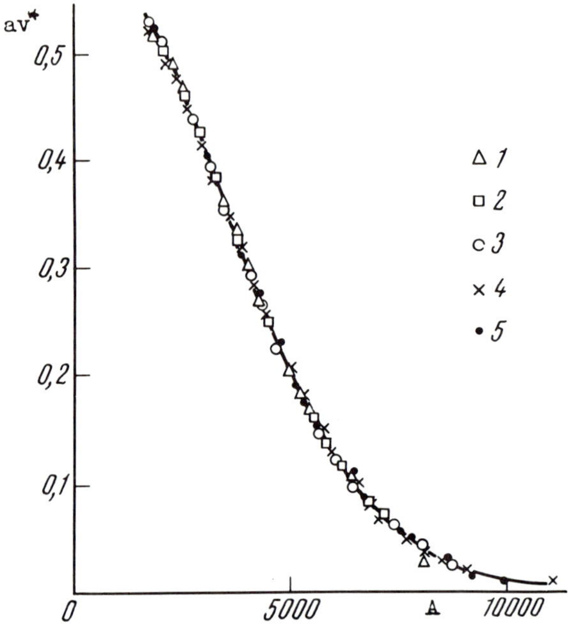

Figure 2. Characteristic curve of benzene adsorption on active carbon SK. The product av is proportional to θ (A, cal/mole; av*, cm³/g). 1 = 20°; 2 = 50°; 3 = 80°; 4 = 110°; 5 = 140°C.*

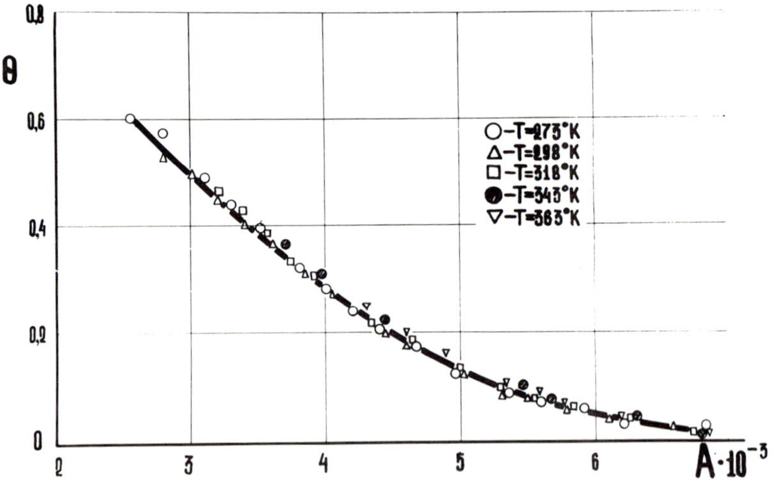

Figure 3. Characteristic curve of CO_2 adsorption on LiX; points correspond to experimental adsorption isotherms for temperatures from 273° to 363°K (A, cal/mole)

sists in the assumption of temperature invariance of the dependence of the differential molar work of adsorption or of the variation in Gibbs' free energy on the degree of filling of the micropores. This assumption is fulfilled approximately only for adsorption in micropores. The nature of the distribution of the degree of filling of micropores over differential molar works of adsorption is approximated by Weibull's statistical equation of distribution. Examples of approximate observance of the abovementioned temperature invariance and of the analytical expression of the dependence studied are in the graphs of Figures 2 to 4, in which the experimental points corresponding to isotherms for different temperatures are denoted by different symbols, while the continuous curves have been calculated on the basis of the parameters of the characteristic equation for each adsorption system. The graph of Figure 2 corresponds to adsorption of benzene on active carbon CK over the temperature range from 20° to 140°C according to experiments of E. F. Polstyanov (*4*), the graph of Figure 3 to adsorption of carbon dioxide on zeolite LiX according to experiments of N. N. Avgul *et al.* (*1*) over the range from 0° to 90°C, and the graph of Figure 4 to adsorption of water on zeolite NaX over the range from 20° to 250°C according to experiments of O. Kadlec

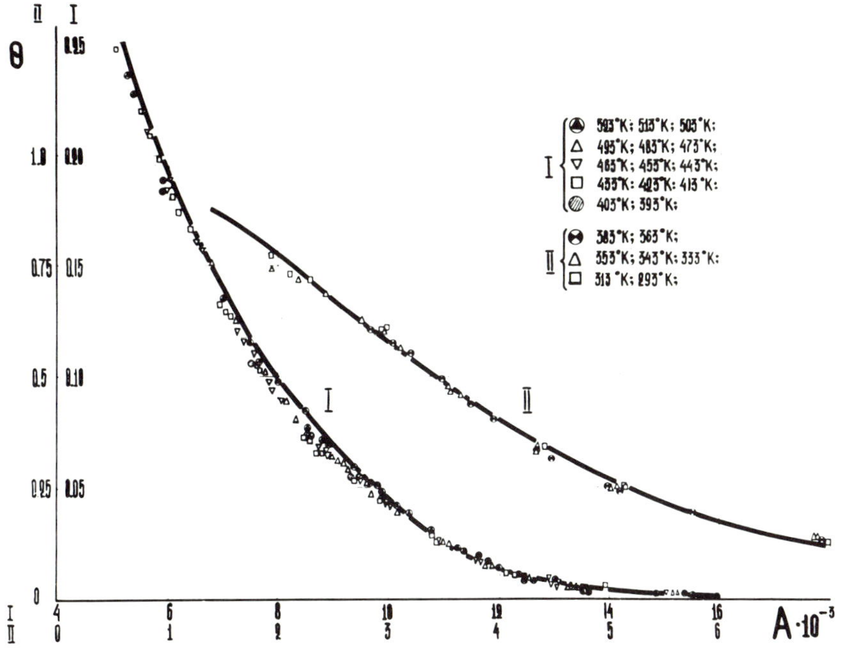

Figure 4. Characteristic curve of H_2O adsorption on NaX; temperature range from 293° to 523°K (A, cal/mole)

and A. Zukal (3). In my paper and the introduction to it, I gave examples of the description of adsorption equilibria and calculating of thermodynamic functions based on the concept of volume filling of micropores, which is in most cases applicable with a satisfactory accuracy within the range of micropore fillings from 0.2 to 1.

Naturally, the concept of volume filling of micropores does not replace, but merely supplements the concepts of adsorption developed at the so-called "molecular level," using, in particular, the molecular–statistical approach. We believe that further development of both approaches will in the long run result in the substantiation of the initial approximate propositions regarding volume filling of micropores, since they express the principal experimental facts.

Literature Cited

(1) Avgul, N. N., Aristov, B. G., Kiselev, A. V., Kurdyukova, L. Y., Frolova, N. N., *Zh. Fiz. Khim.* **1968,** 42, 2682.
(2) Bering, B. P., Zhukovskaya, E. G., Rakhmukov, B. Kh., Serpinsky, V. V., *Izv. Akad. Nauk SSSR, Ser. Khim.* **1967,** 1662.
(3) Dubinin, M. M., Kadlec, O., Zukal, A., *Coll. Czech. Chem. Commun.* **1966,** 31, 406.
(4) Dubinin, M. M., Polstyanov, E. F., *Izv. Akad. Nauk SSSR, Ser. Khim.* **1966,** 793.
(5) Mikos, K. N., thesis, Moscow, 1970.

A. V. Kiselev: Zeolites are porous crystals. This means that we can find the molecular field distribution in their channels. The advantage of describing the adsorption on zeolites using the molecular theory consists in obtaining the constants which have a definite physical meaning (for example, the Henry constant and second virial coefficient). Further development of the theory needs a further improvement of the model based on the investigation of the adsorbate–zeolite systems by the use of modern physical methods.

The other way of treating the experimental isotherm of adsorption is connected with the Polanyi assumption that the adsorbate has a density which is close to the density of the corresponding liquid. The advantage of this method consists in the possibility of calculating the isotherm using only a few constants. The disadvantage of this concept is connected with the empirical character of these constants and with the impossibility of describing by this method the initial part of the isotherm and also the isotherms having an inflexion point.

44

Description of Adsorption Equilibria of Vapors on Zeolites over Wide Ranges of Temperature and Pressure

M. M. DUBININ

Institute of Physical Chemistry, Academy of Sciences of the USSSR, Moscow V-71, USSR

V. A. ASTAKHOV

Bielorussian Technological Institute, Minsk 50, USSR

The distinguishing feature of zeolites as microporous adsorbents is the presence of cations in the micropores. These cations are centers for the adsorption of molecules with a nonuniform electron density distribution. An attempt has been made to develop the theory of volume filling of micropores for approximate description of adsorption equilibria of vapors on zeolites over wide temperature ranges. An adsorption equation has been obtained which takes into consideration, in the general case, both dispersion forces and the forces of interaction of molecules with ions. This equation describes adsorption on the active centers and the filling of the remainder of the adsorption space of the voids after the blocking of the active centers. Several examples of agreement between the results of calculation and experimental data are given.

The description of adsorption equilibria on real microporous adsorbents (*3*)—*i.e.*, active carbons, zeolites, and other fine-pore mineral adsorbents—over wide ranges of temperature and pressure may have 2 objectives: (1) The most accurate possible analytical expression of the aggregate experimental data, which is used to determine the constants of the adsorption equation. In this case, the number of constants is relatively large, and usually most of them acquire a semi-empirical nature. (2) Utilization of minimal experimental information—for instance, adsorption isotherms at 1 or 2 temperatures—for approximate calculation of adsorption equilibria over a sufficiently wide range of temperature and pressure. In this case, the adsorption equation has fewer experimental constants. The following discussion pursues this aim.

It is expedient to base the description of adsorption equilibria for microporous adsorbents on the theory of volume filling of micropores. This theory has been developed mainly for microporous carbonaceous adsorbents—*i.e.*, active carbons—for which the decisive role in adsorption interaction is played by dispersion forces (*4, 5, 6*). The theory is based on the concept of temperature invariance of the characteristic curve expressing the distribution of the degree of filling, θ, of the volume of the adsorption space according to the differential molar work of adsorption, A—an experimental fact which actually was noted by Polanyi with a different interpretation. Determining the differential molar work of adsorption as a decrease in Gibbs' free energy ($A = -\Delta G$):

$$A = RT \ln (p_s/p) = 2.303\, RT \log (p_s/p) \qquad (1)$$

where p_s is the pressure of the saturated vapor of the substance under study (standard reference state) at a temperature T or fugacity, and p is the equilibrium pressure. The characteristic curve equation of the theory may be represented as follows:

$$\theta = \exp [-kA^2] = \exp [-(A/E)^2] \qquad (2)$$

In these 2 equivalent forms of Equation 2

$$\theta = a/a_o \qquad (3)$$

where a is the adsorption at a temperature T and equilibrium pressure p, and a_o is the limiting adsorption value corresponding to the filling of the whole volume of the adsorption space W_o or of the micropore volume, and k or E are parameters of the distribution function ($E = 1/\sqrt{k}$). The limiting adsorption, a_o, depends on the temperature as a result of the thermal expansion of the substance adsorbed. We neglect the temperature changes of W_o. If ρ^* is the density of the substance adsorbed at a limiting micropore filling, which easily can be calculated to a good approximation according to M. M. Dubinin and K. M. Nikolaev from the physical constants of the substance (densities of the bulk liquid to the boiling temperature, the constant b in the van der Waals equation calculated from the critical temperature and pressure) (*10*), then

$$a_o = W_o\, \rho^* \qquad (4)$$

If in Equation 2, we express A as in Equation 1 and θ as in Equation 3, we will obtain the adsorption isotherm equation for the given temperature, T. Reference *4* quotes examples of experimental verification over wide ranges of temperature and equilibrium pressures of these equations of the theory of volume filling of micropores for the adsorption of various vapors on active carbons with different parameters of the microporous structure.

Generalization of the Concept of Volume Filling of Micropores to Adsorption of Gases and Vapors on Zeolites

The generalization under review is based on an analysis of adsorption equilibria of various gases and vapors over wide ranges of temperature and pressure on different types of zeolites. For this purpose, use is made of the authors' own experimental data and of published results of investigations by other workers. This experimental material is analyzed and discussed from a new point of view.

The distinguishing feature of dehydrated zeolites as microporous aluminosilicate adsorbents lies in the presence in their voids—*i.e.*, micropores—of cations. These cations compensate excess negative charges of their aluminosilicate skeletons. The cations form, in the zeolite micropores, centers for the adsorption of molecules with a nonuniform distribution of the electron density (dipole, quadrupole, or multiple-bond molecules) or of polarizable molecules. These interactions, which will be called, somewhat conventionally, electrostatic interactions, combine with dispersion interactions and cause a considerable increase in the adsorption energy. As a result, the adsorption isotherms of vapors on zeolites, as a rule, become much steeper in the initial regions of equilibrium pressures as compared with isotherms for active carbons.

The total amount of Na^+ cations in dehydrated zeolite crystals in passing from NaA ($x = SiO_2/Al_2O_3 = 2$) to the zeolite NaY ($x = 5$) varies from 7.2 to 4.2 mmole/gram. A typical example is zeolite NaX ($x = 2.96$), for which the amount of cations is equal to 5.9 mmole/gram. The principal results of adsorption investigations given below have been obtained by us and by A. V. Kiselev on this sample of zeolite, which was synthesized by S. P. Zhdanov. If we exclude, for this zeolite, cations in position s_1 inside six-membered oxygen bridges, which are inaccessible to the molecules being adsorbed (2 cations per large void), the amount of cations in large zeolite voids will be about 4.7 mmole/gram. Of this amount, 2.4 mmole/gram are localized in six-membered windows of cubeoctahedra and 2.3 mmole/gram are nonlocalized. It is not clear yet whether cations not localized in large voids can be adsorption centers—*i.e.*, localization centers of the molecules adsorbed. Therefore, the amount of active centers in the zeolite NaX under study is at least 2.4 mmole/gram.

The maximal adsorption value, a_0, and the average number, N, of molecules per large void of zeolite ($x = 2.96$) depend on the size of the molecules adsorbed. The experimental values of a_o and the calculated values of N are given in Table I.

Table I. Limiting Adsorption Values

No.	*Substance*	T,°*K*	a_0, *Mmole/Gram*	N
1	Water	293	20.26	33.4
2	Oxygen	90	10.72	18.0
3	Argon	90	10.26	17.0
4	Nitrogen	77	9.71	16.5
5	Benzene	293	3.30	5.5
6	*n*-Pentane	293	2.56	4.2
7	Cyclohexane	293	2.32	3.8

Table I shows that, for the small molecules, substances 1–4, the maximum number of adsorbed molecules considerably exceeds the number of adsorption centers of the zeolite. For the larger molecules, substances 5–7, the maximum number of adsorbed molecules is close to the number of adsorption centers. Therefore, 2 limiting cases are typical for adsorption on zeolites. The first case corresponds to the adsorption of relatively larger molecules (as compared with the void sizes for the zeolite in hand), which is determined to a great extent by the interaction of the molecules adsorbed with the adsorption centers of the zeolite even for the maximal filling of the zeolite voids. In the second case, after the filling of the adsorption centers, there may remain a free space in the zeolite voids for adsorption as a result of the manifestation of both dispersion forces (adsorbent–adsorbate interaction) and the forces of interaction between the molecules adsorbed (adsorbate–adsorbate interaction).

Let us first consider the most common case. An analysis of many adsorption isotherms on zeolites of various vapors with relatively large molecules has shown that the characteristic curves are expressed by an equation similar to Equation 2, but with power (distribution order, n) higher than 2:

$$\theta = \exp\left[-(A/E)^n\right] \tag{5}$$

where n are integers from 3 to 6. In most cases, Equation 5 satisfactorily describes experimental data over the range of fillings, θ, from $\sim$0.1 to 1 as distinct from Equation 2, which is applicable only to the region of high fillings from 0.8 to 1.

Let us now consider the basic properties of the characteristic Equation 5, which is a Weibull distribution equation (*13*). Irrespective of the distribution order, n, the curves expressed by it have 2 points in common: a point corresponding to maximal adsorption $a = a_o$ ($\theta = 1$) and a characteristic point at $a = a_o/e$, where e is the natural logarithm base ($\theta = 0.368$). If the differential molar work of adsorption for the characteristic point is A_o, then, according to Equation 5, $E = A_o$. Thus, the parameter E of Equation 5 is expressed, like A, in units of energy per mole of the substance adsorbed. We will call E the characteristic adsorption energy and express it in cal/mole. A thermodynamical analysis has shown that the characteristic adsorption energy, E, practically coincides with the net differential heat of adsorption for the characteristic point. Table II lists, as an illustration, the characteristic adsorption energy, E, of a number of various temperatures calculated from experimental adsorption isotherms on zeolite NaX ($x = 2.96$) according to experiments of E. F. Polstyanov (*11*) (pentane, benzene, and cyclohexane), B. P. Bering and V. V. Serpinsky (*2*) (propane), and A. V. Kiselev (*8*) (hexane). On the basis of the adsorption values for the characteristic points of the isotherms, which are equal to $a = 0.368\ a_o$ according to Equation 3, the equilibrium pressures, p, were found by interpolation, and $A_o = E$ were calculated from Equation 1. The limiting volumes of the adsorption space, W_o, were determined from the adsorption isotherm for each vapor at the lowest temperature according to Equation 4, and the values of a_o were calculated from the same equation for other temperatures with the condition $W_o =$ constant; that is:

$$a_o = a_o^\circ\ (\rho_o^*/\rho^*) \tag{6}$$

In Equation 6, a_o° is an experimentally determined adsorption value for a temperature T_o, and a_o the calculated limiting adsorption value for a pre-assigned temperature, T. In the case of zeolites, the limiting adsorption volumes, W_o, are perceptibly reduced with an increase in the size of the molecules adsorbed. Thus, W_o acquire the nature of effective values.

As seen from Table II, the deviations of the value of E from the average values for the given substances, with the exception of benzene, are small. The temperature invariance of E implies the temperature invariance of the characteristic Equation 5. For benzene ($n = 4$) over the temperature range from 353° to 413°K, the value of E decreases perceptibly with rising temperature. Deviations from the average value of

Table II. Characteristic Adsorption Energies

Propane		n-*Pentane*		*Cyclohexane*		n-*Hexane*		*Benzene*	
T,°*K*	E	T,°*K*	E	T,°*K*	E	T,°*K*	E	T,°*K*	E
253	4150	273	6080	293	6190	473	6696	353	9250
298	4150	293	6120	323	6160	498	6645	383	9020
313	4160	323	6040	353	6150	423	6605	413	8790
333	4160	353	6080	383	6150	548	6607		
353	4080	383	6020	413	6010	573	6659		
369	4190								

E are equal to ±2.5% and are matched by deviations of the calculated values of fillings, θ, (and accordingly of the adsorption values) by ±9.5% from the value $\theta = 0.368$ for the characteristic point. At lower temperatures, considerable deviations are observed. On the other hand, for the adsorption of acetylene on active centers of zeolite NaA, the value of E remains practically constant ($E = 8090$ cal/mole) over the temperature range from 273° to 423°K.

Naturally, for adsorption on cations in zeolite voids, which we have called electrostatic adsorption, the concept of the invariance of characteristic curves is a rather crude approximation. This is directly indicated by characteristic curves constructed from experimental adsorption isotherms for a sufficiently wide temperature range. However, for the purposes of practical description of adsorption equilibria, such an approximate assumption in a controlled temperature range is reasonable. A general evaluation will be made below.

From Equations 5 and 3, it is easy to obtain the expression

$$n = \ln\ln (a_o/a)/\ln (A/E) \qquad (7)$$

which serves to estimate the value of the exponent n, assuming that a/a_o is not too close to 0.368, *i.e.*, to the characteristic point. Usually, the calculated value of n differs only insignificantly from an integer, which is taken as the order of distribution of n in the characteristic Equation 5.

We now will consider the second case of adsorption on zeolites A and X of relatively small molecules, when, at the initial and final stages of the filling of the adsorption space of zeolites, adsorption is due to the predominant manifestation of various forms of adsorption interactions. Then for each variety of predominant interaction, for instance dispersion or electrostatic interaction, Equation 7 may be represented in the form

$$\theta_n = \exp [-(A/E_n)^n] \qquad (8)$$

Here, the distribution order n, the characteristic energy E_n, and the maximal adsorption value will be different for each interaction. In accordance with Equation 3 for θ and with Equation 8, the total adsorption value may be written thus:

$$a = a_{o2}\,\theta_2 + a_{on}\,\theta_n \qquad (9)$$

where $n = 2$ in the first term of Equation 9. Proceeding from the concept of the volume filling of micropores with a limiting volume of the adsorption space, W_o, the limiting adsorption value (in principle at $p/p_s = 1$ when all $\theta_n = 1$) will be expressed as:

$$W_o\rho^* = a_{o2} + a_{on} \qquad (10)$$

From Equation 10, we obtain for limiting adsorption a_{o2}, considering the dispersion forces,

$$a_{o2} = W_o\rho^* - a_{on} \tag{11}$$

If no adsorption space remains for adsorption under the effect of dispersion forces because of the size of the molecules already adsorbed on the active centers—*i.e.*, $a_{o2} = 0$—we obtain from Equation 10:

$$a_{on} = W_o\rho^* \tag{12}$$

According to the data of analysis of many adsorption systems, the first term in Equation 9 corresponding to the second order appears only when considering adsorption of relatively small molecules. They include molecules of linear shape, such as the diatomic gases, carbon dioxide, carbon monoxide, etc. Experimentally realizable orders, n, are integers from 3 to 6 in the general case. With larger polyatomic molecules, no adsorption space remains in the zeolite voids for final adsorption under the effect of dispersion forces. Then Equation 9 retains only the second term, and a_{on} is expressed by Equation 12.

The number of adsorption centers of zeolite a_{on} can be determined by different methods—for instance, from the adsorption isotherm of water at a temperature of about 300°C. Under these conditions, the contribution of the first term of Equation 9 to the total adsorption value in the initial region of the isotherm is negligibly small. For zeolite NaX, we first adopt the average number of adsorption centers for their possible range estimated above from the zeolite composition, which is about 3.5 mmole/gram, or an adsorption value of about 1.2 mmole/gram for the characteristic point. From Equation 1, we find an approximate value of $E = A_o$ and, using Equation 7, we estimate the exponent, n, which is close to 4. Assuming $n = 4$, we obtain, by the method of successive approximations, $a_{on} = 2.72$ mmole/gram. This value of a_{on} was adopted as the number of the adsorption centers of zeolite NaX and was used with an insignificant correction for cases of adsorption of other gases and vapors with relatively small molecules—for instance, CO_2.

Our experiments on the blocking of adsorption centers (*i.e.*, zeolite cations) by preadsorbed water molecules serve to substantiate the physical meaning of Equation 9. For the adsorption of carbon dioxide on dehydrated crystals of zeolite NaX, $E_2 = 3470$ cal/mole, $n = 3$, $a_{o3} =$ 2.90 mmole/gram, and $E_3 = 5200$ cal/mole, the second term of Equation 9 expressing adsorption on active centers, which amount to 2.90 mmole/gram. Water is adsorbed energetically on active centers ($n = 4$, $E_4 =$ 9150 cal/mole), and as a result of preadsorption of 3.5 mmole/gram of water vapor on zeolite, all the active centers are practically blocked and the adsorption space of the zeolite is reduced only by 19.5%. The equilibrium pressure of preadsorbed water at 20°C is of the order of 0.001 torr and practically does not affect the measurements of equilibrium pressures in subsequent adsorption of carbon dioxide. As a result of the blocking of adsorption centers, adsorption of carbon dioxide is expressed only by the first term of Equation 9 at $E_2 = 3050$ cal/mole. A further increase in the amount of preadsorbed H_2O reduces E_2 only slightly. Thus, if adsorption on cations is excluded, the zeolite becomes an analogue of microporous carbonaceous adsorbents to which Equation 2 is applicable.

Equation 5 at $n = 2$ describes, to a good approximation, the adsorption of vapors on active carbons, for instance, the adsorption of benzene with a variation of the characteristic energy E_2 from 3000 to 6000 cal/mole. However, for active carbons with the finest micropores, when E_2 of benzene substantially exceeds 6000 cal/mole—for example, for active carbon obtained from polyvinylidene chloride ($E_2 = 7240$ cal/mole)—Equation 5 with $n = 2$ is applicable only to $\theta \leqslant 0.5$ with the effective

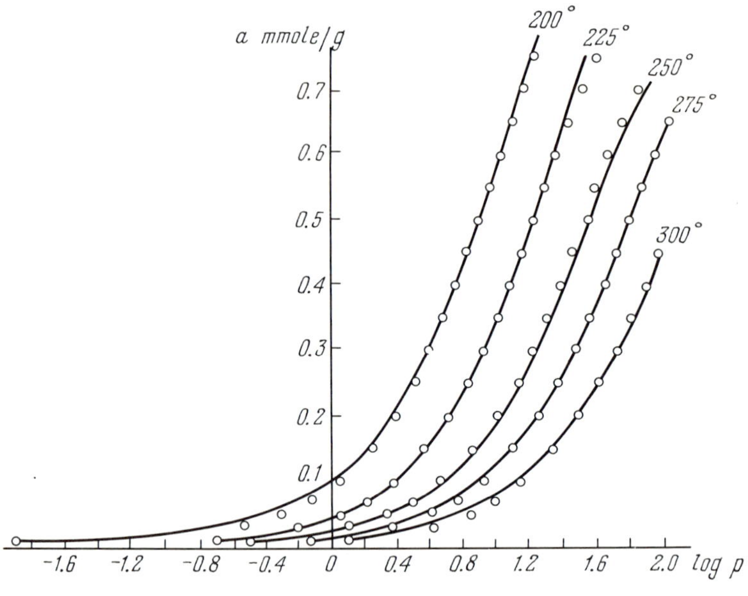

Zhurnal Fizicheskoi Khimii

Figure 1. Adsorption isotherm of n-*hexane vapor on zeolite NaX* (x = 2.96) *at different temperatures* (8)

Solid lines: Equation 5 (using f_s*),* $W_o = 0.226$ *cm³/gram, the values of* a_o *were calculated by Equation 4,* $n = 4$*, and* $E_{4} = 6642$ *cal/mole*
Circles: experimental points

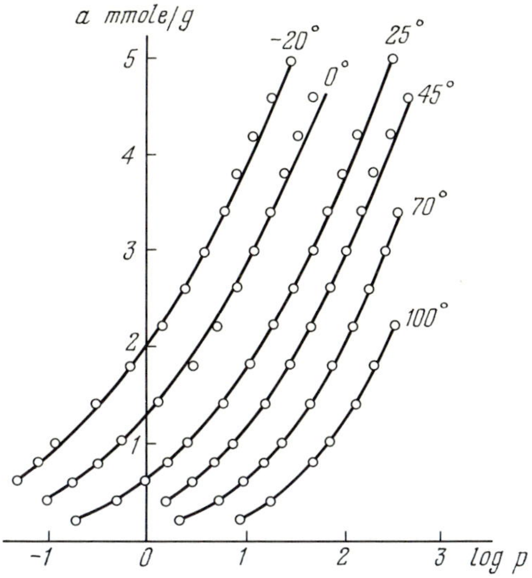

Zhurnal Fizicheskoi Khimii

Figure 2. Adsorption isotherms of carbon dioxide on zeolite NaX (x = *2.96) at different temperatures* (1)

Solid lines: Equation 9, W_o = *0.345 cm³/gram,* E_2 = *3470 cal/mole,* n = *3,* a_{o3} = *2.90 mmole/gram, and* E_3 = *5200 cal/mole*
Circles: experimental points

value of W_o almost 1.5 times the real volume of the carbon micropores. But with $n = 3$, Equation 5 approximates quite satisfactorily the characteristic curve over the range $\theta = \sim 0.1\text{–}1$, with the value of W_o corresponding to the real micropore volume. Thus, for the adsorption of vapors as a result of the manifestation of dispersion forces on active carbons with the finest micropores, the order of the characteristic equation becomes equal to 3. Perhaps for a similar reason, the adsorption of acetylene on molded zeolite NaA [experiments of the Linde Co. (9)] is expressed by an equation similar to Equation 9, but with $n = 3$ for adsorption caused by dispersion forces, *i.e.*

$$a = a_{o3}\ \theta_3 + a_{on}\ \theta_n \tag{13}$$

For adsorption on active centers of zeolite, $n = 5$. In this case, for dehydrated zeolite, $E_3 = 4870$ cal/mole and $E_5 = 8090$ cal/mole. Preadsorption of 5.1% water practically blocks the adsorption centers, the second term of Equation 13 disappears, and the equation for acetylene becomes a one-term equation with $E_3 = 4590$ cal/mole. A further increase in the preadsorbed amount of water to 19.1% is accompanied by a decrease in E_3 to 4030 cal/mole.

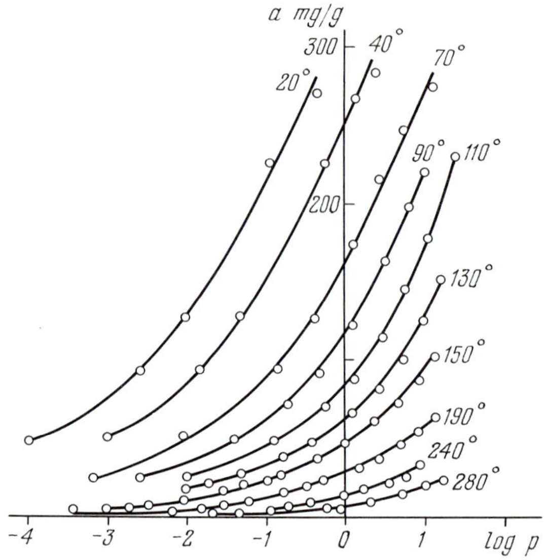

Figure 3. Adsorption isotherms of water vapor on zeolite NaX (x = 2.96) at different temperatures (7); densities of water in adsorbed state ρ were calculated after Ref. 7*

Solid lines: Equation 9, $W_o = 0.365$ cm³/gram, $E_2 = 3660$ cal/mole, $n = 4$, $a_{o4} = 2.72$ mmole/gram = 49.0 mg/gram, and $E_4 = 9150$ cal/mole
Circles: experimental points

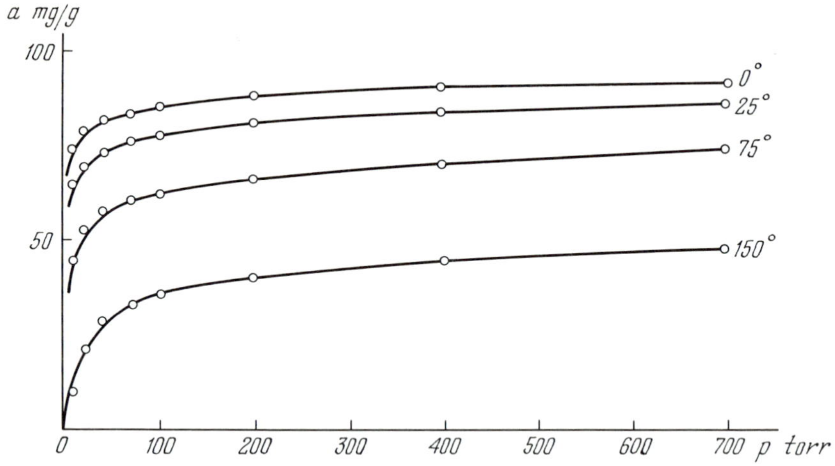

Figure 4. Adsorption isotherms of acetylene on pellets of zeolite NaA [experiments of the Linde Co. (9)]

Solid lines: Equation 13, $W_o = 0.172$ cm³/gram, $E_3 = 4870$ cal/mole, $n = 5$, $a_{o5} = 2.23$ mmole/gram = 58.1 mg/gram, and $E_5 = 8090$ cal/mole
Circles: experimental points

Examples of Description of Adsorption Equilibria on Zeolites

With the aid of a computer, about 40 adsorption systems have been analyzed for equilibrium. Typical examples are presented in the graphs of Figures 1 to 4, where the solid curves represent calculated adsorption isotherms and the circles denote experimental points. The temperatures are expressed in degrees centigrade and pressures in mm of Hg (torr). From above-critical temperatures, for instance, *n*-hexane ($t_c = 235°C$) and acetylene ($t_c = 36°C$), effective values have been obtained by extrapolating the linear dependence of ρ^* on t according to Equation 5 for the temperatures indicated. Effective values of p_s for $t > t_c$ were calculated by the van der Waals equation, which may be written in the form (*12*)

$$p_s = p_c \exp\left[\gamma \left(T - T_c\right)/T\right] \tag{14}$$

the value for *n*-hexane and acetylene being $\gamma = 7.00$ and 6.61 respectively. This equation shows good agreement with experimental data in the range of p_s from 1 atm to p_c.

For a number of systems, for instance, for carbon dioxide and acetylene in the examples quoted, $E_n/n \cong$ constant.

The examples above demonstrate satisfactory agreement between the calculated results and the experimental data. This shows that the initial approximate assumptions are reasonable. In most cases, the one-term Equation 5 is applicable for the description of adsorption equilibria on zeolites, particularly for zeolites with small voids (zeolite L, chabasite, erionite, mordenite) for which, in adsorption of hydrocarbons, $n = 3$ as a rule. The concept of the volume filling of micropores makes it possible to describe adsorption equilibria over sufficiently wide ranges of temperatures and pressures (using f_s instead of p_s) with the use of only 3 experimentally determined (usually from 1 adsorption isotherm for the average temperature) constants, W_o, A, and n. The constant n requires only a tentative estimation, since it is expressed by an integer.

Literature Cited

(1) Avgul, N. N., Aristov, B. G., Kiselev, A. V., Kurdyukova, L. Ya., *Zh. Fis. Khim.* **1968**, 42, 2678.
(2) Bering, B. P., Zhukovskaya, E. G., Rahmukov, B. H., Serpinsky, V. V., *Izv. Akad. Nauk SSSR, Ser. Khim.* **1967**, 1662.
(3) Dubinin, M. M., *Advan. Colloid Interface Sci.* **1968**, 2, 217.
(4) Dubinin, M. M., "Chemistry and Physics of Carbon," 2, 51, Marcel Dekker, New York, 1966.
(5) Dubinin, M. M., *J. Colloid Interface Sci.* **1967**, 33, 487.
(6) Dubinin, M. M., Symposium on Surface Area Determination, Bristol, 1969.
(7) Dubinin, M. M., Kadlec, O., *Czech. Chem. Commun.* **1966**, 31, 406.

(8) Garkavenko, L. G., Dzhigit, O. M., Kiselev, A. V., Mikos, K. N., *Zh. Fiz. Khim.* **1968,** 42, 1033.
(9) Linde Co., Isotherm Data, Sheet No. 43, Acetylene–Molecular Sieve Type 4A Pellets.
(10) Nikolaev, K. M., Dubinin, M. M., *Izv. Akad. Nauk SSSR, Otd. Khim. Nauk* **1958,** 1165.
(11) Polstyanov, E. F., Dubinin, M. M., "Zeolites, Their Synthesis, Properties, and Application," p. 109, Publ. Sci. USSR, Moscow, 1965.
(12) Van der Waals, J. D., *Koninkl. Akad. Belg. Kl. Wetenshap. Verhandel.* **1896,** 5, 248.
(13) Weibull, W., *J. Appl. Mech.* **1951,** 18, 293.

RECEIVED January 23, 1970.

Addendum

Synthetic zeolites of various types differ in the number of cations in their voids which are accessible for direct interaction with the molecules adsorbed. Table I lists, for typical examples of zeolites, the numbers of accessible cations N_a per zeolite void and their number Z in mmole/gram for dehydrated zeolites. When passing from zeolite NaA to zeolite L, the number of accessible cations *Z—i.e.,* the number of adsorption centers in the void—decreases almost by a factor of 10. Therefore, in the case of zeolite L, the relative role of interactions among cations and molecules adsorbed, conventionally called electrostatic, will be approximately one order lower than for zeolite NaA. In adsorption on this zeolite of substances with slightly pronounced nonuniformity of distribution of electron density in molecules—for instance, saturated hydrocarbons—one may expect that electrostatic interactions will not play the decisive role. As a result, we obtain the limiting case of adsorption on zeolites like zeolite L and erionite with a weak electrostatic interaction.

Examples are adsorption equilibria on zeolite L of methane within the temperature range from −117° to −30°C studied in a work of Barrer

Table I. Basic Data on Cations in Zeolites

Zeolite	SiO_2/Al_2O_3	*Cations per Void* Total N	Accessible N_a	*Z, Mmole/Gram*
NaA	2	12	12	7.1
Na–Faujasite	2.2–.5	11.4–6.9	9.4–4.9	5.7–2.9
Na, K–Chabazite	4	4	3	3.8
Na, K–Erionite	6.6	4	2	1.7
Na–Mordenite	10	4	2	1.3
Na, K–Zeolite	6	9	2	0.81

and Lee (*1*). This case is also of general interest because the critical temperature of methane, −82.5°C, lies in this interval.

If we neglect the difference between electrostatic and dispersion interaction energies, then, in accordance with the concept of volume filling of micropores described previously, the equation of adsorption of methane on zeolite L will be expressed by Ref. *3* as

$$a = a_o \exp\left[-(A/E)^3\right] \tag{1}$$

where E is the characteristic adsorption energy, the exponent of Equation 1, $n = 3$ for an adsorbent with fine micropores, and A the differential molar work of adsorption

$$A = RT\ln(f_s/p) \tag{2}$$

equal (with a minus sign) to the variation in Gibbs' free energy.

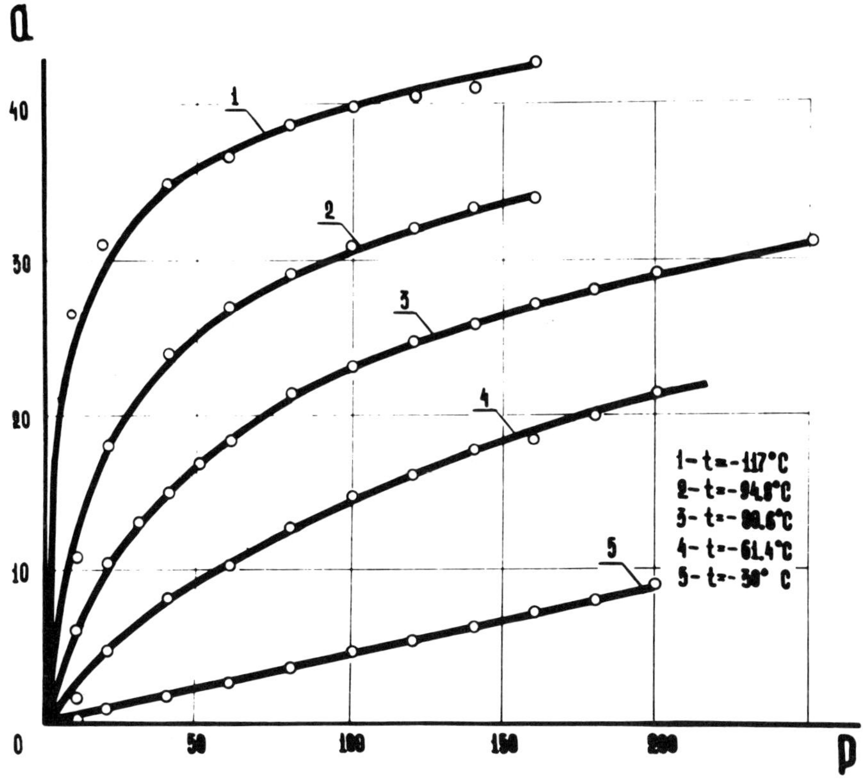

Figure 1. Adsorption isotherms of methane on zeolite L for temperatures: 1, −117.2°; 2, −94.8°; 3, −80.6°; 4, −61.4°; 5, −30°C (a, *Cm*³*NTP/G;* p, *Torr)*

$$A = -\Delta G \qquad (3)$$

We take, as a standard reference state, the state of the bulk liquid methane of the same temperature T in equilibrium with a vapor of fugacity f_s. For supercritical temperatures, we assumed extrapolated effective values of fugacity f_s.

The initial computational data—*i.e.*, the characteristic energy E and the limiting adsorption value $a_o = a_o{}^\circ$—were determined from the graph of a single experimental adsorption isotherm for a mean temperature, $t_o = -80.6°C$, represented in the linear form of Equation 1:

$$\log a = \log a_o - (0.434/E^3)A^3 \qquad (4)$$

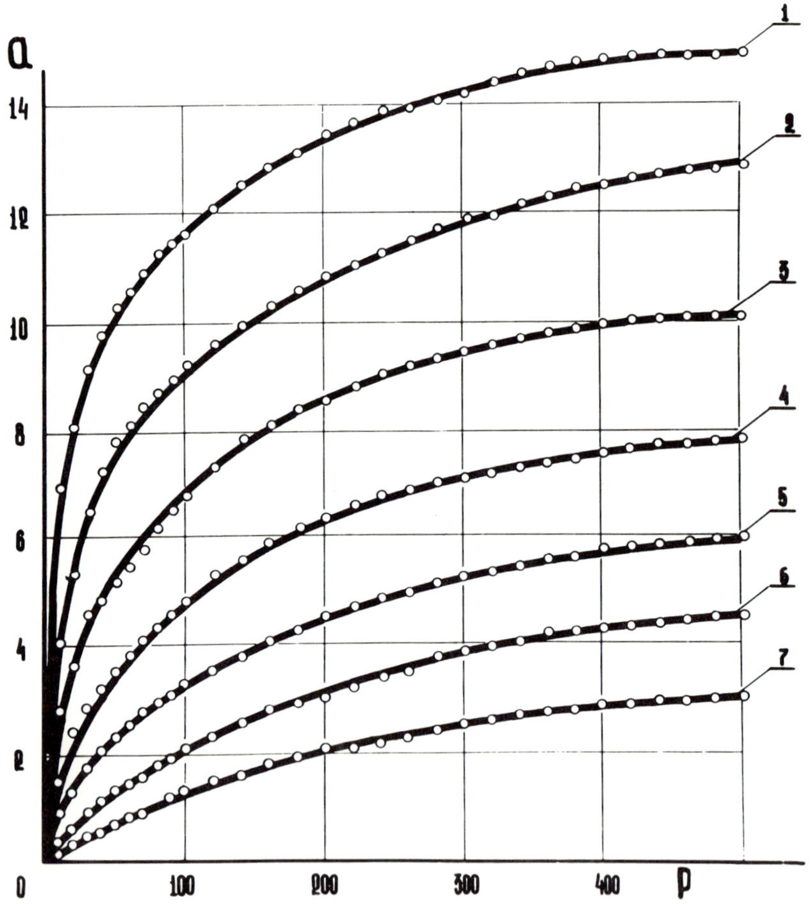

Figure 2. Adsorption isotherms of carbon dioxide on K,Na-erionite for temperatures: 1, 20°; 2, 40°; 3, 60°; 4, 80°; 5, 100°; 6, 120°; 7, 140° (a, %; p, Torr)

We obtained $E = 2350$ cal/mole and the limiting adsorption value, $a_0^\circ = 46.8$ cm^3 NTP/g, at $t_o = -80.6$°C. The values of a_o for other temperatures were calculated with the use of the coefficient of thermal expansion of the adsorbate, $\alpha = 1.51 \times 10^{-3}$ l/deg, computed according to the scheme of the Dubinin–Nikolayev method on the basis of the physical constants of methane (*4*).

$$a_o = a_o^o \exp\left[-\alpha(t - t_o)\right] \tag{5}$$

In Figure 1, the continuous curves depict adsorption isotherms calculated from Equation 4. Experimental points are denoted by circles. The calculation and experimental results are in good agreement. A similar example is illustrated in Figure 2, showing experimental and calculated (from Equation 4) adsorption isotherms of carbon dioxide on Na,K-erionite. The data used in calculation were $E = 5250$ cal/mole and $a_0^\circ = 12.4\%$ at $t_o = 80$°C. Thus, the general nature of gas and vapor adsorption on zeolites at weak electrostatic interactions is similar to adsorption on active carbons with the finest micropores (*3*).

In the case of adsorption in micropores, when the condition of temperature invariance of the characteristic curve is fulfilled, the net differential molar heat of adsorption, q, and the differential molar entropy of adsorption, ΔS, may be expressed (*2*) as

$$q = A - \alpha T\left(\frac{\partial A}{\partial \ln a}\right)_T \tag{6}$$

and

$$\Delta S = \alpha\left(\frac{\partial A}{\partial \ln a}\right)_T \tag{7}$$

Substituting into Equations 6 and 7 the values of the derivative and A calculated from Equation 1, we obtained the following expressions for differential entropy and differential heat of adsorption (*2*):

$$q = E\left[(\ln a_o/a)^{1/3} + \frac{\alpha T}{3}(\ln a_o/a)^{-2/3}\right] \tag{8}$$

and

$$\Delta S = -\frac{\alpha E}{3}(\ln a_o/a)^{-2/3} \tag{9}$$

Note that the degree of filling of the limiting adsorption space is:

$$\theta = a/a_o \tag{10}$$

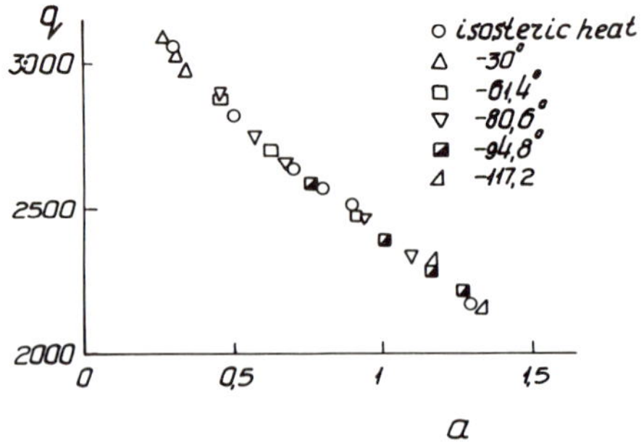

Figure 3. Dependence of net differential heat of adsorption of methane, q, *on adsorption value,* a, *for zeolite* L *(*q, *Cal/Mole,* a, *Mmole/G)*

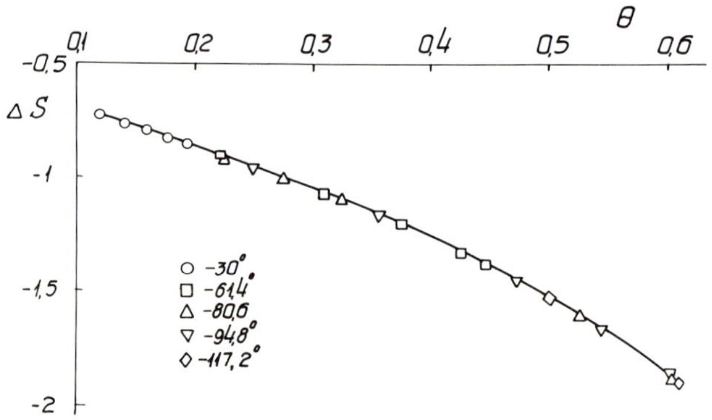

Figure 4. Dependence of differential molar entropy of adsorption, ΔS, *on degree of filling,* θ, *for zeolite* L (ΔS, e.u.)

Figure 3 shows differential heats of adsorption of methane on zeolite L. The circles denote isosteric heats of adsorption (*1*). The other symbols denote differential heats of adsorption for various temperatures corresponding to experimental adsorption isotherms. These heats were computed by Equation 8. The example quoted points of satisfactory agreement between the theoretical and experimental data.

The graph of Figure 4 depicts the dependence curve of the differential entropy of adsorption of methane on zeolite L on the degree of filling, θ. The curve was computed by Equation 9. According to it, temperature has no effect on this curve since the calculated points in the range of experimental adsorption values for various temperatures fall on one and the same curve. Note that in the range of fillings considered, the differential entropy of adsorption varies monotonically.

Literature Cited

(1) Barrer, R. M., Lee, J. A., *Surface Sci.* **1968,** 12, 354.
(2) Bering, B. P., Gordeeva, V. A., Dubinin, M. M., Efimova, L. I., Serpinsky, V. V., *Izv. Akad. Nauk SSSR, Ser. Khim.,* Commun. 4 (in press).
(3) Dubinin, M. M., Astakhov, V. A., *Izv. Akad. Nauk SSSR, Ser. Khim.,* Commun. 1–3 (in press).
(4) Nikolayev, K. M., Dubinin, M. M., *Izv. Akad. Nauk SSSR, Otd. Khim. Nauk* **1958,** 1165.

RECEIVED November 23, 1970.

Discussion

G. V. Tsitsishvili (Academy of Sciences of the Georgian SSR, Tbilisi, USSR): The new method of description of the adsorption isotherms allows one, on the basis of the theory of volume filling of micropores, to characterize adsorption on the microporous adsorbent containing adsorption sites of different nature. For zeolites, there are approximately two adsorption sites: cations and framework. The existence of cations in different positions must be taken into account. It is important to use the Dubinin approach to desorption–adsorption equilibrium on decationized and other forms of zeolites. We have already obtained some good results in this direction.

45

Ultraviolet Spectroscopic Investigation of Molecular Adsorption and Ionization of Molecules Adsorbed on Zeolites

V. I. LYGIN

Department of Chemistry, Lomonosov University, Moscow, USSR

A study was made of the ultraviolet spectra of benzene, alkyl-, amino-, and nitro-derivatives of benzene, diphenylamine, triphenylmethane, triphenylcarbinol, and anthraquinone adsorbed on zeolites with alkali exchange cations, on Ca- and Cu-zeolites, and on decationized zeolites. The spectra of molecules adsorbed on zeolites totally cationized with alkali cations show only absorption bands caused by molecular adsorption. The spectra of aniline, pyridine, triphenylcarbinol, and anthraquinone adsorbed on decationized zeolite and Ca-zeolite are characterized by absorption of the corresponding compounds in the ionized state. The absorption bands of ionized benzene and cumene molecules appear only after uv-excitation of the adsorbed molecules. The concentration of carbonium ions produced during adsorption of triphenylcarbinol on Ca-zeolite and on the decationized zeolite depends on the degree of dehydroxylation of the zeolite.

Ultraviolet spectra make it possible to establish the following: the type of the electron system in the molecules and in exchange cations of transition elements in zeolites which take part in the interaction (*4, 5, 11, 12, 25*); the degree of perturbation of the ground and excited levels (*9, 15*); and the role played by the excited states of the molecules in the molecular adsorption and ionization of the molecules (*10*). In the case of interaction involving charge transfer, ultraviolet spectra can be used to analyze the nature of the ions produced from the adsorbed molecules, and thus the nature of the acid centers in zeolites can be determined (*7, 8, 9, 10*).

In the present work, a study was made of the ultraviolet spectra for a series of aromatic compounds which were either in the molecular state or in the ionized state during adsorption on zeolites.

Experimental

The zeolites used had the following composition:

Zeolite	*Composition*
NaX	$H_{0.03}Na_{0.97}AlO_{4/2}(SiO_{4/2})_{1.48}$
NaY	$H_{0.05}Na_{0.95}AlO_{4/2}(SiO_{4/2})_{2.3}$
CuX	$Na_{0.58}Cu_{0.22}AlO_{4/2}(SiO_{4/2})_{1.82}$
CaX	$H_{0.04}Na_{0.12}Ca_{0.42}AlO_{4/2}(SiO_{4/2})_{1.49}$

The degree of Na^+ exchange NaX for LiX, KX, RbX, and CsX was 59, 66, 63, and 55%, respectively. Decationized zeolites HX and HY were prepared by decomposing ammonium zeolites which had the following composition: $H_{0.03}Na_{0.21}(NH_4)_{0.76}AlO_{4/2}(SiO_{4/2})_{1.48}$ and $H_{0.03}Na_{0.22}(NH_4)_{0.75}AlO_{4/2}(SiO_{4/2})_{2.3}$. The spectra were obtained following thermal treatment of the sample *in vacuo* in quartz cells at 300° and 600°C. The adsorption of the indicator molecules was carried out in the vapor phase.

Samples were irradiated with light from a mercury lamp. The spectra were obtained with the aid of a single-beam SF-4 spectrophotometer which was capable of operating over the wavelength 220–1100 mμ. This spectrophotometer was equipped with a diffusion reflector used in Refs. *8, 9, 10.*

Results

Adsorption on Zeolites Containing Alkali Cations. The spectra of molecules adsorbed on zeolites containing alkali cations were characterized only by the absorption bands of molecules in the molecular state. The small shift of the absorption bands (1–2 mμ) corresponding to π-π^* transitions in the benzene molecule, in its aliphatic derivatives, and in polynuclear aromatic compounds can be explained by the theory of the reaction field (*1*). A greater change in the spectrum of amino-derivatives of benzene—*e.g.,* the shift for aniline was 5–8 mμ—was owing to change in the n-π^* transition caused by the participation in the interaction of the lone pair electron in the substituent group (*8, 9*). The red shift of the absorption band of the adsorbed nitrobenzene (15 mμ) and *p*-nitroaniline (90 mμ) is owing, according to Ref. *21,* to a large perturbation of the excited electron level caused by an increase in the dipole moment of the molecule in the excited state.

The increase in the electrostatic potential of the exchange cation, *e/r*, causes an increase in the shift of the absorption band (Figure 1). The difference in this dependence for *p*-phenylenediamine, aniline, and pyridine is owing mainly to the difference in the properties of the lone electron pair of nitrogen in the molecules of these compounds.

The spectra of the typical acid catalyst indicators (diphenylamine, triphenylmethane, triphenylcarbinol) adsorbed on Na-zeolites are characterized by absorption bands of adsorbed molecules in the molecular state (*7, 8, 9*). The observed spectral behavior is in accordance with the general concepts of adsorption on zeolites (*6*). The spectrum of triphenylcarbinol adsorbed on Na-zeolite after thermal treatment in vacuum at 300°C (Figure 2, curve 1) is characterized by absorption bands at 260,

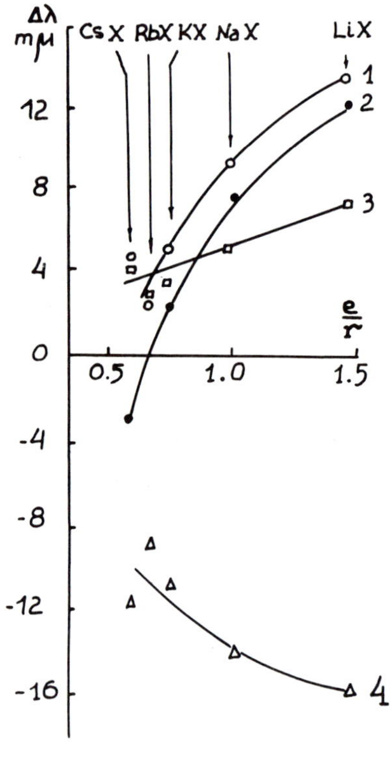

Figure 1. The dependence of the shift of the absorption bands (Δλ) on the electrostatic potential of the exchange cations (e/r), A^{-1}, *in the case of adsorption of (1) aniline, (2)* p-*phenylenediamine, (3) pyridine, (4) nitrobenzene on LiX, NaX, KX, RbX, and CsX zeolites*

282 mμ and by a broad absorption band with a maximum at 350 mμ. The spectrum of triphenylcarbinol adsorbed on the zeolite after thermal treatment in vacuum at 600°C contains the same bands, but they are of lesser intensity (Figure 2, curve 4). The spectrum of anthraquinone adsorbed on Na-zeolite subjected to thermal treatment in vacuum at 300°C contains absorption bands at 255 and 280 mμ and a broad band at 342 mμ (Figure 3, curve 1).

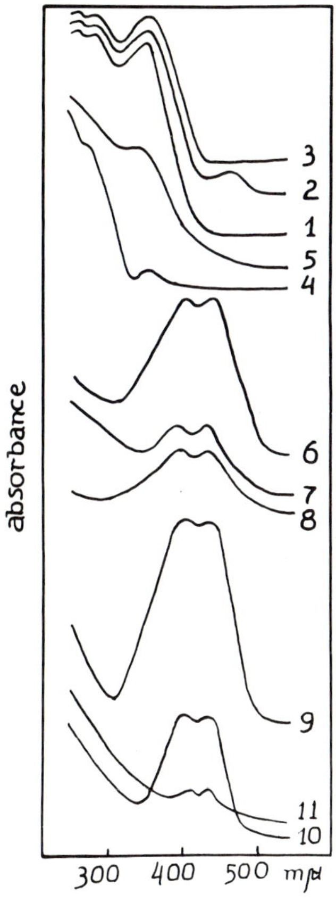

Figure 2. The spectrum of the triphenylcarbinol adsorbed on zeolites:

(1) NaY subjected to thermal treatment in vacuum at 300°C
(2) Sample 1 with adsorbed triphenylcarbinol after ultraviolet irradiation
(3) Sample 1 with adsorbed triphenylcarbinol after adsorption of water
(4) NaY subjected to thermal treatment in vacuum at 600°C
(5) Sample 4 with adsorbed triphenylcarbinol after ultraviolet irradiation
(6) CaX subjected to thermal treatment in vacuum at 300°C
(7) CaX subjected to thermal treatment in vacuum at 600°C
(8) Sample 7 with adsorbed triphenylcarbinol after ultraviolet irradiation
(9) HY subjected to thermal treatment in vacuum at 300°C
(10) HY subjected to thermal treatment in vacuum at 600°C
(11) Sample 10 with adsorbed triphenylcarbinol after adsorption of water

Adsorption on Ca-Zeolites. The adsorption of benzene, toluene, and cumene on CaX zeolite, as in the case of adsorption on NaX zeolite, is accompanied by the appearance of absorption bands characteristic of molecular absorption (Figure 4, curve 1). But the adsorption on Ca-zeolite of aniline (*8*), diphenylmethane, triphenylmethane (*9, 15*), and triphenylcarbinol is characterized by the appearance of new absorption bands. The absorption of triphenylcarbinol at 410 and 440 mμ adsorbed on CaX zeolite (Figure 2, curves 6–8) are characteristic for carbonium

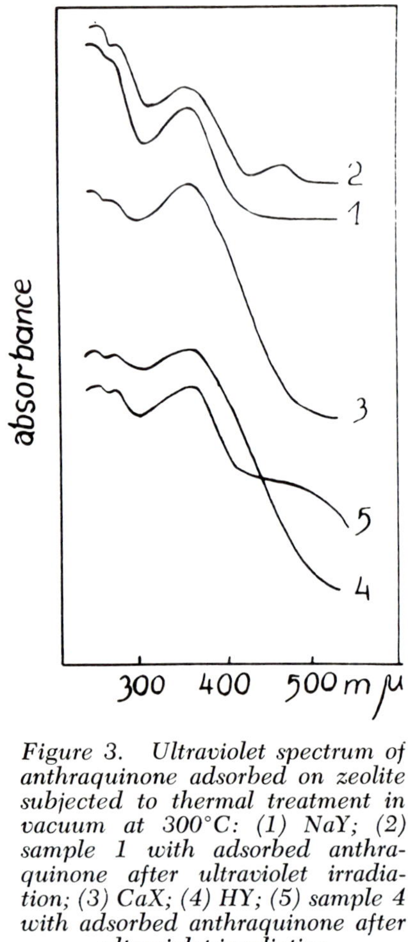

Figure 3. Ultraviolet spectrum of anthraquinone adsorbed on zeolite subjected to thermal treatment in vacuum at 300°C: (1) NaY; (2) sample 1 with adsorbed anthraquinone after ultraviolet irradiation; (3) CaX; (4) HY; (5) sample 4 with adsorbed anthraquinone after ultraviolet irradiation

ions (*3, 14, 16*). The spectrum of triphenylcarbinol adsorbed on Ca-zeolite which has been subjected to thermal treatment in vacuum at 600°C is characterized by the appearance of the same bands, but of lesser intensity (Figure 2, curve 7).

The shifts of the absorption bands in the case of adsorbed anthraquinone (255, 280, and 342 mμ) (Figure 3) relative to the position of these absorption bands in the case of *n*-hexane solution of anthraquinone (250, 270, and 320 mμ, respectively) also indicates that the molecule is ionized (*14*). The shift of the absorption band for pyridine adsorbed on Ca-zeolite (*10*) can be considered, according to Ref. *17*, as an indication of the formation of pyridine ions.

Adsorption on Decationized Zeolites. There is great similarity in the spectral characteristics of adsorption on Ca-zeolites and on decationized zeolites. The adsorption of such molecules as benzene (Figure 4), and cumene (*10*) on these zeolites is characterized only by molecular adsorption.

The spectra of aniline (*8*), diphenylamine, triphenylamine (*9, 15*), triphenylcarbinol (Figure 2) and anthraquinone (Figure 3) adsorbed either on decationized zeolite or on Ca-zeolite show the presencc of new absorption bands caused by ionization of these compounds on the zeolite.

The Role of Excitation in the Ionization of Molecules Adsorbed on Zeolites. Ultraviolet irradiation of benzene and cumene (*10*), adsorbed on Na-zeolite, did not lead to the appearance of absorption bands reflect-

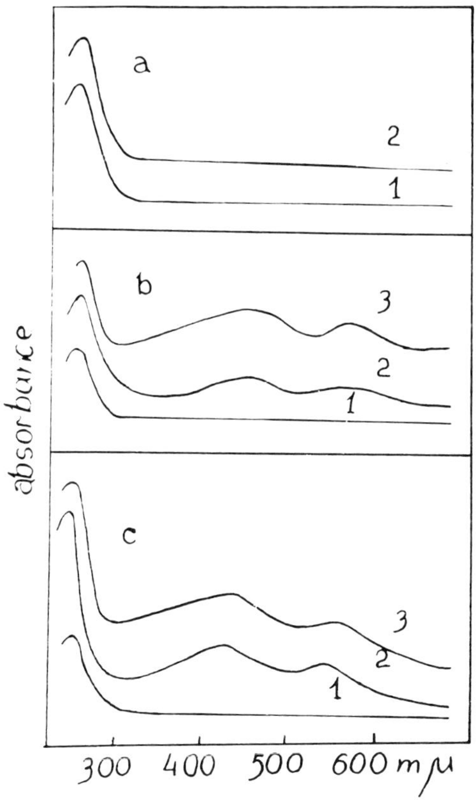

Figure 4. Ultraviolet spectrum of benzene adsorbed on: (a) NaX; (b) HX; (c) CaX; (1) no ultraviolet irradiation and after ultraviolet irradiation for: 2a = 38; 2b = 25; 3b = 46; 2c = 17; 3c = 44 hours

ing ionization (Figure 4a, curve 2). Ultraviolet irradiation of benzene and cumene adsorbed on Ca-zeolite and on decationized zeolite caused the appearance of absorption bands at 450 and 565 mμ (Figure 4b, c). These absorption bands appeared only after ultraviolet irradiation with the light whose wave length corresponded to the absorption molecule itself. The absorption band at 450 mμ is owing to $C_6H_6H^+$ ion and the band at 565 mμ to cation-radical $(C_6H_6\cdot)^+$ (*10*). The irradiation of pyridine adsorbed on Ca-zeolite and on decationized zeolite resulted in a series of subsequent photochemical reactions (*10*, in agreement with Ref. *13*). The triphenylcarbonium ions produced during adsorption on Ca-zeolite and on decationized zeolite did not undergo decomposition following ultraviolet irradiation (Figure 2).

Ultraviolet irradiation of triphenylcarbinol adsorbed on Na-zeolite was accompanied by the appearance of a weak absorption band at 460 mμ (Figure 2, curve 2). This band is near the absorption band of triphenylcarbonium ion. Ultraviolet irradiation of adsorbed anthraquinone resulted in the appearance of new absorption bands at approximately 450 mμ (Figure 3).

The Effect of Water Adsorption on the Spectrum of Ionized Adsorbed Molecules. The adsorption of water molecules on 300°C vacuum treated zeolite with preadsorbed anthraquinone molecules had little effect in the spectrum of anthraquinone. In the case of adsorbed triphenylcarbinol, the adsorption of water caused the disappearance of the bands characteristic for the triphenylcarbonium ion (Figure 2, curves 3 and 11). When the sample was treated in vacuum, the original spectrum of the triphenylcarbonium ion was restored. Analogous behavior was observed (*19*) in the case of triphenylcarbinol adsorbed on silica–alumina.

The Effect of Adsorbed Molecules on the Spectrum of the Cu^{2+} Cations in Zeolites. Figure 5 shows the change in the spectrum corresponding to the transitions between *d*-electron levels of Cu^{2+} during dehydration of the zeolite. The spectrum of completely hydrated zeolite revealed a broad absorption band with a maximum at 12,100 cm^{-1}. Thermal treatment of the zeolite at 100°C resulted in the appearance of a new absorption band at approximately 15,500 cm^{-1}. After vacuum treatment at high temperatures, there appeared in the spectrum an absorption band at 11,200 cm^{-1}. The position of the absorption band due to Cu^{2+} in the spectrum of completely hydrated zeolite is close to that of the $[Cu(H_2O)_6]^{2+}$ complex (12,600 cm^{-1}) (*2*). This indicates that Cu^{2+} enters the hydrated zeolite structure as an octahedral hexaquo-complex. The same conclusion has been reached by other investigators (*4, 18, 20*) on the basis of e.s.r. spectroscopic measurements of the Cu^{2+} cations in completely hydrated zeolites.

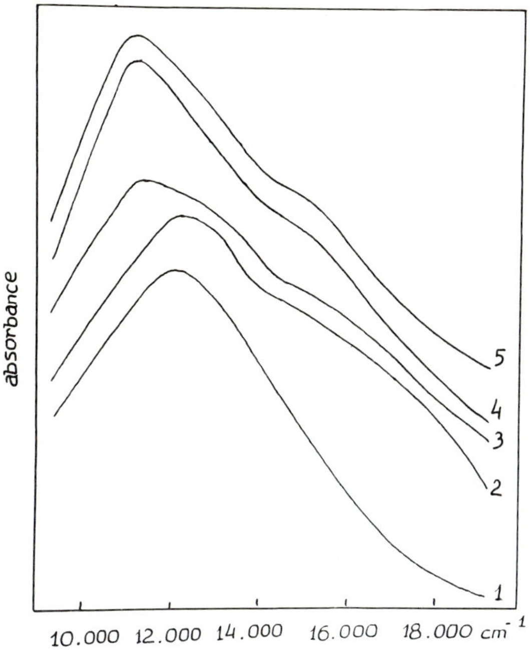

Figure 5. Spectrum of CuX zeolite: (1) completely hydrated zeolite; (2) after following thermal treatment for 3 hours at 100°C; (3) at 200°C; (4) at 300°C; (5) at 400°C

The appearance of the high-frequency absorption band after thermal treatment in vacuum at 100°C (Figure 5) indicates that ions have a lower symmetry of coordination. The theoretical interpretation of e.s.r. spectra of partially dehydrated Cu^{2+} cation in zeolites is based on the model of a tetragonally distorted octahedral or on a square-planar model (*4, 18, 20*). It is assumed that such a state can be reached as a result of the location of the cation at S_{II} sites. Because such distortion of octahedral coordination is accompanied by an increase in the splitting of the $d_{x^2-y^2}$, d_{xz}, and d_{yz} levels of Cu^{2+} (*20*), the high-frequency absorption at 15,500 cm^{-1} (Figure 5) may be caused by transition to those levels. The assignment to a definite type of transition of the low-frequency absorption band (11,200 cm^{-1}), which becomes apparent in the spectrum during more rigorous dehydration, requires further study. When water was adsorbed on zeolites that had been dehydrated at 400°C, the original spectrum of hydrated Cu^{2+} cations was restored. The restoration of the original spectrum of Cu^{2+} ions in the case of CuA zeolite dehydrated at 500°C can be accelerated by keeping the sample in water vapor at 50°C (*11*). If a zeolite containing Cu^{2+} is kept in vacuum at 70°C followed by treatment of NH_3, then in accordance with the rule of spectrochemical series (*2*), the molecules of H_2O in the Cu^{2+} coordination sphere are replaced by NH_3 molecules. In this case, the spectrum shows the appearance of several absorption bands which are characteristic of the distorted octahedral symmetry of the Cu^{2+} coordination. At maximum replacement of H_2O molecules by NH_3 molecules, the spectrum has only one absorption band at 16,000 cm^{-1}. The absorption band of the complex $[Cu(NH_3)_6]^{2+}$ in solution lies at 15,100 cm^{-1} (*2*).

After adsorption of pyridine molecules, an absorption band appeared with 2 maxima of 13,300 and 14,200 cm^{-1}.

Discussion

The spectra of ionized molecules adsorbed on zeolites containing alkali exchange cations showed no apparent absorption. From this it follows that the surface of these zeolites is free of acidic centers. This conclusion also agrees with infrared spectroscopic measurements (*23*) which indicate the absence of structural hydroxyl groups and centers of protonic acidity on the surface of completely cationized zeolites containing alkali cations. A very weak absorption attributed to the carbonium ion was observed only after ultraviolet irradiation of adsorbed triphenylcarbinol (Figure 2, curve 2). This fact may indicate the existence on the zeolite surface containing alkali exchange cations of a small number of proton-donor centers formed by partial decationation (about 3%) of the investigated zeolites.

The spectrum of triphenylcarbinol, adsorbed on zeolites containing alkali cations, is characterized by the appearance of an absorption band at 350 mμ which belongs neither to molecules in the state of molecular adsorption nor to the triphenylcarbonium ion. This indicates that there are adsorbed molecules in the same state of intermediate perturbation. The interpretation of this absorption band requires further investigation.

The dependence of the shift of the band of adsorbed molecules (Figure 1) on the type of the exchange cation indicates that the contribution of the ground and the excited levels of the molecule to the interaction of cations and to the increase in perturbation is significant. This contribution increases with the increase in the magnitude of the electrostatic potential. A similar increase in the degree of polarization of the adsorbed molecules was observed in the investigation of the infrared spectra of pyridine adsorbed on zeolites containing alkali cations (*23*).

The position of the absorption band of pyridine adsorbed on Cu-zeolite (258 mμ) showed a greater perturbation than in the case of adsorption on Na- and a lower perturbation than in the case of adsorption on H-forms of zeolites (252 and 266 mμ, correspondingly) (*10*).

From Figures 2 and 3, it follows that molecules which become ionized during adsorption on Ca-zeolites also become ionized during adsorption on decationized zeolites. These results seem to indicate a similarity in the nature of proton-donor centers of this type of zeolites. According to the existing concepts (*22, 24*), the center of the aprotonic type of acidity is the triply coordinated atom of aluminum, and the center of the proton-donor type is the hydroxyl groups close to the Al atom. These groups are formed as a result of the reaction of the proton with the zeolite, the proton being the decomposition product of ammonium ion during formation of decationized zeolite. Similar surface structures are formed as a result of dissociation of water molecules according to Ref. *23*. In the case of Ca^{2+}, there are cationic vacancies which can produce with adsorbed water the acidic centers of protonic types.

The observed decrease in the concentration of carbonium ions produced during adsorption of triphenylcarbinol on dehydroxylated Ca-zeolite and on decationized zeolite (Figures 2 and 3) indicates that OH-groups played the determining role in the appearance of proton-donor properties of these zeolites.

Thus interactions involving charge transfer between acidic centers on the surface of Ca-zeolites and decationized zeolites and the adsorbed molecules depend on the nature and the state of the molecule. These interactions become more pronounced after ultraviolet excitation of the molecules.

The spectra of Cu^{2+} cations in zeolites show that the state of the exchange cations in zeolites to a large extent depends on the type and on the concentration of molecules that form the coordination with the cation. In the strongly dehydrated state, the spectrum of Cu^{2+} is determined mainly by its interaction with the oxygen atoms of zeolite framework.

Acknowledgment

The samples of zeolites were kindly supplied by S. P. Zhdanov.

Literature Cited

(1) Barachevsky, V. A., "Elementary Photoprocesses in Molecules," p. 359, Nauka, Moscow, 1966.
(2) Day, M. D., Selbin, J., "Theoretical Inorganic Chemistry," Reinhold, New York, 1962.
(3) Hirshler, A. E., *J. Catalysis* **1963**, 2, 428.
(4) Kazansky, V. B., *Kinetics Catalysis* **1970**, 11, 55.
(5) Khodakov, Yu. S., Mikheikin, I. D., Nahshunov, V. S., Shvets, V. A., Kazansky, V. B., Minachev, H. M., *Izv. Akad. Nauk SSSR Ser. Khim.* **1968**, 3, 525.
(6) Kiselev, A. V., ADVAN. CHEM. SER. **1971**, 102, 37.
(7) Kiselev, A. V., Kitiashvili, D. G., Lygin, V. I., *Kinetics Catalysis*, in press.

(8) Kiselev, A. V., Kupcha, L. A., Lygin, V. I., *Kinetics Catalysis* **1966,** 7, 705.
(9) *Ibid.,* **1967,** 8, 475.
(10) Kiselev, A. V., Kupcha, L. A., Lygin, V. I., Shatskii, V. G., *Kinetics Catalysis* **1969,** 10, 449.
(11) Kiselev, A. V., Lygin, V. I., *Siberian Conf. Spectry., 6th, Tomsk,* **1968.**
(12) Klier, K., Ralek, M., *J. Phys. Chem. Solids* **1968,** 29, 951.
(13) Kotov, E. I., Pankratov, A. O., "Elementary Photoprocesses in Molecules," p. 314, Nauka, Moscow, 1966.
(14) Kotsarenko, N. S., Karakchiev, L. G., Dzisko, V. A., *Kinetics Catalysis* **1968,** 9, 158.
(15) Kupcha, L. A., Lygin, V. I., Mineeva, L. V., *Kinetics Catalysis* **1968,** 9, 940.
(16) Leftin, H. P., Hobson, M. C., *Advan. Catalysis* **1963,** 14, 115.
(17) Mason, S. F., "Physical Methods in Heterocyclic Chemistry," p. 318, Academic Press, New York, 1963.
(18) Nicula, A., Stamires, D., Turkevich, J., *J. Chem. Phys.* **1965,** 42, 3684.
(19) Porter, R. P., Hall, W. K., *J. Catalysis* **1966,** 5, 366.
(20) Richardson, J. T., *J. Catalysis* **1967,** 9, 178.
(21) Terenin, A. N., "Photonics of Dye Molecules," Nauka, Moscow, 1967.
(22) Ward, J. W., *J. Catalysis* **1967,** 9, 225.
(23) *Ibid.,* **1968,** 10, 34.
(24) Zhdanov, S. P., Kiselev, A. V., Lygin, V. I., Titova, T. I., *Zh. Fiz. Khim.* **1966,** 10, 1041.
(25) Zhdanov, S. P., Zubareva, N. A., Kiselev, A. V., Lygin, V. I., *Kinetics Catalysis,* in press.

RECEIVED February 11, 1970.

46

Calculation of Some Adsorption Properties of Zeolites by Statistical Methods

P. BRÄUER, A. A. LOPATKIN, and G. PH. STEPANEZ

Moscow State University, Department of Chemistry; Institute of Physical Chemistry, Academy of Science of USSR, Moscow

The calculations of isotherms and heats of adsorption at low filling of the cavities of zeolite with adsorbed molecules have been carried out for the systems Ar–NaA and CH_4–CaA. Although a simple and rather crude model was used, it was possible to obtain results that agree qualitatively with the experimental data. The equilibrium distribution of molecules among the zeolite cavities also was studied.

Because of their regular structure and the presence of more or less separate cavities, the zeolites offer a very convenient system for the application of statistical methods. In this work, an attempt has been made to calculate some adsorption properties of zeolites using statistical thermodynamics.

Let us consider the B cavities of a crystal of zeolite together with the adsorbed substance. Each cavity may contain not more than m adsorbed molecules. We assume that the contribution of the interaction energy of molecules which are in different cavities to the total energy of the system may be neglected. Denoting by n_s the number of cavities which contain s adsorbed molecules ($s = 0, 1, \ldots, m$), we can write for this model the canonical partition function, Q (*1, 4*)

$$Q(N,B,T) = \sum \frac{B!Q_1^{n_1} \ldots Q_s^{n_s} \ldots Q_m^{n_m}}{n_0!\, n_1! \ldots n_s! \ldots n_m!} \tag{1}$$

where N is the total number of molecules within the zeolite cavities and Q_s is the partition function for s molecules adsorbed in one cavity. The summation in Equation 1 is carried out over all distributions ($n_o, \ldots, n_s, \ldots, n_m$) which are compatible with the condition

$$n_1 + 2n_2 + \ldots + sn_s + \ldots + mn_m = N \tag{2}$$

The grand partition function Ξ for this system may be written as

$$\Xi = (1 + Q_1\lambda + \ldots + Q_s\lambda^s + \ldots + Q_m\lambda^m)^B$$

or

$$\Xi = (1 + Z_1\zeta + \ldots + Z_s\zeta^s + \ldots + Z_m\zeta^m)^B \tag{3}$$

where $\lambda = \exp(\mu/\kappa T)$, ζ is the activity, which in the case of an ideal gas is equal to $p/\kappa T$, and Z_s is the configuration integral for the system which consists of s molecules in one cavity. In the quasiclassical approximation, we can write for Z_s.

$$Z_s = \frac{1}{s!} \int_v \exp\left\{-\frac{U_s(\mathbf{r}_1, \ldots, \mathbf{r}_s)}{\kappa T}\right\} d\mathbf{r}_1 \ldots d\mathbf{r}_s \tag{4}$$

where v is the volume of the cavity and U_s is the potential energy of s molecules adsorbed in the cavity. The average number of molecules adsorbed by a crystal is given by

$$<N> = B \frac{Z_1\zeta + 2Z_2\zeta^2 + \ldots + mZ_m\zeta^m}{1 + Z_1\zeta + Z_2\zeta^2 + \ldots + Z_m\zeta^m} \tag{5}$$

At sufficiently low values of ζ, Equation 5 may be written in virial form

$$\theta = Z_1\zeta + (2Z_2 - Z_1^2)\zeta^2 + (3Z_3 - 3Z_1Z_2 + Z_1^3)\zeta^3 + \ldots \tag{6}$$

where $\theta = <N>/B$. The quantity $Z_1/\kappa T$ represents the Henry constant. The virial expansion for the isosteric heat of adsorption also can be obtained by using Equation 5, which we write in the following form

$$-\theta + (1 - \theta)Z_1\zeta + \ldots + (s - \theta)Z_s\zeta^s + \ldots + (m - \theta)Z_m\zeta^m = 0 \tag{7}$$

After differentiating this expression with respect to T at constant θ, we obtain the equation

$$Q_{st} = RT - \frac{(1 - \theta)<U_1>Z_1\zeta + (2 - \theta)<U_2>Z_2\zeta^2 + \ldots + (m - \theta)<U_m>Z_m\zeta^m}{(1 - \theta)Z_1\zeta + 2(2 - \theta)Z_2\zeta^2 + \ldots + m(m - \theta)Z_m\zeta^m} \tag{8}$$

where $<U_s>$ is the average potential energy of the system which consists of s molecules in one cavity. Rearranging Series 6, we can obtain ζ as a function of θ, and inserting this expression into Equation 8 and expanding in powers of θ at $\theta < 1$, we find

$$Q_{st} = RT - <U_1> - 2\frac{Z_2}{Z_1^2}(<U_2> - 2<U_1>)\theta -$$

$$3\frac{Z_3}{Z_1^3}\Big[(<U_3> - 3<U_1>) +$$

$$\frac{1}{Z_3}(Z_1Z_2 - 4\frac{Z_2^2}{Z_1})(<U_2> - 2<U_1>)\Big]\theta^2 + \ldots \qquad (9)$$

It is of interest also to find for this model the equilibrium distribution of the molecules among the zeolite cavities. For this purpose it is convenient to introduce $\rho_s = <n_s>/B$—the average fraction of cavities containing s molecules. It may be shown that

$$\rho_s = \frac{Z_s t_+^s}{1 + Z_1 t_+ + Z_2 t_+^2 + \ldots + Z_m t_+^m} \qquad (10)$$

Z_0 being equal to 1. Here t_+ is the single positive root of the equation

$$-\theta + (1 - \theta)Z_1 t + (2 - \theta)Z_2 t^2 + \ldots + (m - \theta)Z_m t^m = 0 \qquad (11)$$

Comparison of this equation with Equation 7 shows that $t_+ = \zeta$. If Z_s equals to $(Z_1^s)/s!$ (*i.e.*, we do not consider the interaction of adsorbed molecules with each other even when they are in the same cavity), we obtain

$$\rho_s = \frac{\frac{1}{s!}x_+^s}{1 + x_+ + \frac{1}{2!}x_+^2 + \ldots + \frac{1}{m!}x_+^m} \qquad (12)$$

where x_+ is the single positive root of the equation

$$-\theta + (1 - \theta)x + (2 - \theta)\frac{x^2}{2!} + \ldots + (m - \theta)\frac{x^m}{m!} = 0 \qquad (13)$$

and $x = Z_1 t$. At low values of θ and high values of m, Distribution 12 reduces to the Poisson's distribution. The ρ_s *vs.* θ curves at $m = 6$ calculated by means of Equation 12 are shown on Figure 1. If the degree of filling is very low, Equation 10 may be used containing only terms in Z_1 and Z_2 (*i.e.*, $\rho_s = 0$ for all $s > 2$).

Calculations

In the present work, the values of Z_1, Z_2, $<U_1>$ and $<U_2>$ for the systems Ar–NaA and CH_4–CaA were calculated at a number of temperatures, and thus the corresponding coefficients in Expansions 6 and 9 were obtained. The configuration integral, Z_1, was calculated by means of Equation 4. The potential energy, $U_1(r)$, was determined for the great number of positions of a molecule within the zeolite cavity. The details

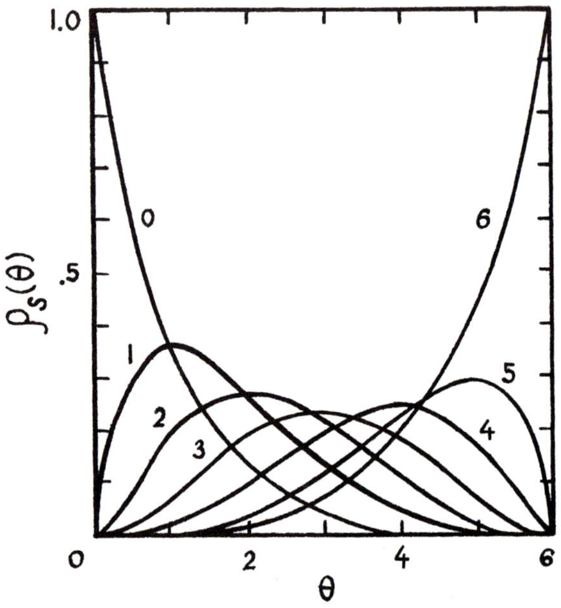

Figure 1. Dependence of ρ_s *on* θ *at* m = 6; *the values of* s *are indicated on corresponding curves*

of calculation of $U_1(r)$ are given in Ref. 8. Thus, it was possible to replace the integral by the corresponding integral sum and to obtain the value of Z_1 with quite acceptable accuracy.

In the calculations of potential energy of molecules adsorbed in the zeolite cavity, Howell's x-ray data (7) of the coordinates of cations and oxygen atoms were used. According to these data, Na^+ in zeolite NaA can take 1 of 3 possible positions near the 6-ring with the same probability, and 1 of 4 possible positions near the 8-ring. It was assumed in the calculations that the cations of the first type are on the 3-fold axes and the cations of second type are in the plane of the 8-rings on the 4-fold axes. If in all 6- and 8-rings there are cations, then in each cavity there will be 11 cations. To take into account the 12th Na^+ and to ensure the electroneutrality of zeolite framework, a statistical weight and charge equal to 12/11 were ascribed to each of 11 cations. These values of weight and charge were used in the calculations of all constituents of potential energy (8).

In the case of zeolite CaA, each elementary cell was assumed to contain 6 Ca^{2+} cations. It was supposed that they occupy the centers of 6-rings in such a way that the 2 unoccupied rings were on the same 3-fold axis. The contribution of Al and Si atoms in the interaction energy has not been taken directly into consideration. It was only through the polarizability of oxygen atoms that this contribution was involved in calculations. Besides, it was assumed that the extra negative charge of AlO_4 tetrahedra is distributed evenly among all oxygen atoms.

The models of zeolites NaA and CaA used in this work were rather crude ones, but no more complete data concerning distribution of cations in zeolites were available.

The average energy of interaction of a molecule with the zeolite was determined using the equation

$$<U_1> = \int_v U_1(\mathbf{r}) \exp\left\{-\frac{U_1(\mathbf{r})}{\kappa T}\right\} d\mathbf{r} \Big/ \int_v \exp\left\{-\frac{U(\mathbf{r})}{\kappa T}\right\} d\mathbf{r} \quad (14)$$

For the calculation of second virial coefficients in Equations 6 and 9, the actual distribution of the adsorption potential in the cavity was replaced by rectangular potential well. The depth of the well was chosen in such a manner that the following relation would be correct.

$$Z_1 = \frac{4\pi}{3} r_0^3 \exp(-U/\kappa T) \quad (15)$$

The meaning of parameters can be seen from Figure 2. Z_1 is the configuration integral calculated as described above and U is a certain effective energy.

The calculation of Z_2 was made by assuming for the interaction of a molecule with the zeolite a rectangular potential well (Figure 2). For

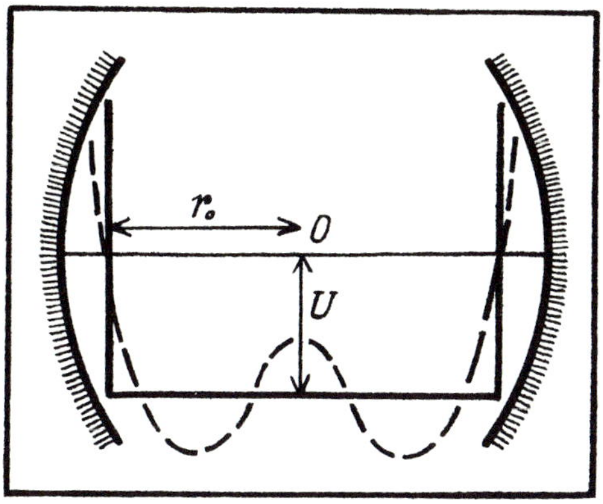

Figure 2. Rectangular potential well substituted for the actual potential curve (dashed line). The point O corresponds to the center of the cavity.

the interaction of 2 molecules with each other, the potential of the following form was used

$$\varphi(r_{12}) = \begin{cases} +\infty, & r_{12} \leq l \\ -w, & l < r_{12} < R_0 l \\ 0, & R_0 l \leq r_{12} \end{cases} \quad (16)$$

where r_{12} is the distance between the centers of interacting molecules. In this case, Z_2 may be expressed explicitly, and for the quantity $2Z_2/Z_1^2$ we obtain

$$2\frac{Z_2}{Z_1^2} = 1 - \frac{l^3}{r_0^3}\left[(e^{w/\kappa T} - 1)R_0^3 f(R_0) - e^{w/\kappa T} f(1)\right] \quad (17)$$

where

$$f(R) = 1 - \frac{9}{16}\frac{l}{r_0}R - \frac{1}{32}\frac{l^3}{r_0^3}R^3 \quad (18)$$

At $w = 0$, Expression 17 corresponds to that for the hard-sphere model. Differentiating ln (Z_2/Z_1^2) with respect to T and multiplying the derivative by RT^2, we obtain

$$\langle U_2 \rangle - 2\langle U_1 \rangle =$$

$$\frac{-w \exp(w/\kappa T) l^3 R_0^3 f(R_0) - f(1)}{1 + l^3[(e^{w/\kappa T} - 1)R_0 f(R_0) - e^{w/\kappa T} f(1)]/r_0^3} \quad (19)$$

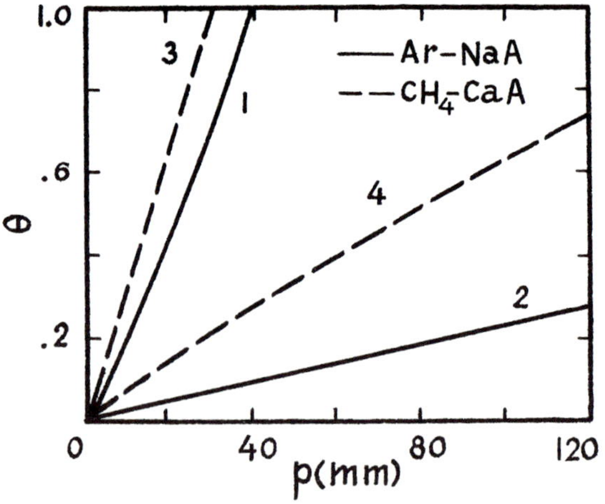

Figure 3. Initial parts of isotherms of adsorption calculated at 1 = 150°K, 2 = 200°K, 3 = 250°K, 4 = 300°K

Table I. Values of $<U_1>$ and q_1, Kcal/Mole

System	T,°K:	50	100	150	200	250	300
Ar–NaA	$<U_1>$	2.81	2.64	2.51	2.43	2.40	2.38
	q_1	0.30	0.15	0.12	0.11	0.10	0.09
CH_4–CaA	$<U_1>$	5.30	5.15	5.02	4.85	4.70	4.55
	q_1	2.32	0.55	0.34	0.27	0.23	0.21

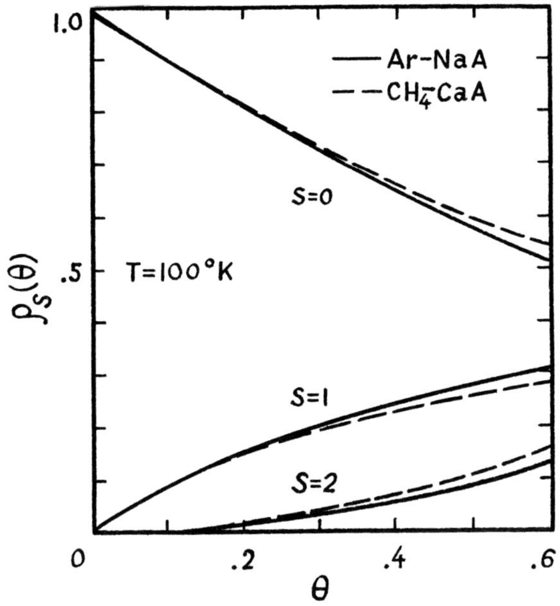

Figure 4. Approximate dependence of ρ_0, ρ_1, and ρ_2 on θ

Calculation of Isotherms and Isosteric Heats of Adsorption

In the calculations by means of Equations 17–19, the following parameters have been used. For Ar: w = 138 cal/mole, R_0 = 1.85, l = 3.16A (*5*). For CH_4: w = 285 cal/mole, R_0 = 1.60, l = 3.35A (*6*). The values of r_0 were estimated from calculated potential curves and are assumed to be equal to 4.30A and 3.81A for the systems Ar–NaA and CH_4–CaA, respectively. The second virial coefficient in Expansion 6 changes the sign from + to − at approximately 200°K for the system Ar–NaA and at 250°K for CH_4–CaA. At these temperatures, the concave isotherms turn into convex ones. Figure 3 shows the initial parts of isotherms for the systems Ar–NaA and CH_4–CaA at different temperatures. The curvature of isotherms at the chosen temperatures is very low. The coefficient at θ in Expansion 9, $q_1 = -(2Z_2/Z_1^2)(\langle U_2 \rangle - 2\langle U_1 \rangle)$, is always positive. Its values as well as the values of $\langle U_1 \rangle$ (in Kcal/mole) are given in Table I.

Distribution of Molecules Among the Zeolite Cavities

As may be seen from Figure 1, even in the case when adsorbed molecules do not interact with each other, the distribution of molecules among the cavities is very irregular. Using Z_1 and Z_2, the approximate values of ρ_0, ρ_1 and ρ_2 at low θ have been calculated by means of Equation 10. The curves ρ_s *vs.* θ for the studied systems are given in Figure 4. At small values of θ, the distribution differs only slightly from that for the noninteracting molecules (*cf.* Figure 1); the interaction between molecules manifests itself more distinctly at higher values of θ.

Conclusion

The isotherms and heats of adsorption calculated for the system Ar–NaA agree with the experimental data (*2*) quantitatively. For the system CH_4–CaA, we have found in the literature only the heats of adsorption (*3, 8, 10*). These experimental data do not agree well with each other. Therefore, it can be said only that for the system CH_4–CaA, the calculated values of isosteric heats are not in contradiction with the experimental ones. Some numerical results of our calculations are given also in Ref. *8*.

Literature Cited

(1) Bakajev, V. A., *Dokl. Akad. Nauk USSR* **1966**, 169, 369.
(2) Bräuer, P., Kiselev, A. V., Lesnik, E. A., Lopatkin, A. A., *Zh. Fis. Khim.* **1969**, 43, 1579.
(3) Greene, S. A., Pust, H., *J. Phys. Chem.* **1958**, 62, 55.
(4) Hill, T. L., "Statistical Mechanics," McGraw-Hill, New York, 1956.
(5) Hirschfelder, J. O., Curtiss, Ch. F., Bird, R. B., "Molecular Theory of Gases and Liquids," Wiley, New York, 1954.
(6) Hoover, A. E., Hagata, J., Leland, T. W., Kobayashi, R., *J. Chem. Phys.* **1968**, 48, 2633.
(7) Howell, P. A., *Acta Cryst.* **1960**, 13, 737.
(8) Kiselev, A. V., ADVAN. CHEM. SER. **1971**, 102, 37.
(9) Kiselev, A. V., Chernenkova, N. L., Yashin, Ya. I., *Neftekhimiya* **1965**, 5, 589.
(10) Turkeltaub, N. M., Zhukhovitzkii, A. A., Porshneva, N. V., *Zh. Prikl. Khim.* **1961**, 43, 1964.

RECEIVED February 11, 1970.

47

Heats of Immersion of Outgassed Ion-Exchanged Zeolites

R. M. BARRER and P. J. CRAM[1]

Physical Chemistry Laboratories, Imperial College, London S.W. 7, England

Heats of immersion in water have been determined for a number of outgassed porous crystals enriched by ion exchange in various cations (zeolites X, Y, A, chabazite, and synthetic ferrierite), and for clinoptilolite and mordenite in their Na-forms, decationated, and in various stages of dealumination. Finally, heats of immersion were determined in NaX, NaY, NaA, and (Ca,Na) chabazite in which the crystals initially contained various known loadings of zeolitic water. From the results, the influence of the exchange cations upon integal heats of sorption of water, ΔH, *and other derived heats have been evaluated and discussed.*

Perhaps the most important of all zeolite–sorbate complexes are those in which water is the guest molecule. Water is essential for the synthesis of zeolites and is present in all the natural and synthetic members of the group, certain of which find application as industrial desiccants. Accordingly, as full an understanding as possible is needed of water–zeolite complexes, especially of the binding energy of the water within the crystals.

Information about this energy has been obtained from isotherm measurements over a temperature range (*3, 7, 8, 9*), by calorimetry (*18, 19*), and by differential thermal analysis (*11*). From the isotherms and by direct calorimetry, the isosteric heats, q_{st}, may be found as functions of the amount of water sorbed. However, some disadvantages may be associated with each procedure. Such is the affinity between water and zeolites that to determine q_{st} for small uptakes may require isotherm measurements at temperatures above 200°C. At these temperatures, lattice breakdown can take place by side reactions involving the water.

[1] Present address: Petroleum Recovery Research Institute, The University of Calgary, Calgary, Alberta, Canada.

Table I. Zeolites Studied[a]

Starting Materials	*Modified Forms*
NaX (Union Carbide)	Li, K, Rb, Cs, Mg, Ca, Sr, Ba, Tl, and PbX
NaY (Union Carbide)	Li, K, Rb, Cs, Mg, Ca, Sr, Ba, Tl, and PbY
NaA (Union Carbide)	Li, K, Rb, Cs, Mg, Ca, Ba, and TlA
(Ca, Na) chabazite	Li, Na, K, Rb, Cs, Ca, and Tl chabazites
Na mordenite (Na-Zeolon, Norton Co.)	H-forms (from 2N, 6N, and 12N HCl-treated Na-Zeolon): H-mordenite (H-Zeolon)
Sr ferrierite (SrD)	Li, Na, and CaD: H-forms (from 0.09*N* HCl-treated SrD, and HD from heated NH_4D)
Na-rich clinoptilolite	H-forms (from 0.25*N*, 0.5*N*, 1.0*N*, and 2.0*N* HCl-treated clinoptilolite)

[a] In Tables I and II, the representation of a zeolite as, for example, RbX, should not be taken to imply 100% exchange of the original Na by Rb.

Secondly, calorimetric measurements from the vapor phase may refer to nonequilibrium distributions of water within the crystals and through the zeolite bed. The very energetic water–zeolite bond, especially for smaller water uptakes, means that water molecules may stick on sites where they first land. Subsequent redistribution can be very slow on the time scale of the experiment, particularly at the low temperatures employed (*19, 21*), 23° and 44°C. Finally, the information derived from differential thermal analysis is qualitative or at best only semiquantitative.

An alternative calorimetric procedure, which in principle may eliminate the redistribution problem of direct calorimetry, consists in measuring heats of immersion as a function of the amount of presorbed liquid. The bulb containing the zeolite and its presorbed water can be heated to a suitably high temperature to promote water migration and then cooled slowly to the experimental temperature. The bulb is broken under liquid water and the heat of wetting measured. This method has been used in the present work for a number of zeolites of differing known structures, with different exchange ions, and also for individual structures decationated and progressively dealuminized.

Experimental

Materials. The zeolites studied are summarized in Table I. The unit cell contents of the dehydrated forms of the starting materials were:

NaX	$87Na^+ [87 AlO_2^- \cdot 105SiO_2]$
NaY	$57 (H_{0.1}, Na_{0.9})^+ [57AlO_2^- \cdot 135SiO_2]$
NaA	$12Na^+ [12AlO_2^- \cdot 12SiO_2]\ 0.4\ NaAlO_2$
Chabazite	$1.6Na^+ 4.4Ca^{2+} [10.4\ AlO_2^- \cdot 25.6\ SiO_2]$
Mordenite	$8 (H_{0.1}\ Na_{0.9})^+ [8AlO_2^- \cdot 40SiO_2]$
Clinoptilolite	$6Na^+ [6AlO_2^- \cdot 30SiO_2]$
Sr ferrierite	Oxide formula: $0.94SrO \cdot Al_2O_3 \cdot 12.3SiO_2$

The oxide formula of Sr ferrierite only is given because the yield of this synthetic zeolite may have been less than 100% (*12*). There are 72 oxygen atoms in the orthorhombic unit cell.

The ion-exchanged forms of sieves X, Y, and A were prepared by shaking a suspension of the starting materials in a solution of the appropriate cation, at room temperature, for more than 6 hours in each treatment. AnalaR chloride solutions were used where suitable. Mg, Tl, and Pb forms were prepared using sulfate, acetate, and nitrate solutions, respectively. The exchanging solutions contained an initial five-fold excess of exchanging ion, except for Rb, Cs, and Tl forms, where the initial excess was about two-fold. Following ion-exchange, the samples were washed thoroughly with nearly boiling water to remove entrained salt. The entire process was repeated at least 4 times. Other zeolite modifications (ferrierites, mordenites, chabazites, and clinoptilolites) were those prepared by other workers in these laboratories (*4, 5, 6, 12, 24*). All the samples studied were dried overnight in a 110°C oven and stored over saturated NH_4Cl solution at room temperature (20° to 23°C) for at least

Table II. Saturation Water Contents, Grams per Gram of Hydrated Zeolite X100

	Ion-Exchanged Forms				
Cation Form	*Zeolite X*	*Zeolite Y*	*Zeolite A*	*Chabazite*	*Ferrierite*
Li	27.8	27.6	22.1	22.0	15.6
Na	25.8	25.9	21.5	19.5	13.4
K	22.5	21.9	18.4	17.3	–
Rb	19.4	20.4	14.2	13.9	–
Cs	17.5	17.6	15.7	12.3	–
Mg	29.3	28.1	27.2	–	–
Ca	27.6	26.1	23.0	22.1, 21.1[a]	14.2
Sr	25.2	25.4	–	–	11.2
Ba	22.2	23.6	–	–	–
Tl	12.3	16.7	8.7	9.5	–
Pb	15.9	–	–	–	–

Decationated and Dealuminized Forms

Clinoptilolite[b]		*Mordenite*[b]		*Ferrierite*	
Natural	14.0	Na-Zeolon	13.2	0.09*N* HCl-treated	15.7
0.25*N* HCl-treated	14.5	2*N* HCl-treated	–	HD from NH_4D	–
0.5*N* HCl-treated	14.3	6*N* HCl-treated	–		
1.0*N* HCl-treated	14.4	12*N* HCl-treated	17.9		
2.0*N* HCl-treated	15.4	H-Zeolon	14.3		

[a] Value for natural (Ca, Na) chabazite.
[b] The fractions of Al removed in the acid treatment are shown in Table IV.

Table III. Heats of Immersion of Cation Forms of Zeolites

Cation Form	q_H, *Cal per Gram*	q_D, *Cal per Gram*	Q_I, *Cal per Mole of Water* × 10^{-3}	Q_{UC}, *Cal per* N_0 *Unit Cells*	Q_{GE}, *Cal per Gram Equiv. of Cation* × 10^{-3}
			Zeolite X		
Li	92.$_5$	128.$_3$	5.9$_7$	15.4 × 10^5	17.7
Na	84.$_3$	113.$_6$	5.8$_9$	15.3 × 10^5	17.6
K	71.$_6$	92.$_3$	5.7$_5$	13.7 × 10^5	15.7
Rb, Na	60.$_3$	74.$_9$	5.5$_9$	12.7 × 10^5	14.6
Cs, Na	52.$_8$	64.$_0$	5.4$_3$	12.6 × 10^5	14.5
Mg	94.$_7$	134.$_1$	5.8$_1$	17.1 × 10^5	19.6
Ca	103.$_4$	143.$_0$	6.7$_3$	18.8 × 10^5	21.$_6$
Sr	94.$_8$	126.$_7$	6.7$_8$	19.3 × 10^5	22.$_2$
Ba	80.$_6$	103.$_6$	6.5$_3$	17.0 × 10^5	19.5
Tl	21.$_2$	24.$_2$	3.2$_1$	7.0$_7$ × 10^5	8.1$_3$
Pb	42.$_4$	50.$_2$	4.7$_9$	10.2 × 10^5	11.8
			Zeolite Y		
Li	63.$_1$	87.$_1$	4.1$_2$	10.3$_4$ × 10^5	18.2
Na	60.$_3$	81.$_5$	4.1$_9$	10.4$_2$ × 10^5	18.3
K	52.$_4$	67.$_1$	4.3$_1$	9.1$_9$ × 10^5	16.1
Rb, Na	47.$_2$	59.$_4$	4.1$_6$	9.1$_2$ × 10^5	16.0
Cs, Na	40.$_8$	49.$_5$	4.1$_6$	8.5$_6$ × 10^5	15.0
Mg, Na	87.$_4$	121.$_6$	5.6$_0$	15.0$_1$ × 10^5	26.$_3$
Ca, Na	73.$_3$	99.$_3$	5.0$_4$	12.2$_9$ × 10^5	21.$_6$
Sr, Na	69.$_5$*	93.$_2$	4.9$_3$	12.7$_1$ × 10^5	22.$_3$
Ba, Na	67.$_3$	88.$_2$	5.1$_3$	12.9$_3$ × 10^5	22.$_7$
Tl, Na	27.$_0$	32.$_4$	2.8$_8$	6.5$_5$ × 10^5	11.5
			Zeolite A		
Li	74.$_9$	96.$_2$	6.1$_1$	14.8 × 10^4	12.3
Na	73.$_9$	94.$_2$	6.1$_9$	16.3 × 10^4	13.6
K	71.$_4$	87.$_5$	6.9$_9$	16.9 × 10^4	14.1
Rb, Na	25.$_3$	29.$_5$	3.2$_1$	6.6$_7$ × 10^4	5.5$_6$
Cs, Na	47.$_7$	56.$_6$	5.4$_8$	13.2 × 10^4	11.0
Mg, Na	98.$_0$	134.$_5$	6.4$_9$	22.$_2$ × 10^4	18.5
Ca	83.$_0$	107.$_7$	6.5$_0$	18.3 × 10^4	15.3
Ba	1.8*	–	–	–	–
Tl	19.$_1$	20.$_9$	3.9$_5$	8.1$_7$ × 10^4	6.8$_1$
			Chabazite		
Li	73.$_2$	93.$_9$	5.9$_9$	21.5$_5$ × 10^4	20.1$_3$
Na	67.$_2$	83.$_5$	6.2$_1$	21.4$_7$ × 10^4	20.0$_6$
K	52.$_2$	63.$_2$	5.4$_4$	16.4$_7$ × 10^4	15.3$_9$
Rb	39.$_1$	45.$_7$	5.1$_1$	14.0$_0$ × 10^4	13.0$_8$
Cs, Ca	30.$_4$	34.$_7$	4.4$_5$	11.7$_6$ × 10^4	10.9$_8$
Ca, Na	76.$_7$	97.$_2$	6.5$_5$	23.5$_5$ × 10^4	22.0$_1$
Ca	80.$_0$*	102.$_5$	6.5$_3$	24.9$_0$ × 10^4	23.2$_7$
Tl	22.$_6$	25.$_0$	4.2$_8$	10.8$_7$ × 10^4	10.1$_6$

Table III. Continued

Cation Form	q_H, Cal per Gram	q_D, Cal per Gram	Q_I, Cal per Mole of Water $\times 10^{-3}$	Q_{UC}, Cal per N_0 Unit Cells	Q_{GE}, Cal per Gram Equiv. of Cation $\times 10^{-3}$
			Zeolite D (Synthetic ferrierite)		
Li	$33._9$	$40._2$	3.9_2	–	17.5
Na	$30._7$	$35._4$	4.1_3	–	15.9
Ca	$35._6$*	$41._5$	4.5_2	–	18.6
Sr	$25._0$*	$28._2$	4.0_2	–	13.3
H (from NH_4^+)	$34._5$	–	–	–	–
H (0.09*N* HCl)	$34._4$*	$40._8$	3.9_5	–	–

2 weeks to ensure that water uptake had reached equilibrium. The saturation water contents, determined by TGA using a Stanton thermobalance, are given in Table II.

Outgassing. Samples were weighed out in thin-walled borosilicate glass bulbs of diameter ~0.8 cm, outgassed at 360°C for 24 hours to a residual pressure $<10^{-5}$ mm of Hg, and sealed off under vacuum. Samples of NaX, NaY, NaA, and (Ca,Na)-chabazite, containing varying amounts of presorbed water, were prepared by partial removal of water from the saturated materials. Equal samples were partially outgassed in pairs through a common tap at temperatures controlled by an oil thermostat between 0° and 190°C for times between 2 and 12 hours. The heat of immersion of 1 sample was measured, and the residual water content of the second was determined by ignition at 1100°C to constant weight in a Pt crucible. The residual water contents of each sample were assumed, in interpreting the heat of immersion, to be the same. To ensure an equilibrium distribution of water prior to the calorimetric measurement, the partially outgassed samples in sealed bulbs were placed in an oven at 100°C for 36 hours and then cooled slowly to room temperature during a period of 6 hours.

Calorimeter. A differential calorimeter, operating at 25.0°C under near-isothermal conditions, was used for all heat measurements. Similar calorimeters, designed for determining heats of ion exchange in zeolites, have been described previously (*5, 6, 14, 15*). The calorimeter was calibrated by measuring the heat of solution of potassium chloride in water. The ratio of the area under the curve traced by the recorder pen to the heat produced was 1.50 ± 0.04 cm^2 per calorie. No heat could be detected when an empty evacuated bulb was broken under water.

Errors. Errors were estimated from duplicate measurements. The calorimetric measurements are accurate to about 4%, the saturation water contents to 1%, and the residual water contents are reproducible to about 3%, except at the lowest coverages, where the error is greater.

Results

Heat of Immersion and Exchange Cation. Heats of immersion in water were determined for the outgassed cationic forms of the zeolites (Table III). The heats given in this table are the following:

a. $q_H =$ calories per gram of hydrated zeolite

b. $q_D = \dfrac{q_H}{1 - w} =$ calories per gram of dehydrated zeolite

c. $Q_I = \dfrac{q_H}{w} \times 18.016 =$ calories per mole of water imbibed

d. $Q_{UC} = q_D \times w_{N_0} =$ calories per Avogadro No. of unit cells

e. $Q_{GE} = \dfrac{Q_{UC}}{n_c} =$ calories per gram equivalent of cations

Here w is the weight of water per gram of hydrated zeolite equilibrated over saturated ammonium chloride solution; w_{N_0} is the weight of an Avogadro number of dehydrated unit cells of zeolite; and n_c is the number of cationic charges per unit cell. The values of q_H were derived directly from the calorimeter measurements. The other quantities are derivable, assuming that all the water is removed by the outgassing at 360°C, and that where no direct chemical analysis was available, ion exchange had reached the limit indicated by published ion exchange isotherms (*5, 6, 14, 15, 31, 33*). All results are the average of 2 or more measurements except those marked with an asterisk, where only 1 measurement was made.

Dealumination and Heat of Immersion. Treatments of mordenite, clinoptilolite, and ferrierite with dilute acid in the first instance remove metallic cations and yield the hydrogen forms (*4, 12, 13*). Such forms also result by heating the ammonium zeolites (*24*). Treatment with stronger acid solutions removes increasing amounts of Al (*4, 13, 26*) (Table IV). The marked effect this has upon q_H is shown in the table.

Heats of Sorption as Functions of Amount Sorbed. The integral ($\Delta\widetilde{H}$) and differential ($\Delta\overline{H}$) heats of sorption of water vapor were obtained from the relationships

$$q_{n_o} - q_{n_s} = (n_s - n_o)\,(\Delta\widetilde{H} - \Delta H_L) \quad (1)$$

where

$$\Delta\widetilde{H} = \frac{1}{(n_s - n_o)} \int_{n_o}^{n_s} \Delta\overline{H}\, dn$$

and so

$$\Delta\widetilde{H} + (n_s - n_o)\left(\frac{\partial\Delta\widetilde{H}}{\partial n_s}\right)_{n_o} = \Delta\overline{H} \quad (2)$$

Here q_{n_s} and q_{n_o} are heats of immersion of the zeolite initially containing n_s and n_o moles of presorbed water per gram, and ΔH_L is the molar heat of condensation of water, taken as 10.51 kcal per mole. The degree of presaturation, θ, is given by

$$\theta = \frac{\text{initial water content}}{\text{saturation water content}} = \frac{n_s}{n_{sat}}$$

where n_{sat} is the number of moles of water needed to saturate the intracrystalline free volume. Smoothed curves of q_{n_s} *vs.* θ were drawn (Figure 1), and the lower limits of θ chosen for the calculations of integral heats were:

Zeolite:	NaX	NaY	NaA	(Ca,Na) chabazite
θ :	0.033	0	0.1	0.133

The values of $-\Delta\widetilde{H}$ and $-\Delta\overline{H}$ derived from the smoothed curves of Figure 1 are shown in Figure 2. At the lowest values of θ, $-\Delta\widetilde{H}$ is very large but diminishes as θ increases. The curves for NaY, NaA, and (Ca,Na) chabazite flatten for higher θ, while that for NaX has a minimum followed by a rising section. For NaX, the water–water self-potential energy, increasing with θ, more than balances the declining values of water–sorbent interactions. After saturation ($\theta \sim 1$), $-\Delta\overline{H}$ and $-\Delta\widetilde{H}$ for all systems would be expected to approach $-\Delta H_L$.

Table IV. Heats of Immersion in Decationated and Dealuminized Zeolites

Sample	*% Al Relative to Original*	q_H, *Cal per Gram*	q_D, *Cal per Gram*	q_I, *Kcal per Mole of Water*	q_{UC}, *Cal. per N_0 Unit Cells $\times 10^{-4}$*
		Clinoptilolite			
Original	100	$41._0$	$47._7$	5.2_7	10.3_3
0.25*N* HCl	58	$41._3$	$48._3$	5.1_1	9.5_3
0.5*N* HCl	33	$32._4$	$37._8$	4.0_8	7.3_3
1.0*N* HCl	7	$27._3$	$31._9$	3.4_2	6.0_7
2.0*N* HCl	0	$23._6$	$27._9$	2.7_6	5.2_9
		Mordenite			
Na Zeolon	100	$46._1$	$53._1$	6.2_9	16.2_1
H Zeolon	87	$34._7$	$40._5$	4.3_7	11.5_7
2.0*N* HCl	49	$38._3$*	–	–	–
6.0*N* HCl	32	$36._7$*	–	–	–
12.0*N* HCl	25	$34._8$*	$42._4$	3.5_0	11.6_1

Discussion

Exchange Ions and Heats of Immersion. Three cationic forms give heats of immersion which are out of line with other results. One of these is BaA, for which q_H is only 1.8 cal per gram. This form can lose much of its structure on outgassing (*32*), and the water uptake and heat of immersion are reduced accordingly. (Rb,Na)A and (Mg,Na)X also may lose some crystallinity during outgassing. Rastrenenko *et al.* (*29*) found q_H for (Rb,Na)A to be only 3 cal per gram, which probably indicates extensive lattice breakdown. Plank (*28*) reported partial breakdown in (Mg,Na)X although this was not confirmed by others (*16, 20, 27, 36*). Because of the abnormalities in q_H observed with BaA, (Rb,Na)A, and (Mg,Na)X, these 3 will be disregarded in the following comparisons. Our values of q_H for Li, (Mg,Na), and Ca forms of zeolite A are 12 to 20% lower than those of Rastrenenko *et al.*, but our value for KA is appreciably higher. However, with the exception of KA, the trends in q_H are the same, and numerical differences may arise from differences in the zeolite sample.

Of the cation forms of chabazite, Cs-chabazite contains about 20% of residual Ca while the other modifications should contain less than 10% of this ion (*6*). In zeolite X, the Rb, Cs, Mg, and Ba forms contain about 30, 30, 27, and 23 residual Na^+ ions per unit cell (*14*) out of a total in our

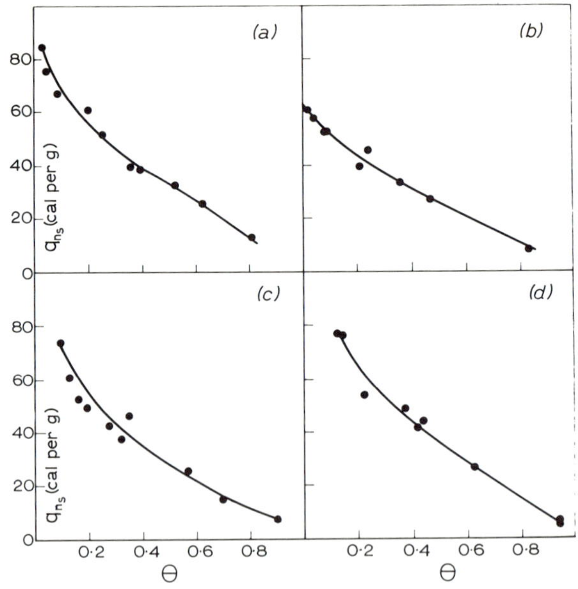

Figure 1. Curves of q_{ns} vs. *θ for (a) NaX, (b) NaY, (c) NaA, and (d) (Ca,Na)chabazite*

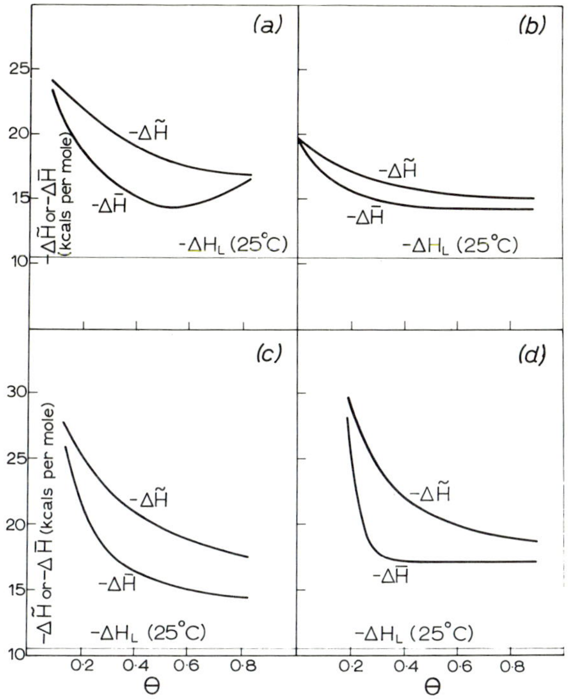

Figure 2. Curves of $-\Delta\bar{H}$ *and* $-\Delta\tilde{H}$ *plotted against* θ *for (a) NaX, (b) NaY, (c) NaA, and (d) (Ca,Na)chabazite*

sample of 87 ions. Of these numbers, 16 Na^+ ions in each case probably occupy sites within the hexagonal prisms and thus are completely shielded from water molecules. The remainder must have some contact with water and thus modify the heat of immersion compared with the homoionic form. In zeolite A, the Rb, Cs, and Mg forms contain about 29, 54, and 33% of Na^+ ions, respectively, all of which can interact with water. In zeolite Y, although the Rb, Cs, Ca, Sr, Ba, and Tl forms each contain about 16 Na^+ per unit cell (*5*), these Na^+ ions are probably within the hexagonal prisms and if so would not interact with water or modify the heats of immersion.

The magnitude of q_H is determined largely by the water content per gram (Table II), with further influence by the number, size, and valence of cations and the framework configuration. Among the alkali metal cations—excluding (Rb,Na)A—there is a regular decrease in q_H and q_D in the sequence.

$$Li^+ > Na^+ > K^+ > Rb^+ > Cs^+$$

For the divalent ions—excluding BaA and (Mg,Na)X—the order is

$$Mg^{2+} > Ca^{2+} > Sr^{2+} > Ba^{2+}$$

while the heat of immersion for a given zeolite containing a divalent exchange ion is uniformly larger than this heat in the same zeolite containing a monovalent ion of similar radius—*e.g.*, Ca^{2+} and Na^{+}—despite the smaller number of divalent ions. The influence of charge density in 2 otherwise identical structures is seen clearly by comparing q_H or q_D for the same exchange ion in zeolites X and Y. In all cases except that of Tl^{+}, the heat of immersion is considerably larger in X. For Tl^{+} the reverse is true, primarily because the water content of TlX is less than that of TlY (Table II).

The effect of polarity of the framework upon the heat of immersion can be shown in a more general way. The lattice-forming units of the anionic framework can be considered as $(Al_x,Si_{(1-x)})O_2^{x-}$ where $0 < x < 0.5$. The heat of immersion of the Na forms of the zeolites in kcal per Avogadro number of lattice-forming units (Q_{N_o}) was plotted against the ratio $\frac{Al}{Al + Si} = x$, which is a measure of the charge density (Figure 3). Despite the diversity of the zeolite structures, there is a correlation between Q_{N_o} and x which shows that intracrystalline charge density is a major factor in determining the magnitudes of these heats. When $x = 0$

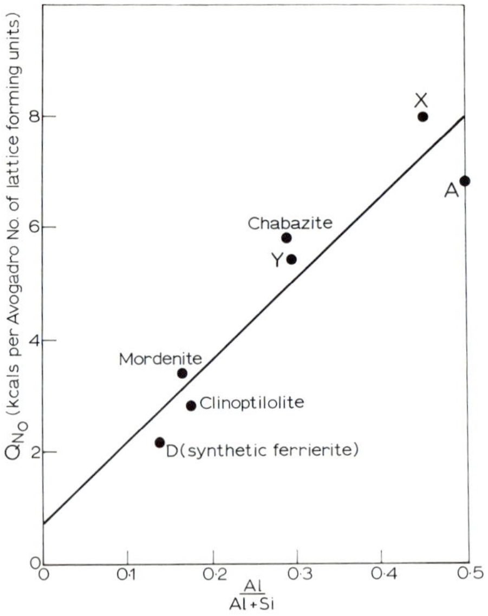

Figure 3. Q_{N_o} *plotted against* $\frac{Al}{Al + Si}$ *for various zeolites*

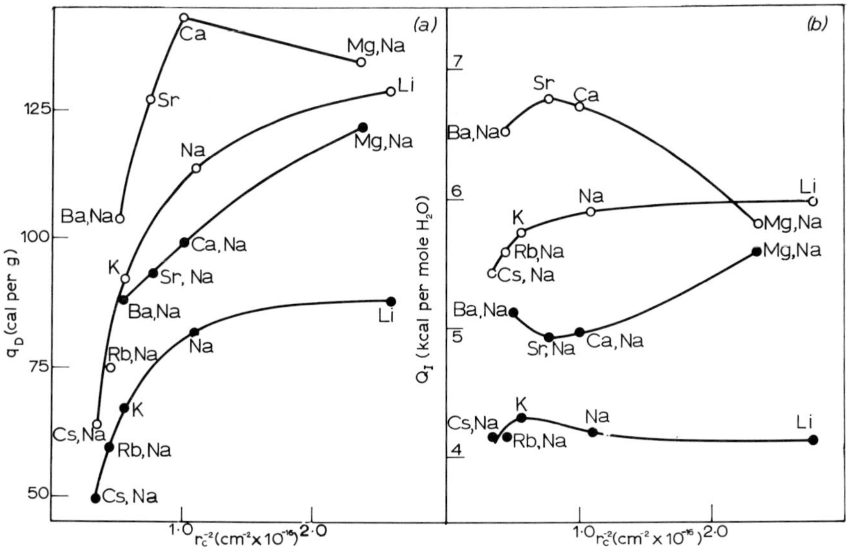

Figure 4. (a) q_D *and (b)* Q_I vs. r_c^{-2} *for zeolites X and Y. Open circles, X; closed circles, Y*

as in a hypothetical porous crystalline silica, Q_{N_o}, obtained by extrapolation, is very small.

The integral molar heat of sorption of liquid water into the zeolites at saturation of the intracrystalline volume is given by $-Q_I$ in column 4 of Table III. The same heat, $\Delta\widetilde{H}$, for water vapor is given by $\Delta\widetilde{H} = -(Q_I + 10.51)$ kcal per mole. This heat is numerically larger in zeolite X than in Y for a common cation, again showing the importance of differing charge densities in otherwise identical frameworks. Nevertheless, $\Delta\widetilde{H}$ does not vary so greatly among the cationic forms of any 1 zeolite (Tl^+ and Pb^{2+} excepted), suggesting that the proportionately more numerous framework oxygens, and water–water interactions, are principal factors in determining such saturation heats.

The integral heat for the saturation of an Avogadro number of unit cells with liquid water is $-Q_{UC}$ (Table III, column 5). It strongly reflects the size of a unit cell and the free volume per unit cell. When, as in a given cation form of X and Y, these factors are the same, then Q_{UC} shows the influence of differing charge density. Finally, $-Q_{GE}$ is the integral heat of immersion per gram equivalent of cations (Table III, column 6). It is influenced for a given zeolite by cation size and valence and also by the average coordination number of the ions towards water. Accordingly, Q_{GE} is larger in zeolite Y than in X. Q_{GE} has its largest values in chabazite, suggesting a particularly big cation contribution in this zeolite.

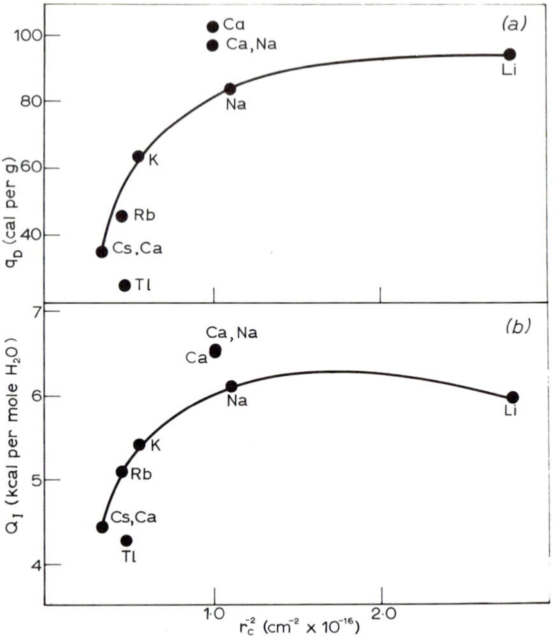

Figure 5. (a) q_D *and (b)* Q_I *plotted against* r_c^{-2} *for chabazite*

Relations between Heats and Cation Radii. For any cation–water pair, the ion–dipole energy should vary inversely as the square of the center-to-center distance between ion and water molecule—*i.e.*, as $(r_c + 1.40)^{-2}$, where r_c is the cation radius in A and 1.40A is taken as the radius of water. Accordingly, correlations might be sought between q_D, Q_I, or Q_{UC} and $(r_c + 1.40)^{-2}$ or r_c^{-2}. Such plots proved qualitatively the same whether $(r_c + 1.40)^{-2}$ or r_c^{-2} was the abcissa. Figures 4 and 5 show several of the observed relations with r_c^{-2} for zeolites X and Y and for chabazite.

For a sequence of ions of the same valence in a given zeolite, q_D—(Mg,Na)X excepted—increases as r_c^{-2} increases. With Q_I, however, maxima or minima are observed which suggest at least 2 opposing influences. These include the extent to which the cations are recessed into and against 4-, 6-, or 8-rings of lattice oxygens and hence the endothermal energy required to detach them wholly or in part when solvation occurs. Secondly, they include the exothermal energy of ion solvation by the entering zeolitic water. Smaller cations may be bonded much more strongly to 4- or 6-rings than large ions; they also are hydrated more energetically. Thus, a balance is likely to be struck involving numbers of ions loosened or detached and the relative energies of detachment and solvation.

The plots of Q_{UC} *vs.* r_c^{-2} also showed maxima or minima, presumably for the reasons given above. Values of Q_{UC}, q_D, or Q_I for Tl and Pb forms did not lie on the correlation curves of the other cations. In Table V, some measured values of Q_{UC} for these 2 ions in several zeolites are compared with interpolated values from the curves for the other ions of the same valency. The differences between experimental and interpolated values of Q_{UC} can be attributed to the greater energy for detaching the highly polarizable Tl^+ or Pb^{2+} from lattice oxygens as compared with ions of comparable radii such as Rb^+ and Sr^{2+}, respectively. The polarizabilities, α, in A^3 per ion, are

Rb^+	1.5	Sr^{2+}	0.9
Tl^+	3.9	Pb^{2+}	3.6

The binding of Tl and Pb to framework oxygen may be partially covalent and should become more energetic as framework polarity increases (chabazite < Y < X < A). The ratios in column 4 of Table V increase in this order. The effect considered is present in the other heats—*e.g.*, q_D, q_H—as seen in Table III.

Influence of Dealumination. Clinoptilolite and mordenites free of metallic cations and with Al/Si ratios which were progressively reduced gave heats of immersion which diminished with the Al content (Table IV). Each Al atom removed reduces the framework charge by 1, according to the reaction (*4*)

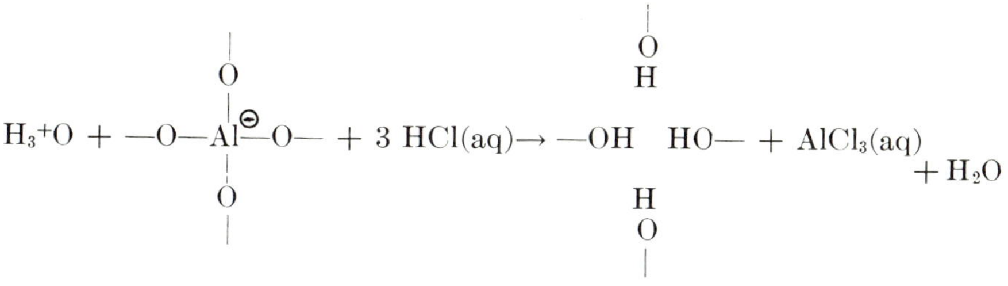

Table V. Experimental and Interpolated Values of Q_{UC} for Tl- and Pb-Forms of Several Zeolites

	Cal per N_0 *Unit Cells*		
Zeolite	Q_{UC}, *Exptl.*	Q_{UC}, *Interpol.*	*Ratio*
Tl chabazite	10.9×10^4	14.6×10^4	1.34
(Tl,Na) Y	$6.5_5 \times 10^5$	$9.1_5 \times 10^5$	1.40
TlX	7.1×10^5	13.0×10^5	1.84
TlA	8.2×10^4	15.2×10^4	1.86
PbX	10.2×10^5	19.3×10^5	1.89

The final product is a crystalline silica rich in hydroxyl defects but having essentially the framework structure of the parent zeolite (*4, 13, 30*). The defects render it polar, but the electrostatic components of the heats of immersion are reduced. The extent to which smooth correlations are obtained for q_D and q_H is shown in Figure 6 for clinoptilolite and mordenite, respectively. Equally clear correlations between Al content and $\Delta\bar{H}$ have been observed for CO_2 in these 2 dealuminized zeolites, whereas for the nonpolar Kr molecule, $\Delta\bar{H}$ remained virtually independent of the

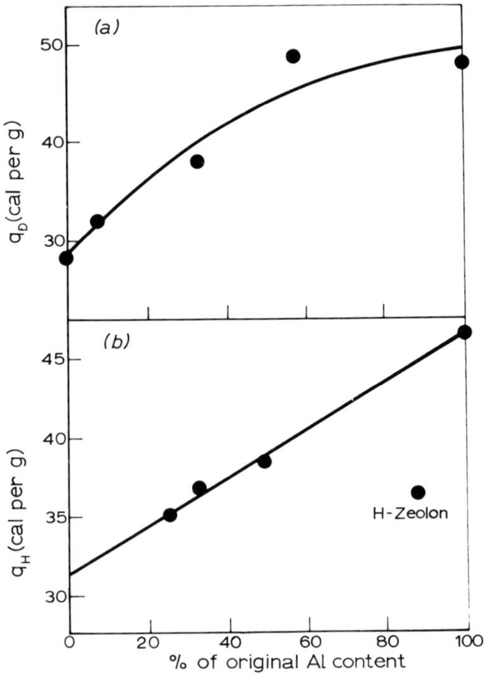

Figure 6. (a) q_D *for clinoptilolite and (b)* q_H *for mordenite, plotted against Al content relative to the original material*

Al content (*4*). These results all exemplify how framework charge density affects heats of sorption of molecules possessing permanent electric moments.

Heats for H-Zeolon, a commercial H-mordenite, fell well below the curve for H-mordenites prepared in our laboratory. The procedure in making H-Zeolon from its Na-form is not known to us.

Differential and Integral Heats of Sorption. $\Delta\bar{H}$ or $\Delta\tilde{H}$ can indicate relative selectivities of zeolites for water, and hence their relative value as desiccants. At low water loadings, it is the initial heats which are important. At $\theta = 0.1$, the order of heats, $-\Delta\bar{H}$, is

(Ca, Na) chabazite	>30 kcal per mole
NaA	30 kcal per mole
NaX	$22._7$ kcal per mole
NaY	17 kcal per mole

For high water loadings, integral heats at the largest measured value of θ common to all the zeolites may indicate the relative usefulness as desiccants at such loadings. For $\theta = 0.8$, $-\Delta\bar{H}$ has the values

(Ca,Na) chabazite	18.9 kcal per mole
NaA	17.7 kcal per mole
NaX	16.9 kcal per mole
NaY	15.2 kcal per mole

This order is the same as that of the initial heats. Retentivity for water at high loadings also depends upon the intracrystalline free volumes which are (*1*) about 0.54 cm^3 per cm^3 for NaY and NaX, and about 0.46 cm^3 per cm^3 for NaA and chabazite.

A comparison of water retentivity is given in Figure 7 in which θ(NaY), θ(NaA), and θ(chabazite) are plotted against θ(NaX). In each case, starting with saturated crystals, outgassing was conducted for the 4 zeolites under identical conditions. Curves shown in Figure 7 are not

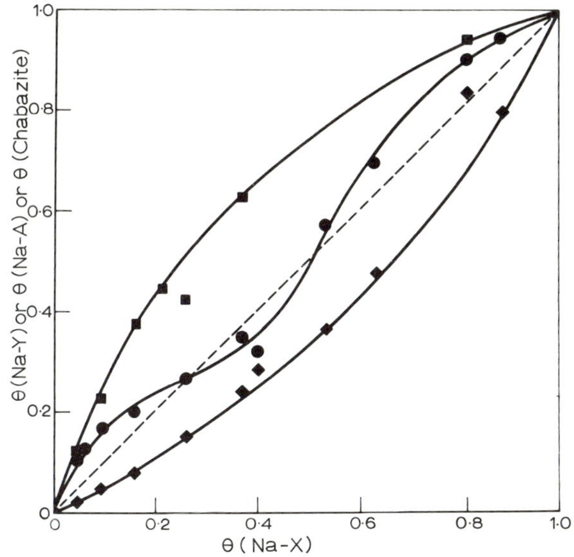

Figure 7. θ for NaY (◆), NaA (●), and chabazite (■) plotted against θ for NaX, following identical outgassing conditions

at equilibrium so that factors such as crystallite size may have an influence. Nevertheless, with the exception of an anomalous region for NaA in the range $0.25 < \theta < 0.5$, the retentivities are in the order expected from the sequence of heats.

Comparisons of our curves of $-\Delta\bar{H}$ *vs.* θ with those of others are made in Figures 8 and 9. Considerable differences are revealed. Apart from experimental difficulties which are considerable with polar, strongly sorbed water discussed above, several special factors are significant:

(a) Samples of NaX used by the Russian workers (*18, 19*) (curves III and IV, Figure 8) and by Barrer and Bratt (*3*) (curve II, Figure 8) contained 77 and 82 Na^+ ions per unit cell, respectively, compared with 87 Na^+ ions in our material. Thus, the rather higher over-all values of $-\Delta\bar{H}$ obtained by us are to be expected.

(b) Dzhigit *et al.* (*19*) (curve IV, Figure 8) and Frohnsdorff (*21*) (curve III, Figure 9) measured $-\Delta\bar{H}$ by adsorption calorimetry. It is

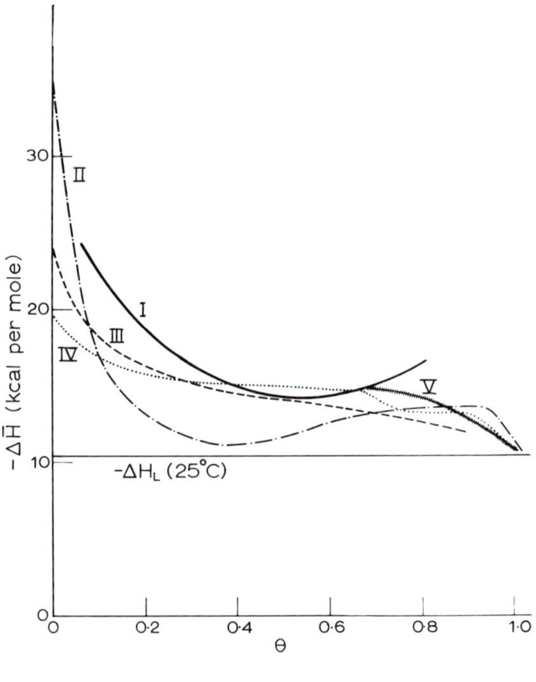

Figure 8. Comparison of curves of $-\Delta\bar{H}$ vs. θ *for the NaX–water system obtained by various workers*

————	*Curve I. This work*
—·—·—	*Curve II.* (3)
– – – – –	*Curve III.* (18)
. . . .	*Curve IV.* (19)
-|-|-|-|-|-|-	*Curve V.* (7)

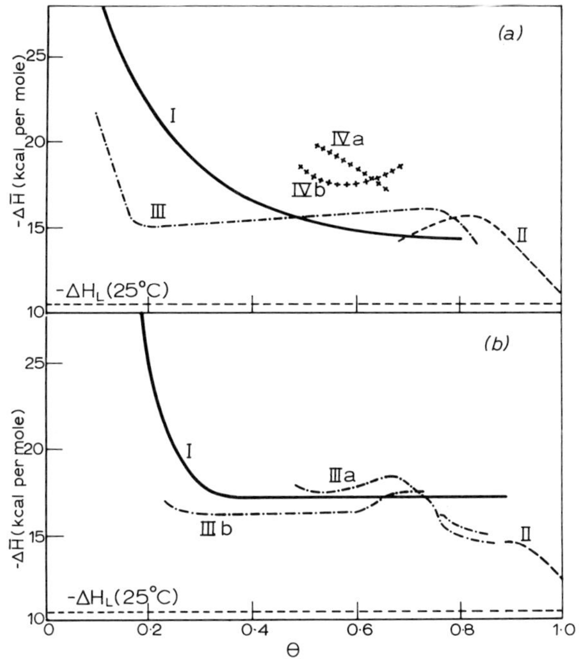

Figure 9. Comparison of $-\Delta \bar{H}$ vs. θ *for*

(a) NaA–water
———— *Curve I. This work*
– – – – – *Curve II.* (7)
—·—·— *Curve III.* (21)
-|-|-|-|-|- *Curve IVa, NaA and Curve IVb, CaA* (17)
(b) (Ca,Na)chabazite–water systems
———— *Curve I. This work*
– – – – – *Curve II* (8)
—·—·— *Curves III a and b, respectively, more and less Ca-rich chabazites* (35)

difficult, especially at low coverages, to ensure uniform distribution of water through the sample.

(c) Residual water contents may have varied on account of differing outgassing conditions. For example, Frohnsdorff outgassed at 200°C, which is a lower temperature than those used by other workers.

(d) Application of the Clapeyron-Clausius equation to determine heats from low coverage high temperature isotherms will yield values of $(\Delta \bar{H} - \Delta H_L)$ which are smaller than our values because ΔH_L changes with temperature—$(\Delta H_L{}^{280°C} - \Delta H_L{}^{25°C}) = 3$ kcal per mole.

(e) Slight hysteresis has been detected in some water–zeolite isotherms (7, *26*).

Clearly, in the light of Figures 8 and 9, it would be useful to compare results obtained by different methods on a single zeolite sample.

Interpretation of Heats at Small Coverage. From the shapes of curves of $-\Delta\bar{H}$ *vs.* θ, the first molecules of water are sorbed by a notably exothermal process upon a small number of sites. Barrer and Bratt (*3*) suggested that these might be a few very well exposed cations. Considerable success was achieved later in determining the sequence of initial heats for ammonia and carbon dioxide in various cationic forms of zeolite X without assuming abnormally exposed cation sites (*9, 10*). However, Habgood (*22*) suggested recently how exposed cation sites may arise, thus reviving the view of Barrer and Bratt. The Si/Al ratio in X exceeds 1 so that a few exchange ions will have positive charge balanced by negative charge on oxygens located elsewhere in the lattice. Such local imbalances expose the cation charge. This situation will arise also in other zeolites where the Si/Al ratio is nonintegral—*e.g.*, NaY, (Ca,Na) chabazite. The Si/Al ratio in NaA is unity, but in this structure 4 Na^+ ions may be associated with 3 8-ring windows (*34*). The possibility therefore exists of at least 1 very energetically sorbing site per unit cell.

For these reasons, as an alternative limiting case to that considered by Barrer and Gibbons (*9, 10*), we consider initial heats in terms of isolated ion–water interactions. Thus, energy terms of electrical origin are maximized, but contributions from lattice oxygens are omitted. The total interaction energy, ϕ, between a water molecule and an isolated ion (Na^+ for X, Y, and A, and Ca^{2+} for chabazite) was calculated as the sum of ion–dipole ($\phi_{F\mu}$), ion–quadrupole ($\phi_{\dot{F}Q}$), polarization (ϕ_P), dispersion (ϕ_D), and repulsion (ϕ_R) energy terms. These are assumed to be given by the expressions

$$\phi_{F\mu} = \frac{-Ze\mu}{r^2};\ \phi_{\dot{F}Q} = \frac{-QZe}{2r^3};\ \phi_P = \frac{-\frac{1}{2}\alpha(Ze)^2}{r^4};\ \phi_D = \frac{-A}{r^6};\ \text{and}\ \phi_R = \frac{B}{r^{12}}$$

where Z and e are the ion valency and electronic charge (4.80×10^{-10} e.s.u.) and α is the polarizability. μ and Q are the dipole and quadrupole moments of water ($\mu = 1.84 \times 10^{-18}$ e.s.u.; $Q = 2 \times 10^{-26}$ e.s.u.). The expressions for $\phi_{F\mu}$, $\phi_{\dot{F}Q}$, and ϕ_P refer to point charges, dipoles, and quadrupoles. The dispersion energy constant, A, was calculated from the Kirkwood-Muller expression (*23, 25*). Polarizabilities, α, and the several diamagnetic susceptibilities, χ, and radii, r, were taken as

	α, *Cm³ per Ion*	χ, *CGS units*	r, *Cm*
Na^+	0.19×10^{-24}	6.96×10^{-30}	0.95×10^{-8}
Ca^{2+}	0.47×10^{-24}	22.1×10^{-30}	0.99×10^{-8}
H_2O	1.48×10^{-24a}	21.5×10^{-30}	1.38×10^{-8}

[a] Cm³ per molecule.

Table VI. Interaction Energies at $r = r_e$, Kcal per Mole[a]

Ion-Water Pair and Repulsion Coeff.		r_e, A	$-\phi$	$-\phi_{F\mu}$	$+\phi_{\dot{F}Q}$	$-\phi_P$	$-\phi_D$	$+\phi_R$
Na^+–H_2O	B_1	1.80	42.5	39.4	12.0	23.2	6.2	14.3
	B_2	2.33	21.8	23.3	5.4	8.1	1.5	5.7
	B_3	2.13	27.0	28.5	7.1	12.1	2.3	8.8
	B_4	2.33	27.5	23.3	5.4	8.1	1.5	0
Ca^{2+}–H_2O	B_1	1.80	122.5	78.7	24.2	94.0	15.7	41.7
	B_2	2.37	53.8	46.0	10.6	31.1	2.5	15.2
	B_3	2.26	60.0	50.7	12.0	37.6	3.7	20.0
	B_4	2.37	69.0	46.0	10.6	31.1	2.5	0

[a] Molecule orientation is determined by the dipole. This makes the quadrupole energy endothermic (*3*).

The repulsion energy constant, *B*, the evaluation of which presents difficulties (*9, 10*), was estimated in 4 ways as follows.

(a) $B_1 = \frac{A}{2} \cdot r_o^6$ where r_o is the equilibrium separation of the ion and water molecule in the presence of dispersion and repulsion forces only; r_o was taken as the sum of the crystal radii of the ion and the water molecule.

(b) $B_2 = \frac{A}{2} r_e^6 \left[1 + \frac{\alpha(Ze)^2 r_e^2}{3A} + \frac{Q(Ze) r_e^3}{4A} + \frac{(Ze)\mu r_e^4}{3A} \right]$ where r_e, the equilibrium distance of separation under the influence of all forces, is set equal to r_o.

(c) $B_3 = \frac{A}{2} (r_o')^6$ where r_o' is the radius of the water molecule plus the radius of the cation in a hypothetical nonelectrostatic environment. This latter radius was taken as the radius of the inert gas atom having the same electronic configuration (Ne, 1.60A; Ar, 1.92A).

(d) The hard sphere model with $B_4 = \infty$ for $r \leqslant r_e$ and $B_4 = 0$ for $r > r_e$. r_e was again set equal to r_o.

The maximum interaction energy, corresponding with the equilibrium distance of separation, r_e, was obtained from the plots of $\phi = (\phi_{F\mu} + \phi_{\dot{F}Q} + \phi_P + \phi_D + \phi_R)$ against *r*. The values of r_e according to the value of *B* taken, and the corresponding values of the components of ϕ, are shown in Table VI. The values of ϕ and r_e associated with B_1 seem unrealistic, so an average value for ϕ was calculated from the other 3 values. Thus, the maximum energies of interaction are:

Na^+–H_2O	−25 kcal per mole
Ca^{2+}–H_2O	−61 kcal per mole

The relationship between ϕ and $\Delta\bar{H}$ at 0°K is $(\phi - \phi_o) = \Delta\bar{H}$ where ϕ_o is the zero point energy. In view of the approximations made, ϕ_o is neglected, and it is assumed also that $\Delta\overline{H}_{298°K} = \Delta\overline{H}_{0°K}$. At the lowest coverages examined, the values of ΔH are

NaX	$\theta = 0.05$	$\Delta\overline{H} = -24$ kcal per mole
NaY	$\theta \sim 0$	$\Delta\overline{H} = -20$ kcal per mole
NaA	$\theta = 0.1$	$\Delta\overline{H} = -29$ kcal per mole
(Ca,Na) chabazite	$\theta = 0.15$	$\Delta\overline{H} = -32$ kcal per mole

The heats of sorption at low coverages for the Na-forms thus are comparable with the calculated value. For NaX, NaA, and (Ca,Na) chabazite, $-\Delta\overline{H}$ at $\theta = 0$ presumably would be somewhat greater. This may explain in part why the measured heat for chabazite is substantially less than that calculated for an isolated Ca^{2+}–H_2O pair. A further possibility is that the Na^+ ions in natural chabazite occupy more exposed positions than the more numerous Ca^{2+} ions.

The numerically largest exothermal term at r_e is $\phi_{F\mu}$ (Table VI). A direct method of estimating the relative importance of the energy components $(\phi_D + \phi_R + \phi_P)$ and $(\phi_{F\mu} + \phi_{\dot{F}Q})$ has been developed (2). The initial heats of sorption for a number of nonpolar, structurally simple molecules were plotted against their polarizability, α, for a given sorbent. In this way, a characteristic curve giving the contribution to $\Delta\overline{H}$ from $(\phi_D + \phi_R + \phi_P)$ as a function of α was obtained. This contribution can be interpolated then for molecules of intermediate polarizabilities which also have permanent electric moments. By difference, the contribution of $(\phi_{F\mu} + \phi_{\dot{F}Q})$ to $\Delta\overline{H}$ can be found. This method demonstrates again the dominance of the electrostatic energy components in water–zeolite systems, as is apparent from Table VII. The calculated values of $(\phi_{F\mu} + \phi_{\dot{F}Q})$, averages for the values of B_2, B_3, and B_4 (Table VI), are -19.1 kcal per mole for Na^+ and -36.5 for Ca^{2+} and are reasonably near the actual values. As noted previously (3), the formula for ϕ_P appears to exaggerate this energy term, as seen when $(\phi_D + \phi_R + \phi_P)$ in Table VII

Table VII. Contributions of $(\phi_D + \phi_R + \phi_P)$ and of $(\phi_{F\mu} + \phi_{\dot{F}Q})$ to Initial Values of $\Delta\overline{H}$

Zeolite	θ	$-\Delta\overline{H}$, *Kcal per Mole*	*Contributions, Kcal per Mole*	
			$-(\phi_D + \phi_R + \phi_P)$	$-(\phi_{F\mu} + \phi_{\dot{F}Q})$
NaX	0.05	24	2.7	21.3
NaY	0	20	2.7[a]	17.3
NaA	0.1	29	—[b]	—
(Ca,Na) chabazite	0.15	32	5.8	26.3

[a] The same value as in the iso-structural NaX is assumed.
[b] No characteristic curve has been constructed.

is considered along with this sum and its individual components in Table VI.

Conclusion

This study of heats of immersion has shown a direct relation between the heat, Q_{N_0}, per Avogadro number of lattice-forming units, $(Al_x, Si_{1-x})O_2^{x-}$, and $x = \frac{Al}{Al + Si}$ for a range of zeolite structures. This indicates 1 general connection not sensitive to lattice configuration between the framework charge per unit and the heat, Q_{N_0}. Relations between the cation radii and $\Delta\widetilde{H}$ are equally apparent but are less simple and include such factors as extent of cation screening and solvation by lattice oxygens, extent of re-siting during hydration, and cation valence, number, and polarizability. High initial heats, $\Delta\bar{H}$, and the rapid decline of $\Delta\bar{H}$ with increasing uptake of water are compatible with a small proportion of relatively unscreened cations. These produce initially large electrostatic energy contributions to $\Delta\bar{H}$.

Literature Cited

(1) Barrer, R. M., *Ber. Bunsengesell. Phys. Chem.* **1965,** 69, 786.
(2) Barrer, R. M., *J. Colloid Interface Sci.* **1966,** 21, 415.
(3) Barrer, R. M., Bratt, G. C., *J. Phys. Chem. Solids* **1959,** 12, 146, 154.
(4) Barrer, R. M., Coughlan, B., "Molecular Sieves," p. 141, Society of the Chemical Industry, London, 1968.
(5) Barrer, R. M., Davies, J. A., Rees, L. V. C., *J. Inorg. Nucl. Chem.* **1968,** 30, 3333.
(6) *Ibid.,* **1969,** 31, 219.
(7) Barrer, R. M., Denny, A. F., *J. Chem. Soc.* **1964**, 4684.
(8) Barrer, R. M., Fender, B. E. F., *J. Phys. Chem. Solids* **1961,** 21, 1.
(9) Barrer, R. M., Gibbons, R. M., *Trans. Faraday Soc.* **1963,** 59, 2569.
(10) *Ibid.,* **1965,** 61, 948.
(11) Barrer, R. M., Langley, D. A., *J. Chem. Soc.* **1958,** 3817.
(12) Barrer, R. M., Lee, J. A., *Surface Sci.* **1969,** 30, 111.
(13) Barrer, R. M., Makki, M. B., *Can. J. Chem.* **1964,** 42, 1481.
(14) Barrer, R. M., Rees, L. V. C., Shamsuzzoha, M., *J. Inorg. Nucl. Chem.* **1966**, 28, 629.
(15) Barrer, R. M., Rees, L. V. C., Ward, D. J., *Proc. Roy. Soc.* **1963, A,** 273, 180.
(16) Barry, T. I., Lay, L. A., *J. Phys. Chem. Solids* **1966,** 27, 1821.
(17) Breck, D. W., Eversole, W. G., Milton, R. M., Reed, T. B., Thomas, T. L., *J. Am. Chem. Soc.* **1956,** 78, 5963.
(18) Dubinin, M. M., Kadlec, O., Zukal, A., *Coll. Czech. Chem. Comm.* **1966,** 31, 406.
(19) Dzhigit, O. M., Kiselev, A. V., Mikos, K. N., Muttik, G. G., *Zh. Fiz. Khim.* **1964,** 38, 1791.
(20) Eberly, P. E., Jr., *J. Phys. Chem.* **1962**, 66, 812.
(21) Frohnsdorff, G. J. C., Ph.D. Thesis, London University, 1959.
(22) Habgood, H. W., *Chem. Eng. Prog. Symp. Ser.* **1967,** 63 (73), 45.
(23) Kirkwood, J. G., *Phys. Zeit.* **1932,** 33, 57.

(24) McDaniel, C. V., Maher, P. K., "Molecular Sieves," p. 186, Society of the Chemical Industry, London, 1968.
(25) Muller, A., *Proc. Roy. Soc.* **1936, A,** 154, 624.
(26) Murphy, E. V. T., Ph.D. Thesis, London University, 1968.
(27) Pickert, P. E., Rabo, J. A., Dempsey, E., Schomaker, V., *Proc. Intern. Congr. Catalysis, 3rd, Amsterdam* **1964,** 1, 714.
(28) Plank, C. S., *Proc. Intern. Congr. Catalysis, 3rd, Amsterdam* **1964,** 1, 727.
(29) Rastrenenko, A. I., Antonovskaya, S. N., Neimark, I. E., *Kolloidn. Zh.* **1965**, 27, 269.
(30) Sand, L. B., "Molecular Sieves," p. 71, Society of the Chemical Industry, London, 1968.
(31) Sherry, H. S., *J. Phys. Chem.* **1966,** 70, 1158.
(32) *Ibid.,* **1966,** 70, 1332.
(33) Sherry, H. S., Walton, H. F., *J. Phys. Chem.* **1967,** 71, 1457.
(34) Smith, J. V., Dowell, L. G., *Z. Krist.* **1968,** 126, 135.
(35) Tiselius, A., Brohult, S., *Z. Physik. Chem.* **1934, A,** 168, 248.
(36) Tsitsishvili, G. V., Andronikashvili, T. G., Chumburidze, T. A., *Dokl. Akad. Nauk SSSR* **1965**, 164, 1104.

RECEIVED November 13, 1969.

Discussion

J. R. Katzer (University of Delaware, Newark, Del. 19711): Boudart and Hwang have considered a similar problem recently and have reported their results in *Ind. Eng. Chem. Fundamentals,* I believe. They calculated heats of adsorption for argon adsorption on type Y zeolite in the sodium and calcium forms. They then compared the calculated values with experimentally determined heats and found good agreement for the sodium form but a value which was much lower than the calculated value for the calcium form. From this they concluded that the Ca cation did not exhibit a full +2 charge but that there was charge shielding for the Ca cation reducing the effective charge. Would this not explain the disagreement which you found for the Ca form you studied?

P. J. Cram: I am not familiar with Boudart and Hwang's work. We believe that an overestimate of the polarization energy term contributes to the disagreement.

M. M. Dubinin (Academy of Sciences of the USSR, Moscow, USSR): In the work under discussion, the authors for the first time undertook an extensive and systematic investigation of heats of immersion into water of various zeolites in different cation-exchange forms containing varied amounts of preadsorbed water. On the basis of their experiments, they calculate the dependence of differential molar heats of adsorption on the adsorption values of water. In principle, assuming that equilibrium states are reached, the curves obtained should coincide with similar curves determined calorimetrically in adsorption of water vapors or with curves of isosteric heats of adsorption. The authors analyze possible difficulties and shortcomings of the methods mentioned above. These are mainly owing to extremely slow attainment of equilibrium states in adsorption of vapors of various substances on zeolites, in particular of water vapors. This effect results from the pronounced energetic nonuniformity of the adsorption space of zeolites.

In a joint work with A. A. Isirikyan with the participation of G. U. Rakhmat-Kariyev, we carried out direct measurements of differential heats of adsorption of water vapors on crystalline and molded zeolite NaA at 22°C using a Tian–Calvet-type calorimeter. The calorimetric installation enabled us to measure thermal effects for each point of the adsorption isotherm for a period of 300 hours and more (Figure 1). The squares and circles in the upper part of the graph denote our data for

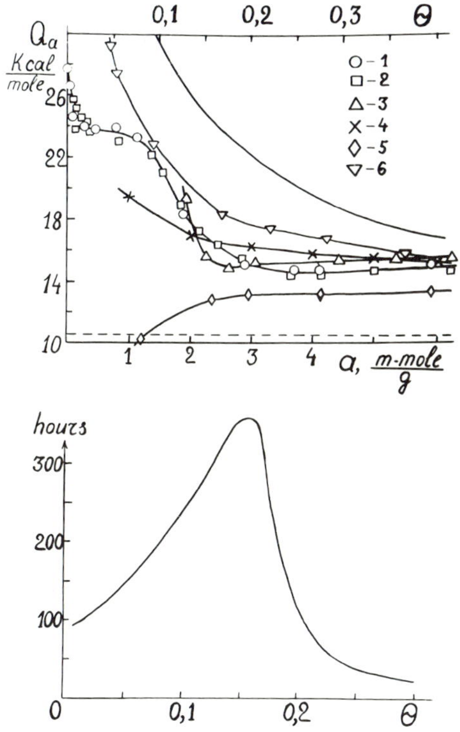

Figure 1. Dependence of differential heats of adsorption, Q_a, of water vapor on adsorption values, a, for zeolite NaA after different investigators (above); the curve illustrates the kinetics of adsorption (below)

crystalline and molded zeolite NaA (*2*). In the latter case, the adsorption values are given per one gram of crystals contained in the pellets. The region of high differential heats of adsorption, which amount to approximately 24 kcal/mole, corresponds to a degree of filling of the zeolite voids of around 0.1; *i.e.*, an average of 3 molecules of water per one large void. In the filling range from 0.1 to 0.15, there is an abrupt drop in the differential heats of adsorption, which vary insignificantly, however, on further filling up to about 0.8. The measurements of Frohndorff and Kington (*3*), which are denoted by triangles in the graph, almost coincide with our data. In their experiments, however, the most interesting region of small fillings below 0.1 was not studied. The curves of differential heats of adsorption of ammonia on type A zeolite, which were also determined with the aid of a modified Tian-Calvet calorimetric installation and are depicted in Figure 2 of the paper by Sichart *et al.* submitted to this conference (*5*), are similar to our curve as well.

The curve in the lower graph expresses the dependence of the time of attainment of equilibrium (in hours) on filling. For this time, the increment of heat release in a calorimetric experiment becomes immeasurably small. The graph shows that for fillings of up to 0.2, the time of attainment of adsorption equilibrium is from 100 to 350 hours, the slowest kinetics of adsorption being typical of fillings of about 0.15 corresponding, on the average, to 4–5 molecules of the water adsorbed per one large void.

In the light of the results obtained by us, the calorimetric experiments of Avgul *et al.* (*1*) (denoted by the letter x in the graph) conducted on a calorimeter with constant heat exchange and a maximum length of the "main period" of the release of the heat effect amounting to 4–5 hours for each point of the isotherm exhibit very considerable divergencies, especially for the small-filling region. The isosteric heats of adsorption of water on zeolite NaA calculated by Morris (*4*) (denoted by diamonds in the graph) are considerably lower than the calorimetric heats determined by us for the entire region of fillings below 0.4, and the shape of the curve is in qualitative contradiction with the results considered above.

The inverted triangles in the upper graph denote the curve of differential heats of adsorption calculated from data on heats of immersion recorded by Vergnaud *et al.* (*6*); these greatly exceed our data throughout almost the entire filling region. And finally, the uppermost curve without any points has been borrowed from Barrer and Cram's paper. The heats of adsorption corresponding to it are unusually high, and the causes of this should be thoroughly discussed.

It is quite natural that we consider our data the closest to the actual fact, and the conclusions following from the nature of the curve obtained and concerning the highest interaction energy of three mobile cations of sodium with three adsorbed molecules of water and the favorable position of these binary complexes (H_2O . . . Na^+) at the windows of zeolite NaA leading into large voids also seem to be plausible. To the fourth (and last) mobile cation of sodium, together with the fourth water molecule that has entered the void, we assign, as distinct from Barrer and Cram, the intermediate region of the curve from 22 to 17 kcal/mole, since all the energetically advantageous places at the entrance windows are already occupied by the preceding three complexes.

However, the curves of differential heats of adsorption presented in the upper graph and obtained by different investigators exhibit such considerable divergencies that, as it appears to me, we should first of all establish a method of investigation providing results most closely reflecting the actual fact, and only then draw conclusions of fundamental importance.

Literature Cited

(1) Avgul, N. N., Kiselev, A. V., *et al., Zh. Fiz. Khim.* **1968,** 42, 1474.
(2) Dubinin, M. M., Isirikyan, A. A., *et al., Izv. Akad. Nauk SSSR, Ser. Khim.* **1969,** 2355.
(3) Kington, G. L., "The Structure and Properties of Porous Materials," p. 247, Butterworths, London, 1958.
(4) Morris, B., *J. Colloid Interface Sci.* **1968,** 28, 149.
(5) Sichhart, K. H., Kolsch, P., Schirmer, W., ADVAN. CHEM. SER. **1971,** 102, 132.
(6) Vergnaud, J. M., *et al., Bull. Soc. Chim. France* **1965,** No. 5, 1279.

A. V. Kiselev (Lomonosov State University, Moscow, USSR): I believe that the most important source of the differences between the values of the differential heat of specific adsorption of water which were indicated in the paper by Barrer and Cram is the difference in the cation concentration in the samples of zeolite used. Indeed Figure 1 shows that the heat of water adsorption at small adsorption values definitely diminishes with the diminishing of cation concentration in the case of two Li^+-containing samples of zeolite X: LiX-1 and LiX-2 (for the compositions, see p. 185). These data were obtained by O. M. Dzhigit, A. V. Kiselev, K. N. Mikos, G. G. Muttik, and T. A. Rakhmanova (paper accepted for publication in *Trans. Faraday Soc.*). Figure 2 shows a similar picture for NaX zeolites. The heat of water adsorption diminishes with dropping cation concentration from 87 Na^+ per unit cell (calculated by R. M. Barrer and P. J. Cram from heat of wetting measurements) to 76 and 75 Na^+ per unit cell in our samples (measured in a calorimeter by O. M. Dzhigit, A. V. Kiselev, K. N. Mikos, G. G. Muttik, and T. A. Rakhmanova). The curve for 87 cations in Figure 2,b represents the

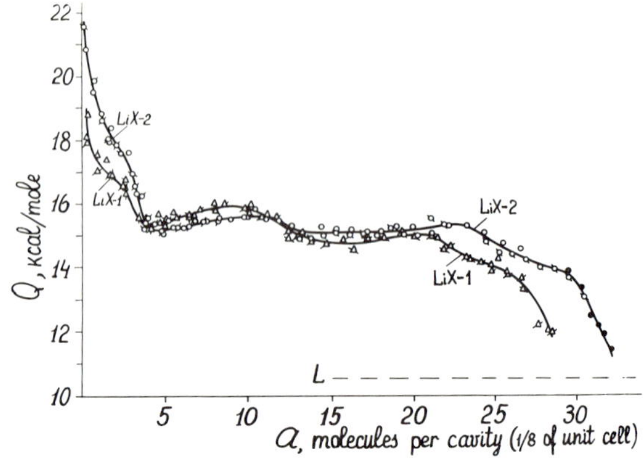

Figure 1. Dependence of calorimetrically measured differential heat of adsorption of water vapor on the adsorption amount. Crossed triangles and circles represent the repeated runs of measurements.

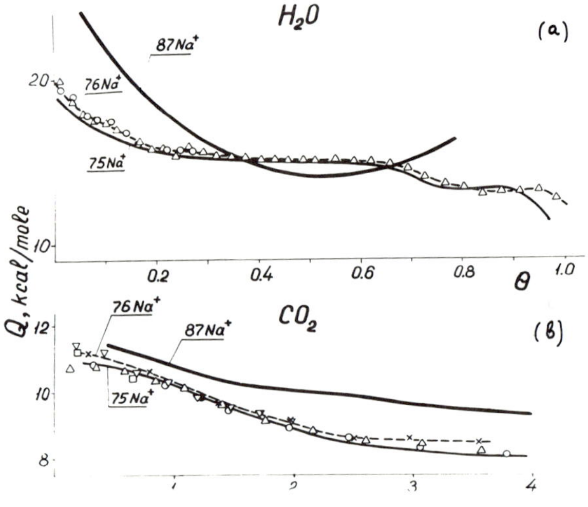

Figure 2. (a) Dependence of the heat of adsorption of water vapor on the adsorption amount for NaX zeolites with different exchange cation concentrations. (b) Dependence of the heat of adsorption of CO_2 on the adsorption amount for the same zeolites.

isosteric heats determined by R. M. Barrer and B. Coughlan; curves for 76 and 75 cations were determined by N. N. Avgul, B. G. Aristov, A. V. Kiselev, and L. Ya. Kurdyukova: points = calorimetric measurements; filled curve = isosteric heats. Similar results (Figures 2 and 3) were obtained in the case of specific adsorption of CO_2 on the same zeolites (R. M. Barrer and R. M. Gibbons, *Trans. Faraday Soc.* **1965,** 61, 948; R. M. Barrer and B. Coughlan, "Molecular Sieves," p. 233, Society of the Chemical Industry, London, 1968; N. N. Avgul, B. G. Aristov, A. V. Kiselev, and L. Ya. Kurdyukova, *Zh. Fiz. Khim.* **1968,** 42, 2678). In Figure 3, for

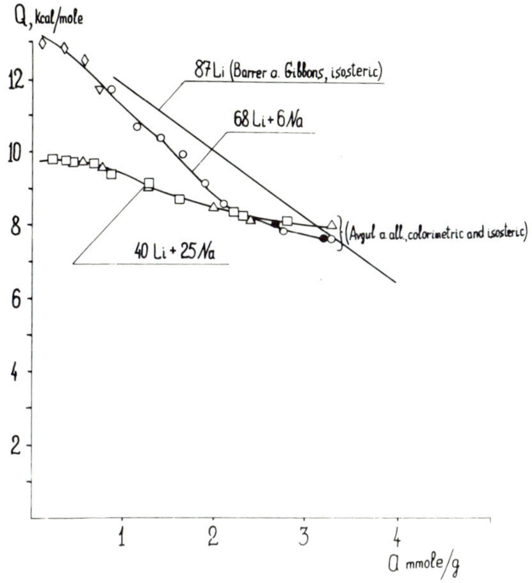

Figure 3. Dependence of the heat of CO_2 adsorption on the adsorption amount for LiNaX zeolites with different exchange cation concentrations (indicated)

87 Li^+ per unit cell, isosteric heats were determined by R. M. Barrer and R. M. Gibbons; for 68 $Li^+ + 6\ Na^+$, calorimetric measurements were made by N. N. Avgul, E. S. Dobrova, and A. V. Kiselev; for 40 $Li^+ + 25\ Na^+$, calorimetric (points) and isosteric (filled curve) heats were obtained by N. N. Avgul, B. G. Aristov, A. V. Kiselev, L. Ya. Kurdyukova, and N. V. Frolova. In the case of CO_2 adsorption, the calorimetric heat values coincide with the isosteric. These examples clearly show that the physicochemical constants calculated from experiments (Henry constant, second virial coefficient, corresponding heat of adsorption, etc.) are influenced by the zeolite structure and chemical composition. Therefore, it is quite necessary to indicate this composition in the representation and discussion of the thermodynamic results. Uncertain results were often obtained for zeolites having a binding material.

48

Calorimetric Measurements of the Systems Zeolite–Ammonia and Zeolite–*n*-Heptane in a Range of Temperature from 0° to 300° C

K.-H. SICHHART, P. KOLSCH, and W. SCHIRMER

Department of Adsorption of the Central Institute of Physical Chemistry of the German Academy of Science, Berlin, East Germany

*Using a conduction calorimeter (improved type Tian-Calvet), we measured the differential enthalpies of adsorption $-\Delta H$ (kcal/mole) of the system zeolite NaCaA–NH_3 within a temperature range from 23° to 300°C. We observed a step-like dependence of the values of $-\Delta H$ from the degree of pore volume filling of the zeolite. This result leads to the assumption that certain structures are formed between the NH_3 molecules and the cations present in the zeolite cages. A drop calorimeter was used to measure the heat capacities of the system NaCaA zeolite–*n*-*heptane in the temperature range 25°–240°C. The heat capacity shows a high maximum at approximately 100°C, especially for very small values of pore volume filling. This behavior may be explained by cooperation between defect structures of the adsorbent and configuration effects of the hydrocarbon chain.*

For the determination of thermodynamic adsorption parameters, we used a drop calorimeter to measure specific heats and a conduction calorimeter to measure enthalpies of adsorption.

In general, 2 types of calorimeters are used for measurement of specific heats, the adiabatic heated and the drop calorimeter. We chose the latter type for our study of specific heats because we needed the enthalpy function for our thermochemical calculations. This is the integral of the specific heat up to the experimental temperature, and cannot be measured directly by an adiabatically heated calorimeter. At higher temperatures the integration error will be considerable, so that it is better to use the drop method which provides the enthalpy differences directly.

The filled sample container is heated to temperature T. Then the sample is thrown into the calorimeter at temperature T_K. After a blind test, the enthalpy difference H_T-H_{T_K} of the investigated substance can be calculated.

The calorimeter which we constructed has a precision of 0.02–0.1%. Measurements are possible between 25° and 1000°C. The calibration constant reduced to 25°C is 2533.08 ± 0.25 J/grd with a standard deviation of m = 1.02 J/grd (*6*).

We built a conduction calorimeter of the Tian-Calvet type to measure the heat of adsorption of gases on zeolites. Figure 1 shows the construction of this calorimeter. The metal block reaches a temperature constancy of 0.01°–0.03°C. At about 300°C, we obtain the same values with an automatic adjustment. The calorimeter attains a time constancy

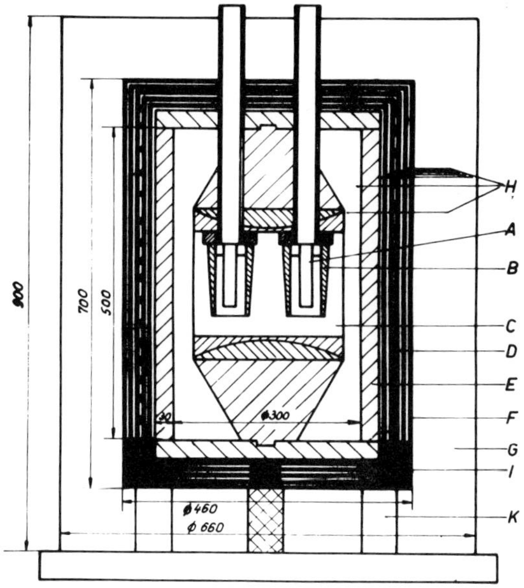

Figure 1. Calorimeter construction

A: Ag vessel
B: External boundary
C: Al block
D: Electric heater
E: Al cylinder
F: 5 Al jackets
G: Glass wool isolation
H: Air isolation
I: Asbestos isolation
K: Ceramic props

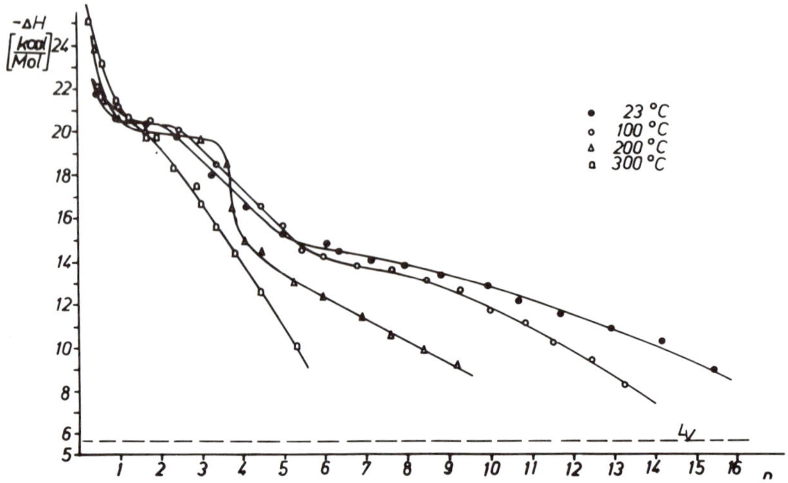

Figure 2. Differential heats of adsorption of ammonia on zeolite NaCaA at 4 temperatures

of 83 seconds and a sensitivity of 2.12 $\times$ 10^{-4} cal/sec mm; the precision is 0.4% (*8*). The differential heats of adsorption of ammonia on zeolite of the type NaCaA (66% ion exchange) have been measured at 23°–300°C by this calorimeter.

Figure 2 shows the results of these studies for 23°, 100°, 200°, and 300°C. We observed heats of adsorption, 22–25 kcal/mole, at the lowest values of ammonia content and a decrease of this value to 20–21 kcal/mole at one molecule of ammonia per large cage. The heat of adsorption remains constant at this value up to the adsorption of 3.5 molecules/cage. This can be seen most distinctly at a measuring temperature of 200°C. Between 3.5 and 4 molecules/cage, the heat of adsorption decreases to 14–15 kcal/mole.

The heat curves at 23° and 100°C also show these steps but in this case the decrease of heat of adsorption occurs in the range 2.5–5 molecules/cage. The measurements at 300°C show no steps, and the values decrease almost linearly from 25 to 10 kcal/mole.

We interpret these results by assuming that the high values for the heats of adsorption pertain to specifically active centers, such as lattice holes (*7*). Dielectric relaxation measurements (*4*) and KMR studies (*2*) substantiate such conclusions.

The adjacent step represented by values of 20–21 kcal/mole leads to the assumption that the first molecules of ammonia are adsorbed at the 4 Ca^{2+} cations in the cage. This is shown by measurements of the adsorption of ammonia in several NaX zeolites partially containing transition metal cations and by studies of the dielectric relaxation of the systems H_2O–NaCaA and H_2O–CaA (*4*).

A value of 23 kcal/mole (*7*) for the interaction of ammonia with the Ca^{2+} cation was calculated from the energy of adsorption of ammonia on zeolite NaCaA by common potential methods (analogous to Ref. *1*). In Ref. *3* other authors found several ranges of adsorption which are accounted for by interactions with the cations. The interpretation in our paper (*7*) thus is confirmed.

In the transition range of temperatures, no extreme drop is observed. This can be explained by diffusional inhibition of the molecules which are adsorbed at these temperatures (*1*). A further step follows. Our explanation for the steep slope of the heats of adsorption at 300°C in this case is that the adsorbed ammonia molecules have a high mobility and do not occupy fixed positions. These measurements have been carried out with a precision of ±2%.

Direct measurements of the heat capacities of the system zeolite NaCaA–*n*-heptane have been made in the temperature range 25°–240°C with the drop calorimeter described above.

In Figure 3, the heat capacity is shown as a function of temperature at 3 coverages, with correction for the heat of adsorption. The quantity

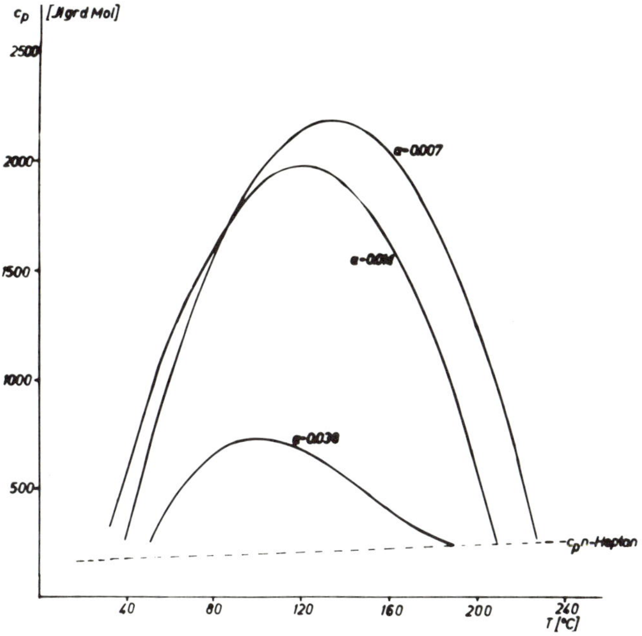

Figure 3. Heat capacity of the system n-*heptane–zeolite NaCaA*

of adsorbed *n*-heptane, *a,* is shown in mmole/gram zeolite. The heat capacity goes through a maximum between 80°–120°C, rising to values much larger than can be attributed to the pure *n*-heptane. This maximum in the curves was expected after having measured isosteres of the system zeolite NaCaA–*n*-heptane (5). The height of the maximum decreases with loading.

This effect can be explained only by regarding the system zeolite–adsorpt as a whole. Certain defect structures of the adsorbent and additional configuration effects of the chain of the hydrocarbon must coincide to give these results. We are preparing a quantitative estimation of this problem for publication.

Literature Cited

(1) Barrer, R. M., Gibbons, R. M., *Trans. Faraday Soc.* **1963,** 59, 2569.
(2) Bulow, M., Kohler, D., Spangenberg, H. J., *Z. Chem.* **1969,** 9, 317.
(3) Dubinin, M. M., Isirikian, A. A., Sarachow, A. I., Serpinksi, V. V., *Izv. Akad. Nauk, SSSR Ser. Khim.* **1969,** 2355.
(4) Lohse, U., Stach, H., Hollnagel, M., Schirmer, W., in press.
(5) Meinert, G., Thesis, Humboldt-University, Berlin, 1969.
(6) Muscheites, K., Thesis, Humboldt-University, Berlin, 1967.
(7) Schirmer, W., Grossmann, A., Fiedler, K., Sichhart, K.-H., Bulow, M., *Chem. Tech.,* in press.
(8) Sichhart, K.-H., Thesis, TH Merseburg, 1967.

RECEIVED February 18, 1970.

Discussion

L. V. C. Rees (Imperial College, London): When zeolites have increasing amounts of polar molecules sorbed in them, the cations show a great increase in mobility when certain low fillings are obtained. Is it not possible that the decrease in the heat of sorption of NH_3 at low isotherm temperatures represents the point where the cations loosen their attachment to the framework? This would be an endothermic movement of a positive charge from the negative framework. At 300°C, the cations would have sufficient thermal energy to mask this effect.

W. Schirmer: I should suppose that the loosening process of the cations would take place at higher coverages, where the decrease in heat of adsorption is rather low. If this process takes place—we took it into consideration, too—then we must get, for a certain range of coverage, an approximately horizontal part of the A *vs.* θ curve (all the Ca cations being equal in energy)—a phenomenon which we did not observe until now.

N. G. Parsonage (Imperial College, London): If one integrates under the curve for $a = 0.0007$ in Figure 3, one obtains ~60 kcal/mole for the excess heat. This seems to be much too large for a process concerned with physical sorption and suggests that a chemical process is involved. At these temperatures on the sieve, a chemical reaction would appear to be quite possible. Would you agree?

W. Schirmer: Although we did not yet finish our investigation of this topic, we should like to conclude that there must be a chemical reaction, perhaps between some OH groups and the hydrocarbon molecule. We shall use other techniques (infrared, NMR) to solve the problem.

49

Low-Temperature Calorimetric Study of Methane in Linde 5A Sieve

H. J. F. STROUD and N. G. PARSONAGE

Imperial College, London, S.W. 7, England

Heat capacities over the range 20° to 300°K have been measured for empty 5A sieve and for the same sample containing 2 different amounts of CH_4. These measurements are more reliable than previous calorimetric results for sorption systems in that the calorimeter is not connected by a metal filling tube to the apparatus at room temperature. Sorption isotherms and isosteric heats of CH_4, C_2H_6, and Kr on the same sample have been determined from 194° to 300°K. A Monte Carlo evaluation has been made of a "cell" model, assuming Lennard-Jones potentials for all interactions. The isotherm and q_{st} *predictions for CH_4 are good, but the* C_s vs. T *curve fails to show the large "hump" near 150°K which is the main feature of the experimental results.*

Measurement of *Cp* of the sorbed phase (C_s) provides a very sensitive test of theories both because of the wide range of temperature which can be covered and because of the nature of the quantity itself.

Calorimetric *Cp* measurements of sorption systems always have been carried out with a tube connecting the sample to the room-temperature parts of the apparatus (*7*). This leads to large heat leaks along the tube and to uncertainty about the temperature which is to be assumed for the gas which is desorbed from the sample during the experiment. Studies have been made of the latter point (*7*), but in the present work we have avoided both the leak and the ambiguities by making the sample container a closed system, although this leads to very considerable experimental difficulties. Correction for the heat associated with desorption into the small amount of dead-space in the sample container was made using sorption isotherms determined for the same sample of zeolite.

For the theoretical treatment of the system, we have used Monte Carlo (MC) methods, which have been very successful for the many-body problems of fluids (9). The present paper contains, as far as we are aware, the first application of such a technique to a sorbed phase.

Low–Temperature Calorimetry

Measurements were made from 20° to 300°K using an adiabatic calorimeter. Before each filling of the container, the zeolite was baked out for at least 1 week while pumping on the system. The amount of sorbate sealed into the platinum–iridium sample container was determined gasometrically, and a small known amount of He also was added to facilitate thermal equilibrium as is normal practice in low-temperature calorimetry. Since the sample container was baked at $T > 300°C$, it was necessary to use glass-coated wires and pyromellitimide varnish for the heater of the sample container. The zeolite was a sample of 5A sieve given by Union Carbide Corp. and had the composition: 0.23 Na_2O, 0.77 CaO, Al_2O_3, 1.89 SiO_2.

Results

Run A: dehydrated zeolite = 23.4338 grams, CH_4 = 0.02597 mole, He = 0.00026 mole. Contribution of sorbate to heat capacity of sorbent + sorbate was 7.7% at 50°K and 4.7% at 250°K. Run B: dehydrated zeolite = 23.4338 grams, CH_4 = 0.01283 mole, He = 0.00026 mole. The average number of CH_4 molecules per large cavity ($\langle n \rangle$) was 1.808 in

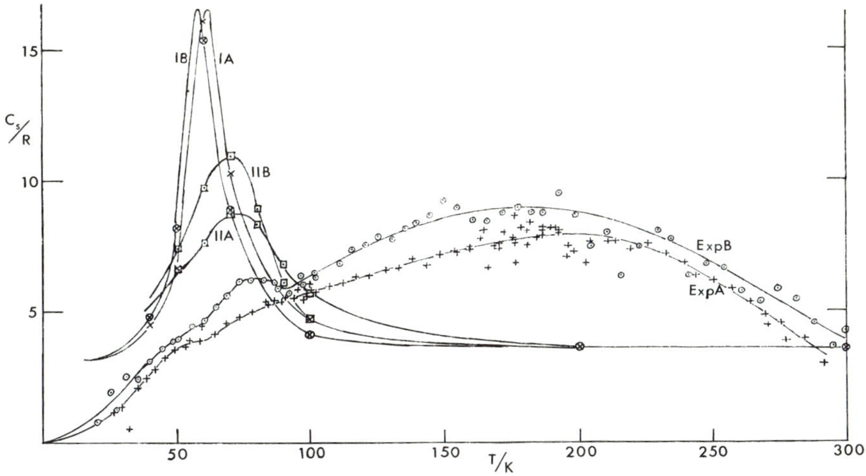

Figure 1. Molar heat capacity of the sorbed CH_4 (C_S) vs. temperature. Experimental results for $\langle n \rangle = 1.808$: Exp A (+). Experimental results for $\langle n \rangle = 0.8933$: Exp B (⊙). Corresponding calculated results from the spherical cell model: IA and IB (for 6-12 potential), IIA and IIB (for 6-18 potential).

Run A and 0.8933 in Run B. A run with the same amount of zeolite and He also was carried out. On subtracting the latter results from those from Runs A and B and dividing by the number of moles of methane, C_s, the molar heat capacity of sorbed methane for the 2 samples, was found (Figure 1, Exp. A, Exp. B).

The above C_s values are derived from measurements for which the initial cooling was slow (from room temperature to 80°K in ~40 hours). For lower temperatures the further cooling was faster, *e.g.*, 80° to 50° or 20°K over 100 minutes. In one set of experiments of Run A (not included in Figure 1), the sample was "quenched" deliberately from ~300° to ~80°K in ~30 minutes. Cp measurements were then normal until ~98°K, above which large, warming drifts were observed (~0.0015 deg minute^{-1}). At 112.6°K, the drift remained apparently constant over 2 hours. This behavior is similar to that found when a system is "frozen" first into a metastable state, from which it escapes with evolution of energy when thermal excitations become sufficiently large to permit it. An estimate of the amount of frozen-in energy was made by heating a quenched sample from 96.6°K, below the temperature at which annealing occurs, to 123.9°K, at which rapid equilibration occurs. The amount of heat required was compared with the amount calculated from the true Cp curve. This gave ~5.40 KJ (mole CH_4)$^{-1}$ for the frozen-in heat.

The trivial explanation that the "quenching" phenomena can be ascribed to spatial inhomogeneity of the CH_4, induced by the temperature gradient during cooling, and subsequent relaxation can be dismissed on the following grounds. For this explanation to be possible, it would be necessary for the graph of q_{st} *vs.* $\langle n \rangle$ to be sharply concave to the $\langle n \rangle$ axis. In fact, our measurements showed that q_{st} was independent of composition over the range of our isotherm measurements. Indeed, even if q_{st} *vs.* $\langle n \rangle$ was as curved as it is for C_2H_6, it would be virtually impossible to explain the magnitude of our observations in this way.

In normal experiments, using the slow-cooling method, 2 regions were found for which the rate of equilibration after heat input was unusually long. These were near 30°K (~50 minutes; normal, ~10 minutes) and at 155°–200°K (~50 minutes; normal, ~15 minutes). In the latter region, this caused a scatter in the total Cp of up to 0.4%.

Sorption Isotherm Results

For $\langle n \rangle$ between 0.2 and 3, q_{st} was 17.76 and 22.36 KJ mole^{-1} for Kr and CH_4, respectively, these results being independent of composition and temperature. For C_2H_6, on the other hand, the behavior was more complex. Thus, between 230° and 280°K, it was 25.19, 29.37, 32.26, and 34.10 KJ mole^{-1} at $\langle n \rangle = 1$, 2, 3, and 4, respectively.

Discussion

Bakaev (*1*) has shown how the Grand Partition Function for the sorbed phase may be expressed in terms of the Canonical Partition Functions for single cavities containing 1,2,3 . . .,*n* molecules. The only assumption necessary is that there is no interaction between particles in different cavities. The problem reduces itself to that of evaluating the configuration integrals, $Z_n = \int e^{-U/kT} d\tau_n$.

Spherically Symmetric Model

Because of the high symmetry of the main cavities of sieve A, it might seem reasonable, in the spirit of the cell theory of liquids (*5*), to assume that the various charged species of the zeolite are distributed with complete spherical symmetry about the center of each cavity. This, of course, leads to zero electric field throughout the cavity and hence to zero contribution from ion-induced dipole and similar forces. The remaining terms, dispersion and repulsion, are assumed to be of the Lennard-Jones form and are taken to be additive. This theory then becomes essentially the same as the cell theory of liquids, except that we permit several molecules per cell. The total P.E. is given as $U_{LJ} = \sum_i \varphi(r_i) + \sum_{i>j} u(r_{ij})$, where $\varphi(r_i)$ is the P.E. for the *i*th molecule–wall interaction when the molecule is r_i from the center and $u(r_{ij})$ is the P.E. for the interaction between the *i*th and *j*th particles. The difficult process now is to evaluate the Z_n. The Importance Sampling Monte Carlo Method (*6, 9*) is relevant to this problem, but it can yield values only for average quantities such as $\langle X \rangle = \int X \exp(-U_{LJ}/kT) d\tau_n / \int \exp(-U_{LJ}/kT) d\tau_n$; it does not give, in particular, a value for the denominator itself. To overcome this, we have used a coupling parameter method (*3*). We introduce cut-offs into the potential such that if either any 2 particles are less than r_o apart or any particle is less than r_o' from the wall, then P.E. $= \infty$. Writing the P.E. as $U(\xi) = U_{HS} + \xi U_{LJ}$, where $U_{HS} = 0$ if all $r_{ij} \geqslant r_o$ and $(a - r_i) \geqslant r_o'$ but $U_{HS} = \infty$ otherwise, we are interested in $U(1)$; or, putting $Z_n(\xi) = \int \exp(-U(\xi)/kT) d\tau_n$, we require $Z_n(1)$. It can be shown that $\ln Z_n(1) = \ln Z_n(0) - \int_0^1 \langle U_{LJ} \rangle_\xi \, d\xi/kT$ where $\langle U_{LJ} \rangle_\xi = \int U_{LJ} \exp(-U(\xi)/kT) d\tau_n / \int \exp(-U(\xi)/kT) d\tau_n \cdot Z_n(0)$, which refers to the fraction of all configurations of the hard spheres which are allowed, was evaluated numerically. $\langle U_{LJ} \rangle_\xi$ was evaluated by the above Monte Carlo method for about 10 values of ξ between 0 and 1, the number of trial moves ranging from 98,000 for 5 particles to 46,000 for 8 particles. For cells with 4 or less particles, the functions were evaluated directly by a simple MC method. The potential parameters for CH_4–CH_4 were those of Ref. *4*. For CH_4–O, the attractive potential was calculated by the Slater-Kirkwood equation (*8*), and the repulsive coefficient was chosen so as to make the potential minimum occur at the van der Waals' distance. For the compositions of Runs A and B, q_{st} decreases from 21.01 to 20.33 KJ mole^{-1} over the range 100° to 300°K. The errors in ΔG and ΔH are ~ -1.6 and ~ 2.0 KJ mole^{-1}, respectively, at 300°K. This is satisfactory, as we have not used any adjustable parameters. The calculated curves of C_s *vs.* T for $\langle n \rangle = 1.808$ and 0.8933 (Figure 1,IA,IB) are, however, clearly of the wrong form. The peaks predicted near 60°K mark disproportionation leading to a mixture of empty and full cells. Even if this process is thermodynamically favored, it would be unlikely to occur at such low temperatures within a reasonable time. Also shown are the curves assuming a 6–18 potential for CH_4–CH_4 (*2*), and corresponding 6–15 potentials for CH_4 with the wall (Figure 1,IIA,IIB).

The Spherically Symmetric Model is clearly unsatisfactory for predicting C_s values. Calculations are now in progress with the Na, Ca, O, Si, and Al atoms treated as charged point centers of force.

The small humps near 60°K in the experimental C_s *vs.* T curves may result from escape of He from solution in the sorbed CH_4. The run with empty zeolite showed a similar hump near 30°K, which we ascribed to desorption of helium.

Literature Cited

(1) Bakaev, V. A., *Dokl. Akad. Nauk SSSR* **1966,** 167, 369.
(2) Byrne, M. A., Jones, M. R., Staveley, L. A. K., *Trans. Faraday Soc.* **1968,** 64, 1747.
(3) Hill, T. L., "Statistical Mechanics," p. 180, McGraw-Hill, New York, 1956.
(4) Hirschfelder, J. O., Curtiss, C. F., Byrd, R. B., "Molecular Theory of Gases and Liquids," p. 1110, Wiley, New York, 1954.
(5) Lennard-Jones, J. E., Devonshire, A. F., *Proc. Roy. Soc., Ser. A* **1937,** 163, 53.
(6) Metropolis, N., Rosenbluth, A. W., Rosenbluth, M. N., Teller, A. H., Teller, E., *J. Chem. Phys.* **1953,** 21, 1087.
(7) Pace, E. L., "The Solid-Gas Interface," E. A. Flood, Ed., Vol. 1, Ch. 4, Dekker, New York, 1966.
(8) Pitzer, K. S., *Advan. Chem. Phys.* **1959,** 2, 59.
(9) Wood, W. W., "The Physics of Simple Liquids," H. N. V. Temperley, J. S. Rowlinson, G. S. Rushbrooke, Eds., Ch. 5, North Holland, Amsterdam, 1968.

RECEIVED February 4, 1970.

Discussion

K. Klier (Lehigh University, Bethlehem, Pa. 18015): The authors have chosen the cell method, which seems to be quite successful in explaining properties of liquids in the range of densities which adsorbing molecules attain in the zeolite cavities. It would be interesting to know whether, complementary to the Monte Carlo method used, analytical solutions could be obtained in terms of partition function being a function of average density of the sorbate. Such solutions would have the value of obtaining the isotherm as $p = -kT(\partial \log[\text{p.f.}]/\partial v)_T$. The model suggested here would be to assign one cell to one molecule within the cavity, not to the cavity itself.

N. G. Parsonage: I can see no possibility of obtaining an analytical expression for the partition function for cavities containing several molecules. For cells containing n molecules, we would need to integrate a complicated function over $\sim 3n$ coordinates. Lennard-Jones and Devonshire considered only the case of one molecule per cell, and for this needed only one coordinate, the radial distance. Even so, they had to evaluate their integrals numerically.

If one were to redefine the "cells" so that each contained only one molecule, then one would find great difficulty since the new "cells" would not correspond with the real cavities. The advantage of the present treatment, that the cells have real physical existence, would be lost.

H. A. Resing (Naval Research Laboratory, Washington, D. C. 20390): What do your experiments show about molecular rotation?

N. G. Parsonage: The experiment does not distinguish between rotational and other contributions to the heat capacity. Information on rotational freedom must depend on a theoretical analysis of the data. In our treatment, we assume that rotation is free over the whole temperature range and so gives a contribution of 3/2 R. This must be wrong at very low temperatures, but the evidence from the rather similar system methane—quinol clathrate, for which a good analysis is available, is that the rotation remains virtually free down to 50°K. At this temperature, our treatment will already be failing with respect to other degrees of freedom because it is classical and so does not show the quantum fall-off of the heat capacity.

50

Theoretical Predictions of Equilibrium Properties and Diffusivities of Carbon Dioxide in Type A Molecular Sieves

R. W. H. SARGENT and C. J. WHITFORD

Imperial College, London, England

Equilibrium sorption isotherms, heats of sorption, and diffusivities have been calculated for carbon dioxide in type A zeolites, using an idealized model of the molecular potential field. Good agreement with practical results are obtained for heats of sorption, but results are poor for equilibrium isotherms, and the simple approximation used for diffusivity is quite inadequate. Nevertheless, useful insight is obtained on the basic assumptions.

Models of the interaction on a molecular scale between gases and zeolite crystals have been considered by other workers, notably Barrer (*2–5*) and Kiselev and Lopatkin (*13*). The results presented here are developed from work initiated by Dworjanyn (*9*) and are described in more detail by Whitford (*19*). The difficulties encountered arise from 2 sources, lack of precise knowledge of the various types of interaction and the complexity of the computations required to produce results which can be compared directly with experiment. With the advent of powerful computers the second difficulty is alleviated, providing results which can throw light on the basic physical assumptions. Thus, calculations have been made of the equilibrium isotherms, the heat of sorption, and the self-diffusion coefficient of carbon dioxide in Ca-A and 5A molecular sieves.

Molecular Potential Model

The sieve structure was taken from Broussard and Shoemaker (*6*). All water in the sieve was neglected, and it also was assumed that aluminum and silicon ions contributed no charge to the field; the charge of

12*e* to balance the cations was shared equally between the 48 oxygen atoms of the unit cage. To allow for cation mobility, the 6 calcium ions of the Ca-A sieve were replaced by 8 "pseudo-ions" with appropriately scaled properties, one in each 6-0 window, and similarly, 8 composite Na–Ca pseudo-ions were used for the 5A sieve. A few calculations showed that the silicon and aluminum ions contributed little to the total interaction potential, so to save computing time their contribution also was approximated by treating them as composite pseudo-ions. In the calculations, all the sieve ions were assumed to remain fixed in their lattice positions.

The interaction of the carbon dioxide molecule with the sieve includes electrostatic, induction, dispersion, and repulsion contributions. The CO_2 molecule was assumed to be capable of free rotation, so that the directional interactions could be averaged over all orientations using a Boltzmann weighting factor (*11*); this causes the electrostatic ion–quadrupole interaction to depend on the temperature. Mean values were used for the polarizability (α), the diamagnetic susceptibility (χ), and the equilibrium radius of the CO_2 molecule. Using vector summation for the total electric field at the CO_2 molecule, the total potential, $\epsilon(r)$, at a given position *r* is given by:

$$\varepsilon(r) = \sum_i \frac{A_i}{r_i^{12}} - \sum_i \frac{B_i}{r_i^6} - \frac{\alpha}{2} \sum_j \left(\sum_i \frac{C_i x_{ij}}{r_i^3} \right)^2 - \frac{Q^2}{20kT} \sum_i \frac{C_i^2}{r_i^6} \quad (1)$$

Here C_i is the charge of the ion in question, and r_i (with components x_{i1}, x_{i2}, x_{i3}) is its separation from the CO_2 molecule; *Q* is the quadrupole moment of the CO_2 molecule. The constant B_i for the dispersion contribution can be obtained from a variety of empirical formulae (*12, 14, 15, 16, 18*); that of Kirkwood and Muller was selected, expressing B_i in terms of the polarizabilities and susceptibilities of the 2 molecules, but since this gives a larger value than most other formulae, it may be expected to overestimate the dispersion contribution. The constant A_i in the repulsion term was chosen to make the pair interaction potential stationary at the equilibrium separation of the 2 molecules. Equilibrium radii, polarizabilities, and susceptibilities were obtained from the literature and also calculated using screening constants (*11*); comparative calculations were made where the data conflicted, and the final values which gave the best results are given in Table I.

The potential, $\phi(r_i, r_j)$, between 2 CO_2 molecules has contributions owing to repulsion, dispersion, and quadrupole–quadrupole interaction:

$$\phi\,(r_i, r_j) = \frac{A'}{r^{12}} - \frac{B'}{r^6} - \frac{7}{40}\,\frac{Q^4}{kT} \cdot \frac{1}{r^{10}} \quad (2)$$

Table I.

	CO_2
Polarizability (α cm$^3 \times 10^{25}$)	26.5(*3*)
Magnetic susceptibility ($-\chi$ cm$^3 \times 10^{30}$)	30.9(*3*)
Equilibrium radius (A)	2.56(*3*)
Quadrupole moment (Q esu. $\times 10^{26}$)	4.10(*7*)
Lennard-Jones constants: σ A	4.12(*11*)
ε/K °K	189 (*11*)

where the constants A' and B' are evaluated by the same methods as above.

Complete maps of $\epsilon(r)$ as a function of the CO_2 molecule position in the unit cell were calculated for 12 temperatures ranging from 173° to 573°K, and a typical result showing the map for the *acd* or *abd* faces at 298°K is given in Figure 1. This shows the open passages through the 8-O windows at *B* and the high-interaction region along the direction of the 6-O windows at *C*. A point of interest is that the CO_2 molecule always is excluded from the small cage.

The maps were obtained by computing potentials at each point of a grid with spacing $a/20$ (where $a = 12.31$ A, the length of side of the unit cube) covering a tetrahedral space 1/48th of the volume of the cube; the complete map is obtained by appropriate rotations and reflections. The ion positions were generated similarly for as many cages as required for the summations in Equation 1. A few calculations were made of all contributions for ion distances up to 5.5*a* from the CO_2 molecule (1331 cages), but the effective range was much less than this. Table II gives

Table II. Molecular Potentials at 298°K (Ergs $\times 10^{15}$)

	ε (*Repulsion*)	$-\varepsilon$ (*Dispersion*)	$-\varepsilon$ (*Induction*)	$-\varepsilon$ (*Electrostatic*)	ε (r)
Cage center	3.6	67.4	3.4×10^{-10}	169.6	−233.4
Center, 8–0 window	675	457.6	0.014	414.2	−196.3
Peak interaction near cation	255.9	242.7	15.3	557.8	−559.9
Range (99%) of interaction	1.5*a*	2.5*a*	3.5*a*	2.5*a*	—

the ranges required to obtain 99% of the total contribution for each type of interaction, together with typical values for 3 positions of interest. The induction contribution is always unimportant, but it is interesting to note the large contribution of the electrostatic interaction.

Molecular Data

$O^{-1/4}$	$\frac{3}{4}\ Ca^{+1.5}$	$NaCa^{+1.5}$	$AlSi^{+3.5}$
39.0/16.5(*3*)	3.5(*4,5*)	2.84(*4,5*)	0.66
20.9(*3,4,5*)	16.6(*4,5*)	14.5(*4,5*)	3.76
1.98(*4,5*)	1.78(*4,5*)	1.77(*4,5*)	1.52
—	—	—	—
—	—	—	—
—	—	—	—

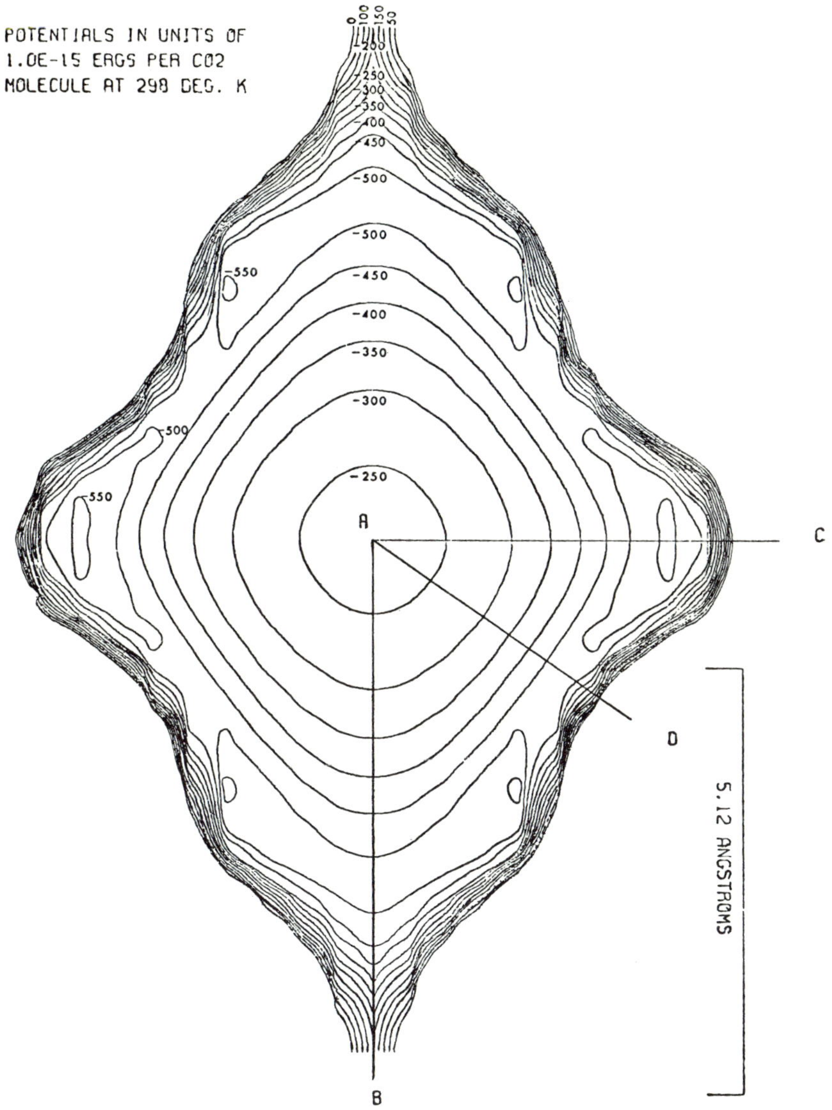

Figure 1. Sieve potentials in the Y = Z plane

Statistical Calculations of Equilibrium Properties and Diffusivities

Consider a large volume, V, of zeolite crystal consisting of cells, each of volume V_c, and containing N gas molecules at uniform temperature. The probability, $P(r^N, p^N)$ of the positions r^N and momenta p^N of the N molecules is given by the canonical distribution

$$P(r^N, p^N) = \frac{1}{N!h^{3N}Z_N} \exp\left[- \frac{\Phi(r^N) + \sum_{i=1}^{N} ||p^i||^2/2m}{kT} \right] \quad (3)$$

where m is the mass of a gas molecule, $\Phi(r^N)$ is the potential energy, and the normalizing factor Z_N is the partition function, given by

$$Z_N = \frac{1}{N!h^{3N}} \int_{-\infty}^{+\infty} dp^N \int_V dr^N \exp\left[- \frac{\Phi(r^N) + \sum_{i=1}^{N} ||p^i||^2/2m}{kT} \right] \quad (4)$$

Thus, the probability, P_N, of molecule N being in a certain cell at time t and passing to a given adjacent cell within the interval δt is given by

$$P_N = \int_{-\infty}^{+\infty} dp^{N-1}\, dp_{1N}\, dp_{2N} \int_0^{\infty} dp_{3N} \int_V dr^{N-1} \int_{S_c} dr_N \cdot \frac{p_{3N}}{m} \delta t \cdot P(r^N, p^N) \quad (5)$$

where S_c is the surface of the face between the 2 cells and p_{3N} is the component of momentum normal to this surface.

It is assumed that the volume V is large enough to neglect boundary effects, that the gas is dilute enough for only two-body interactions to be significant, and that the potential function is of the form

$$\Phi(r^N) = \sum_{i=1}^{N} \{\varepsilon(r_i) + \sum_{j<i} \phi(r_i, r_j)\} \quad (6)$$

Also, since the zeolite cells are all identical, the potential $\epsilon(r_i)$ depends only on the position of molecule i relative to the cell and not on its absolute position. With these assumptions, the integrals in Equations 4 and 5 factorize, and if the integrations over momenta are carried out, the following equations are obtained

$$Z_N = \frac{1}{N!\lambda^{3N}} \left(\frac{V\, V_o}{V_c}\right)^N \sum_{r=0}^{N/2} \frac{N!\, \xi^r}{r!(N-2r)!N^r}; \quad \lambda = \left(\frac{h^2}{2\pi mkt}\right)^{1/2} \quad (7)$$

$$P_N = \frac{S_o \cdot \delta t}{N!\lambda^{2N} Z_N} \cdot \left(\frac{kT}{2\pi m}\right)^{1/2} \cdot \left(\frac{V\, V_o}{V_c}\right)^{N-1} \cdot \left\{ \sum_{r=0}^{\frac{N-1}{2}} \frac{(N-1)!\xi^r}{r!(N-1-2r)!N^r} + \right.$$

$$\frac{2\eta}{N}\sum_{r=0}^{\frac{N-2}{2}}\frac{(N-1)!\xi^r}{r!(N-2-2r)!N^r}\Big\} \tag{8}$$

where

$$V_o = \int_{V_c} \exp\left[-\varepsilon(r)/kT\right] \cdot dr; \quad S_o = \int_{S_c} \exp\left[-\varepsilon(r)/kT\right] \cdot dr \tag{9}$$

$$\xi = \frac{1}{2}\cdot\frac{NV_c}{V\,V_o^2}\cdot\int_V\int_{V_c}\exp\left[-\frac{\varepsilon(r_1)+\varepsilon(r_2)}{kT}\right]\cdot \left\{\exp\left[-\frac{\phi(r_1, r_2)}{kT}\right]-1\right\}\cdot dr_1\, dr_2 \tag{10}$$

$$\eta = \frac{1}{2}\cdot\frac{NV_c}{V\,V_oS_o}\cdot\int_V\int_{S_c}\exp\left[-\frac{\varepsilon(r_1)+\varepsilon(r_2)}{kT}\right]\cdot \left\{\exp\left[-\frac{\phi(r_1, r_2)}{kT}\right]-1\right\}\cdot dr_1\, dr_2 \tag{11}$$

Now it can be shown (*10*, p. 263) that for a dilute gas ($\xi << 1$) the sum in Equation 7 is approximated adequately by $(1 + \xi)^N$, and the sums in Equation 8 may be approximated similarly, to yield the results

$$Z_N = \frac{1}{N!\lambda^{3N}}\left\{\frac{V\,V_o}{V_c}(1+\xi)\right\}^N;$$

$$P_N = \frac{V_cS_o\cdot\delta t}{V\,V_o\,(1+\xi)}\cdot\left(\frac{kT}{2\pi m}\right)^{1/2}\cdot\left\{1+\frac{2\eta}{(1+\xi)}\right\} \tag{12}$$

The same analysis applies to a gas in free space simply by putting $\epsilon(r_i) = 0$, and the corresponding quantities will be denoted by the suffix G.

It is convenient to introduce molar quantities and "effective molar densities" (n' and n'_G) by the relations

$$R = k\,N_A; \qquad M = m\,N_A; \qquad n = N/V\,N_A$$

$$B = -\,V_o\xi/V_c\,n; \quad B_s = -\,V_o\eta/V_c\,n; \quad B_G = -\,\xi_G/n_G \tag{13}$$

$$n' = \frac{nV_c/V_o}{1 - B\cdot nV_c/V_o}; \qquad n'_G = \frac{n_G}{1 - B_Gn_G}$$

where N_A is Avogadro's number. From Equations 10, 11, and 13

$$B = -\frac{1}{2} \cdot \frac{N_A}{V_o} \int_V \int_{V_c} \exp\left[-\frac{\varepsilon(r_1) + \varepsilon(r_2)}{kT}\right] \cdot \left\{\exp\left[-\frac{\phi(r_i, r_j)}{kT}\right] - 1\right\} dr_1\, dr_2 \quad (14)$$

$$B_s = -\frac{1}{2} \cdot \frac{N_A}{S_o} \int_V \int_{S_c} \exp\left[-\frac{\varepsilon(r_1) + \varepsilon(r_2)}{kT}\right] \cdot \left\{\exp\left[-\frac{\phi(r_i, r_j)}{kT}\right] - 1\right\} dr_1\, dr_2 \quad (15)$$

$$B_G = -\frac{1}{2} \cdot \frac{N_A}{V_c} \int_V \int_{V_c} \left\{\exp\left[-\frac{\phi(r_1, r_2)}{kT}\right] - 1\right\} dr_1\, dr_2 \quad (16)$$

Because of the limited effective range of the gas interactions, the integrals in Equations 14–16 become independent of V, and in evaluating them it is only necessary to take V just large enough to include all significant interactions.

Using Stirling's formula, the definition of the Helmholtz free energy: $F = -kT \ln Z_N$, and standard thermodynamic relations, several useful formulae may be deduced from the expression for Z_N in Equation 12. The chemical potential (μ) is given by

$$\mu = RT \ln N_A \lambda^3 + RT \ln n' + RTBn' \quad (17)$$

This expression may be equated to the corresponding expression for the gas phase to yield a simple equilibrium adsorption isotherm

$$n' \exp(Bn') = n'_G \exp(B_G n'_G) \quad (18)$$

The equation of state for the gas is deduced in the form

$$p = -\left(\frac{\partial F}{\partial V}\right)_N = n'_G RT \simeq n_G RT (1 + B_G n_G) \quad (19)$$

Since the second virial coefficient, B_G, of the gas is small over the range of the approximations, Equation 18 is closely approximated by the more convenient form

$$p = \frac{RT\, n' \exp(Bn')}{1 + B_G n' \exp(Bn')} \quad (20)$$

The internal energy of sorption per mole is given by:

$$\frac{U - U_G}{RT^2} = (1 + Bn') \frac{\partial \ln V_o}{\partial T} - n' \frac{\partial B}{\partial T} + n'_G \frac{\partial B_G}{\partial T} \qquad (21)$$

and the heat of sorption ($H - H_G = \Delta H$) is obtained readily from Equations 18, 19, and 21.

An "effusion" formula for the crystal diffusivity (D_c) is obtained from the expression for P_N in Equation 12. The average absolute flux of molecules across the face S_c is $NP_N/\delta t$, with an equal flux in the opposite direction at equilibrium. For self-diffusion the equilibrium is not disturbed, and if the difference in number-fraction of "labelled" molecules between adjacent cells is Δx, the definition of diffusivity yields:

$$\text{Net flux of labelled molecules} = \frac{NP_N}{\delta t} \cdot \Delta x = S_c D_c \cdot \frac{N\Delta x}{Va} \qquad (22)$$

where a is the center-to-center distance between the cells. From Equations 12, 13, and 22, and using the relations: $V_c = a^3$, $S_c = a^2$

$$D_c = a^2 \left(\frac{RT}{2\pi M}\right)^{1/2} \frac{S_o}{V_o} (1 + Bn') (1 - 2B_s n') \qquad (23)$$

Discussion of Results

There is not space to give extensive results, but the main results are summarized in Table III, and representative curves are shown in Figure 2.

Because of the two-body interaction assumption, one would expect the results to be applicable only to dilute systems, a probable upper limit corresponding to an average of 2 carbon dioxide molecules per cage (about 5 wt %). Over this range, the heat of sorption ($-\Delta H$) is dominated by the first term in Equation 21, the second order effects amounting to about 18% of the total, and the $p\Delta v$ term never rising above 8%. At higher concentrations, the second order contribution increases rapidly, but almost exactly compensates the decrease in the term in V_o so that ΔH remains independent of concentration up to at least 25 wt %. In fact, the value of V_o depends critically on 2 assumptions concerning the molecular potentials: the degree of ionization of the oxygen atoms and the equilibrium radius of the CO_2 molecule. Use of the "kinetic" radius (1.65 A) for the latter gives values of V_o too high by a factor of 10^{24}, while the assumption of complete ionization gives values too high by a factor of about 10. The values in Table III are based on partial ionization and an equilibrium radius equal to the half-length (2.56 A) of the CO_2 molecule. Barrer and Coughlan's (*1, 8*) experimental results for ΔH vary significantly with amount sorbed, but their values for Ca-A lie in the

Table III. Summary

T,°C	30	60
B_{GLJ}, liter/mole	−0.120	−0.098
B_G, liter/mole	0.044	0.048
$\alpha_0 = 39.0 \times 10^{-25}\ cm^3$		
V_o/V_c	158000	19300
B, liter/mole	1404	151
$-\Delta H$ kcal/mole	15.53	13.97
$\alpha_0 = 16.5 \times 10^{-25}\ cm^3$		
V_o/V_c	858000	72160
B, liter/mole	7.47×10^4	6291
B_{LJ}, liter/mole	700	98.5
$-\Delta H$ kcal/mole	18.66	16.52
D_c cm²/sec	1.05×10^{-7}	2.64×10^{-7}

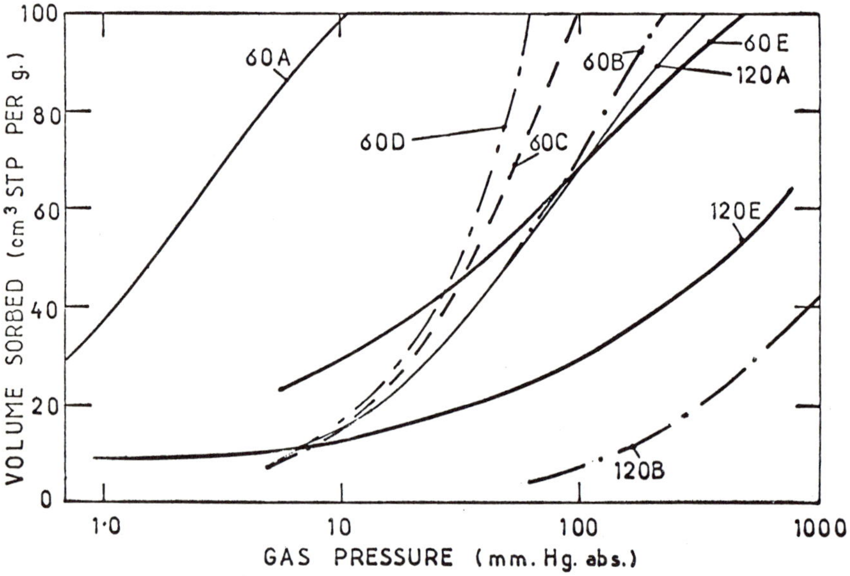

Figure 2. Equilibrium isotherms

———	E	*Experimental*
———	A	*Calculated* ($\alpha = 39.0$)
— · —	B	*Calculated* ($\alpha = 16.5$)
— — —	C	*Calculated* ($\alpha = 16.5$ *using* B_{LJ})
— - —	D	*Calculated* ($\alpha = 16.5$ *using* B $= 0$)

Numbers on curves refer to temperatures in °C

Table IV. Effect of

T,°C		30	60
$-\Delta H$, kcal/mole	(Oα = 39.0)	15.53	13.97
	(Oα = 16.5)	18.66	16.52
D_c, cm²/sec	(Oα = 16.5)	8.74×10^{-8}	1.99×10^{-7}

of Results

90	120	150	190
−0.080	−0.065	−0.052	−0.038
0.051	0.054	0.056	0.058
4070	1240	489	188
39.0	13.2	5.10	1.74
12.46	11.30	10.55	10.12
11100	2580	805	237
968	226	71.0	21.2
11.2	7.30	3.08	1.18
14.95	13.79	13.15	12.95
5.24×10^{-7}	9.15×10^{-7}	1.48×10^{-6}	2.60×10^{-6}

range 8 to 15 kcal/mole, which agrees surprisingly well with the theoretical results.

The value of the oxygen polarizability also affects the molecular potentials, but has a minor effect on ΔH (Table IV). However, it has a large effect on the equilibrium isotherms, as can be seen in Figure 2. Agreement with experimental data (*1*, *8*) is poor with both values of the polarizability, and it is evident that the theory predicts too large a variation with temperature in both cases. To give an indication of the extent of second-order effects, a first-order curve at 60°C and the lower polarizability (corresponding to $B = B_G = 0$) is included.

The molecular potentials also can be used to obtain values of the gas virial coefficient, B_G, but the results in Table III show that this is unsuccessful. In fact, a very good fit of Equation 2 to experimental values is obtained over the whole temperature range using a CO_2 radius of 1.81 A, but unfortunately this gives grossly high values of V_o, as noted above. Values of B and B_G were therefore recalculated, using the Lennard-Jones 12-6 potential to replace Equation 2, and the results for the lower polarizability are given also in Table III as B_{LJ} and B_{GLJ}. Of course, B_{GLJ} now agrees with experimental values, but B_{LJ} gives a much smaller second-order correction of the equilibrium isotherm (Figure 2). These difficulties probably are associated with the assumption of free rotation of the CO_2 molecule; there are almost certainly strongly preferred orientations in some regions of high interaction in the sieve-cage where the effective

Oxygen Polarizability

90	120	150	190
12.46	11.30	10.55	10.12
14.95	13.79	13.15	12.95
3.86×10^{-7}	6.62×10^{-7}	1.56×10^{-7}	2.77×10^{-7}

radius is the half-length, while in the gas the free rotation results in a smaller effective mean radius. Further investigation of this point would require much more complex calculations.

The simple theory used to estimate the diffusivity is obviously quite inadequate, for the predicted values are about 10,000 times larger than values obtained by experiment (*17*). One would expect the predicted values to be unnaturally low in view of the assumption that the ions in the 8-O window do not relax to allow passage of the CO_2 molecule, but remain fixed in their lattice positions. However, the question of preferential orientation is probably even more important in this case, and the effusion model itself is too crude an assumption to make further conjecture profitable.

Literature Cited

(1) Barrer, R. M., Coughlan, B., "Molecular Sieves," Society of the Chemical Industry, London, 1968.
(2) Barrer, R. M., Gibbons, R. M., *Trans. Faraday Soc.* **1963,** 59, 2569.
(3) *Ibid.*, **1965,** 61, 948.
(4) Barrer, R. M., Stuart, W. I., *Proc. Roy. Soc.* **1957,** A 242, 172.
(5) *Ibid.*, **1959,** A 249, 464.
(6) Broussard, L., Shoemaker, D. P., *J. Am. Chem. Soc.* **1960,** 82, 1041.
(7) Buckingham, A. D., Disch, R. L., *Proc. Roy. Soc.* **1963,** A 273, 275.
(8) Coughlan, B., Ph.D. Thesis, University of London, 1964.
(9) Dworjanyn, L. O., Ph.D. Thesis, University of London, 1962.
(10) Fowler, R. H., Guggenheim, E., "Statistical Thermodynamics," Cambridge University Press, 1939.
(11) Hirshfelder, J. O., Curtiss, C. F., Bird, R. B., "Molecular Theory of Gases and Liquids," Wiley, New York, 1954.
(12) Kirkwood, J. G., *Physik. Z.* **1932,** 32, 57.
(13) Kiselev, A. V., Lopatkin, A. A., "Molecular Sieves," Society of the Chemical Industry, London, 1968.
(14) Lennard-Jones, J. E., reported in Ref. *11*, page 964.
(15) London, F., *Z. Physik. Chem.* **1930,** B 11, 222.
(16) Muller, L. A., *Proc. Roy. Soc.* **1936,** A 154, 624.
(17) Sargent, R. W. H., Whitford, C. J., ADVAN. CHEM. SER. **1971,** 102, 155.
(18) Slater, J. C., Kirkwood, J. G., *Phys. Rev.* **1931,** 37, 682.
(19) Whitford, C. J., Ph.D. Thesis, University of London, 1969.

RECEIVED February 13, 1970.

51

Diffusion of Carbon Dioxide in Type 5A Molecular Sieve

R. W. H. SARGENT and C. J. WHITFORD

Imperial College, London, England

The self-diffusion of carbon dioxide in single pellets of commercial type 5A molecular sieve has been studied using $C^{14}O_2$ as a tracer. Experiments were carried out at atmospheric pressure between +25° and −25°C. Using a simple model of the pellet structure, it was possible to deduce effective diffusivities for both pore and crystal diffusion. Ordinary gas diffusion occurs in the pores; crystal diffusivities have values of the order of 10^{-11} cm^2/sec.

Diffusion in porous pellets is often the rate-limiting process in industrial adsorption or catalytic processes. Much useful work in this field has been done by Smith and coworkers (*3, 5*), but for molecular sieve pellets the situation is complicated by diffusion in the zeolite crystal itself, as well as through the pores formed between the crystals. Few studies have been made of zeolite crystal diffusion, but Barrer and Brook (*1*) reported some results on diffusion of simple gases in various cation-substituted mordenites, and Wilson (*7*) gives some indirect results from the study of separation of CO_2 from air using a fixed bed of type 4A zeolite pellets. In the present work, results have been obtained by studying self-diffusion of CO_2 in a single pellet of type 5A zeolite under controlled conditions. The experimental results were fitted satisfactorily by a very simplified model of the pellet structure, which made it possible to deduce approximate values of the self-diffusion coefficients for both pore and crystal diffusion.

Experimental Apparatus and Procedure

Full details of the experimental work are given by Whitford (*6*), and a diagram of the diffusion cell arrangement is shown in Figure 1. Commercial ⅛-inch diameter pellets were selected from a batch to fit

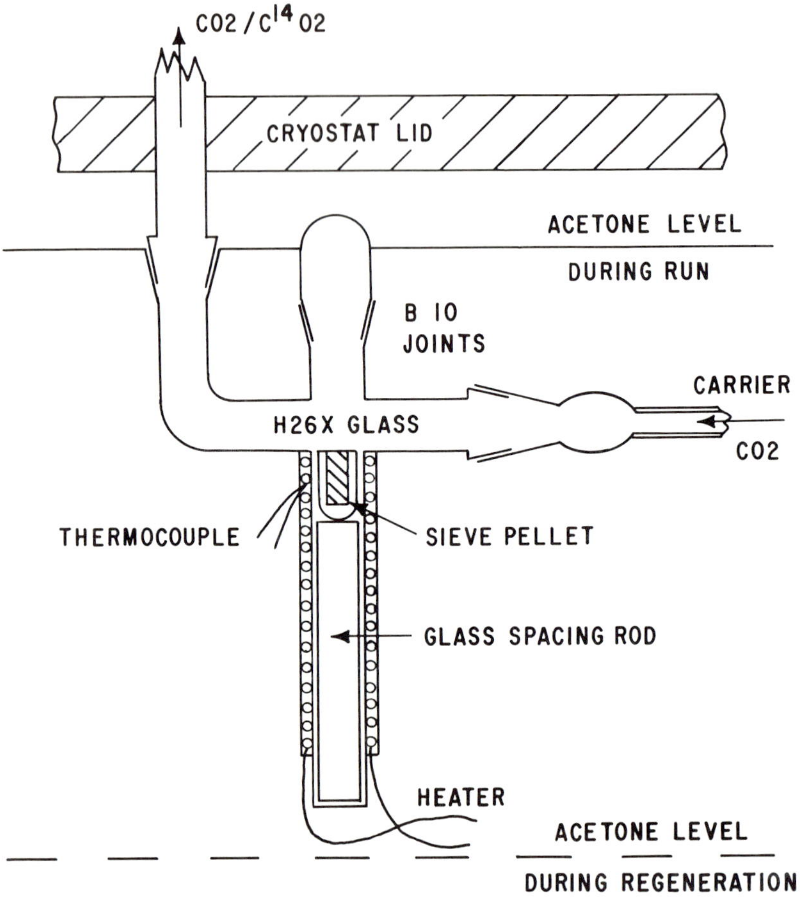

Figure 1. Schematic arrangement of the diffusion cell

tightly into a Viridia tube, closed at one end with the squared-off end of the pellet flush with the other open end. This tube was in turn a close fit in the closed side-arm of a capillary tube T-piece, with the end of the pellet again flush with the mouth of the side-arm. The side-arm was wound with an electric heater for degassing the pellet. The through-section of the T-piece was connected at its inlet to the gas preparation and flow-control system and at its outlet to a small-volume scintillation counter. This counter used anthracene crystals mounted on glass discs and viewed by a photomultiplier; the response time of the counter was 1.3 nanoseconds. A thermostat bath could be raised to immerse the complete test section during a trial, or lowered to allow regeneration, without disturbing the apparatus or the thermostat temperature. A high-vacuum system made it possible to evacuate the test section to 10^{-5} mm mercury absolute, but when degassing the pellet at temperatures above

300°C, the pressure did not fall below 10^{-3} mm; degassing was assumed to be complete when these conditions had been maintained for 3 hours.

After degassing, the pellet was saturated with $C^{14}O_2$ under the desired conditions, the volume of gas admitted being measured accurately with a gas burette. A stream of unlabelled CO_2 was passed through the test section, and both count rate and cumulative count were recorded. After about 3 hours in most cases, the count rate had dropped to the background value; the thermostat was lowered and the pellet heated to 300°C to expel the remaining gas, which also was counted to give an accurate end-point value for the total amount sorbed and to cross-check the initial value. Background count rates were determined separately, and blank runs were made under identical conditions, but with the pellet tube replaced by a solid glass blank. This made possible measurements of dead space and amounts sorbed on glass surfaces, as well as correction of the start-time to the time at which unlabelled gas actually arrives at the pellet surface.

The Pellet Model

The pellet is saturated with CO_2 throughout a test run, but $C^{14}O_2$ is replaced gradually by $C^{12}O_2$ by diffusion, first through the "pores," or interstices between the crystals and binder, and then through the crystals themselves. This is analogous to the situation in a fixed-bed adsorption process, where the crystals are analogous to the porous pellets, except that the transfer process in the interstices is a diffusive flow rather than a bulk gas flow. Smith (5) considered complex detailed models for the various flow paths in a fixed-bed process, but as there is inevitably some approximation involved in describing the structure of the interstices, a much simpler approach has been used here.

It is assumed that each crystal is surrounded completely by gas at a uniform concentration, which means that direct transfer between crystals is neglected and that all the crystal surface is accessible to the gas. The 2 assumptions are self-compensating, but one could expect serious errors if there is significant crystal-binder agglomeration. It is assumed further that all the crystals are uniformly-sized spheres, mainly for mathematical simplicity. Of course, 5A crystals are in fact cubic, but with the corners cut off, so that the spherical approximation is not unreasonable. The uniform size is a more drastic assumption, which is discussed below. The diffusion processes in both crystals and pores are assumed to be governed by Fick's law. For the pores, an effective diffusivity (D_p) is used, which can be related to the bulk gas diffusivity (D_G) by the relation:

$$D_p = \frac{\varepsilon D_G}{\tau} \tag{1}$$

where ϵ is the pellet porosity and τ is the tortuosity. Because of the simple geometry of the experimental system, pore concentration (v) varies only with distance (x) from the open end, and for self-exchange of CO_2 there are no thermal effects and the whole system is isothermal.

With these assumptions, the following equations describe the system

Pore Diffusion: $$D_p \frac{\partial^2 v}{\partial x^2} = (1 - \varepsilon) \cdot \frac{3D_c}{R} \left(\frac{\partial w}{\partial r}\right)_{r = R} \tag{2}$$

Crystal Diffusion: $$\frac{\partial w}{\partial t} = D_c \left(\frac{\partial^2 w}{\partial r^2} + \frac{2}{r}\frac{\partial w}{\partial r}\right) \tag{3}$$

where: v, w are the concentrations (mole/liter) of $C^{12}O_2$ in the pore and crystal phases, respectively; D_p, D_c are the corresponding effective diffusivities; t is time; and r is the radial coordinate inside a crystal.

It is assumed that instantaneous thermodynamic equilibrium exists at the crystal surface, and that both $C^{12}O_2$ and $C^{14}O_2$ have the same equilibrium isotherm

Pore–crystal surface: $$(w)_{r = R} = Kv \tag{4}$$

where K is a constant obtained from the equilibrium isotherm. There is, of course, symmetry at the center of the crystal

$$t > 0 \quad (\partial w/\partial r)_{r = 0} = 0 \tag{5}$$

The 2 boundary conditions for the pores are

Closed end: $$t > 0 \quad (\partial v/\partial x)_{x = L} = 0 \tag{6}$$

Open end: $$t > 0 \quad (v)_{x = 0} = v_o \tag{7}$$

where v_o is the molar density (mole/liter) of $C^{12}O_2$ under test conditions.

Initially the pellet contains no $C^{12}O_2$, so we have

$$\text{At } t = 0\text{: } v = 0,\ 0 \leq x \leq L,\ w = 0,\ 0 \leq r \leq R \tag{8}$$

The equations were put in dimensionless form, using the variables

$$\left.\begin{aligned} x' = x/L;\ r' = r/R;\ v' = v/v_o;\ w' = w/v_o \\ t' = tD_p/L^2;\ P = D_pR^2/D_cL^2 \end{aligned}\right\} \tag{9}$$

They were solved then for given P, K, and ϵ by use of finite-difference approximations, using the following algorithm for each time-step:

1. Assume a set of values for $v' = v'(x)$ at the end of the current time-step.
2. Compute crystal-surface gradients $(\partial w'/\partial r')_{r'=1}$ using Equation 2.

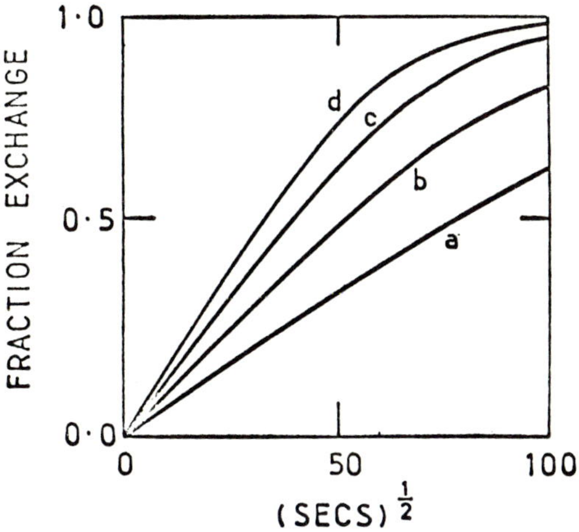

Figure 2. Curves for P = *0*

Curve:	*a*	*b*	*c*	*d*
D_p *(cm²/sec)*	*0.005*	*0.010*	*0.015*	*0.020*

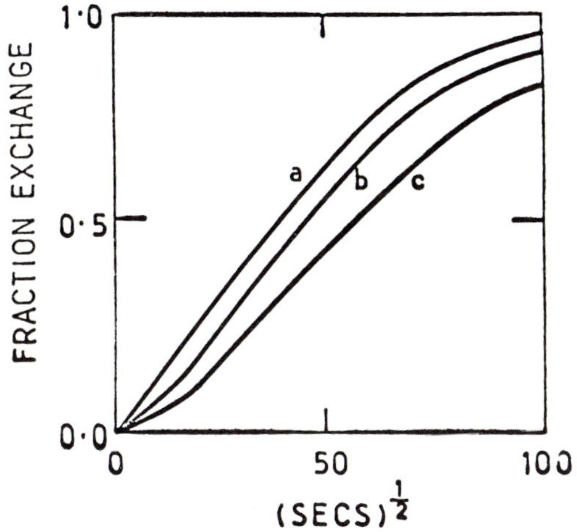

Figure 3. Effect of P

Curve:	*a*	*b*	*c*
P	*0*	*150*	*600*

3. Solve Equation 3 for $w' = w'(r')$ for each x at the end of the current time-step (Equation 3 becomes a set of linear simultaneous equations).
4. Compute new values of $v' = v'(x')$ from Equation 4.
5. Combine these new values with the assumed set using the successive over-relaxation technique (*4*), and repeat from step 2 to convergence.

The initial estimate in step 1 conveniently can be the initial values for the time-step, but simple extrapolation rules can shorten the computation. From the resulting solution, it is easy to compute the fraction of $C^{14}O_2$ exchanged (E) as a function of the parameters:

$$E = f(t', \varepsilon, K, P) \tag{10}$$

The parameter P is a measure of the relative importance of the pore and crystal diffusion processes; $P = 0$ corresponds to an infinite value of D_c, and it is easy to see from the above equations that the crystals are in equilibrium with the local pore concentration and the exchange process is described by the simple diffusion equation

$$\frac{\partial v}{\partial t} = \frac{D_p}{\varepsilon + (1 - \varepsilon)K} \frac{\partial^2 v}{\partial x^2} \tag{11}$$

Typical solutions of this equation are plotted in Figure 2 for a range of values of D_p. The effect of the parameter P for a typical value of D_p is shown in Figure 3.

Results

Three sorption runs and 3 blank runs were made for each set of conditions. After correction of the initial times, values of cumulative count were taken from the records at equal increments of $\sqrt{t}$; the sorption values were corrected using the data from the blank runs and used to compute values of the fraction of $C^{14}O_2$ exchanged (E), which finally were averaged over the 3 runs. The over-all scatter attributable to all causes was about 1% on the blank runs and 4% on the sorption runs. Three typical sets of data are given in Figure 4, where the circles denote experimental points.

In order to compute E from the model and plot it against time (t), it is necessary to have values for the parameters ϵ, K, P, D_p, L. The dimensions of the pellet tube were known accurately, and the tube was weighed before and after inserting the pellets. Thus, values for L (1.03 cm) and ϵ (0.375) were determined directly. The equilibrium constant (K) was calculated from the experimental data on the total amounts sorbed and desorbed in each run. The results are given in Table I; they agreed with published data (2) to within 2%.

Hence, if values for D_p and P are assumed, the curve of E *vs.* t can be computed from the model. Computed curves were fitted to the experimental curves by finding the values of D_p and P which minimized the sum of squares of errors in E at the experimental points; these curves are shown as continuous lines in Figure 4.

Table I. Results

Temperature, °C	*−25*	*0*	*+25*
K	166.3	167.2	181.8
P	100.5	17.19	10.0
D_p, cm²/sec	0.0136	0.0134	0.0148
Mean square error (29 points)	9.05×10^{-6}	1.13×10^{-5}	1.26×10^{-4}
D_G, cm²/sec	0.0617	0.0675	0.0729
τ	1.71	1.88	1.84
D_c, cm²/sec	1.28×10^{-12}	7.36×10^{-12}	1.39×10^{-11}

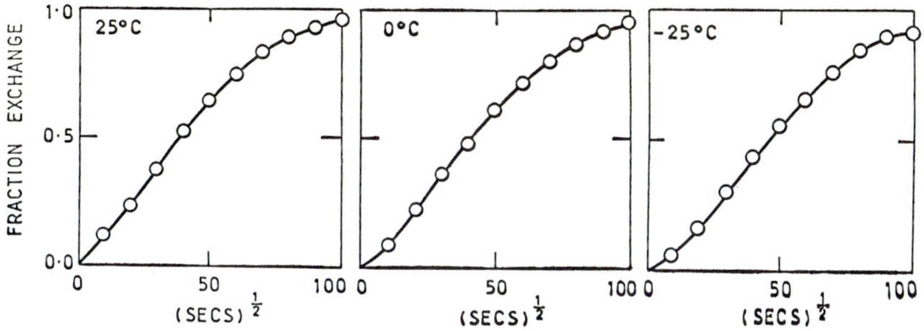

Figure 4. Typical data

○ *Experimental points*
— *Computed curves*

The fit obtained is excellent. Only at the highest temperature is a simple diffusion model (Equation 11) adequate to describe the curve, and although the curves are only slightly sigmoid in shape, the value of P is quite well determined by the least-squares fitting process in each case. The conclusion is that crystal diffusion plays a role in determining over-all rates of sorption, and the effect becomes more significant as the temperature decreases.

Fitted values of D_p and P are given in Table I, together with values of the tortuosity (τ) determined from Equation 1. The tortuosity is reasonably constant, as it should be for a geometric factor, and has a value typical of beds of spheres. Therefore, it can be concluded safely that the transport mechanism in the pores is ordinary bulk gas diffusion, and in particular, there is no evidence of surface diffusion.

To make practical use of diffusion results obtained in this way, for example in designing a fixed-bed separation process, it suffices to obtain the ratio D_c/R^2 from the fitted values of P and D_p. This ratio is characteristic of the gas–zeolite system and the size-distribution of the crystals in the pellet, and to obtain the value of D_c itself it is necessary to estimate an appropriate mean value of R. This cannot be done rigorously without a knowledge of the crystal size-distribution, but one can infer from the form of the model solution that the mean will not be far below the maximum size; the small crystals come to equilibrium quickly but represent a small fraction of the total adsorptive capacity. A limited set of measurements by electron microscopy indicated that the crystal size ranged from about 0.1μ to just above 2μ; the order of magnitude of D_c is obtained therefore by using $R = 1\mu$, and results based on this value are given in Table I.

The diffusivities obtained by Barrer and Brook (*1*) ranged from 5×10^{-12} cm²/sec for sorption of argon in Li-mordenite to 5×10^{-15} for nitrogen in Ca-mordenite. A quantitative comparison of results for such different systems is not possible, but considering the relative effective sizes of the gas molecules and zeolite windows, the comparison with the present results is not unreasonable.

Literature Cited

(1) Barrer, R. M., Brook, D. W., *Trans. Faraday Soc.* **1953,** 49, 1049.
(2) Linde Co., Adsorption Equilibrium Data Sheets No. 22 and 24, 1959.
(3) Masamune, S., Smith, J. M., *A.I.Ch.E. J.* **1962,** 8, 217.
(4) Varga, R. S., "Matrix Iterative Analysis," Prentice-Hall, Englewood Cliffs, N. J., 1962.
(5) Wakao, N., Smith, J. M., *Chem. Eng. Sci.* **1962,** 17, 825.
(6) Whitford, C. J., Ph.D. Thesis, University of London, 1969.
(7) Wilson, K. B., Ph.D. Thesis, University of London, 1966.

RECEIVED February 13, 1970.

Discussion

C. N. Satterfield (Massachusetts Institute of Technology, Cambridge, Mass. 02139): Sargent mentioned that he found about a 10^2 difference between the self-diffusion of CO_2 in one system and the counterdiffusion of CO_2 and N_2 in a second system. A brief elaboration of these studies would be interesting.

R. W. H. Sargent: The CO_2–N_2 (or rather CO_2–air) values were obtained indirectly from studies on a fixed-bed separation process for removal of CO_2 from air. These studies were made on a bed of pellets 4 ft long and 4 inches in diameter; two pellet sizes were used, and a range of air flow rates was covered at pressures from 1 to 25 atm and from +15° to −40°C. All the results were successfully predicted by a model assuming both pellet–pore and intracrystalline diffusion under isothermal conditions; the two relevant diffusivities were used as adjustable parameters to fit the experimental break-through curves by least squares for one set of conditions at each temperature, and these values then predicted the results for all other conditions at the same temperature. As with the single-pellet results presented in the paper, it was found that a model assuming only one diffusion process was inadequate and the two diffusivities were well defined by the fitting procedure. The fitted pore diffusivity varied correctly with temperature and pressure to give a virtually constant tortuosity of about 1.8 but the fitted crystal diffusivity differed from the single-pellet self-diffusion results by a factor of about 100. The results of these fixed-bed studies will be published in the *Transactions of the Institution of Chemical Engineers*, London.

J. R. Katzer (University of Delaware, Newark, Del. 19711): Regarding the previous question, could you please clarify for me which of the two situations had the higher diffusion coefficient?

R. W. H. Sargent: The CO_2–air diffusivities were about 100 times higher than the self-diffusion values, which is in the right direction but only partially accounted for by the correction factors given by Barrer in his paper.

E. F. Kondis (Mobil Research & Development, Paulsboro, N. J. 08066): Please describe the technique used to determine the two parameters in your nonlinear model.

R. W. H. Sargent: The method used is given in the Results section of the paper, page 160. The errors between the model and experimental results at 29 points were computed and the sum of squares of the errors minimized by adjusting D_p and P using Powell's conjugate gradient method (*Computer Journal*, 1964).

I. Zwiebel (Worcester Polytechnic Institute, Worcester, Mass. 01609): Why did you assume rapid transport within the pore; specifically, why did you use the steady state model in Equation 1?

R. W. H. Sargent: In the fixed-bed work already referred to, we did include the term for rate of accumulation in the gas phase in the pores but this was always negligible (about 1000 times smaller) compared with the other terms. In the later single-pellet work, we therefore omitted the term from the beginning.

R. B. Anderson (McMaster University, Hamilton, Ontario, Canada): Could the higher diffusivity in practical adsorber be attributed to the temperature of the bed being higher than that of the gas?

R. W. H. Sargent: The maximum concentration of CO_2 in the fixed-bed tests was 1000 ppm and there was no measurable rise in gas temperature along the bed. We are currently doing tests with much higher CO_2 concentrations and have extended the model to deal with thermal effects. Calculations with this model for the original tests confirm that temperature gradients are negligible, both in the gas phase and in the interior of the pellets.

52

Kinetics of Adsorption on A Zeolites. Temperature Effects

JAMES D. EAGAN, BRUNO KINDL, and ROBERT B. ANDERSON

Department of Chemical Engineering, McMaster University, Hamilton, Ontario

Maximum temperatures measured in the adsorbent during the adsorption of nitrogen on 4A and propane on 5A zeolite, both at −78°C, were 15° and 50°C above the bath temperature. Finite difference calculations, taking into account the generation and loss of heat and changes in diffusivity and equilibrium adsorption with temperature, reproduced the pertinent features of the rate and temperature data. When the temperature maximum occurs late in the adsorption process, the rate curve is drastically different from that expected for isothermal adsorption.

Some rate curves for the adsorption of gases on zeolitic molecular sieves have unusual shapes (*5, 6*). A possible cause of some of these phenomena is the changing temperature of the sample during adsorption owing to the heat of adsorption (*5*). A group of experiments has demonstrated that samples overheat substantially during kinetic measurements. Temperatures higher than 15°C above the bath temperature have been measured during kinetic experiments in both volumetric and gravimetric apparatuses. Finite difference calculations, which approximate experimental results, indicate that these temperature differences do not change the shape of the initial part of the rate curve substantially, but in some cases cause major changes at longer times.

Experimental and Computing Methods

Commercial Linde 4A and 5A powders were used after evacuation at 450°C for 15 hours. Experiments with N_2 on 4A powder were made in a conventional glass volumetric system equipped with a graduated burette and a manostat (*3*). A 0.675-gram sample of 4A powder was contained in a spherical bulb of inside diameter 12 mm. Inserted in the powder was a 36-gauge chromel-alumel thermocouple that passed through the top of the sample tube at a point above the cold bath by a commercial glass–metal seal. The reference junction was immersed in the cooling bath.

Tests with propane on 5A were performed in a Cahn R.G. electrobalance system. In the gravimetric apparatus, the initial disturbance caused by introducing the gas to the evacuated sample persisted for about 6 seconds and adsorption measurements could not be made in this period. In separate experiments, 0.3- and 4-mm layers of zeolite on the balance pan were used and respective adsorption rate curves were recorded. For the thicker layer in another experiment, the balance pan was supported on a 32-gauge thermocouple, the junction of which was immersed in the zeolite; the temperatures were determined during a similar adsorption experiment in which weight changes were not measured. In both apparatuses, the measured temperature was probably not equal to the actual particle temperature but it seems reasonable to assume that they differ only slightly. In all experiments, the cooling bath was dry ice–acetone through which a slow stream of CO_2 was bubbled to maintain an atmosphere of CO_2 under all conditions.

Finite difference calculations were made on a CDC6400 computer to determine the effect of the temperature rise of the powder on the adsorption rate. Here the particles were assumed to be spheres of 1-micron diameter for convenience; the actual sample consisted largely of cubes with wide variation in size, 0.1 to 6 microns. Calculations indicated that the temperatures throughout the particle could be assumed uniform. Heat was assumed to be removed from the surface of the particle by natural convection. The largest heat transfer resistance resulted from slow conduction through the powder. Thus, during adsorption, temperature gradients existed within the sample bed but it was assumed to have a uniform average temperature which could be considered the temperature of an "average" particle. The heat transfer characteristics of the single particle were made equivalent to those of the actual sample bed in the following way: The thermal conductivity of the powder was assumed to be the same as diatomaceous earth (*4*), and the temperature response of the bed was calculated for a given change in external temperature. The heat transfer coefficient of the single particle was calculated so that its cooling rate was the same as that of the average particle in the powder. This value, 5.0×10^{-8} cal/sec-cm^2-°C, was used in the finite difference calculations.

The particle was divided into 30 to 100 shells for the calculation and Fick's Law was assumed to hold. Heat generated was taken as rate of adsorption times the heat of adsorption, and the latter quantity was assumed to be independent of amount adsorbed. With increasing tem-

Table I. Parameters for Computed Rate Curves

Data in Figure	*Heat of Adsorption, Kcal/Mole*	*Activation Energy, Kcal/Mole*	D/R^2, Min^{-1}	H, $Cal/Sec\text{-}Cm^2\text{-}°C$
1	5.0	5.5	5.0×10^{-3}	5.0×10^{-8}
2	8.0	3.0	0.4	5.0×10^{-8}
3	5.0	5.5	5.0×10^{-3}	Variable

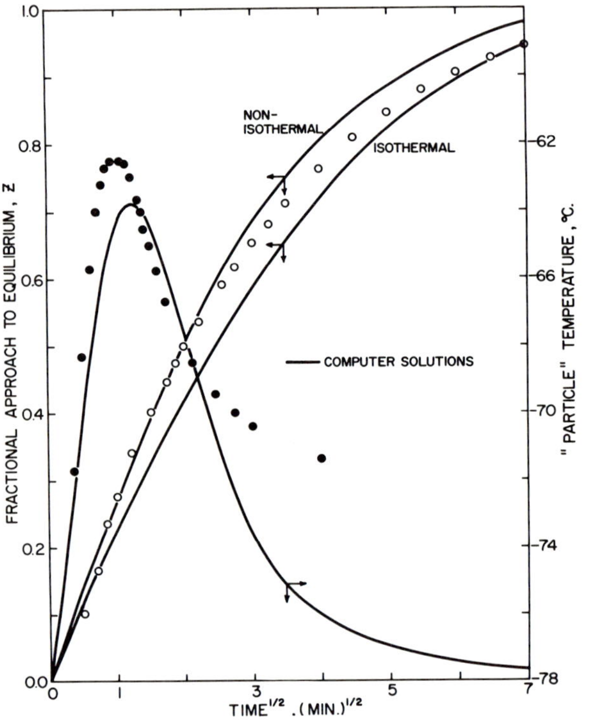

Figure 1. Adsorption of nitrogen on 4A zeolite powder at −78°C and 760 torr.
Open circles denote adsorption and solid circles the temperature. Solid lines are computed adsorption and temperature data.

perature, the diffusivity increases according to the Arrhenius equation but the equilibrium amount adsorbed decreases. The change in equilibrium adsorption was calculated using the heat of adsorption and the Freundlich isotherm with pressure exponents from adsorption studies. The exponent of the Freundlich equation was assumed to vary linearly with temperature between experimental temperatures.

The computer program calculates in each time step a new concentration profile, temperature, diffusivity, and amount adsorbed at equilibrium. At convenient intervals, the concentration profile is integrated using Simpson's rule to give the amount adsorbed. This quantity was divided by the equilibrium adsorption at the bath temperature to give the approach to equilibrium factor Z.

Appropriate values of activation energies and heats of adsorption from our laboratory or from the literature (2) were chosen, and diffusivities at −78°C were estimated from our rate curves; these values are given in Table I.

In the present calculations, we have attempted only to obtain rate and temperature curves reasonably approximating the experimental data, and no attempt was made to improve the predicted results by adjustment of constants.

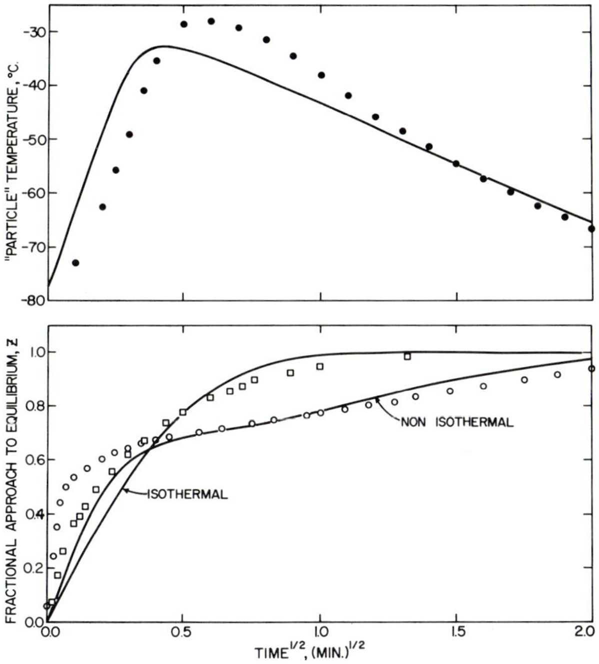

Figure 2. Adsorption of propane on 5A zeolite powder at −78°C.

Open squares denote adsorption at 21 torr for an 0.3-mm layer. Open and solid circles denote adsorption and temperatures, respectively, at 30 torr in a 4-mm layer. Solid curves are calculated adsorption and temperatures.

Experimental Results and Discussion

Figure 1 shows the rate of adsorption of N_2 on 4A powder and concurrent temperature measurements. Figure 2 gives the rate of adsorption of propane on 5A with a thin (0.3-mm) and a thick (4-mm) layer of zeolite on the balance pan; temperatures were measured in a separate experiment in the thick layer.

For nitrogen on 4A, Figure 1, the nonisothermal rate data despite the heating could be represented reasonably by usual isothermal Fick's law equations (*1*), if D/R^2 is taken as 7.5×10^{-3} min^{-1}. Thus, the value of D/R^2 calculated from the isothermal equation was 50% larger than that used to derive the nonisothermal curve in Figure 1. Here the temperature maximum occurs at low amounts adsorbed, and the increased rate owing to increased diffusivities is nearly compensated by the decreased equilibrium adsorption at the observed temperatures. Propane

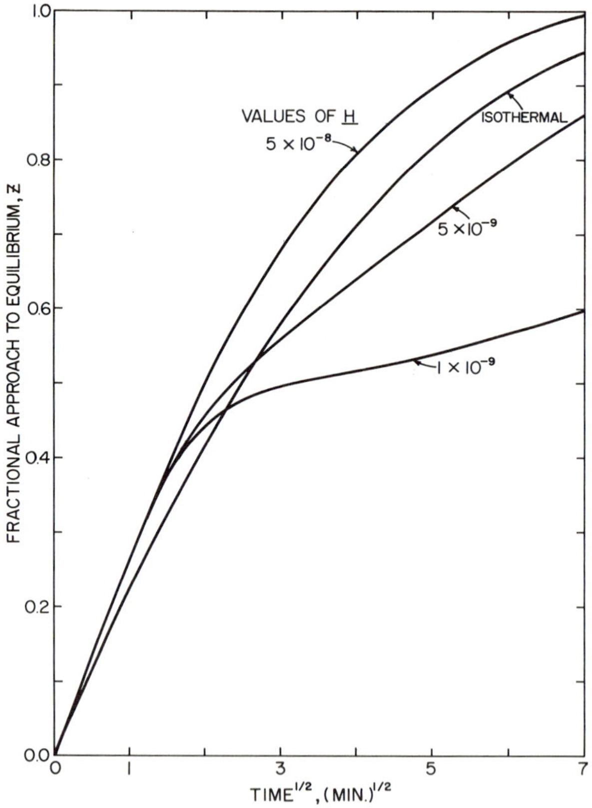

Figure 3. Calculated adsorption data for different heat transfer coefficients

on 5A (Figure 2) was chosen as an example of a rapid adsorption, where the rate and heat generation were greater. For the thick sample, the amount adsorbed approached equilibrium slowly compared with the thin sample. The temperature maximum for the thick sample occurred at 0.7 of the equilibrium adsorption. In this case, the adsorption is rapid at short times because of the increased diffusivity, and equilibrium at the particle temperature is approached at about the time corresponding to the maximum temperature. The amount adsorbed then increases with time and the cooling of the samples—*i.e.*, reflecting increasing equilibrium adsorption.

The calculated curves in Figures 1 and 2 reproduce all of the pertinent features of the experimental rate and temperature data, although there are some quantitative differences. For example, at longer times, experimental measurements of adsorption lie below computed curves for the single particle. The actual powder consisted of particles of different sizes, and at large times the small particles, having already reached equilibrium, no longer contribute to the rate. Calculated rate curves for the same values of diffusivity at the bath temperature, activation energy, and heat of adsorption as for nitrogen, but different heat transfer coefficients, are given in Figure 3. These curves approximate all of the types of rate curves described in the present paper. Changing the heat transfer coefficient by a factor of 50 produced profound effects on the last part of the rate curve but has virtually no effect on the early portion.

Our data show that large temperature changes can occur during measurements of adsorption kinetics. When the temperature maximum occurs early in the process, no pronounced effect on the rate curve is observed, and the unwary experimenter may conclude that his data were obtained isothermally.

Acknowledgment

The authors are pleased to acknowledge fellowship and operating funds provided by The National Research Council of Canada.

Literature Cited

(1) Anderson, R. B., Bayer, J., Hofer, L. J. E., *Ind. Eng. Chem. Process Design Develop.* **1965**, 4, 167.
(2) Eberly, P. E., Jr., *Ind. Eng. Chem. Prod. Res. Develop.* **1969**, 8, No. 2, 140.
(3) Harper, R. J., Stifel, G. R., Anderson, R. B., *Can. J. Chem.* **1969**, 47, 4661.
(4) Kreith, F., "Principles of Heat Transfer," 2nd ed., p. 594, International Textbook Co., Scranton, Pa., 1965.
(5) Satterfield, C. N., Margetts, W. G., *62nd Annual Meeting, A.I.Ch.E., Washington, D. C., Nov. 1969*, Reprint 56d.
(6) Stifel, G. R., Master's thesis, McMaster University, 1967.

RECEIVED February 4, 1970. Resubmitted September 1, 1970.

Discussion

C. N. Satterfield (Massachusetts Institute of Technology, Cambridge, Mass. 02139): Recent experimental results of Margetts (Satterfield and Margetts, *A.I.Ch.E. J.*, in press; preprint, A.I.Ch.E. Meeting, November 1969) are consistent with the observations and calculations of Anderson. Margetts studied the gas phase sorption of CH_4 or n-C_4H_{10} on Na-mordenite crystals at 25°C and pressures of about 10 to 80 torr. The calculated effective diffusivities showed a maximum in the general region of $M_t/M_\infty \sim 0.5$. Calculations and experimental observations indicated this was caused by a temperature increase within the sample, occurring during the first few seconds.

R. B. Anderson: Thanks for the supporting evidence.

Y. H. Ma (Worcester Polytechnic Institute, Worcester, Mass. 01609): What is the diffusion coefficient you obtained? Could the temperature rise be owing to the adiabatic compression?

R. B. Anderson: The heat of adsorption was the cause of the temperature increase. The adiabatic compression of gas introduced to the evacuated sample should lead to a negligible temperature increase.

E. F. Kondis (Mobil Research & Development Corp., Paulsboro, N. J. 08066): We confirmed some of the trends in nonisothermal results which you report. Our data are in a flow system and hence discount the possibility of heat effects caused by expansion of gas using your Cahn instrument.

R. B. Anderson: Thanks.

Kinetics of Ethane Sorption on 4A Molecular Sieve Crystal Powder and Pellets

EDWARD F. KONDIS[1] and JOSHUA S. DRANOFF

Department of Chemical Engineering, Northwestern University, Evanston, Ill. 60201

The sorption of ethane from dilute mixtures with helium by 4A sieve crystal powder and pellets made without binder has been studied with a microbalance in a flow system at temperatures between 25° and 117°C. Results show clearly that intracrystalline diffusion is the rate-controlling process and that it is represented well by a Fick's law diffusion model. Transient adsorption and desorption are characterized by the same effective diffusivity with an activation energy of 5660 cal/gram mole.

The objective of the work reported here was to characterize clearly the kinetics of ethane sorption by 4A molecular sieves under isothermal conditions. Determination of the significant phenomena at work in adsorption and desorption is important for the basic understanding of the sorption process, as well as for the rational analysis and design of equipment for its application. The present study was undertaken to settle the question of the relative importance of micropore and macropore diffusion for one characteristic adsorbate and molecular sieve adsorbent.

The experimental technique adopted for this work was the measurement of weight gain and loss as a function of time when small amounts of molecular sieve are exposed to gas streams of constant composition in a flow system. Comparison of experimental data with curves based on a suitable mathematical model permitted the determination of effective diffusivities for ethane. Equilibrium data also were obtained from these experiments. The present technique was chosen in preference to fixed-bed studies because of the increased sensitivity offered by the "single particle" approach.

[1] Present address: Mobil Oil Corp., Paulsboro, N. J.

Experimental

Equipment. The basic equipment used in this study was a microbalance (Thermo-Grav, American Instrument Co.). It consists essentially of a small sample holder connected to a fine calibrated spring, the motion of which is followed by an electrical transducer and automatically recorded. The sample holder is contained in a small borosilicate glass vessel (sample tube) through which gas may be circulated continuously and which may be heated easily to high temperatures by an electrical furnace for regeneration purposes or submerged in a circulating oil bath for temperature control during sorption experiments. The apparatus was completed by the addition of suitable equipment for gas flow measurement and control and system calibration.

Materials. The gases used were research grade ethane and helium, both 99.99 mole % pure. Impurities in the helium were not significant in this work. However, those present in the ethane, principally ethylene, were of some concern. The adsorbent was standard pure crystalline Linde 4A molecular sieve in powder and pelletized form (Linde Lot Number 441079). The cylindrical pellets were formed from pure crystals in a single pellet press and were 1/8 inch in diameter and 1/16 inch long. These pellets had a density of 1.06 grams/cc (dry and regenerated). Since they contained no binding clays, the pellets could withstand only mild stress but this was not a problem in the present work. Pellets were made from the pure crystal powder as received from Linde and from such powder after ellutriation to remove the finer particles. Samples used in sorption runs ranged from 0.25 to about 6 grams.

Procedure. The basic pattern for these experiments consisted of regeneration of the absorbent sample followed by adsorption and desorption runs. Regeneration was accomplished by heating the sample to 550°C (at a rate of 20°C per minute) and maintaining it at that temperature under a helium purge until a stable weight was reached (about 1 hour). The sample was allowed then to cool to room temperature after which the sample tube was submerged in the constant-temperature bath and brought to bath temperature while still under helium purge. The weight of the sample was monitored continuously during this process to assure the absence of contamination. An adsorption experiment was initiated by replacing the helium stream by a pre-set mixture of ethane and helium. When the sample was saturated with ethane at the experimental conditions, a desorption experiment was carried out by replacing the feed gas by pure helium. Desorption was continued until the sample weight returned to its initial value.

Experiments were performed at 25.2°, 73.8°, and 116.8°C, with ethane concentration in the feed gas of 2, 4, and 8 volume %, and at atmospheric pressure. The recorded traces were corrected to account for buoyancy and drag and converted to plots of weight gain and loss *vs.* time.

Special care was required to eliminate a number of undesired effects uncovered in preliminary experiments. These included sample contamination from impurities in the ethane, which was eliminated by placing a small amount (about 200 mg) of 4A sieve in the base of the sample tube. This material, which was regenerated automatically each time along with the sample, did not interfere with the main sorption experiments but did permit reproducible operation of the system. In addition, the problems of initiating a sharp step change in gas composition in the sample tube and the elimination of gas phase mass transfer effects were solved by careful redesign of the sample tube and an annular sample holder, and verified by a series of experiments at various gas flow rates and compositions.

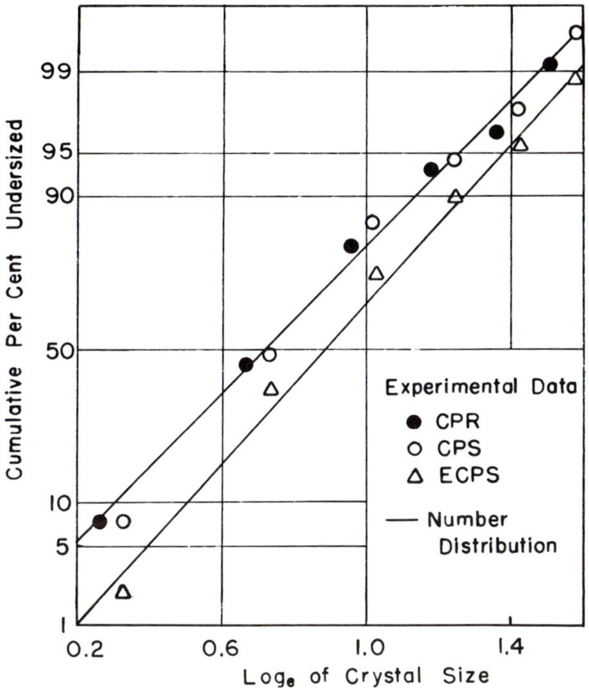

Figure 1. Log-normal size distributions for 4A crystals

Crystal Size Measurement. The size distribution of the sieve crystals used also was measured in some auxiliary experiments. This was done by first suspending the sieve sample (crystal powder or crushed pellets) in a sodium hydroxide solution (to help disperse the easily-formed crystal agglomerates) and then examining the solution with a calibrated optical microscope at a magnification of 5000. About 250 particles picked at random on a slide were measured for each sample. The particles, which have a shape somewhere between that of a sphere and a cube, were treated as cubes with the length of a side equal to $2a$ microns.

More complete details of equipment and experimental techniques are presented by Kondis (7).

Results and Discussion

Crystal Size Distribution. The measured crystal size distribution followed a log-normal form, suggested as characteristic for most small particles by Herdan (*6*). Figure 1 shows results obtained with the 4A crystal powder and with the 2 types of pellets formed from it. Here and below, the Linde crystal powder as received, the pelletized powder, and the pellets formed from the elutriated powder will be referred to as CPR, CPS, and ECPS, respectively. Clearly, the pelletizing process did not affect the size distribution of the original material. Furthermore, the elutriated particles do have a somewhat larger average size.

For analysis of sorption data, the particles were represented as spheres with radius equal to the hydraulic radius of the particles. For cubes, this leads to an equivalent spherical radius equal to one-half the length of the cube side or a. The weight average radius may be found from the previous data and the relation among weight and number average radii and the standard deviation of the distribution given by Herdan (*6*).

$$\ln a_g = \ln a_g' - 3 (\ln \sigma_g)^2 \tag{1}$$

The results are listed in Table I.

Sorption Kinetics. The adsorption and desorption data were analyzed in terms of a model based on the following main assumptions. Micropore diffusion within the sieve crystals is the rate-controlling process. Diffusion may be described by Fick's law for spherical particle geometry with a constant micropore diffusivity. The helium present in the system is inert and plays no direct role in the sorption or diffusion process. Sorption occurs under isothermal conditions. Sorption equilibrium is maintained at the crystal surface, which is subjected to a step change in gas composition. These assumptions lead to the following relation for the amount of ethane adsorbed or desorbed by a single particle as a function of time (Crank, *4*).

$$\frac{Q - Q_0}{Q_1 - Q_0} = 1 - 6 \sum_{n=1}^{\infty} \frac{1}{n^2\pi^2} \exp\left(- \frac{n^2\pi^2 D_c}{a^2} t\right) \tag{2}$$

The corresponding solution for a cubic particle with sides $2a$ in length is in very close agreement with Equation 2, deviating at most by a few per cent.

Table I. Average Crystal Particle Size

Sample	*Weight Average Radius (Microns)*
CPR, CPS	1.39
ECPS	1.60

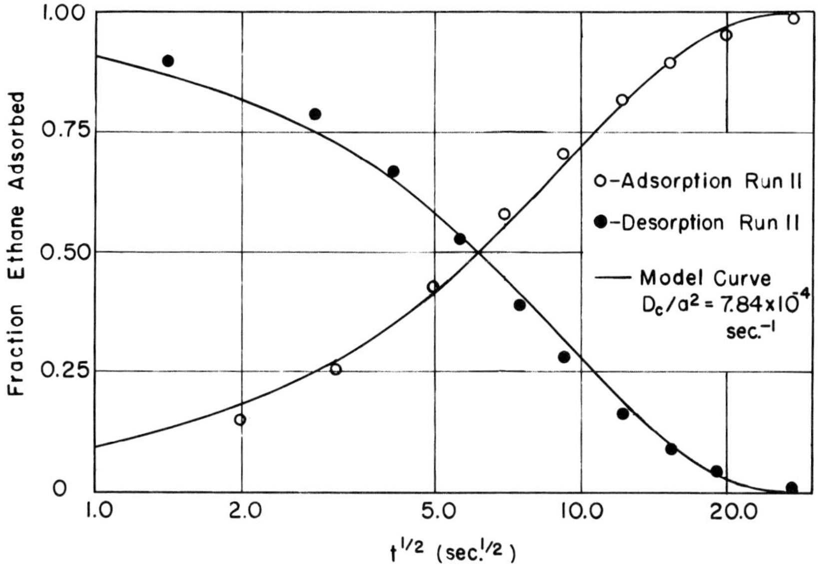

Figure 2. Adsorption and desorption curves for 4A crystal pellets at 73.8°C

Experiments were first performed using sieve pellets containing an imbedded thermocouple and with gas concentrations up to 10% ethane. These showed temperature variations of less than 0.5°C during sorption, thus confirming the isothermal assumption.

Thereafter, the experimental data were fitted to the above model equation by a graphical superposition technique. The data and model curve were plotted separately as fraction adsorbed or desorbed against the log of the square root of time. The experimental curve was moved horizonally until the best fit was obtained, thereby determining the appropriate value at D_c/a^2. This method uses all of the data, as opposed to some approaches based on the values at early times which have been used by others (*1, 2, 5*). It was applied easily in this work because of the excellent agreement of the model with the data obtained.

Figure 2 presents results for pure crystal pellets (CPS) at 73.8°C, which are typical of all the data obtained. Clearly, the model provides an excellent fit to these data, with the same diffusivity applying to both adsorption and desorption runs. There is some slight deviation at early times, but this is well within limits of experimental error. Results at other temperatures and with all 3 adsorbent types show the same kind of model fit.

The corresponding values of D_c/a^2 are presented in Table II and are plotted against temperature on the Arrhenius plot in Figure 3. Three

Table II. D_c/a^2 Values Determined from Experimental Data

Run	Adsorbent	Temp., °C	Ethane Concentration,[a] Vol. %	D_c/a^2 (Sec^{-1}) × 10^4 Adsorption Cycle	Desorption Cycle
9	CPS	73.8	4	7.84	7.84
10	CPS	73.8	2	8.41	7.84
11	CPS	73.8	8	7.84	7.84
12	CPS	116.8	2	22.1	20.3
13	CPS	116.8	4	26.0	22.1
14	CPS	116.8	8	23.0	23.0
19	CPS	25.2	4	2.40	2.56
20	CPS	25.2	2	2.25	2.40
21	CPS	25.2	8	2.72	2.40
21A	CPS	73.8	8	7.84	—
30	CPS	25.2	4	2.72	—
31	CPS	25.2	4	2.56	2.56
33	CPR	25.2	4	2.10	2.40
34	CPR	25.2	8	2.40	2.40
36	CPR	73.8	4	7.29	7.29
37	CPR	73.8	8	7.29	8.40
38	CPR	116.8	4	21.2	22.1
39	CPR	116.8	8	22.1	22.1
44	ECPS	25.2	4	1.56	1.56
45	ECPS	25.2	8	1.69	1.69
53	ECPS	73.8	4	5.11	5.11
54	ECPS	73.8	8	5.11	5.3
55	ECPS	116.8	4	15.2	15.2
56	ECPS	116.8	8	13.7	13.7

[a] Concentration for adsorption cycle. For desorption cycle, concentration is zero.

principle results are apparent from these data. First, there is essentially no difference between the values of D_c/a^2 obtained from adsorption or desorption experiments, indicating that the same mechanisms are at work in both processes. Secondly, the fact that identical results are obtained with pure crystal powder and pellets made from that powder indicates that D_c/a^2 is independent of over-all particle size and therefore that micropore or crystal diffusion is the rate-controlling process. Finally, the values for the elutriated crystal pellets are lower by a factor of 1.44 compared with the other results. Thus, the diffusion measurements imply that the value of a for these pellets is 1.2 times the average radius for the pure crystals. This compares quite favorably with the measured radius ratio of 1.15 indicated by the data of Table I, thus providing further confirmation of the model.

The effect of the crystal size distribution on these results was investigated using the observed distribution function. The prediction of the model for the average particle radius was compared with a prediction weighted according to the distribution. The results showed negligible difference over the entire range of interest, thus validating the use of an average particle radius.

The D_c values found in this work may be compared with the value of 4.8×10^{-12} cm²/sec reported previously by Brandt and Rudloff (*3*), who studied ethane sorption by 4A crystals at 22.9°C but in the absence of helium carrier gas. The present data at 25.2°C indicate (for an average radius of 1.39 microns) that D_c is 4.6×10^{-12} cm²/sec, in excellent agreement with the earlier work. However, the activation energy reported by Brandt and Rudloff was 7.4 Kcal/mole, as compared with the value of 5.66 Kcal/mole found here.

Equilibrium Data. Finally, equilibrium isotherms were established from the measurements made in this study. These are plotted in Figure 4. Clearly, the results indicate the isotherms to be essentially linear except at 25.2°C. Subsequent work at higher ethane concentrations has con-

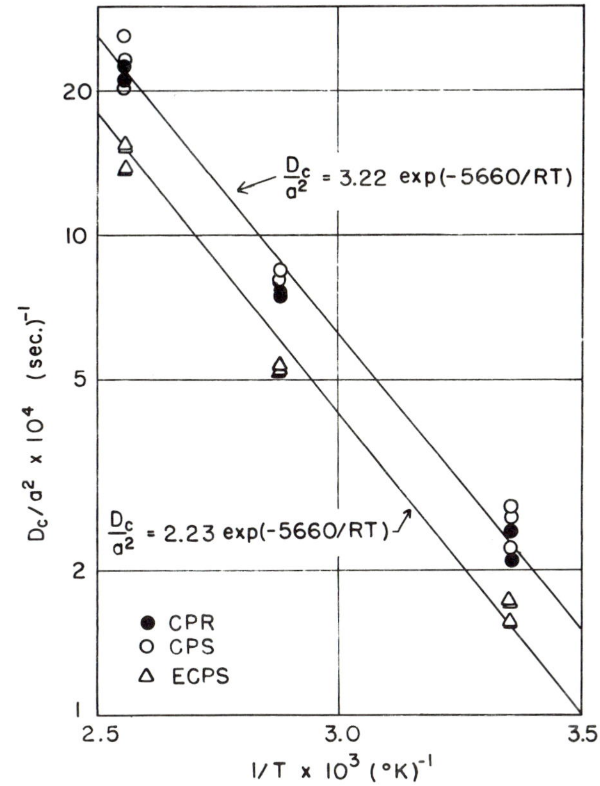

Figure 3. Arrhenius plot for experiments micropore diffusivity data

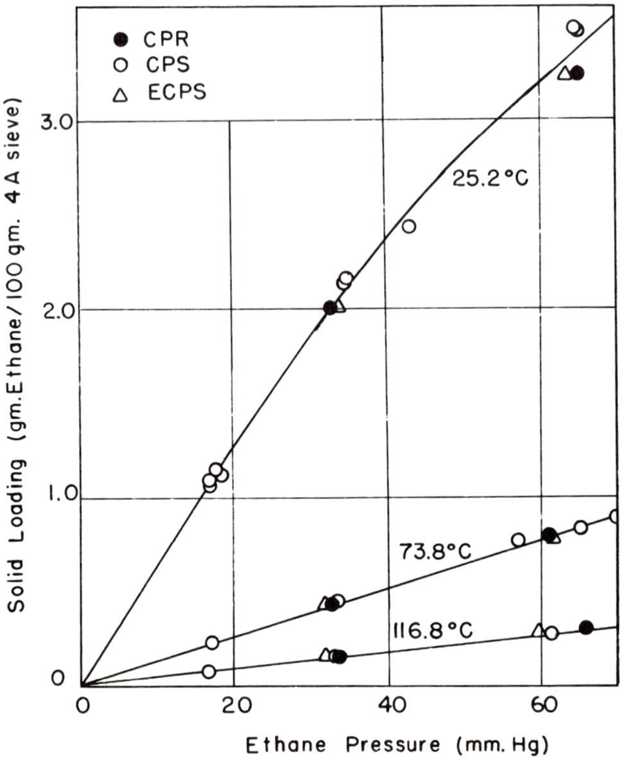

Figure 4. Equilibrium isotherms at low ethane concentrations for 4A crystal pellets and powder

firmed that the isotherms follow the usual Langmuir form. As expected, data for all 3 adsorbents show the same equilibrium behavior. The heat of adsorption was determined from these and more complete equilibrium data (7) and found to be 7.04 Kcal/mole.

Conclusions

The results in this study have demonstrated clearly that the rate of adsorption and desorption of ethane at low concentrations on 4A molecular sieves in the absence of binder is controlled by intracrystalline diffusion of the ethane. Furthermore, the diffusion process may be characterized by Fick's law and an effective diffusivity dependent only on temperature, and applicable to both adsorption and desorption. It may be expected, therefore, that such micropore diffusion also determines the rates of ethane sorption with commercial 4A pellets containing clay binders since the same microporous structure should exist in these adsorbents.

Acknowledgment

The authors thank F. G. Dwyer of Mobil Oil Corp. for providing the samples used in this work, and D. L. Johnson of Northwestern University for assistance in crystal size measurement. The financial assistance of the Mobil Oil Corp. Incentive Fellowship Program is gratefully acknowledged.

Nomenclature

- a = Effective crystal radius, cm
- a_g = Number average particle size, cm
- a_g' = Weight average particle size, cm
- D_c = Intracrystalline diffusivity, cm^2/sec
- Q = Average ethane content of adsorbent, grams/gram of solid
- Q_0 = Initial average ethane content of adsorbent, grams/gram of solid
- Q_1 = Final average ethane content, grams/gram of solid
- t = Time, sec
- σ_g = Standard deviation of size distribution, cm

Literature Cited

(1) Barrer, R. M., *Trans. Faraday Soc.* **1949,** 45, 358.
(2) Barrer, R. M., Fender, B. E. F., *J. Phys. Chem. Solids* **1961,** 21, 12.
(3) Brandt, W. W., Rudloff, W., *J. Phys. Chem. Solids* **1965**, 26, 741.
(4) Crank, J., "The Mathematics of Diffusion," Oxford University Press, Oxford, England, 1956.
(5) Habgood, H. W., *Can. J. Chem.* **1958,** 36, 1384.
(6) Herdan, G., "Small Particle Statistics," Butterworths, London, England, 1960.
(7) Kondis, E. F., Ph.D. Dissertation, Northwestern University, Evanston, Ill., 1968.

RECEIVED February 13, 1970.

Discussion

D. M. Ruthven and **K. F. Loughlin** (University of New Brunswick, Fredericton, N. B., Canada): The data of Kondis and Dranoff illustrate a number of important points relating to the problem of calculating diffusion coefficients from sorption curves. For the analysis of sorption curves, the crystal size distribution should be expressed on a weight (or volume) fraction basis. The type A zeolite crystals are cubic and by plotting the cumulative volume fraction

$$\left(\frac{\sum_{d_i = 0}^{d_i} N_i d_i^3}{\sum_{d_i = 0}^{\infty} N_i d_i^3} \right)$$

against crystal diameter d_i on arithmetic probability paper, it may be shown that the crystal size distribution data of Kondis (*1*) are well represented by a normal distribution function with mean crystal size 2μ = 2.70 micron and standard deviation 2σ = 1.08 micron; $s = \mu/\sigma = 2.5$. Similar crystal size distribution data were obtained by Ruthven and Loughlin (*2*) for Linde 5A zeolite.

Figure 1 shows the typical sorption curve presented by Kondis and Dranoff (run 11) together with the theoretical curve for uniform spherical particles, calculated from Equation 2 of the preceding paper, using the value $D/a^2 = 7.84 \times 10^{-4}$ sec^{-1}. The deviation of the experimental points from the theoretical curve is clearly apparent. By summing the contributions of the individual size fractions of particles it may be shown that, for a system of cubic particles with an approximately normal distribution of size (on a weight or volume fraction basis), the proper expression for the sorption curve is given by (*2*):

$$W = \frac{512s}{\pi^6\sqrt{2\pi}} \sum_{l=1}^{\infty} \sum_{m=1}^{\infty} \sum_{n=1}^{\infty} \int_{y=0}^{\infty}$$

$$\frac{\exp\left\{-\tfrac{1}{2}s^2(y-1)^2 - \frac{\pi^2}{4y^2} \cdot \frac{Dt}{\mu^2} \cdot [(2l-1)^2 + (2m-1)^2 + (2n-1)^2]\right\} dy}{(2l-1)^2 (2m-1)^2 (2n-1)^2}$$

where $W = 1 - \dfrac{Q_t}{Q_\infty}$

Q_t = Mass of sorbate adsorbed or desorbed during time t

Q_∞ = Mass of sorbate adsorbed or desorbed as $t \rightarrow \infty$

D = Diffusion coefficient

μ = Mean half side of cubic particle

$s = \dfrac{\mu}{\sigma}$

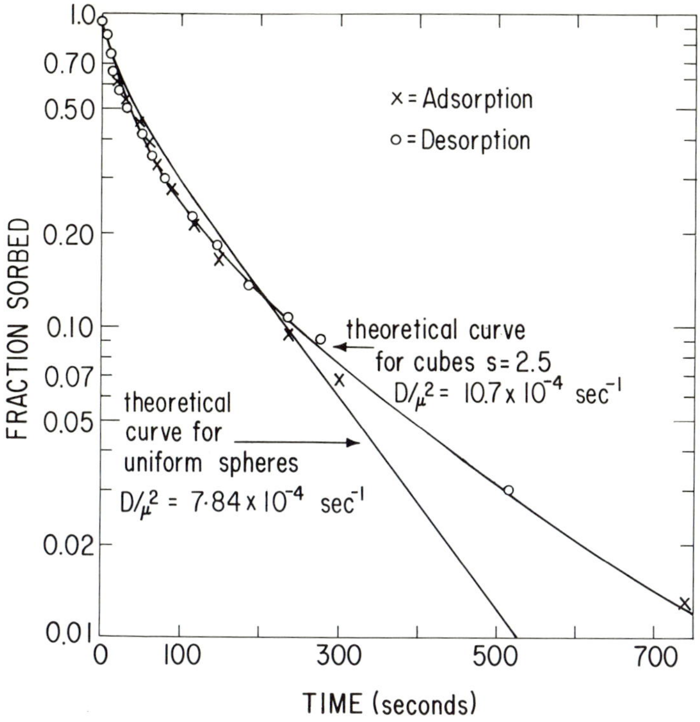

Figure 1. Sorption curve for ethane in 4A zeolite; data of Kondis (Run 11)

When the value of s is known from measurements of crystal size distribution, this expression may be evaluated numerically to give W as a function of Dt/μ^2. The value of D/μ^2 may then be obtained by matching the experimental data to the theoretical curve.

When this analysis is applied to the data of run 11, with $s = 2.5$, a value of $D/\mu^2 = 10.7 \times 10^{-4}$ sec^{-1} is obtained, compared with the value $D/\mu^2 = 7.84 \times 10^{-4}$ sec^{-1} obtained from the equation for uniform spherical particles.

The corresponding theoretical curve is shown in Figure 1. It is apparent that this curve fits the experimental data well, whereas the theoretical curve calculated assuming a mean equivalent spherical radius gives only a rather poor fit. In this type of system in which there is present a significant range of crystal sizes, the assumption of a mean equivalent spherical radius is a rather poor approximation which can lead to significant errors in the calculated diffusivities.

Literature Cited

(1) Kondis, E. F., Ph.D. thesis, Northwestern University, 1969.
(2) Ruthven, D. M., Loughlin, K. F., *Chem. Eng. Sci.*, in press.

E. F. Kondis and **J. S. Dranoff:** We agree with Loughlin and Ruthven that crystal shape and size distribution are important considerations for adsorption on zeolites. Moreover, we have previously evaluated their effects in our work. Dr. Kondis has pointed out at this symposium and elsewhere (*1*) that for our experimental data and for our method of obtaining the diffusion coefficients, these two parameters caused about a 10 to 20% variation in D_c/a^2. We accepted this variation as being within the accuracy of our results. Loughlin and Ruthven in their Figure 1 show that by accounting for size distribution and shape they obtain a variation of 35% in D_c/a^2. However, they used a different technique to obtain the diffusion coefficient. Moreover, their method of presentation accents a much different portion of the data.

In Figure 1 we show a comparison of our data and method of analysis with that of Loughlin and Ruthven using the same coordinate system as presented in our paper at this symposium. It is quite obvious that the model used by Loughlin and Ruthven gives a much poorer fit to the experimental data over the first 50–60% sorbed or desorbed. In fact, their model requires $D_c/a^2 \cong 7 \times 10^{-4}$ sec^{-1} to fit the data in this region (not 10.7×10^{-4} sec^{-1} as shown in their Figure 1). Their method gives a

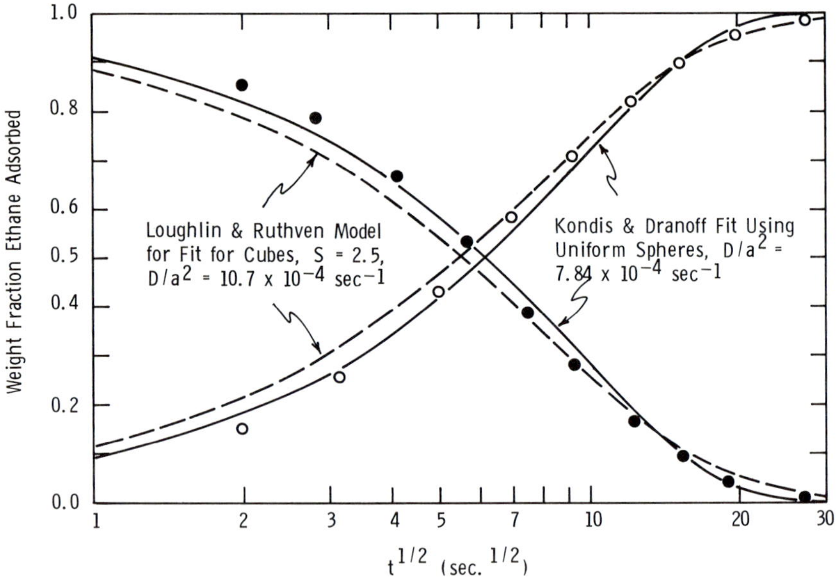

Figure 1. A comparison of the experimental data and model fit of Kondis and Dranoff with the proposed model of Loughlin and Ruthven

Experimental Data of Kondis & Dranoff
○ Run 11 Adsorption Data
● Run 11 Desorption Data

better fit primarily over the last 10% of the sorption or desorption curve (see their Figure 1). However, we have cautioned against the use of the data in this region (*1*) because small amounts of contamination in the ethane gas can shift this portion of the sorption curve quite markedly.

We have also calculated the value of D_c/a^2 using the $\sqrt{t}$ method of Barrer (*2*) as another check of our data. By this method, D_c/a^2 is about 7×10^{-4} sec^{-1}, which is in fair agreement with our result of 7.84×10^{-4} sec^{-1} but also differs markedly from the Loughlin and Ruthven result. In this case, it is quite obvious the method of analysis by Loughlin and Ruthven completely ignores the first 50% of the adsorption or desorption curve. Had they considered this data region to be equal in importance to the tailend of the adsorption or desorption curve, they would have obtained a value of D_c/a^2 within 10–20% of our value.

Literature Cited

(1) Kondis, E. F., Ph.D. dissertation, Northwestern University, 1969.
(2) Barrer, R. M., *Trans. Faraday Soc.* **1949**, 45, 358.

J. D. Sherman (Union Carbide, Tarrytown, N. Y. 10591): How were the pelleted crystal samples prepared? In your introduction, you mentioned a desire to compare your results for binderless forms with results for clay-bonded pellets. Have you indeed made such comparisons, and could you comment on your results?

E. F. Kondis: Samples were carefully prepared in a single pellet press. Binderless pellets show values of D_c/a^2 roughly 6 to 8 times greater than commercially made pellets with binder. Detailed experimental results (to be published in *Ind. Eng. Chem. Proc. Design Develop.*) attribute the reason for the change in D_c/a^2 to high-temperature steaming of the 4A crystals. Micropore diffusion remains the rate-controlling process.

54

Heats of Adsorption on X-Type Zeolites Containing Different Alkali Metal Cations

N. N. AVGUL, A. G. BEZUS, and O. M. DZHIGIT

M. V. Lomonosov State University of Moscow and Institute of Physical Chemistry, USSR Academy of Sciences, Moscow

Heats of adsorption of ethane, ethylene, and water on LiX, NaX, KX, RbX, and CsX zeolites as well as heats of adsorption of the homologous series of n-*alcohols* C_1–C_4 *on NaX zeolite have been investigated. The contribution of specific interactions to the total energy of adsorption of ethylene on zeolites decreases from LiX to CsX. The curves of the heat of adsorption of water have a wavelike form, especially for KX zeolite. At small filling, the heat diminishes from LiX to CsX zeolite but at intermediate fillings it goes through a maximum for KX. The increments of the heat of adsorption for the* CH_3, CH_2, *and OH groups in investigated homologous series are constant.*

If the molecules with π-bonds or polar functional groups—*i.e.*, the molecules of B and D groups according to classification (*14, 18*)—are adsorbed on zeolites, they interact strongly with the exchange cations (*12, 13, 16*). The value of this additional contribution to the total interaction energy depends on the type, charge, radius, and concentration of the exchange cations, degree of decationization, type of zeolite, and structure of adsorbed molecules. The particular facets of the structure that are important are the nature of the π-bonds, the value and localization of dipole and quadrupole moments, and the presence of a free electron pair in the adsorbed molecule. In this work, we have investigated the changes of the heats of adsorption of ethane, ethylene, and water on X-type zeolites containing alkali cations, as well as the changes of heats of adsorption in a homologous series of *n*-alcohols on NaX zeolite.

Experimental

The composition of the zeolites studied, which were kindly supplied by S. P. Zhdanov, is shown in Table I. Differential heat of adsorption of water was measured with an isothermal constant heat exchange calorimeter (*17*) at 23°C. The heat of adsorption of the alcohols was measured with a variant of this calorimeter (*1*). The heat of adsorption of ethane and ethylene was determined from isosteres. The zeolites were pretreated by evacuating at 400°C for 30 hours.

Table I. Composition of X-Type Zeolites Studied

Zeolite	*Composition*	*Degree of Decationization, %*
LiX-1	$Li_{0.54}Na_{0.33}(AlO_2) \cdot (SiO_2)_{1.35}$	13
LiX-2	$Li_{0.91}Na_{0.08}(AlO_2) \cdot (SiO_2)_{1.35}$	1
NaX-1	$Na_{0.94}(AlO_2) \cdot (SiO_2)_{1.38}$	6
NaX-2	$Na_{0.97}(AlO_2) \cdot (SiO_2)_{1.48}$	3
KX	$K_{0.66}Na_{0.26}(AlO_2) \cdot (SiO_2)_{1.31}$	8
RbX	$Rb_{0.63}Na_{0.29}(AlO_2) \cdot (SiO_2)_{1.31}$	8
CsX	$Cs_{0.55}Na_{0.37}(AlO_2) \cdot (SiO_2)_{1.34}$	8

Results and Discussion

The heats of adsorption of ethane and ethylene at zero coverage are presented in Table II. These heats are obtained by means of the virial equations (*12*). The heats of adsorption of ethylene exceed those of ethane because of specific interactions (*7, 8*). As the dimensions and polarizability of the ethane and ethylene molecules are similar, the energy of nonspecific interaction of these molecules with the zeolite surface must be approximately equal (*6*). Therefore, the differences of the heats of adsorption of ethane and ethylene, ΔQ, on the same zeolite is considered to be approximately the contribution of the specific interaction of the π-bonds of ethylene with the corresponding cation. The values of these differences for all systems studied are presented also in Table II.

Table II. Heats of Adsorption, Q_1, of Ethane and Ethylene on X-type Zeolites and Their Differences, ΔQ

Zeolite	*Cation Radii,* r, *A*	Q_1, *Kcal/Mole* C_2H_6	C_2H_4	$\Delta Q = Q_{C_2H_4} - Q_{C_2H_6}$, *Kcal/Mole*
LiX-1	0.78	5.5	8.9	3.4 (*7,8*)
NaX-2	0.98	6.2	9.2	3.0 (*7,8*)
KX	1.33	6.4	7.8	1.4
RbX	1.49	6.6	7.8	1.2
CsX	1.65	6.7	7.7	1.0 (*7,8*)

The energy of nonspecific interaction of ethane and other saturated hydrocarbons (*4, 9, 11*) increases from zeolite LiX to CsX. Conversely, the energy of specific interaction, ΔQ, of ethylene decreases sharply as cation radii r are increased. The character of dependence ΔQ on r are shown in Figure 1. In this figure, the similar dependence on r of the energy of specific interaction of nitrogen with LiX, NaX, and KX zeolites also is represented. In the last case, the contribution ΔQ is determined from the differences of heats of adsorption of nitrogen and argon. Similar results were obtained earlier for ethers (*9*). Unfortunately, the zeolites used were of different degrees of decationization. The influence of decationization on adsorption is being studied separately.

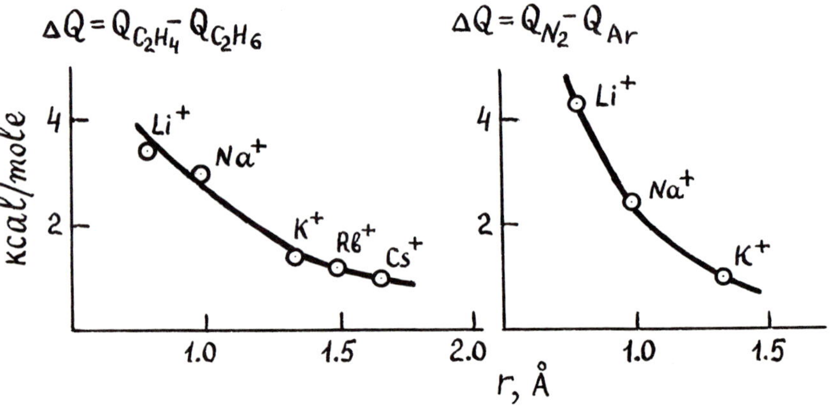

Figure 1. Dependence of the contribution of the specific interaction, ΔQ, in total heat of adsorption of ethylene and nitrogen on cation radius r *(obtained together with B. G. Aristov, Z. Sedlachek, and Pham Quang Du)*

The heats of adsorption of water by zeolites are mainly caused by specific interaction (*4*). The water molecules can interact specifically with the exchange cations, with negatively charged oxygen framework of zeolite and with each other. The heats of adsorption of water on zeolites containing alkali cations are shown in Figure 2. These curves have a wavelike form which is particularly pronounced for KX zeolite (*10*). Probably there are several energetically different groups of sites of specific adsorption in the cavities of zeolites, the sites of any particular group being energetically similar.

At small values of *a*, adsorption probably takes place mainly on cations occupying S_{III} sites. The heat of adsorption of water in this region decreases from LiX to CsX in agreement with the order of sequence of the heats of hydration of these ions. This indicates that the interaction is mainly between water and cations. With increasing coverage, adsorption probably takes place mainly on cations occupying S_{II} sites. In this region, the heat of adsorption of water is greatest on KX zeolite. K^+ is the first cation which cannot be placed in the plane of the six-membered oxygen ring at the S_{II} site and is displaced to a great extent into the large cavity. The particular form of the curve of the heat of adsorption *vs.* amount adsorbed probably is connected with this peculiarity of the K^+ cation. As the size of exchange cations increases from K^+ to Cs^+, the contribution of water–cation interaction and the energetical differences of location of these cations on sites S_{II} and S_{III} in zeolite become smaller.

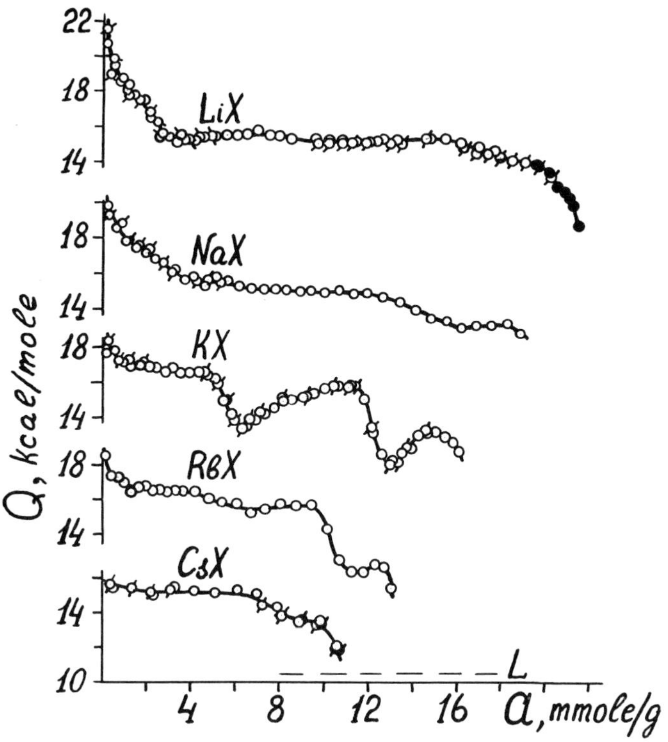

Figure 2. Heats of adsorption of water vapor on zeolites LiX-2, NaX-1, KX, RbX, and CsX (obtained together with K. N. Mikos and T. A. Rakhmanova)

Therefore, the heat of adsorption of water on the CsX zeolite practically does not depend on coverage. The last "waves" of the curves shown in Figure 2 are related to the filling of the central part of the large cavities by the formation of mutual hydrogen bonds as in Ref. *15*.

The dependence of the differential heats of adsorption of *n*-alcohols by zeolite NaX on adsorption level *a* is shown in Figure 3. The initial values are very large (20–26 kcal/mole). Therefore, at room temperature, it is difficult to determine the heat of adsorption of alcohols at small

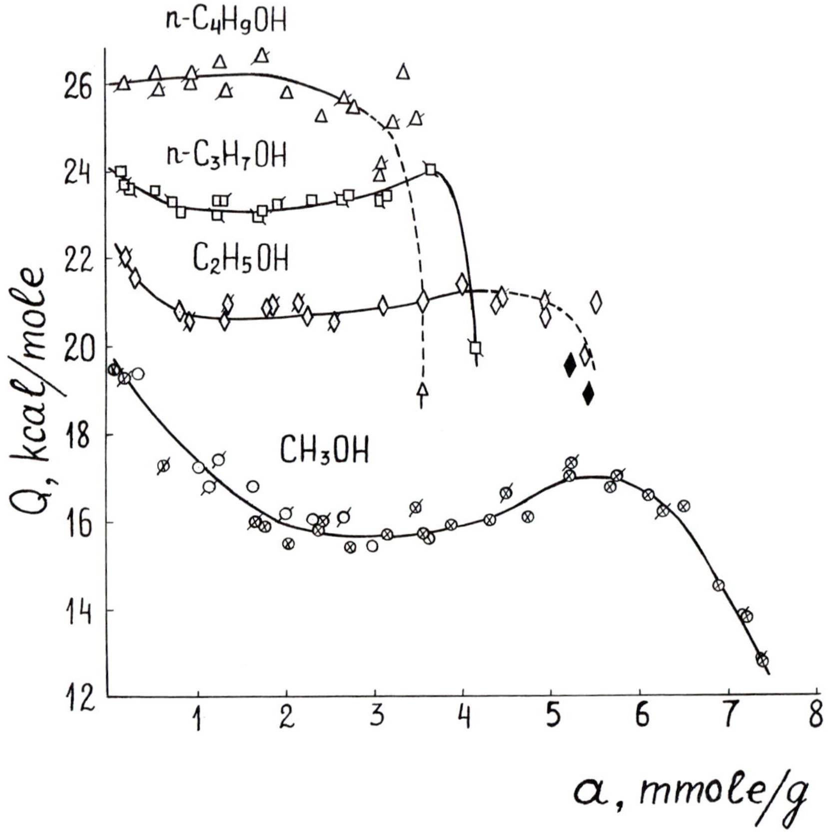

Figure 3. Heats of adsorption of n-*alcohol vapors on zeolite NaX-2 (obtained together with E. S. Dobrova)*

coverages. The heat of adsorption of methanol decreases with increasing *a* in the low-coverage region (*2, 3*). This curve is typical for strong specific adsorption of small molecules with the NaX zeolites (*16*). In the homologous series of *n*-alcohols, the decrease of heats of adsorption at low coverage becomes smaller and the heat of adsorption of 1-butanol does not depend to any large extent on *a*. It is connected with increasing lateral interaction between the hydrocarbon chains of adsorbed molecules, as in the case of hydrocarbons (*9, 11*).

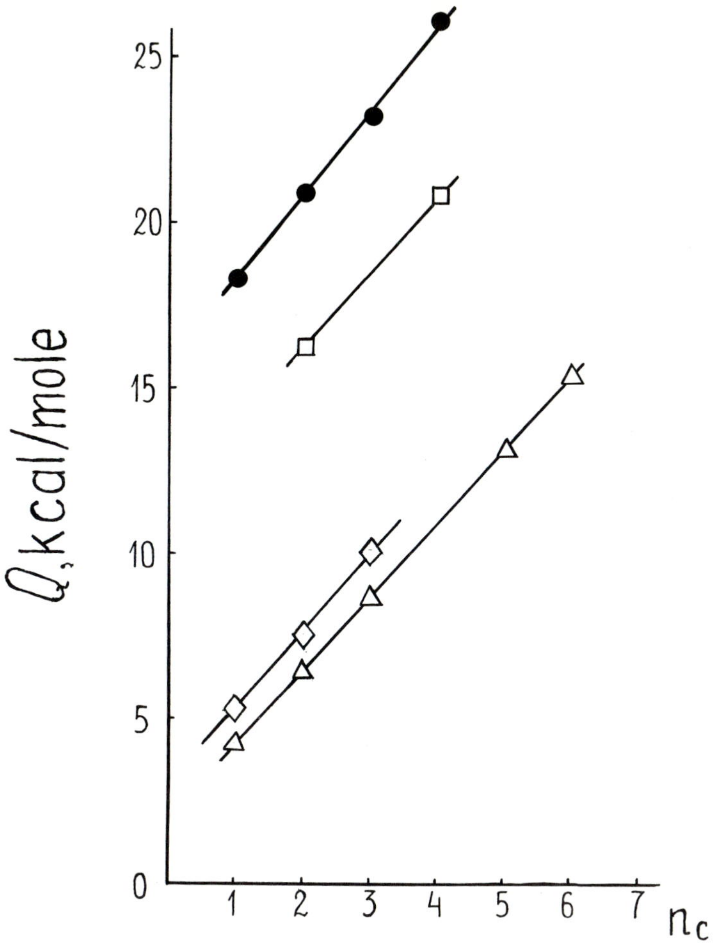

Figure 4. Dependences of the heats of adsorption, Q, of n-*alcohols (●), ethers (□), perfluoroalkanes (◇), and* n-*alkanes (△) on zeolite NaX on the number of the carbon atoms,* n, *in their molecules*

Table III. Increments, *q*, of the Heats of Adsorption of Different Functional Groups

Functional Groups	q, *Kcal/Mole*
CH_3 in *n*-alkanes	3.2
CH_2 in *n*-alkanes	2.3
CH_2 in ethers	2.3
CH_2 in alcohols	2.5
O in ethers	10.0
OH in *n*-alcohols	15.2

In Figure 4, the dependences of the heat of adsorption on number of the carbon atoms, *n*, in a molecule for the homologous series *n*-alkanes (*9*), perfluoroalkanes (*5*), ethers (*9, 11*), and *n*-alcohols on the zeolite NaX are shown. All data in Figure 4 and Table III relate to $a = 1$ molecule per cavity. In each case, there is a linear dependence of Q on n.

It was possible to obtain the increments of the heat of adsorption, q, for the CH_3 and CH_2 groups from the heats of adsorption of the *n*-alkane series (*9*). The increments for the ether's O and alcohol's OH groups were obtained by comparing the heats of adsorption of the particular *n*-alkane with those of the ether and alcohol, respectively, having the same value of n (Figure 4). These increments are presented in Table III. They are defined more precisely in comparison with (*16*). The increments of the heats of adsorption on NaX zeolite for functional groups containing oxygen and nitrogen (*16*) are very large because of the specific interactions of lone electron pairs of the oxygen and nitrogen atoms with cations of the zeolite. The increments of the alcohol hydroxyl group are particularly large owing to the additional interaction with a zeolite through the hydrogen atom of this group.

Conclusion

Determination of the heats of adsorption of substances of different structure demonstrates the dependence of the interaction energy with a zeolite on the type of the exchange cation as well as on the structure of the adsorbed molecule. It is possible to evaluate the heat of adsorption of a complicated molecule by means of the increments of the functional groups forming this molecule. These measurements are still needed for molecules with branched chains, with π-bonds, and with conjugated bonds on zeolites of different structure and composition.

Literature Cited

(1) Avgul, N. N., Berezin, G. I., Kiselev, A. V., Lygina, I. A., Muttik, G. G., *Zh. Fiz. Khim.* **1957**, 31, 1111.
(2) Avgul, N. N., Bgazhba, N. K., Dobrova, E. S., Kiselev, A. V., *Z. Physik. Chem.*, in press.
(3) Avgul, N. N., Kiselev, A. V., Kurdyukova, L. Ya., Serdobov, M. V., *Zh. Fiz. Khim.* **1968**, 42, 188.
(4) Barrer, R. M., *J. Colloid Interface Sci.* **1966**, 21, 415.
(5) Barrer, R. M., Reucroft, P. J., *Proc. Roy. Soc.* **1960**, A258, 431.
(6) Bezus, A. G., Dreving, V. P., Kiselev, A. V., *Zh. Fiz. Khim.* **1964**, 38, 947.
(7) Bezus, A. G., Kiselev, A. V., Sedlachek, Z., *Zh. Fiz. Khim.* **1969**, 43, 1224.
(8) Bezus, A. G., Kiselev, A. V., Sedlachek, Z., Pham Quang Du, *Trans. Faraday Soc.*, in press.
(9) Dzhigit, O. M., Karpinsky, K., Kiselev, A. V., Melnikova, T. A., Mikos, K. N., Muttik, G. G., *Zh. Fiz. Khim.* **1968**, 42, 198.

(10) Dzhigit, O. M., Kiselev, A. V., Mikos, K. N., Muttik, G. G., Rakhmanova, T. A., *Trans. Faraday Soc.*, in press.
(11) Dzhigit, O. M., Zhdanov, S. P., Kiselev, A. V., Melnikova, T. A., Mikos, K. N., Muttik, G. G., *Zh. Fiz. Khim.* **1967**, 41, 1431.
(12) Kiselev, A. V., ADVAN. CHEM. SER. **1971**, 102, 37.
(13) Kiselev, A. V., *Discussions Faraday Soc.* **1965**, 40, 228.
(14) Kiselev, A. V., *Zh. Fiz. Khim.* **1967**, 41, 2470.
(15) Kiselev, A. V., Kubelkova, L., Lygin, V. I., *Zh. Fiz. Khim.* **1964**, 38, 2408.
(16) Kiselev, A. V., Lopatkin, A. A., "Molecular Sieves," p. 252, Society of the Chemical Industry, London, 1968.
(17) Kiselev, A. V., Muttik, G. G., *Zh. Fiz. Khim.* **1961**, 35, 2153.
(18) Kiselev, A. V., Yashin, Ya. I., "Gas-Adsorption Chromatography," Plenum, New York, 1969.

RECEIVED February 11, 1970.

Discussion

A. V. Kiselev (Lomonosov State University, Moscow, USSR): Figure 1 shows that the difference in positions of alkali cations in sites S_{II} and S_{III} changes very much from Li^+ to K^+ and to a less extent from K^+ to Cs^+. This illustrates the difference in the electrostatic field distribution which produces the difference in the heats of adsorption of water indicated on

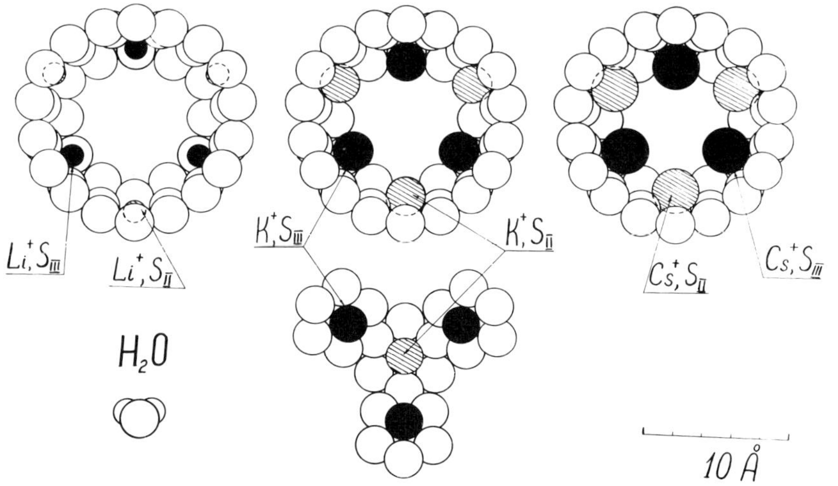

Figure 1. Scheme of the arrangement of the cations Li^+, K^+, and Cs^+ at S_{II} sites (hatched circles) and S_{III} sites (filled circles). The other projection of that part of the lattice of the KX zeolite containing a six-membered oxygen ring as well as a model of a water molecule are also shown on the same scale (indicated).

Figure 2, p. 187. It seems that the intermediate "wave" on the heat of adsorption *vs.* amount of adsorption curve for KX zeolite is connected with the change in the distribution of water molecules in the zeolite cavities. A detailed discussion of these results is given in the paper by O. M. Dzhigit, A. V. Kiselev, K. N. Mikos, G. G. Muttik, and T. A. Rakhmanova, accepted for publication in *Trans. Faraday Soc.*

55

Counterdiffusion of Liquid Hydrocarbons in Type Y Zeolites

CHARLES N. SATTERFIELD and JAMES R. KATZER[1]

Massachusetts Institute of Technology, Cambridge, Mass.

The nature of the cations present in a zeolite can have a marked effect upon the rate of intracrystalline counterdiffusion, as shown by studies with several selected aromatic hydrocarbons in a series of ion-exchanged forms of the type Y zeolite. For 1-methylnaphthalene diffusing from type Y into bulk cumene, the desorptive diffusion coefficients vary by 2 orders of magnitude over different ion-exchanged forms in the order:

$$\text{NaY} < \text{CaY} < \text{SK–500} < \text{CeY} < \text{HY}$$

The results are interpreted in terms of the size of the diffusing molecule and the effect of the cation upon the pore size of the zeolite. Counterdiffusion of the molecules studied occurs readily in the various forms of type Y zeolite, but molecule–molecule interactions between the counterdiffusing molecules have a pronounced effect upon the diffusion rate.

In most applications of zeolites, it is necessary for molecules to be able to diffuse into or out of their fine pore structure, and in many of these applications, particularly catalysis, the counterdiffusion of at least 2 different kinds of molecules occurs. The rates of these diffusion processes can have a profound effect upon the apparent activity and selectivity of zeolitic catalysts (*21*) and upon such characteristics as dispersion and sharpness of separation in the use of zeolites in separation and purification processes. The state of knowledge of intracrystalline diffusion in zeolites is reviewed by Barrer in a paper for this symposium (*4*). Little is known about unidirectional diffusion in zeolites of substances of indus-

[1] Present address: University of Delaware, Newark, Del.

trial interest such as hydrocarbons, and no information seems to be available on counterdiffusion except for the self-diffusion of water (*4, 5*). Limited information is available on diffusion in zeolites when contacted with a liquid (*1, 2, 3, 19*) and apparently none in the type X and Y zeolites, which are of importance in catalysis.

The pore structure of type Y is described, for example, by Breck and Flanigen (*9*). It consists essentially of a three-dimensional array of large cavities with a diameter of about 12 A interconnected by pore apertures with diameters of about 8 A. These dimensions vary slightly with the Si/Al ratio in the zeolite and the number and kinds of cations present.

The main objective of this study was to determine the counterdiffusion characteristics of selected liquid hydrocarbons—benzene, cumene, 1-methylnaphthalene, and 2-ethylnaphthalene—in several ion-exchanged forms of the type Y zeolite and thus to begin to provide an understanding of the role which diffusion may be playing in various applications of zeolites. This is a part of the larger, theoretical problem of understanding the physics of molecular motion inside pores with diameters which approach the diameter of the molecules diffusing in them.

Experimental Materials and Methods

Materials Used. The NaY zeolite and an ion-exchanged form of it, SK-500, were supplied by Union Carbide Corp., Linde Division, in the form of uncalcined powder. The SK-500 (Lot Number 12506-39) is a rare earth–ammonium exchanged type Y zeolite and had not been activated previously or calcined in its preparation. The calculated unit cell formula was

$$(Re^{3+})_{8.8}\ (NH_4^+)_{21.1}\ (Na^+)_{8.3}\ [(AlO_2)_{55.7}\ (SiO_2)_{136.3}]\ ZH_2O$$

According to the analysis supplied by Linde, the rare earths were predominantly lanthanum. An independent sodium analysis agreed with that supplied by Linde.

The NaY zeolite powder (Lot Number 12, 119-69) had a calculated formula of

$$Na_{57}\ [(AlO_2)_{57}\ (SiO_2)_{135}]\ \cdot\ ZH_2O$$

The NaY was divided into 4 equal portions. One of these was used as supplied; the other 3 were ion-exchanged to either the calcium, cerium, or ammonium form. Ion exchange was carried out by a combination batchwise–continuous procedure at about 80°C with a large excess of exchange solution. These were prepared from reagent grade chlorides and comprised 5 wt% $CeCl_3$, 10 wt% $CaCl_2$, or 10 wt% NH_4Cl, respectively, in distilled water. After exchange, the zeolites were washed with distilled water until no trace of the chloride ion appeared in the wash water. The degree of exchange based on the number of sodium cations replaced per unit cell was 89% for CaY, 97% for NH_4Y, and 70% for CeY. The low degree of exchange for cerium probably resulted from the inability of the hydrated cerium ion to enter the sodalite units through the six-membered oxygen rings and replace the 16 sodium cations per unit cell residing there (*22*).

After the standard activation procedure (described below), the NaY and SK-500 remained highly crystalline. The degree of crystallinity as observed by J. F. Charnell at the Mobil Oil Co. equalled the maximum found there for other samples of each zeolite. Neither sample contained appreciable amorphous matter (*26*). The benzene sorption capacity after ion exchange and activation of each of the type Y zeolites was essentially the same and equal to the value reported for NaX (*8*), further indicating that these procedures did not result in any marked changes in the crystal structure. However, use of a slightly larger molecule such as triethylamine would have been a more critical test.

The diffusion characteristics of a hydrocarbon can be affected markedly by traces of impurities, so considerable care was devoted towards obtaining and maintaining materials of highest available purity. Several sources of supply were examined for each material, and the availability of a particular compound in high purity was a criterion in the choice of diffusants. Each hydrocarbon was examined for impurities by gas chromatography and then stored in the dark over a mixture of freshly activated 4A and 13X zeolite pellets. The benzene, from Fischer, was thiophene-free and 99.944 mole % pure, the only measurable impurities being methylcyclopentane and an unknown compound of higher molecular weight. The cumene was from Matheson, Coleman and Bell and was 99.953% pure, with the only measurable impurity being a trace of benzene. Both substituted naphthalenes were from Columbia Organic Chemicals Co. The 2-ethylnaphthalene (2-EN) was 99.92 mole % pure and contained only 1 impurity appearing before the 2-EN peak and a second impurity some distance beyond the 2-EN peak. The 1-methylnaphthalene (1-MN) was 99.39 mole % pure and contained 3 impurities appearing before the 1-MN peak and 1 appearing after it. No attempt was made to identify these impurities.

Apparatus and Procedure. The zeolite in the form of single crystals was activated to remove water, and also ammonia in the case of the SK-500 and NH_4Y, from the pore structure. The activation was carried out by slowly heating the zeolite spread out in a layer about 5 mm thick to 500°C at 0.5°C per minute either in a vacuum or in a stream of predried air (total activation time = 18 hr). The SK-500 was always activated in an air stream, whereas all other zeolites were activated in a vacuum. The NH_4Y thus activated is converted to an HY form. After activation, the zeolite was saturated from the vapor phase with the hydrocarbon whose desorptive diffusion rate was to be measured, saturation conditions being such that essentially complete saturation was achieved. With benzene and cumene, the saturation was carried out at room temperature, whereas with 1-methyl and 2-ethylnaphthalene, the temperature was 46°C because of their low vapor pressure.

Each diffusion run was carried out in an apparatus consisting of a stirred, 500-ml Morton flask in a constant temperature bath (± 0.2°C). At zero time, the zeolite saturated with the desired hydrocarbon was placed in the stirred flask which contained a known quantity of a second, liquid hydrocarbon. Samples of about 1 ml were removed from the stirred flask at predetermined time intervals by a hypodermic needle and syringe, and the zeolite was immediately removed from the mixture by forcing it through a Millipore filtering unit. Precautions were taken to ensure that the sample withdrawn was representative of the mixture in the flask (*14*). The composition of the hydrocarbon phase was determined by gas chromatography.

The desorptive diffusion process was assumed to follow Fick's second law in spherical geometry. Diffusion can occur in 3 dimensions in this pore structure, and electron micrographs showed that the particles were nearly spherical in shape. The experimental results were fitted to the solution of the diffusion equation for the proper initial and boundary conditions (*4, 10, 12, 14, 20*), and point values of D_{eff}, the effective diffusion coefficient, were determined for the initial portion of the data and at 25, 60, and 75% of the approach to equilibrium. Details are given by Katzer (*14*). The ratio of the value at 25% of equilibrium to that at 75% gives a measure of the degree of departure from Fick's law. The type Y zeolite particles had a rather narrow particle size distribution, with three-fourths of the particles falling in the range of particle diameter of 0.65 to 1.25 microns. The arithmetic mean radius of 0.55 micron was used for the characteristic diffusion length in the diffusion equation. There was no discernible difference between the particle size distribution of the NaY and that of the SK-500.

Results and Discussion

Counterdiffusion of cumene and 1-MN occurred readily in type Y zeolite, as shown by several studies. The 1-MN is selectively adsorbed relative to cumene; thus, when the zeolite was initially saturated with cumene and placed in 1-MN, essentially 100% of the cumene diffused out; but when the zeolite was saturated with 1-MN and placed in cumene, only about 74% of the 1-MN diffused out. The same end point was also reached when SK-500 saturated with cumene was placed in a mixture of 1-MN and cumene in the proper ratio. This selective adsorption equilibrium value was essentially independent of temperature and, except for the cerium form of type Y, was independent of the nature of the cation

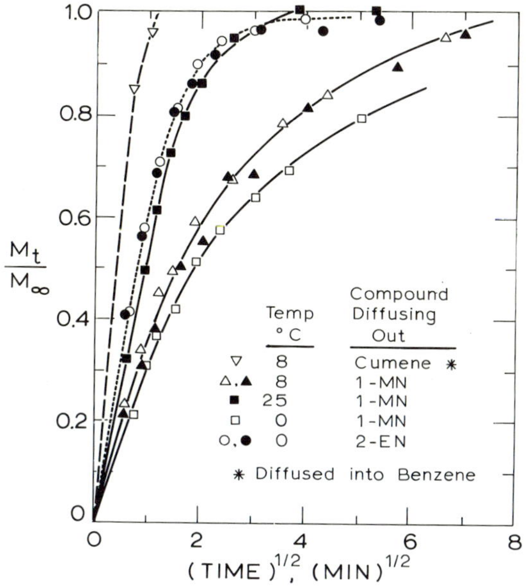

Figure 1. Diffusion rates from SK-500 into cumene

within the structure. This behavior is in direct contrast to the system of benzene–cumene in H-mordenite, in which counterdiffusion of benzene and cumene does not readily occur in the H-mordenite pores (*14*).

Diffusion Coefficients. Figure 1 presents diffusion rates from SK-500 of cumene into benzene and of 1-MN and 2-EN into cumene. The curves show the reproducibility of the data and the effect of temperature and of the nature of the counterdiffusing species upon the diffusion rate. Table I summarizes calculated effective diffusivities in SK-500 from these and other runs. The rate of diffusion of benzene or cumene from SK-500 into the opposite hydrocarbon was too rapid to allow an accurate measurement to be made, and the quoted value of 1.4×10^{-11} cm²/sec is an approximation. Larger molecules such as the substituted naphthalenes diffuse out of type Y much more slowly. The rate of diffusion of 1-MN at 0° or 25°C (into cumene) is of the right order of magnitude to allow the effects of cation exchange to be discerned readily. The experimental results indicated, as did also mass transfer calculations (*14*), that mass transfer rates from the particle surface to the bulk liquid were an insignificant resistance and therefore that diffusion within the type Y zeolite particles themselves is the rate-controlling step. Zeolite crystals tend to agglomerate in many liquids, but diffusional resistance through the macropores between the crystals in an agglomerate was apparently negligible.

Table I. Desorptive Diffusion

Compound Desorbing	*Saturated Zeolite Placed In*	*Temp., °C*
Cumene	Benzene	8
1-MN	Cumene	25
1-MN	Cumene	8
1-MN	Cumene	0
2-EN	Cumene	8
2-EN	Cumene	0

At the stirring rate of about 600 rpm, visual observations including the rate of settling after stirring indicated that the agglomerates were less than 0.1 mm in size, and for this size essentially complete interchange through the macropores will occur in a time much less than that required for the diffusion effects observed here. Furthermore, there seemed to be no appreciable effect of the nature of the hydrocarbon on degree of agglomeration, although it had a marked effect on the diffusion rates.

Figure 2 presents diffusion rates from NaY of cumene into benzene or into 1-MN. These and other results were used to calculate the diffusion coefficients reported in Table II. Tables I and II show that the

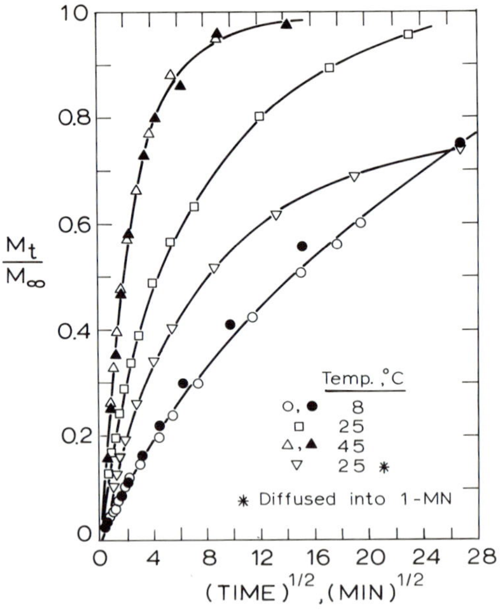

Figure 2. Cumene diffusion rates from NaY into benzene

Coefficients from SK-500

$D_{eff} \times 10^{13}$, Cm^2/Sec				
$M_t/M_\infty \rightarrow 0$	$M_t/M_\infty = 0.25$	$M_t/M_\infty = 0.60$	$M_t/M_\infty = 0.75$	$D_{0.25}/D_{0.75}$
			~140	
16.6	16.6	16.5	14.2	1.17
4.31	4.06	3.07	2.66	1.52
3.09	2.96	1.72	1.12	2.64
40.3	40.3	40.3	39.2	1.03
23.8	23.8	22.8	21.1	1.13

desorptive diffusion coefficients for cumene diffusing into benzene vary by over 3 orders of magnitude between the forms of type Y represented by NaY and SK-500. Likewise, for 1-MN diffusing into cumene, the coefficient varies by nearly 2 orders of magnitude between NaY and HY.

Except for situations in which the diffusion rate was very rapid, the maximum deviation in the diffusion coefficients was about 10%. The ratio $D_{0.25}/D_{0.75}$ is the ratio of the effective diffusion coefficient at $M_t/M_\infty = 0.25$ to that at 0.75. Rudloff (*20*) has shown that for a particle size distribution he studied in which 85% of the particles fell within a range in which the particle diameter varied by a factor of 2—slightly narrower than our distribution for type Y (*14*)—the ratio of the effective diffusion coefficient (that observed for the group of particles) at $M_t/M_\infty = 0.25$ to that at $M_t/M_\infty = 0.75$ is 1.20. The value of the ratio for these 2 M_t/M_∞ points ($D_{0.25}/D_{0.75}$) for SK-500 as found here is close to this value (except for the low-temperature run), which indicates that our desorptive diffusion studies with SK-500 fit the Fick's law diffusion model fairly well. The same is true for those with HY and CeY, but the deviations from ideality become quite large for those with CaY and NaY (*see* Table II). It is apparent that Fick's law is followed much more closely when faster diffusion rates are being measured, whether the faster rate is caused by a higher temperature or by the particular combination of diffusant and form of zeolite. Plausibly, when the size of the diffusing molecules is very close to the size of the passageways, a very slight degree of irregularity of structure or slight variation in the local degree of interaction between diffusant molecule and zeolite can have a marked blocking effect. For those situations in which the diffusion rate is quite rapid—*e.g.*, 2-EN desorption from SK-500 at 0° and 8°C and 1-MN at 25°C—the value of M_t/M_∞ at the first data point was quite high, resulting in forced inaccuracies in $D_{0.25}$ and therefore in $D_{0.25}/D_{0.75}$.

Effect of Cation in Zeolite. The diffusion coefficients for 1-MN into cumene for the ion-exchanged zeolites vary by about 2 orders of magnitude from NaY to HY in the following order:

$$NaY < CaY < SK\text{–}500 < CeY < HY$$

The change in the nature of the cations within the zeolite structure is undoubtedly the cause of this large variation in the diffusion rate. The important structural parameters, bond lengths, and bond angles are essentially the same for dehydrated NaY (*13*), rare earth faujasite (*17*), and hydrogen faujasite (*16*). Olson (*15*) reported significant variations in the lattice parameters of CaX, variations which should result in a reduced pore diameter; and the type X zeolite undergoes a significant reduction in the effective pore diameter upon calcium exchange and activation. However, Breck and Flanigen (*9*) found that the type Y zeolite does not undergo a similar reduction in the effective pore size upon calcium exchange and activation. It thus appears that even for CaY the extent of change in the structural parameters is small.

In dehydrated NaY, the population parameter of Na in the type II sites is 0.947 (*13*). There is therefore 1 sodium cation resting in almost every six-membered ring of the free hexagonal faces of the sodalite units (type II sites) and probably protruding into the main sorption cavity near the pore aperture. The presence of such cations could hinder the diffusion of molecules through the pore apertures and result in a diffusion rate less than it would be without the cations. In dehydrated calcium faujasite, the type II sites have a population parameter of 0.50 (*18*): that is, about one-half as many calcium cations as sodium cations are present in positions which protrude into the main sorption cavities near the pore apertures. Since calcium has approximately the same cation radius as sodium, this should result in a reduced resistance to diffusion through the pore apertures and an increased diffusion rate, as found. Rare earth cations exchanged into the type Y zeolite move into the sodalite cages upon activation (*17, 23*). If complete exchange has occurred, the main

Table II. Desorptive Diffusion Coefficients

Zeolite	*Compound Diffusing*	*Saturated Zeolite Placed In*	*Temp., °C*
NaY	Cumene	Benzene	8
NaY	Cumene	Benzene	25
NaY	Cumene	Benzene	45
NaY	Cumene	1-MN	25
NaY	1-MN	Cumene	25
NaY	1-MN	Cumene	0
CaY	1-MN	Cumene	0
CeY	1-MN	Cumene	0
HY	1-MN	Cumene	0

pore structure of the type Y zeolite should be free of all rare earth cations, thus permitting diffusion to occur more rapidly and more easily than when cations are present. The NH_4Y upon heating becomes HY, and it is reasonable to assume that the hydrogen cations do not protrude far out into the pore structure since they probably bond quite closely and tightly to the framework atoms and are small in size. They thus should not greatly obstruct the diffusion of molecules.

The experimental results are explained well in terms of the above considerations concerning structure. For 1-MN desorbing into cumene, the diffusion coefficients in NaY are about one-tenth those for SK-500, which contains mainly hydrogen cations and rare earth cations, and the activation energy for diffusion from NaY is considerably higher than that for diffusion from SK-500. For CaY, the diffusion coefficients are over 5 times those for NaY. For the rare earth exchanged Y in which all of the rare earth cations should have moved from the main pore structure, the diffusion coefficients are more than 5 times greater than those for CaY, and NH_4Y (HY) shows an even greater rate of diffusion.

SK-500 contains predominantly hydrogen cations, and it is not clear why the diffusion coefficients in it were much less than in CeY or HY. All the other forms of Y were prepared from the same batch of NaY, whereas the SK-500 was from a different batch, so batch variations may have been significant. Barry and Lay (*6, 7*) have shown that the positions of cations within the structure of X and Y zeolites depend upon the relative proportions and types of other cations within the structure and upon the prior activation conditions. Since the CeY, HY, and SK-500 each had a different cation balance, this could have caused the rates to differ. Although the SK-500 was activated in a stream of predried air rather than in a vacuum as were the other zeolites, this probably was

from NaY and Ion-Exchanged Zeolites

$D_{eff} \times 10^{13}$, Cm^2/Sec				
$M_t/M_\infty \rightarrow 0$	$M_t/M_\infty = 0.25$	$M_t/M_\infty = 0.60$	$M_t/M_\infty = 0.75$	$D_{0.25}/D_{0.75}$
0.126	0.101	0.0553	0.0406	2.49
1.25	0.983	0.374	0.289	3.40
4.73	4.68	3.39	2.31	1.98
0.489	0.392	0.075	0.016	24.2
2.95	2.91	2.22	1.80	1.62
0.360	0.242	0.105	0.079	3.07
1.84	1.32	0.718	0.608	3.04
10.7	10.7	9.98	8.13	1.32
20.0	20.0	17.1	10.6	1.87

not a significant variable. Diffusion coefficients varied by only 20% for 1-MN diffusing from SK-500 into cumene when it was activated according to the 2 different procedures and saturated at 100° rather than 46°C. The cations in the type Y zeolite do not affect greatly the effective pore diameter as determined by adsorption capacities (*9*), but diffusion rates are probably extremely sensitive to slight variations in structure when the diffusing molecule is nearly the same size as the minimum pore aperture.

Effect of Nature of Diffusing Molecule. The data in Table I also allow comparison to be made of the effect of the nature of the diffusing molecule on the intracrystalline diffusion coefficient. Benzene diffuses from SK-500 into cumene as rapidly as does cumene into benzene. The large variations in the diffusion coefficient for benzene, 2-EN, and 1-MN, each into cumene, seem to be caused by the variation in the critical diameter of the desorbing molecule. The calculated value for benzene and cumene is 6.29 A, that of 2-EN is 6.96 A, and that of 1-MN is 7.70 A (*14*) based on bond lengths, bond angles, and ionic radii. Other methods of calculation will give slightly different values, but the order remains the same. However, some factors other than critical diameter seem to be involved. In SK-500 at 8°C, cumene diffuses into benzene much more rapidly than does 1-MN into cumene, yet in NaY at 25°C (or 8°C) cumene diffuses into benzene more slowly than does 1-MN into cumene at 25°C (or even 0°C) (*see* Tables I and II). Small amounts of impurities in the 1-MN may interact in a different fashion with different forms of Y sieve to give rise to various blocking effects.

Activation Energies. Table III lists activation energies from Arrhenius plots of values of D_{eff} calculated from the initial rate or at $M_t/M_\infty = 0.60$. The considerable deviations from Fick's law that exist in some cases are reflected in a considerable variation in calculated activation energy for different portions of the desorption process, so one must be cautious in interpreting the values obtained. Substances sorbed in zeolites exhibit somewhat liquid-like properties, but it is apparent that the activation energies here are much higher than those for bulk liquid-phase diffusion. Further, the highest activation energies are found with the systems having the lowest diffusion rates, which would be expected with a highly restricted form of diffusion. Tuang *et al.* (*24*) have reported the heat of adsorption of cumene vapor on NaY to be 19.3 kcal/mole, which indicates a value of about 11 kcal/mole for liquid. Since the activation energy for desorptive diffusion of cumene is considerably greater than the heat of adsorption, corrected to liquid-phase conditions, this again suggests that the activation energy observed reflects primarily a diffusion process rather than one of desorption.

Table III. Activation Energies

		Activation Energy, Kcal/Mole	
Zeolite	*Diffusing System*	M_t/M_∞–*Initial*	$M_t/M_\infty = 0.60$
SK-500	1-MN into cumene	11.3 ± 1.5	15.0 ± 1.5
SK-500	2-EN into cumene	9.95	11.3
NaY	Cumene into benzene	17.4	19.8
NaY	1-MN into cumene	13.8	19.4

Other Comments. Cheng (*11*) has recently shown that the rate of sorption of pure, liquid cumene into freshly activated NaY is extremely rapid even at −15°C. Sorption equilibrium was essentially achieved by Cheng in less than 30 seconds at −15°C, whereas in this study more than 1 week was required for desorptive counterdiffusion to reach equilibrium (cumene desorbing from NaY into benzene at 8°C) which is equivalent to a ratio of the diffusion coefficient for sorption to that for the desorptive counterdiffusion of cumene greater than 2000. This value is much larger than the ratio of the calculated diffusion coefficient for the adsorption of water to that for the self-diffusion of water in chabazite, reported by Barrer and Fender (*5*) to be 38, and it therefore appears that there is no relation between the rate of adsorption and that of counterdiffusion. This large reduction in the rate from pure sorption to counterdiffusion was also found for 1-MN and appears to be the result of large interactions between the counterdiffusing molecules. This hypothesis is further supported by the fact that the rate at which the preadsorbed hydrocarbon desorbs is affected by the nature of the species diffusing into the pore structure (counterdiffusing to it). This is clearly shown by the 2nd and 4th entries in Table II, in which all conditions were the same except for the hydrocarbon into which the zeolite saturated with cumene was placed. The same effect occurred with cumene diffusing from SK-500. This behavior is contrary to Knudsen diffusion, in which each flux is independent. It is also different from normal surface diffusion, for which Whang (*25*) has shown that the flux in one direction is independent of the flux in the opposite direction for simple molecules surface diffusing on 96% silica glass.

Conclusions

Type Y zeolite has a much more open pore structure than most other zeolites, but counterdiffusion through its pore structure of small aromatic hydrocarbons can still be quite slow and depends strongly upon the nature of cations present. Counterdiffusion in the type Y zeolite appears to be modeled best as diffusion over periodic high energy barriers which are the pore apertures joining the supercages. The activation energy is that required for a molecule to jump through the pore aperture from one supercage to another. The presence of cations near these pore apertures increases the resistance to diffusion through them, increasing the height of the energy barrier and resulting in slower counterdiffusion rates and higher activation energies. The nature of the species diffusing in one direction greatly affects the flux of the species counterdiffusing to it. The initial presence of molecules within the zeolite pore structure impedes the diffusion rate of a second substance by orders of magnitude below those observed when the pore structure is initially empty.

The ramifications of these findings are many. Unlike Knudsen difusion, the rate of diffusion in one direction is affected markedly by the opposite flux. Adsorption or desorption measurements cannot be used to approximate counterdiffusion rates; these must be determined independently. They are a function of the nature of the zeolite, the type of cation within the pore structure, and the nature of the counterdiffusing species.

Acknowledgment

The authors acknowledge the financial support of the National Science Foundation under Grant GK-1707 and that of the Hertz Foundation to JRK. Linde Division of the Union Carbide Corp. supplied the zeolite, and J. Charnell of Mobil Oil Corp. performed the x-ray diffraction studies reported.

Nomenclature

D_A = diffusion coefficient at $M_t/M_\infty = A$
D_{eff} = effective diffusion coefficient, cm^2/sec
2-EN = 2-ethylnaphthalene
M_t = gram-moles of material diffused out of zeolite at time t
M_∞ = gram-moles of material diffused out of zeolite after infinite time
1-MN = 1-methylnaphthalene
t = time, sec

Literature Cited

(1) Aleksandrov, G. G., Larionov, O. G., Chmutov, K. V., *Russ. J. Phys. Chem.* **1967**, 41, 805.
(2) Aleksandrov, G. G., Larionov, O. G., Chmutov, K. V., Yudilevich, M. D., *Russ. J. Phys. Chem.* **1967**, 41, 807.
(3) Aleksandrov, G. G., Larionov, O. G., Chmutov, K. V., *Russ. J. Phys. Chem.* **1967**, 41, 1107.

(4) Barrer, R. M., ADVAN. CHEM. SER. **1971,** 102, 1.
(5) Barrer, R. M., Fender, B. E. F., *J. Phys. Chem. Solids* **1961,** 21 (112), 1.
(6) Barry, T. I., Lay, L. A., *J. Phys. Chem. Solids* **1968,** 29, 1395.
(7) Barry, T. I., Lay, L. A., *Nature* **1965,** 208, 312.
(8) Breck, D. W., *J. Chem. Educ.* **1964,** 41, 678.
(9) Breck, D. W., Flanigen, E. M., "Molecular Sieves," p. 47, Society of Chemical Industry, London, 1968.
(10) Carman, P. C., Haul, R. A., *Proc. Roy. Soc., Ser. A* **1954,** 222, 109.
(11) Cheng, C. S., M.I.T., Cambridge, Mass., personal communication, 1969.
(12) Crank, J., "The Mathematics of Diffusion," Oxford University Press, London, 1956.
(13) Eulenberger, G. R., Shoemaker, G. P., Keil, J. G., *J. Phys. Chem.* **1967,** 71, 1812.
(14) Katzer, J. R., Ph.D. Thesis, M.I.T., Cambridge, Mass., 1969.
(15) Olson, D. H., *J. Phys. Chem.* **1968,** 72, 4366.
(16) Olson, D. H., Dempsey, E., *J. Catalysis* **1969,** 13, 221.
(17) Olson, D. H., Kokotailo, G. T., Charnell, J. F., *J. Colloid Interface Sci.* **1968,** 28, 305.
(18) Pickert, P. E., Rabo, J. A., Dempsey, E., Shoemaker, V., *Proc. Intern. Congr. Catalysis, 3rd, Amsterdam, 1964* **1965,** 1, 714.
(19) Roberts, P. V., York, R., *Ind. Eng. Chem., Process Design Develop.* **1967,** 6, 516.
(20) Rudloff, W. K., Ph.D. Thesis, I.I.T., Chicago, Ill., 1965.
(21) Satterfield, C. N., "Mass Transfer in Heterogeneous Catalysis," M.I.T. Press, Cambridge, Mass., 1970.
(22) Sherry, H. S., *J. Colloid Interface Sci.* **1968,** 28, 288.
(23) Smith, J. V., Bennett, J. M., Flanigen, E. M., *Nature* **1967,** 215, 241.
(24) Tuang, Ho Chi, Romanovskii, B. V., Topchieva, K. V., Piguzova, L. I., *Kinetics Catalysis* **1967,** 8, 594.
(25) Whang, H. Y., Sc.D. Thesis, M.I.T., Cambridge, Mass., 1961.
(26) Wise, J. J., Mobil Oil Corp., Paulsboro, N. J., personal communication, 1968.

RECEIVED February 13, 1970.

Discussion

L. V. C. Rees (Imperial College, London): Have you studied counterdiffusion in X with exchange of Na^+ by Ca^{2+}? If not, can you give an explanation for the reduction in the effective window size in CaX compared with that of NaX?

J. R. Katzer: No, we have not studied counterdiffusion in the type X zeolite, but I am now in the process of doing so. The effect appears to be produced by the cations within the pore structure and resting near the pore apertures. Olson (Olson, D. H., *J. Phys. Chem.* **1968,** 72, 4366) indicated that structural variations occurred in the CaX form which careful examination shows could reduce the pore aperture. However, he has recently indicated (Olson, D. H., personal communication, 1970) that the pore aperture remains essentially the same upon exchange from the sodium to the calcium form. The effect must therefore be the result of the cations within the structure blocking the pore. Note that there are more calcium cations in the X than in the Y zeolite because of the lower Si/Al ratio. Nivikova *et al.* (*Kolloidn. Zh.* **1967**, 29, 573) have shown that diffusion in CaX is slower than in NaX.

O. M. Fuller (McGill University, Montreal, Quebec, Canada): What were the methods you used to determine that the resistance to mass transport between the crystal and the fluid was negligible?

J. R. Katzer: First variations in stirring speed had no effect upon desorption rate, but this would probably not have an effect for particles this small unless they agglomerated seriously. Observation indicated that they did not do so. The large span of rates observed, the very high activation energies (*e.g.*, 10–20 kcal/mole), and the fact that system changes such as changing the zeolite cation form or the hydrocarbon type produced marked changes in the diffusion rate observed all indicated that mass–transfer limitations between the particle surfaces and the bulk liquid were not rate-limiting. Furthermore, mass–transfer calculations assuming that the particles followed the streamlines of flow showed mass transfer from the particle surfaces to the bulk liquid was not rate-limiting, the calculated rate of mass transfer being greater than 1000 times the fastest rate observed.

P. E. Eberly, Jr. (Esso Research Laboratories, Baton Rouge, La. 70821): In evaluating diffusivities, one frequently finds that the parameters D and M_∞ interact. Thus, by changing M_∞, one can affect the value of D. In your results comparing the various forms of Y zeolite, could changes in M_∞ have affected the diffusivity values?

J. R. Katzer: The answer is no. The amount of preadsorbed hydrocarbon which desorbed from the presaturated zeolite at equilibrium, M_∞, was constant at a value of 73 ± 2% of the initially preadsorbed hydrocarbon in all cases except for the cerium form, for which the value was 64% of the initially preadsorbed material. Thus, the changes in M_∞ were not large enough to produce significant effects in D_{eff} value. Furthermore, the M_∞ used was the true equilibrium value as determined experimentally in all cases, and thus the value of D_{eff} found should be the same as that which should also be found if the run conditions were changed for a given zeolite such that at equilibrium the value of M_∞ was indeed changed from its previous value.

R. M. Barrer (Imperial College, London): In counterdiffusion systems such as you have examined, the volume flow of Component 1 into the crystals is not necessarily equal to the volume flow of the displaced Component 2 into the external solution, because the degree of filling of the crystals may change with composition of the sorbed fluid. Thus, for partially filled crystals, both flows should be measured, and two diffusion coefficients are involved.

J. R. Katzer: We were concerned with the question of equal volume rates of flow during counterdiffusion but did not attempt to measure the two flow rates; this would be extremely difficult in a freely and quite rapidly counterdiffusing system as was the case here. Note, however, that the total volume flow in each direction should have been essentially the same unless there were some strange phenomena which occurred at intermediate composition because it was found that the volume of pure compound sorbed at saturation based on the liquid density at sorption temperature was essentially the same for all components studied here.

R. B. Anderson (McMaster University, Hamilton, Ontario, Canada): Does the diffusivity vary with "driving force?" For example, suppose a sieve filled with an aromatic is placed in a large amount of nonaromatic hydrocarbon?

J. R. Katzer: I do not believe that the diffusivity would vary with "driving force" to any larger extent than the variation with M_t/M_∞ which we observed in this work and noted in the paper. This could easily be checked by starting with the zeolite pore structure equilibrated with a mixture of the two hydrocarbons involved and then measuring the approach to equilibrium for this system as was done in the above work. It would not be possible to change the hydrocarbon type on the outside and effect only a change in the driving force because, as we have clearly shown in this paper, the rate of diffusion out of the structure is highly dependent upon the nature of the hydrocarbon diffusing into the zeolite. Placing the saturated zeolite into a nonaromatic hydrocarbon changes the nature of the hydrocarbon diffusing into the structure in addition to possibly altering the driving force.

D. M. Ruthven and **K. F. Loughlin** (University of New Brunswick, Fredericton, N. B., Canada): We would like to re-emphasize the significance of the effects of crystal size distribution in the analysis of sorption curves. The variations in diffusivity which are reported in this paper show precisely the trend which is to be expected if the effect of crystal size distribution is considered. We feel sure that if crystal size distribution measurements are available, it should be possible to obtain the correct values of diffusivity using the type of analysis which we presented in our comment on the paper of Kondis and Dranoff.

J. R. Katzer: You are correct in your statement; as we indicated in our paper, the crystallite size distribution is responsible for the majority of the variation found for SK-500 under most conditions. However, your approach would not eliminate all of the variation found in the systems studied here and is a small correction relative to the far more significant variations found in the systems studied.

J. D. Sherman (Union Carbide Corp., Tarrytown, N. Y. 10591): The heat of adsorption (ΔH_{ads}) of interest is for high degrees of filling and is for the difference between the ΔH's of different molecules at high degrees of filling. Interpretation can be facilitated by comparison of E_{diff} and ΔH_{ads}—*i.e.*, the activation energies for diffusion and heats of adsorption for a series of pairs of counterdiffusing species.

J. R. Katzer: This comment is in response to a comment by Paul Eberly noting that the rates observed increase in zeolites exhibiting a higher heat of adsorption. However, Khudier *et al.* (*Bull. Acad. Sci. USSR, Div. Chem. Sci.* **1968**, 4, 694) found the differential heat of adsorption of C_6H_6 on HY to be lower than that on NaY over the entire range. In response to the first point of the above comments, I do think that heats of adsorption for low degrees of filling are probably also of importance here because a few molecules adsorbed on sites exhibiting a high heat of adsorption (strong adsorption) could greatly affect the diffusion rate. Your second comment, however, is well taken.

D. P. Roelofsen (Technological University, Delft, Netherlands): The sequence of the investigated molecular sieves regarding increasing desorption velocities is roughly similar to the expected sequence of decreasing adsorption affinities of the sieves for these large aromatic compounds. For instance, NaY, with its exposed cations in S_{III} sites, would be expected to adsorb naphthalenes stronger than CaY with its shielded cations in S_{II} positions. Therefore, I would like to suggest that the differences in desorption velocities may be partly explained by the differences in affinities of the sieves for the diffusing molecules. These differences could be studied, for instance, by measuring the adsorption equilibria of a dilute solution of a substituted naphthalene in a saturated hydrocarbon solvent and the different sieves.

J. R. Katzer: There should be no cations in the type III sites you refer to in the dehydrated form of either the NaY or CaY zeolite. Your comments on diffusion rates being affected by adsorption affinity is a very valid one, however, and Barrer has already noted in this conference that the stickiness of water is important to its diffusion rate. I think that there is no question but what sorbate–zeolite interactions of the nature you point out and others have a very large effect on the diffusion rate and should be studied in more detail.

56

Multicomponent Adsorption Isotherms on Hydrogen–Mordenite

JAMES I. JOUBERT and IMRE ZWIEBEL

Chemical Engineering Department, Worcester Polytechnic Institute, Worcester, Mass. 01609

Adsorption isotherms of oxygen, nitrogen, carbon dioxide, and sulfur dioxide on hydrogen–mordenite were measured at several temperatures in the range of 0°–100°C. The SO_2 and CO_2 had considerably greater affinity for the adsorbent than the O_2 and N_2. Using the pure-component data, multicomponent isotherms were predicted and compared with experimental results. The more strongly adsorbed species completely overwhelm the lesser adsorbed components (e.g., *SO_2* vs. *N_2). Wherever 2 species of approximately equal affinity are adsorbed* (e.g., *$CO_2 + SO_2$), the temperature sensitivity of the individual components influences the extent of the competition.*

The literature contains a relatively meager amount of multicomponent adsorption data for inorganic gases. The early literature was summarized by Young and Crowell (*8*), and since then, the bulk of the published data has dealt with mixtures of organic vapors or inorganic adsorbates at liquid nitrogen temperatures.

Motivated by the air pollution combat, our research was directed at the components making up a simulated exhaust gas, namely nitrogen, oxygen, carbon dioxide, and sulfur dioxide. The experiments were restricted to the 0°–100°C range at 1 atm absolute pressure.

First, pure-component isotherms were measured. As a consequence of preliminary screening in which such adsorbents as several commercial grade 5A sieves, 13X sieve, H-mordenite and activated carbon were compared, the hydrogen form of mordenite was selected as the adsorbent for the multicomponent experiments. This decision was based upon capacity, selectivity, and durability criteria, since practical application under rather severe conditions was one of the objectives.

Experimental

The apparatus used is a modification of the equipment employed by Lewis *et al.* (*6*). For pure components, it is a constant-volume unit; for gas mixtures, it operates at constant total pressure (nearly constant volume). It utilizes an internal recirculation device to expose the adsorbent continuously to a gas mixture of approximately known composition, and thus avoid localized concentration variations which could arise as a result of differences in the diffusivities of the various species.

The pure-component isotherms were determined by admitting successive increments of adsorbate into the chamber and measuring the pressures. For the multicomponent measurements, the system pressure is maintained constant by a manostat. While equilibration takes place, the gas is circulated in a loop containing the adsorption chamber and a large reservoir where the concentration is approximately constant. At the termination of the experiment, the adsorption chamber is isolated, and the final composition in the reservoir is measured by withdrawing a sample and analyzing it with a mass spectrometer. Then, by desorbing, measuring the total amount of desorbate, analyzing it, and correcting for the void space gas content, the desired adsorbed phase quantities are obtained.

The adsorbents are activated *in vacuo* below 10^{-5} torr at 350°–400°C for 16 hours. Desorptions are carried out by repeatedly connecting the adsorption chamber to an evacuated vessel from which the desorbate is collected in a mixing chamber. Desorptions usually are carried out at the adsorption temperature, and completeness of removal is checked by material balance; it is usually within 2%. Occasionally, to speed up the procedure, desorptions may be carried out at elevated temperatures.

The hydrogen–mordenite (unit cell; hydrated $H_8Al_8Si_{40}O_{96} \cdot 24H_2O$) used in this study was provided by the Norton Co., Worcester, Mass., in the form of 1/16-inch pellets fabricated without a binder. This material is characterized by parallel 12-membered rings of silica–alumina tetrahedra forming pores with effective diameters of 7–9A; smaller cavities occur in the walls of the large channels. Mordenite has reported B.E.T. surface area of 400 to 500 m^2/gram (*3*); synthesis and other characteristics of this material are described well elsewhere (*1, 5*).

Correlations

While ultimately the objective of our project will be the theoretical analysis of the behavior of complex mixtures, at this point we only can correlate our results with previously developed multicomponent models. With the objective of predicting the total amount adsorbed and the adsorbed phase composition from no more than pure-component data, only a few specially selected systems have been used in previous experimental programs to verify the various models. Hence, these relationships, which usually were derived from theoretical considerations, have not been tested sufficiently yet. Thus, they all can be classified as empirical models, especially as they have been modified to make them readily applicable (*2, 4, 6, 8*).

Correlations with the model based on thermodynamic principles developed by Myers and Prausnitz (*7*), hereafter referred to as the ideal solution model, are presented in this paper. The model is based upon the assumption that Raoult's law is applicable to the adsorbed phase, and the spreading pressure of the mixture and those of the pure components are set equal at the same surface coverage. Unlike some of the other models, this one does not require that the pure components adhere to any specific pure-component behavior. Hence, for example, the SO_2, which was correlated by the Langmuir equation, and CO_2, which best fitted the Freundlich equation, could be combined readily in predicting mixture characteristics. This model could be extended to higher coverages than any of the other relationships; however, the present work is restricted to the low coverage range. Finally, this work presents adsorption data on inorganic adsorbates with widely different characteristics in complex mixtures.

Results

Pure Components. Pure-component isotherms were measured for 4 gases, oxygen, nitrogen, carbon dioxide, and sulfur dioxide. The first 3 exhibited reversible sorption, while the SO_2 showed a rather broad hysteresis loop.

Relatively speaking, the affinity of the hydrogen–mordenite is greatest for the SO_2, then in decreasing order the CO_2, N_2, and O_2 follow. This can be explained by differences in the electronic configurations of these adsorbates; no "sieving" would be expected in any of these systems since the molecular diameter of these gas molecules is well below the published pore diameters. The sorption of all the species is likely to be physical, since the calculated isosteric heats are all below 5 kcal/mole.

The oxygen and nitrogen exhibited similar characteristics; they were both correlated well by the Freundlich isotherm. Their similar molecular structure might account for this, with differences in their polarizability being the major reason for their relative adsorbability.

The CO_2 and SO_2, having critical temperatures of 31.1° and 157.2°C, respectively (*vs.* −118.8°C for O_2 and −147.1°C for N_2), are more likely to adsorb by the so-called multilayer mechanism, and hence, in the region of the current experiments exhibit considerably higher capacities than the O_2 and N_2—as much as 15 times higher. In the experimental region, the CO_2 isotherms could be correlated by the Freundlich equa-

Table I. Correlation of Pure-Component Isotherms

P in torr, W in mmol/gram

Temp., °C	*0°*	*25°*	*36°*	*56°*	*78°*	*100°*
Oxygen; $W = KP^m$						
$K \times 10^4$	7.44	3.31	2.62	1.52	1.42	0.824
m	0.909	0.943	0.946	0.970	0.924	0.950
Nitrogen; $W = KP^m$						
$K \times 10^4$	17.4	7.56	5.28	2.55	1.78	0.836
m	0.841	0.870	0.891	0.936	0.919	0.977
Carbon Dioxide; $W = KP^m$						
$K \times 10^1$	5.51	3.70	2.17	1.13	0.491	0.220
m	0.257	0.277	0.341	0.405	0.492	0.565
Sulfur Dioxide; $W = WmbP/(1 + bP)$						
$b \times 10^3$	11.95	9.13	8.28	7.02	6.80	5.80
Wm	4.18	3.90	3.78	3.48	3.00	2.66

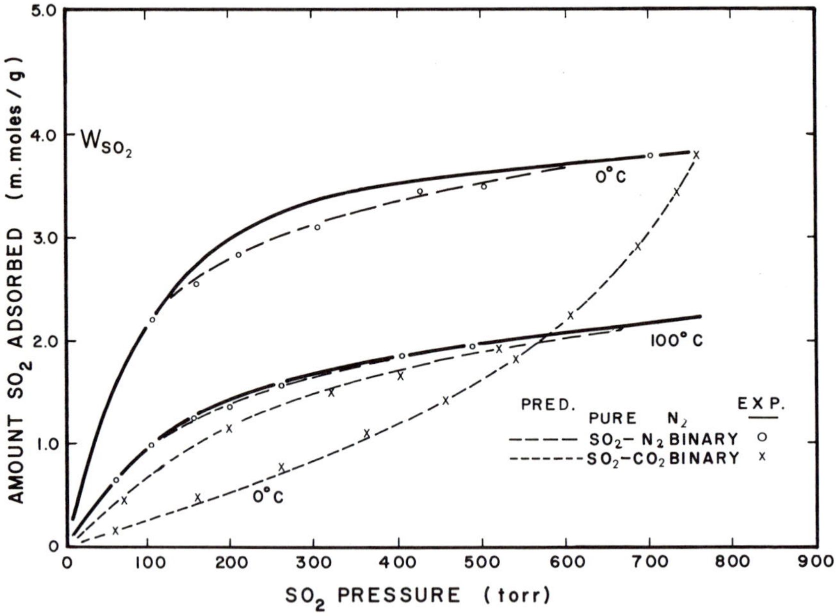

Figure 1. SO_2 adsorption isotherms

tion, although there were some deviations at low pressures. The SO_2 data seemed to be fitted best by the Langmuir equation. Attempts to fit the other pure components to the Langmuir equation or the B.E.T. equation were not successful; hence, the mixed Langmuir model or the B.E.T.

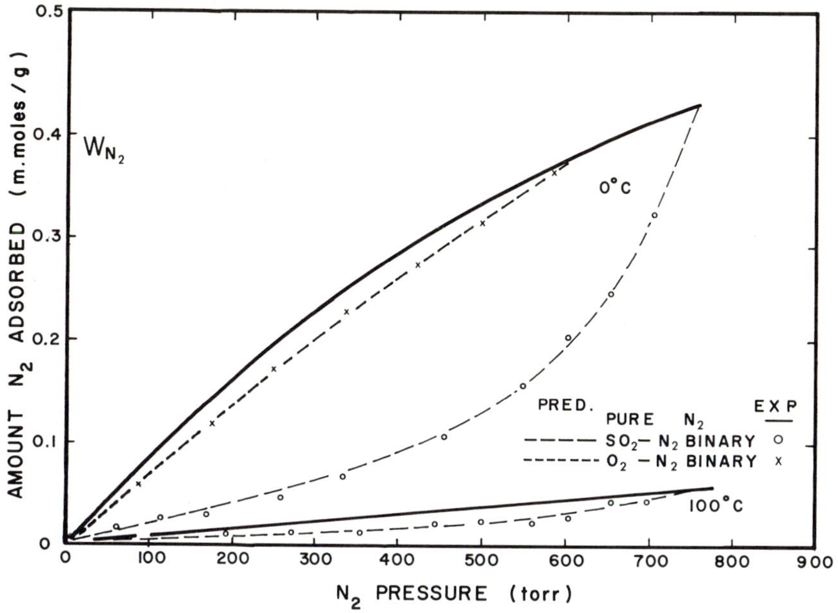

Figure 2. N_2 adsorption isotherms

mixture model was not used in this work. The results of these pure-component correlations are summarized in Table I.

Multi-Component. Only a representative selection of the many possible combinations of multicomponent equilibria involving the 4 gases used in the current program is presented in this paper.

First, the effects of the lightly adsorbed component on the more strongly adsorbed species are viewed. No significant effect was predicted, and this was confirmed experimentally. Figure 1 illustrates the amount of SO_2 adsorbed. The SO_2–N_2 binary system hardly deviates from the pure SO_2 isotherm. On the other hand, a very strong reverse effect was predicted. As shown in Figure 2, the presence of SO_2 significantly reduces the amount of nitrogen adsorbed. This is quite pronounced even at the very low SO_2 partial pressures where the amounts adsorbed of the 2 pure components is approximately equal. Similar behavior has been observed in the CO_2–N_2 and CO_2–O_2 binary systems.

The SO_2–CO_2 system is a bit more interesting (Figure 1); here the competition for adsorption sites is more intense, and deviations from the pure-component curves is far more significant. With an increase of temperature, the selectivity for SO_2 is greatly increased, so much so that around 30°C a reversal occurs and SO_2 becomes the preferred species.

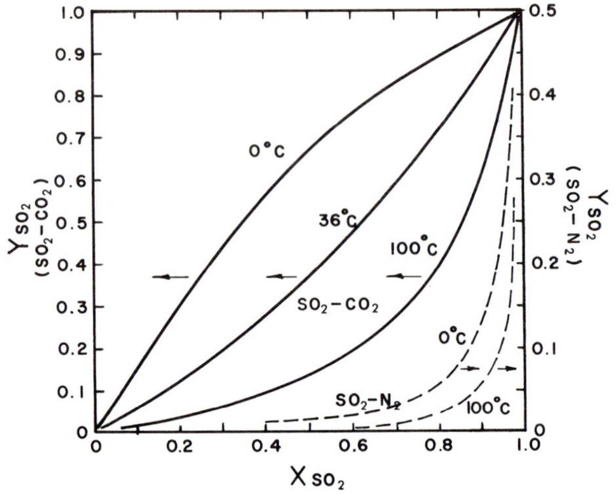

Figure 3. Equilibrium composition diagrams for SO_2–CO_2 and SO_2–N_2 binary mixtures

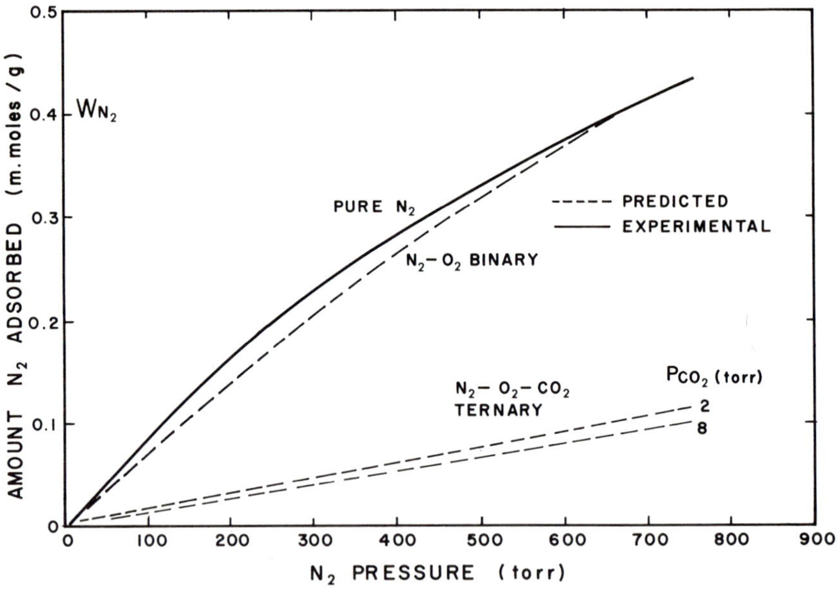

Figure 4. N_2 adsorption isotherms at 0°C

This is apparently because of the temperature dependence of the various components; the CO_2 is much more affected than any of the other components tested.

It can be noted from *Y-X* diagrams (Figure 3) that with increased temperatures the nitrogen adsorption selectivity is decreased with respect to SO_2. Similar decrease in selectivity was noted for the CO_2 with respect to the SO_2. This is encouraging with regard to air pollution combat, since exhaust gases contain large quantities of CO_2 and N_2 in addition to the undesirable SO_2.

Finally, Figure 4 shows an example of the O_2–N_2–CO_2 ternary isotherms at 0°C. The solid curve is pure N_2, the dashed curve is the binary O_2–N_2 system, while the remaining curves are the ternary cases with the amount of CO_2 indicated. Note that the CO_2 effect, even at very low concentrations, is far more significant than the O_2 effect on the N_2 sorption. The O_2–N_2–SO_2 system behaves similarly; the CO_2–SO_2–N_2, or the CO_2–SO_2–O_2, and the quaternary systems are not much different from the SO_2–CO_2 binary system. This is probably owing to the overbearing effect of the 2 "active" components with respect to O_2 and N_2.

Conclusions

Experimentally measured pure-component adsorption characteristics of O_2, N_2, CO_2, and SO_2 on H-mordenite were correlated to predict the behavior of multicomponent mixture of these gases. These correlations, based upon the relationships developed by Myers and Prausnitz, were successfully substantiated experimentally. The CO_2 and SO_2, which are the predominantly adsorbed components, controlled the fate of the multicomponent sorption. This prevailed even at the concentration levels where the pure-component data indicate comparable affinity for both the "strongly" and the "weakly" adsorbed species. Hence, indications are that adsorption may be effectively useful in exhaust gas cleanup processes. The temperature sensitivity of the pure components contributes significantly to the selectivity of the sieve for the various components, and the data obtained indicate that this also tends to favor the desired applications in pollution combat.

Nomenclature

P Pressure, torr
W Amount adsorbed, mmoles/gram
X Mole fraction of adsorbed phase
Y Mole fraction of gas phase
Subscripts denote components

Literature Cited

(1) Barrer, R. M., Peterson, D. L., *Proc. Roy. Soc.* **1964**, 280, 466.
(2) Danner, R. P., Wenzel, L. A., *A.I.Ch.E.* **1969**, 15, 515.
(3) Eberly, P. E., *J. Phys. Chem.* **1963**, 67, 2404.
(4) Hill, T. L., *J. Chem. Phys.* **1946**, 14, 268.
(5) Keough, A. H., Sand, L. B., *J. Am. Chem. Soc.* **1961**, 83, 3526.
(6) Lewis, W. K., Gilliland, E. R., Chertow, B., Cadogan, W. P., *Ind. Eng. Chem.* **1950**, 42, 1319.
(7) Myers, A. L., Prausnitz, J. M., *A.I.Ch.E. J.* **1965**, 11, 121.
(8) Young, D. M., Crowell, A. D., "Physical Adsorption of Gases," Butterworths, London, 1962.

RECEIVED February 26, 1970.

57

Influence of Zeolite Cation Nature on Adsorption and Chromatographic Properties

G. V. TSITSISHVILI and T. G. ANDRONIKASHVILI

Institute of Physical and Organic Chemistry, Academy of Sciences of the Georgian SSR, Tbilisi, USSR

For zeolites with 1-, 2-, and 3-charge cations, the specificity of adsorption interaction is determined by nature of the cations replacing sodium ions, by the degree of exchange, by population of single-cation positions, and by other factors. A more pronounced influence of cations replacing sodium ions on sorption kinetics has been established for zeolites of type A. Chromatographic data show that when the number of cations in a unit cell of a faujasite type zeolite decreases, the values of retention volumes of hydrocarbon gases and carbon monoxide decrease. The nature of alkali metal cations of type X zeolites and their positions in the crystal lattice essentially influence values of the retention volumes of the compounds studied.

Studies of new perspective fine porous crystal zeolites have shown that their properties depend strongly on cations compensating the negative charge of the aluminosilicate framework.

Within one type of zeolite, adsorption, chromatographic, catalytic, and other properties are determined to a certain extent by the nature of ion exchange cations, by their number per unit cell, by the degree of population of single centers (positions) in the crystal framework, and by other things.

The crystal structure of zeolites A, X, and Y is comparatively well studied by x-ray crystal analysis (*5, 6, 7, 11, 24*). The existence of single centers of cation localization is established for faujasite group zeolites, and their populations have been determined for a number of cation forms (*7, 9, 10, 18, 23, 26*).

This is very important in analysis of the influence of ion exchange cations on different properties of zeolites, and it was taken into account in our work.

This paper presents some results of the studies of the influence of ion exchange cations on adsorption, kinetics, and chromatographic characteristics of zeolites of types A, X, and Y.

K. E. Avaliani, D. N. Barnabishvili, M. S. Shuakrishvili, E. P. Grigolia, M. G. Adolashvili, Sh. D. Sabelashvili, Z. I. Koridze, and S. S. Chkheidze have taken part in this work.

Different cation forms of zeolites have been obtained from original sodium zeolites by means of ion exchange and by the technique worked out earlier (*28*). The composition was determined by chemical analysis, and stability of the zeolite crystal lattice by x-ray technique.

The studies of adsorption capacity of vapors were made by microbalance technique, and kinetic characteristics were determined on a dynamic apparatus. The statics of vapor adsorption was studied at 20°C.

The specimens were dehydrated at 350°C and for adsorption studies the vacuum corresponded to 10^{-6}mm of mercury. For chromatographic studies, zeolite granules with alkali metal cations, before they were placed into a chromatographic column, had been heated at 450°C with thermal treatment in the flow of the carrier gas at 300°C.

It is better to give experimental data not only in the usual form of isotherms, where the adsorption value is expressed in millimoles per gram of adsorbent as a function of relative (reduced) pressure, $a \frac{\text{mmole}}{\text{gram}} = f\left(\frac{P}{P_0}\right)$, but also in the form of isotherms when the value of adsorption is determined by the number of the adsorbed molecules per unit cell depending on the relative pressure, $a \frac{\text{mol}}{\text{u.c.}} = f\left(\frac{P}{P_0}\right)$. The number of unit cells corresponding to a unit mass of zeolite was determined by M. M. Dubinin (*12, 13, 15*). Such an approach eliminates the influence of cation masses, which is felt particularly in heavy cations. During the analysis of adsorption data, one should pay the main attention to the portion of the isotherm corresponding to low pressures. The influence of cations in zeolites on interaction with molecules of the adsorbed substances at low coverages is clearly seen while studying chromatographic processes (*8, 16, 17, 19, 20, 21, 37, 39*).

Study of vapor adsorption on potassium–sodium zeolites of the faujasite type (the ratio of silica to alumina $n = 2.49$) has shown that adsorption capacity for water is decreased and for benzene increased with the degree of exchange $\alpha = 0.44 \left(\alpha = \frac{K^+}{Na^+ + K^+}\right)$, to compare it with the sodium form. The observed character of the adsorption process may be caused by weakening of electrostatic interaction and strengthening of dispersion.

Almost linear decrease of adsorption capacity has been observed with thallium–sodium zeolites of type A ($n = 1.98$) for water vapor with an increase of the degree of exchange (Figure 1). This is a result of a decrease of the cation electric field which causes a decrease of the adsorption layer density.

Using thallium–sodium zeolites of type X ($n = 2.30$), we observed an increase of adsorption capacity for water vapor and benzene at 38% replacement of sodium ions by thallium ions, and then its decrease with an increase in the degree of exchange. As thallium ions replace sodium ions in the position S_{III} (*25*) we may assume redistribution of cations at dehydration under the conditions of high vacuum and thermal treatment at 350°C, and stronger chemical bonds of thallium atoms in screened positions in comparison with sodium atoms.

The replacement of sodium ions by ions of calcium and strontium leads to a decrease in the adsorption value per unit cell. A similar picture is observed in studies of adsorption of benzene vapor on calcium forms.

Properties of calcium and strontium zeolites apparently are caused by the fact that ions of calcium (strontium) during the process of exchange occupy the screened positions S_I and S_{II}, and the number (density) of cation centers per unit cell decreases (*32, 34*).

In principle, hydrolitic exchange (*35, 36*) also might influence the adsorption capacity; however, such an assumption is difficult to bring into agreement with the data of x-ray crystal studies of calcium and

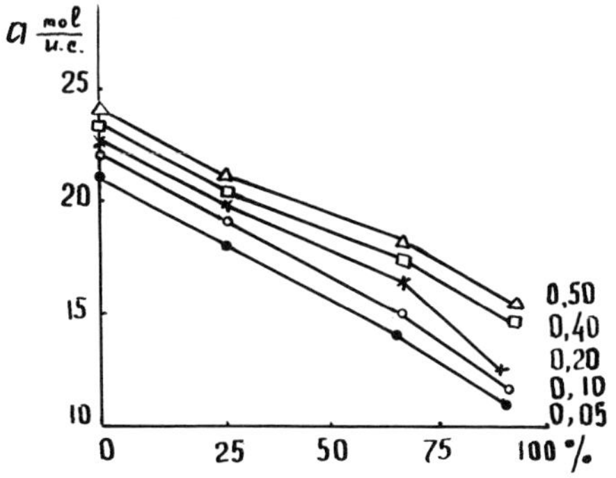

Figure 1. Dependence of the number of adsorbed water molecules per unit cell on the degree of exchange of sodium ions by thallium ions at different relative pressures from 0.05 to 0.50

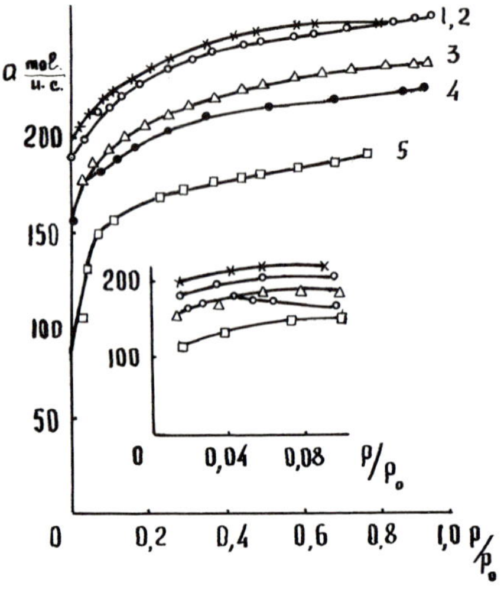

Figure 2. Isotherms of adsorption of water vapors on original sodium zeolite X (4) and lanthanum forms (1,2,3, and 5). (1) $\alpha = 0.44$; (2) $\alpha = 0.22$; (3) $\alpha = 0.50$; (5) $\alpha = 0.91$.

strontium zeolites (7). Barium zeolites are characterized by properties close to those of calcium and strontium forms.

Our studies have shown that replacement of 29% of sodium ions by lanthanum ions in an A zeolite little affects its adsorption capacity for water vapor. Obviously, lanthanum ions occupy comparatively screened positions in a crystal. The presence of lanthanum ions up to the degree of exchange $\alpha = 0.44$ and $\alpha = 0.64$, respectively, in zeolites X ($n = 2.44$) and Y ($n = 4.16$) promotes an increase of water vapor adsorption (Figure 2). At degrees of exchange exceeding 0.5–0.6, adsorption probably decreases owing to weakening of the lattice. That is shown by roentgenograms of the corresponding lanthanum zeolites.

The data show that lanthanum ions in zeolites X and Y, even in comparatively screened positions, interact with polar molecules of water. Such a point of view is confirmed by the results of our studies of water vapor adsorption on sodium and lanthanum hydrosodalites.

Consideration of isotherms of benzene vapor adsorption on specimens of lanthanum zeolite X shows that up to 50% replacement of sodium ions leads to some increase of adsorption capacity, and that deeper exchange promotes decrease of adsorption capacity. Adsorption of benzene vapor is decreased on lanthanum and yttrium zeolites of type Y in comparison with sodium zeolites (Figure 3). Such a difference in properties of zeolites X and Y apparently is caused by the fact that lanthanum ions occupy strongly screened positions in zeolite Y.

The kinetics of water vapor adsorption has been studied at 25°C and the relative concentration of the water vapor $\frac{C}{C_0} = 0.10$ ($C_0 = 23.1$ mg/l), using a sodium zeolite of type A ($n = 1.95$), and obtained from it lithium (degree of exchange $\alpha = 0.60$), potassium ($\alpha = 0.40$), rubidium ($\alpha = 0.22$), and cesium ($\alpha = 0.27$) forms. The kinetic curve $a = f(t)$ for the sodium form is the highest, while the lowest is for cesium zeolite. Calculations of the effective diffusion coefficients according to D. P. Timofejev (*27*) show that these values gradually decrease from the sodium zeolite to the cesium one.

	NaA	LiNaA	KNaA	RbNaA	CsNaA
$D_e \cdot 10^7$ cm²/sec	9.0	7.0	6.5	6.3	5.8

The data show that an appreciable decrease of sorption and its rate take place with increase in the cation radius. We obtained a smaller value for the diffusion coefficient of the lithium form than for sodium zeolite. Such a result may be caused by a strong hydration of lithium ion that does not allow, under the usual conditions of dehydration, complete elimination of water molecules in a lithium zeolite. Maybe the data (*38*) indicating that the effective diameter of windows in lithium

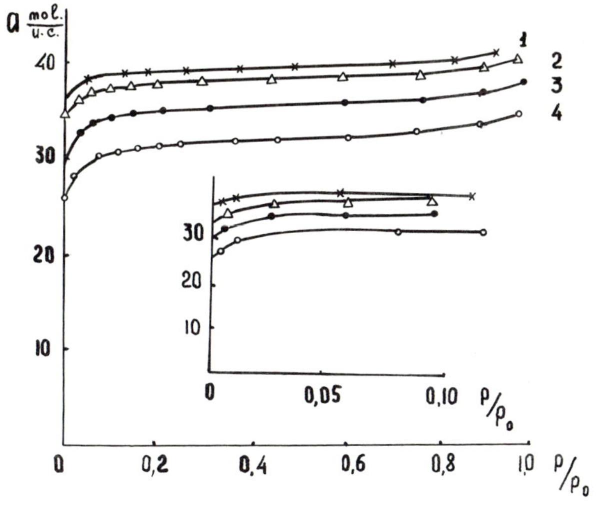

Figure 3. Isotherms of adsorption of benzene vapors on the original sodium zeolite Y (1) and lanthanum forms (2,3, and 4). (2) $\alpha = 0.31$; (3) $\alpha = 0.59$; (4) $\alpha = 0.64$.

zeolites type A is close to 3A is caused by this fact. We have not observed such an appreciable influence on sorption rate for type X zeolites when sodium ion is replaced by cations of other alkali metals. Apparently, this is determined by an essential difference in the diameters of entering "windows" of zeolites of type A (3–5A) and type X (8–9A).

Cations occupying positions in the entering "windows" of zeolite A probably decrease their effective diameter (*14*), causing a decrease of the diffusion coefficients.

A mixture of hydrocarbon gases C_1–C_4, carbon monoxide, and hydrogen was used as a model mixture for chromatographic studies.

Zeolites of type NaX and NaY have practically identical structure, but differ from each other by the number of cations per unit cell. A decrease of cation density in a zeolite takes place because of decationization of the positions S_{III}.

As our data show (Table I), the values of the retention volumes of all studied compounds were decreased with a decrease of the total number of cations per unit cell. There are 84 Na^+ per unit cell for NaX and 62 Na^+ for NaY. The values of the retention volumes for components on hydrogen and decationized forms also are decreased (*3*).

The maximum replacement of Na^+ by NH_4^+ was 75%. At such an exchange, replacement of cations apparently takes place mainly on the sites S_{II} and S_{III} (*25*).

The values of the retention volumes on the hydrogen form with 35% replacement of Na^+ by NH_4^+, where the S_{III} sites are probably vacant, approach the values obtained for zeolite NaY. The higher the replacement of Na^+ by NH_4^+ is, the greater decrease of the retention volumes for all studied components is observed. Similar results are obtained in Ref. *22*. Such a decrease of retention volumes on these forms may be caused by a decrease of the cation concentration per unit cell (the sites S_{II} and S_{III} become vacant) and perhaps by a partial destruction of the zeolite structure.

Table I. Specific Retention Volumes of Hydrocarbon Gases C_1–C_4 and CO, Expressed at Flowmeter Temperature, 20°C

Components	*Column Temp., °C*	*NaX*	*NaY*	*HX, 35%*	*DcX, 35%*	*HX, 75%*	*DcX, 75%*
Methane	25	25.9	14.3	14.6	16.4	3.9	5.0
Ethane	80	56.0	25.4	29.2	32.0	8.1	8.7
Propane	140	67.0	20.8	34.0	40.8	9.1	10.1
Butane	140	194.0	118.9	131.2	148.1	37.2	37.4
Carbon monoxide	25	63.8	25.7	27.0	31.0	4.9	7.5
Ethylene	140	62.9	20.8	26.4	32.4	5.2	6.7
Propylene	200	71.1	47.3	54.3	59.2	–	–

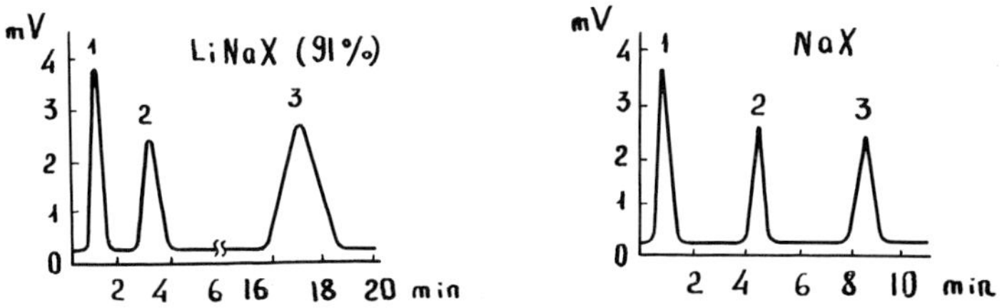

Figure 4. Order of elution of methane and carbon monoxide on type X zeolites containing either sodium or lithium; 1 = hydrogen, 2 = methane, 3 = carbon monoxide. The carrier gas is air; rate is 100 ml/min; column temperature is 25°C.

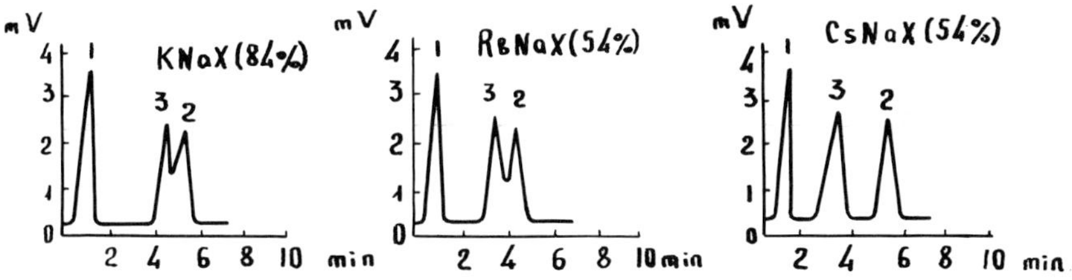

Figure 5. Order of elution of methane and carbon monoxide on potassium, rubidium, and cesium type X zeolites; 1 = hydrogen, 2 = methane, 3 = carbon monoxide. The carrier gas is air; rate is 100 ml/min; column temperature is 25°C.

An increase in the values of retention volumes of compounds on decationized zeolites may be explained both by a stronger dehydration of the remaining sodium cations and by a possible formation of active three-coordinate aluminum groups.

An increase of the dispersion part of interactions takes place with enrichment of the cation electron shell, while the smaller the cation radius is—*i.e.*, the higher the concentration of the positive charge—the more pronounced the electrostatic interaction is. This may be illustrated taking the example of separation of the mixture hydrogen–methane–carbon monoxide on type X zeolites with alkali metal cations.

A methane molecule, in contrast to that of carbon monoxide, has no dipole and quadrupole moment. For lithium and sodium forms of a zeolite, the following order of elution is characteristic: methane–carbon monoxide (Figure 4).

On the specimens with a high degree of replacement of Na^+ by K^+, a partial separation of the components takes place, but with the inverse sequence of elution. Separation of methane and carbon monoxide proceeds also with inverse elution on zeolites containing either rubidium or cesium. More complete separation of these compounds is characteristic for cesium form (Figure 5).

Both the extent of cation exchange and the temperature of chromatographic column heating (*1, 2, 4, 29, 30, 31, 33*) influence the order of elution of different components. Temperature rise leads to an appreciable decrease of ion–dipole interactions. Thus, on a specimen with the replacement of Na^+ by K^+ equal to 26% at 25°C, the value of the retention volume for carbon monoxide is greater than for methane, the same as on NaX. However, separation of the components does not take place. The rise of the column temperature to 40°C leads to alignment of the values of the retention volumes. Further rise of the column temperature leads to inversion of elution order for these compounds, with their separation only at 80°C. On the specimen with replacement of Na^+ by K^+ (48%), the inverse elution sets up at room temperature; however, it is followed by separation only at 60°C. As to the form with a high content of potassium cations, at 16°C the retention volume of methane is considerably higher than the retention volume of carbon monoxide, and therefore separation may take place at the temperature mentioned above (Figure 6) and higher.

The nature of a cation influences the sequence of elution of different pairs of hydrocarbons; for example, ethylene on NaX is eluted after propane at the column temperature of 80°–100°C. When temperature rises to 120°–140°C, there is no appreciable separation of these components. At 160°C and higher, separation of the mixture propane–

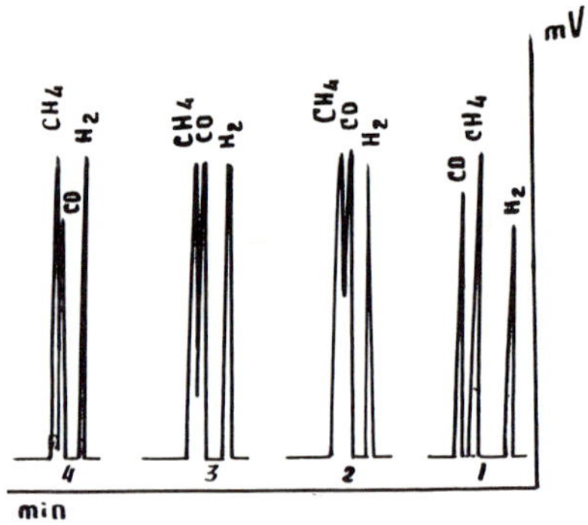

Figure 6. Separation of the mixture of hydrogen–methane–carbon monoxide on type X potassium zeolites. The carrier gas rate is 100 ml/*min; column temperature is 80°C. 1 = NaX, 2 = KNaX (degree of exchange of Na^+ by K^+ is 26%), 3 = KNaX (degree of exchange of Na^+ by K^+ is 48%), and 4 = KNaX (degree of exchange of Na^+ by K^+ is 84%).*

ethylene proceeds with the inverse sequence of elution. On a zeolite with 91% replacement of Na^+ by Li^+ at any temperature of the column heating, propane is eluted first and then ethylene.

On potassium, rubidium, and cesium forms, after each saturated compound an unsaturated one is eluted with the same number of carbon atoms in a molecule. The following sequence of elution is observed: methane–ethane–ethylene–propane–propylene–butane–butylene.

Such cations as sodium and lithium show a greater ability to retain a compound characterized by π-bonds than larger cations (potassium, rubidium, cesium).

The cation action is not always very strong. It depends on the arrangement of cations in the crystal lattice of a zeolite and hence both on the degree of cation screening by oxygen atoms of the lattice and on the degree of its hydration. Thus, for example, on a zeolite containing lithium cations at 47% replacement of Na^+ by Li^+, a decrease of the retention volumes and heats of adsorption takes place for unsaturated hydrocarbons and carbon monoxide instead of the expected increase. Only an increase of the degree of exchange to 87% and higher leads to a gradual increase of the above quantities, which at the maximum Li^+ content exceeds the values corresponding to these quantities on NaX.

Table II. Ratio of Heats of Adsorption $Q_{C_2H_4}/Q_{C_2H_6}$; $Q_{C_3H_6}/Q_{C_3H_8}$

Degree of Replacement of Na^+ by Li^+, %	0	5	22	43	87	91	NaX	LiX
$Q_{C_2H_4}/Q_{C_2H_6}$	1.45	1.38	1.37	1.36	1.75	1.88	1.44 (*20*)	1.24 (*20*)
$Q_{C_3H_6}/Q_{C_3H_8}$	1.38	1.26	1.29	1.29	1.49	1.83		

In Table II, the ratios of adsorption heats $Q_{ethylene}/Q_{ethane}$ and $Q_{propylene}/Q_{propane}$ are given on NaX and zeolites with different content of Li^+, as well as the data obtained in (*20*). The ratio of heats is higher for the specimens with high Li^+ content in comparison with NaX.

At high degree of exchange, Li^+ cations are located in the least hydrated sites (*25*). Such sites may be S_I, S_I', S_{II}' (*26*). Because of their location in the screened positions, they interact weakly with molecules of adsorbed substances but, as under the accepted conditions of thermal activation other cations are surrounded by water molecules, one may assume that just these cations form the main centers responsible for the increase of both retention volumes and heat adsorption of unsaturated compounds and carbon monoxide.

From the above and other data on physico-chemical properties of zeolites, it follows that for molecular sieves with one-charge cations, one observes both some similarity and some discrepancy. The same conclusion may be made for zeolites with two- and three-charge cations.

Each cation form of a zeolite, in the region of stability of their structure and composition, is an individual fine porous body representing considerable interest for detailed studies.

Literature Cited

(1) Andronikashvili, T. G., Sabelashvili, Sh. D., *Mater. All-Union Conf. Zeolites, 1st, Akad. Nauk SSSR, Moscow,* **1962,** 65.
(2) Andronikashvili, T. G., Sabelashvili, Sh. D., Tsitsishvili, G. V., *Neftekhimiya* **1962,** 248.
(3) Andronikashvili, T. G., Tsitsishvili, G. V., Sabelashvili, Sh. D., *Bull. Acad. Sci. Georg. SSR* **1969,** 56, 113.
(4) Andronikashvili, T. G., Tsitsishvili, G. V., Sabelashvili, Sh. D., Chumburidze, T. A., *Mater. All-Union Conf. Zeolites, 2nd, Nauka, Moscow-Leningrad,* **1965,** 179.
(5) Barrer, R. M., Meier, W. M., *Trans. Faraday Soc.* **1958,** 54, 1074.
(6) Baur, W. H., *Am. Mineralogist* **1964,** 49, 697.
(7) Bennet, I. M., Smith, I. V., *Mater. Res. Bull.* **1968,** 3, 633.
(8) Bosacek, V., *Symp. Mol. Sieves, London,* April 1967.
(9) Breck, D. W., *J. Chem. Educ.* **1964,** 41, 678.
(10) Breck, D. W., Flanigen, E. M., *Symp. Mol. Sieves, London,* April 1967.
(11) Broussard, L., Shoemaker, D. P., *J. Am. Chem. Soc.* **1960,** 82, 1041.
(12) Dubinin, M. M., *Dokl. Akad. Nauk SSSR* **1961,** 138, 866.
(13) *Ibid.,* **1964,** 159, 166.
(14) *Ibid.,* **1966,** 168, 860.
(15) Dubinin, M. M., *Izv. Akad. Nauk SSSR, Ser. Khim.* **1961,** 1192.
(16) Eberly, P. E., Jr., *J. Phys. Chem.* **1961,** 65, 69.
(17) *Ibid.,* **1962,** 66, 812.
(18) Eulenberger, G. R., Shoemaker, D. P., Keil, I. G., *J. Phys. Chem.* **1967,** 71, 1812.
(19) Habgood, H. W., *Can. J. Chem.* **1964,** 42, 2340.
(20) Habgood, H. W., *Chem. Eng. Progr. Symp. Ser.* **1967, 63, 45.**
(21) Kiselev, A. V., Cheren'kova, Yu. L., Yashin, Ya. I., *Neftekhimia* **1965,** 1, 141.
(22) Neddenriep, R. S., *J. Colloid Interface Sci.* **1968,** 28, 293.
(23) Olson, D. H., Kokotailo, G. T., Charnell, J. F., *Natl. Colloid Symp., 41st, Buffalo, N. Y.,* **1967.**
(24) Reed, T. B., Breck, D. W., *J. Am. Chem. Soc.* **1956,** 78, 5972.
(25) Sherry, H. S., *J. Phys. Chem.* **1966,** 70, 1158.
(26) Smith, J. V., Bennett, I. M., Flanigen, E. M., *Nature* **1967,** 215, 5098, 241.
(27) Timofejev, D. P., Erashko, I. T., *Izv. Akad. Nauk SSSR, Ser. Khim.* **1961,** 1192.
(28) Tsitsishvili, G. V., Andronikashvili, T. G., *Mater. All-Union Conf. Zeolites, 1st, Akad. Nauk SSSR, Moscow,* **1962,** 117.
(29) Tsitsishvili, G. V., Andronikashvili, T. G., Sabelashvili, Sh. D., Chkheidze, S. S., *Neftekhimiya* **1967,** 305.
(30) Tsitsishvili, G. B., Andronikashvili, T. G., Sabelashvili, Sh. D., Koridze, Z. I., *Bull. Akad. Sci. Georg. SSR* **1967,** 46, 611.
(31) Tsitsishvili, G. V., Andronikashvili, T. G., Sabelashvili, Sh. D., Urotadze, S. L., *Neftekhimiya* **1969,** 790.

(32) Tsitsishvili, G. V., Bagratishvili, G. D., Avaliani, K. E., Andronikashvili, T. G., Barnabishvili, D. N., *Intern. Congr. Pure Appl. Chem., 20th, A23, USSR, Moscow,* **1965.**
(33) Tsitsishvili, G. V., Krupennikova, A. Yu., Andronikashvili, T. G., Urotadze, S. L., *Bull. Akad. Nauk Georg. SSR* **1969,** 54, 581.
(34) Tsitsishvili, G. V., "Surface Phenomena of Aluminosilicates," p. 5–15, Mecniereba Publ., Tbilisi, 1965.
(35) Ward, J. W., *J. Catalysis* **1967,** 9, 225, 396.
(36) *Ibid.,* **1968,** 10, 34.
(37) Wolf, F., Furtig, H., *Symp. Mol. Sieves, London,* April 1967.
(38) Wolf, F., Furtig, H., *Tonind. Ztg. Keram. Rundschau* **1966,** 90, 297.
(39) Wolf, F., Hadicke, U., *Tonind. Ztg. Keram. Rundschau* **1967**, 91, 45.

RECEIVED January 23, 1970.

Discussion

G. C. Blytas (Shell Development Co., Oakland, Calif. 94623): Your data of retention volumes refer to low degree of coverage of sorbent. Would you care to comment on the separation factors, for example between ethylene and ethane, at higher coverages?

G. Tsitsishvili: Our studies of zeolite chromatographic properties indicated that the retention volumes, separation factors, and other characteristics are strongly dependent on cation nature, degree of exchange, and temperature. We have observed the inversion of the sequence of elution by temperature alteration.

J. D. Eagan (McMaster University, Hamilton, Ontario, Canada): Is there an explanation for the decrease in $\frac{Q_{C_2H_4}}{Q_{C_2H_6}}$, for low degrees of Li^+ exchange, with increasing Li^+ exchange?

G. Tsitsishvili: Interaction of Li^+ with molecules of unsaturated hydrocarbons and CO is very sensitive to the presence of water molecules in a zeolite and to column temperature increase. On a zeolite containing hydrophilic lithium cations, with most of the substitutions corresponding to Li^+ in open positions S_{II} and S_{III}, the adsorption heat of the studied compounds is lower than with the sodium form. In lithium forms with a high percentage of lithium ions, the Li^+ in nonhydrated positions influence adsorbate molecules, causing an increase of adsorption heat in comparison with NaX zeolite. Migration of Li^+ ions from S_I to other positions is also possible.

J. R. Katzer (University of Delaware, Newark, Del. 19711): I was interested in the comment in your paper that you observed a variation in the rate of sorption with a change in the cation present in the type X zeolite. How much variation did you find, and how were your rates of adsorption determined?

G. Tsitsishvili: The variation was about 20–30% at the transition from the sodium to other forms. Rates of adsorption were measured by following the weight of pellets during uptake.

F. Wolf (Martin Luther University, Halle/Salle, Germany, DDR): Tsitsishvili's results that there are maxima and minima in the curves for the adsorption amounts as functions of the degree of ion exchange were reported at the Molecular Sieves Conference in London, 1967. This seems to be the general behavior of molecular sieves when adsorbing molecules.

58

Kinetics of Sorption Processes as a Basis for Estimating Cation Distribution in Unit Cells of Zeolites

FRIEDRICH WOLF, FLORIN DANES, and KURT PILCHOWSKI

Chair of Chemical Technology, Martin Luther University Halle-Wittenberg, Halle/Saale, German Democratic Republic

Kinetic studies of ion exchange on partially ion-exchanged type A zeolites of Mg^{2+}, Ca^{2+}, and Mn^{2+} revealed that minimums and maximums characterize the differential coefficients of internal diffusion for every exchange of 2 Na^{+} ions for one divalent cation per unit cell of the zeolite. On the basis of these observations, assuming definite interactions between the cations and the zeolite lattice, predictions can be made concerning the distribution and arrangement of cations in the unit cells of a type A zeolite. Research on liquid phase adsorption of n*-alkanes on partially ion-exchanged type A zeolites indicated that the differential diffusion coefficients for alkane adsorption are influenced likewise by cation distribution in the unit cells of the zeolite.*

Since the natural zeolites were discovered and introduced as adsorbents, numerous investigations have been devoted to the sorption behavior and ion exchange properties of zeolites (*1, 2, 5, 14*). Sorption behavior of zeolites can be influenced powerfully by ion exchange, partly in the uptake capacity of the zeolites and partly in sorption rate since these may undergo decisive changes. The present study undertakes, on the basis of kinetic research in ion exchange and in alkane adsorption on type A zeolites after partial ion exchange, to show what influence is exerted by cation distribution on sorption by type A zeolites.

Kinetics of Ion Exchange

Although the kinetics of ion exchange on type A zeolites has been discussed previously in several papers (*4, 10, 12, 15, 19*), no systematic

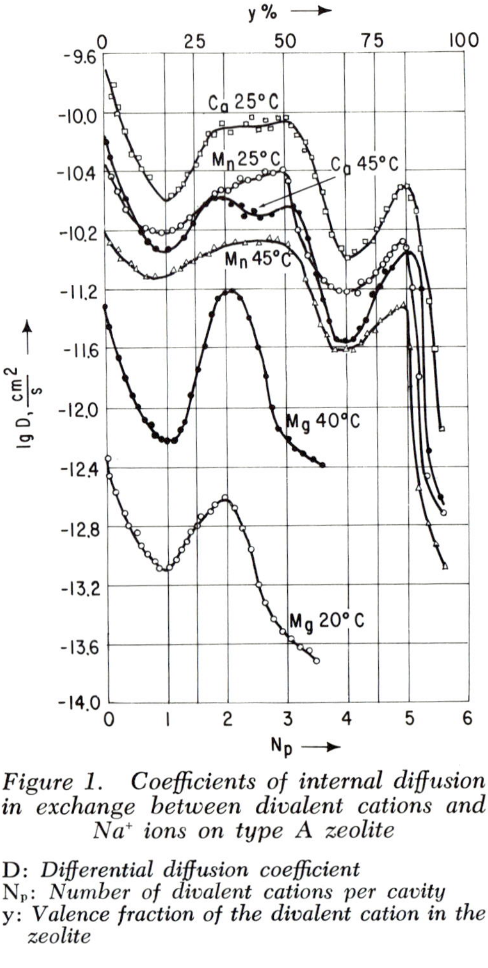

Figure 1. Coefficients of internal diffusion in exchange between divalent cations and Na^+ ions on type A zeolite

D: *Differential diffusion coefficient*
N_p: *Number of divalent cations per cavity*
y: *Valence fraction of the divalent cation in the zeolite*

research has been reported on ion exchange kinetics as influenced by cation uptake of the zeolite. Accordingly, we studied the kinetics of ion exchange with Mg^{2+}, Ca^{2+}, and Mn^{2+} ions on partially ion-exchanged type A zeolites. Our starting material was Zeosorb 4 A, a synthetic zeolite of Volkseigener Betrieb (State-Owned) Chemiekombinat Bitterfeld. The ion exchange study was performed in a batch process with 0.1 *N* aqueous salt solution. Internal diffusion (particle diffusion) was the rate-determining exchange step under the hydrodynamic conditions employed here (shape of container and agitator, agitator speed, concentration of the solution, quantities of solution and of zeolite). Changes in concentration generally were followed by complexometric titration, and radioactive isotopes of Ca^{2+} and Mn^{2+} were used for very fast exchange processes.

For each cation, the total uptake range of the zeolite was divided into 12 zones and the kinetic study of ion exchange was individual in each zone. For this purpose, the final uptake (equilibrium charge) of each zone corresponded to the initial uptake of the next zone, so that the total uptake range was covered. The requisite equilibrium isotherms for calculating diffusion coefficients also were determined experimentally in the batch process.

The effective diffusion coefficients were calculated from the experimentally observed data (time, amount of cation exchanged, temperature), using Paterson's solution of Fick's second law, or published approximate solutions (*8, 16*). Taking into consideration particle shape and particle size distribution, the differential coefficients of internal diffusion in ion exchange can be ascertained by a method previously described (*9*).

Figure 1 shows the differential diffusion coefficients in ion exchange as a function of zeolite uptake for the 3 investigated ion exchange pro-

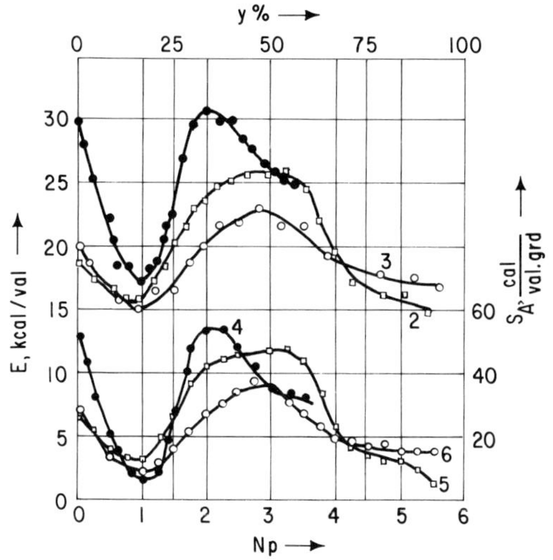

Figure 2. Activation energies and activation entropies of internal diffusion in cation exchange on type A zeolite

E: *Activation energy*
N_p: *Number of divalent cations per cavity*
S_A: *Activation entropy*
y: *Valence fraction of the divalent cation in the zeolite*

1: E *for the interchange* Mg^{2+}/NaA
2: E *for the interchange* Ca^{2+}/NaA
3: E *for the interchange* Mn^{2+}/NaA
4: S_A *for the interchange* Mg^{2+}/NaA
5: S_A *for the interchange* Ca^{2+}/NaA
6: S_A *for the interchange* Ca^{2+}/NaA

cesses. The substantial differences among the investigated ion exchange systems are caused primarily by differences in activation energy. Figure 2 shows activation energies as calculated from the Arrhenius equation and the activation entropies of ion exchange, calculated by the theory of absolute reaction rates, as a function of zeolite uptake. The observed downward trend of activation energies and activation entropies presumably is attributable to partial stripping away of the aqueous sheath on cations entering the zeolite.

The influence of cation uptake of the zeolite (*see* Figures 1 and 2) on diffusion coefficients, activation energies, and activation entropies in ion exchange is much more pronounced than that of changes in diffusion magnitudes from one cation to another. With increasing zeolite uptake, the differential diffusion coefficients tend to drop sharply, the curve being interrupted by maximums and minimums. While there have been previous reports of falling diffusion coefficients with rising zeolite uptake (*3, 11, 17, 20*), the curves which we obtained show extremes in diffusion coefficients for zeolite uptakes which are multiples of 1/6, namely 1/6, 2/6, 3/6, 4/6, 5/6, and 6/6.

Existence of at least 2 diffusion mechanisms must be assumed, taking into consideration the structural peculiarities of type A zeolites, for interpreting these results. It is a familiar fact that diffusion of cations or molecules into zeolites is influenced by the narrow channels. Rate of diffusion depends on the ratio d_p/d_k, where d_p is diameter of the cations or molecules and d_k is channel diameter. When $d_p/d_k << 1$, diffusion is not influenced by pore size; when $d_p/d_k \sim 1$, however, a small increase in d_p/d_k sharply increases activation energy, along with a saltatory decrease in rate of diffusion. During diffusion of an ion pair into a zeolite, the change d_p with zeolite uptake (y) is insignificant; pore diameter may be notably smaller when a cation is present in the pore. Hence, blocking of a zeolite pore is governed by cation uptake, especially since when uptake varies there are changes in the unit cell as to cation number and distribution and in the number and distribution of unoccupied cation positions.

Cation Distribution in Type A Zeolite

There are 14 cation positions in the large cavity in a unit cell of a type A zeolite (*6, 13*). Of the 14 possible cation positions, 6 are near the 8-membered oxygen rings within the zeolitic cavity (octahedral cation positions), and 8 are near the 6-membered oxygen rings situated in the corners of the zeolitic cavity (cubic cation positions). Whereas the cations in the octahedral pore openings partially block these openings and thus influence interdiffusion of cations, cations in the cubic oxygen rings are significant in relation to motion of cations in the zeolitic cavity.

Octahedral and cubic cation positions may be occupied by univalent and/or divalent cations, or they may be unoccupied; *i.e.*, they may remain as "holes." In our opinion, the number and distribution of unoccupied cation positions are responsible for deceleration or acceleration of cation diffusion. Designating as l_o and l_c, respectively, the number of unoccupied octahedral and cubic cation positions per cavity, as m_o and m_c the number of univalent cations, as p_o and p_c the number of divalent cations situated in the octahedral and cubic cation positions, and as N_l, N_m, and N_p the total number of unoccupied cation positions, of monovalent and of divalent cations, then for uptake y on a type A zeolite we have:

$$N_l = l_o + l_c = 14 - N_p - N_m = 2 + 6\,y$$
$$N_m = m_o + m_c = 12(1 - y)$$
$$N_p = p_o + p_c = 6\,y$$

and

$$l_o + m_o + p_o = 6$$

In addition to these 4 equations with 6 unknowns, the following additional conditions must be met for determining cation distribution:

(a) So far as possible, symmetrical cation distribution is attained, since in the stable state the cations are farthest apart because of their repulsion.

(b) Divalent cations are preferred, when they do not run counter to the first condition, because of the high polarizing activity of the smaller cubic oxygen ring. This assumption also is supported by the fact that introducing divalent cations in a type A zeolite actually widens pore size from 4 to 5 A. This pore expansion would not be understandable if these cations were taking the places of Na^+ ions at pore entrances.

(c) When completely symmetrical distribution is impossible, at least a partial symmetry is achieved by having an unoccupied cation position adjacent to each divalent cation, owing to the equilibrium in charge distribution. Accordingly, for each exchanged Ca^{2+} ion there is an unoccupied octahedral cation position ($p_c = l_o$) because the nearest neighbors of a cubic cation position are 3 octahedral cation positions.

(d) Univalent cations preferentially occupy the octahedral cation positions, but this tendency is not extremely pronounced. For example, it was found (*6, 7, 13, 18*) for the dehydrated zeolite NaA that x-ray analysis showed the Na^+ ions preferentially taking up positions vicinal to the 8 cubic cation positions. Yet the fact that a type A zeolite with univalent cations does not adsorb any molecule having $d_p > 4$ A can be explained by occupation of the octahedral oxygen rings (pore openings) by univalent cations.

Cation distribution in type A zeolites can be calculated by using the equations above, taking these conditions into consideration. Thus, for example, with 4 Ca^{2+} and 4 Na^+ ions per zeolite cavity—*i.e.*, $N_m = N_p = 4$—condition (a) is met if the 4 Ca^{2+} and 4 Na^+ ions occupy all 8 cubic cation positions so that all the octahedral cation positions remain free:

$$p_c = 4;\ m_c = 4;\ l_c = 0;\ l_o = 6;\ p_o = 0;\ m_o = 0.$$

This distribution has been confirmed by x-ray analysis (7).

Table I shows the corresponding distribution of ions, Na^+ and Ca^{2+} or Mn^{2+}, as well as unoccupied cation positions as a function of zeolite uptake.

Table I. Cation Distribution and Unoccupied Cation Positions at the Octahedral and Cubic Oxygen Rings in Na,Ca and Na,Mn Type A Zeolites

y	*0*	*1/6*	*2/6*	*3/6*	*4/6*	*5/6*	*1*
N_p	0	1	2	3	4	5	6
p_o	0	0	0	0	0	0	6
p_c	0	1	2	3	4	5	0
m_o	6	5	4	3	0	1	0
m_c	6	5	4	3	4	1	0
l_o	0	1	2	3	6	5	0
l_n	2	2	2	2	0	2	8

For a Na,Mg zeolite, the distribution is different at $y = 0.50$ and $y = 0.83$ because of the higher polarizing activity of Mg as compared with Ca or Mn. For these values of y there should be, as compared with the corresponding Na,Ca or Na,Mn zeolites, one additional Na^+ ion at the cubic cation positions, so that there is one more unoccupied position remaining at the octahedral cation positions.

Thus, it is necessary for unhampered diffusion of cations in a zeolite that at least 1/3 of the octahedral and 1/4 of the cubic cation positions shall be unoccupied simultaneously, since every cubic cation position is adjacent to 3 octahedral positions, and every octahedral position is adjacent to 4 cubic cation positions—*i.e.*, $l_o = 6/3 = 2$ and $l_c = 8/4 = 2$. If $l_o < 2$ and/or $l_c < 2$, cation diffusion by this mechanism is impossible. In that case, it proceeds by simultaneous position interchange between 2 cations. For Ca zeolites and Mn zeolites, the resulting maximums are at $N_p = 0$, 2, 3, and 5, or for Mg zeolites at $N_p = 0$ and 2, while the minimums for Ca zeolites and Mn zeolites are at $N_p = 1$, 3, and 6, or for Mg zeolites at $N_p = 1$, 3, 4, 5, and 6, confirming these assumptions (*see* Figure 1).

To explain this $D(y)$ relation, it is necessary to assume overlapping of 2 diffusion mechanisms: a "pore–fissure" diffusion and a diffusion by simultaneous position interchange between cations. In both cases, cation motion would be along the walls of cavities, since cations are strongly attracted by the negatively charged zeolite lattice. Rate of diffusion is smaller in the interchange mechanisms than in pore–fissure diffusion.

But cation diffusion always can occur by the interchange mechanism, whereas in pore–fissure diffusion it depends on the presence of unoccupied cation positions in a certain number and a definite distribution. Accordingly, the coefficients of internal diffusion are maximal when both diffusion mechanisms are operating at the same time, and minimal when diffusion is occurring only by the interchange mechanism.

Kinetics of Alkane Adsorption

Kinetic research on adsorption of *n*-decane from a solution of *n*-decane in toluene to partially ion-exchanged type A Mg and Ca zeolites in batch processes showed that the rate of adsorption of the *n*-paraffin molecule on zeolite 5A depends on the degree of cation exchange. *n*-Decane was not adsorbed below a cation exchange of 33%. The rate of diffusion of *n*-decane molecules increases with increasing cation exchange; for the Mg zeolite, it reaches a maximum at about 49% cation exchange and then drops off somewhat, whereas it rises rapidly for the

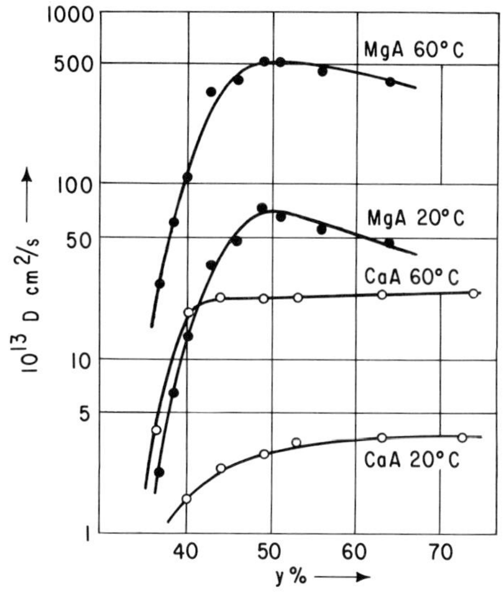

Figure 3. Coefficients of internal diffusion in adsorption of n-*decane from* n-*decane–toluene solution by type A magnesium zeolites and calcium zeolites at a zeolite uptake of* q_t *in grams* n-*decane/grams zeolite = 0.120*

D: *Differential diffusion coefficient*
y: *Valence fraction of the divalent cation in the zeolite*

Ca zeolite to about 53% cation exchange and thereafter changes only very slightly as cation exchange proceeds further (Figure 3).

This phenomenon must be attributed to cation distribution in zeolitic cavities and to interactions between *n*-decane molecules and divalent cations in the strong adsorption field in zeolite cavities. With increasing proportions of divalent cations in the zeolite, pore openings are free of cations to an increasing extent; this favors diffusion of *n*-decane molecules. On the other hand, even with increasing extent of cation exchange, the interactions between *n*-decane molecules and divalent cations in zeolitic cavities increase, a situation in which particular consideration must be given to the strong polarizing activity of divalent cations in connection with the very small polar anionic lattice. Here again, diffusion is impeded. Both of these effects produce diffusion rates for alkane molecules which are governed essentially by the distribution and properties of cations in the unit cells of the zeolite.

Literature Cited

(1) Barrer, R. M., *Ber. Bunsenges. Physik. Chem.* **1965,** 69, 786.
(2) Barrer, R. M., *Brennstoff-Chem.* **1954,** 35, 325.
(3) Barrer, R. M., Rees, L. V. C., *J. Phys. Chem. Solids* **1964,** 25, 1035.
(4) Bukata, S. J., Marinsky, J. A., *J. Phys. Chem.* **1964,** 68, 994.
(5) Breck, D. W., *J. Chem. Educ.* **1964,** 41, 678.
(6) Breck, D. W., Eversole, W. G., Milton, R. M., Reed, T. B., Thomas, T. L., *J. Am. Chem. Soc.* **1956,** 78, 5963.
(7) Broussard, L., Shoemaker, D. P., *J. Am. Chem. Soc.* **1960,** 82, 1041.
(8) Danes, F., Pilchowski, K., Wolf, F., *Ber. Bunsenges. Physik. Chem.* **1969,** 73, 99.
(9) *Ibid.*, in press.
(10) Dyer, A., Fawcett, J. M., *J. Inorg. Nucl. Chem.* **1966,** 28, 615.
(11) Helfferich, F., Plesset, M. S., *J. Chem. Phys.* **1958,** 28, 418.
(12) Hoinkis, E., Lewi, H. W., *Z. Naturforsch.* **1967,** 22 a, 226.
(13) Howell, P. A., *Acta Cryst. (Copenhagen)* **1960,** 13, 737.
(14) Nikolina, V. Ya., Neimark, Ee., Piontkovskaya, M. A., *Usp. Khim.* **1960,** 29, 1088.
(15) Panchenkov, G. M., Tolmachev, A. M., *Kinetika i Kataliz* **1963,** 4, 855.
(16) Pilchowski, K., Danes, F., Wolf, F., *Kolloid-Z., Z. Polymere* **1969,** 230, 328.
(17) Rao, A., Rees, L. V. C., *Trans. Faraday Soc.* **1966,** 62, 2505.
(18) Reed, T. B., Breck, D. W., *J. Am. Chem. Soc.* **1956,** 78, 5972.
(19) Wolf, F., Furtig, H., *Chem. Tech.* **1966,** 18, 1.
(20) *Ibid.*, **1966,** 18, 20.

RECEIVED February 10, 1970.

Discussion

R. M. Barrer (Imperial College, London): Can you indicate the way in which you were able to obtain the differential diffusion coefficients from Paterson's solution (which refers to a constant *D*) when *D* is a function of concentration? Have you considered your kinetic results in terms of a rapid exchange throughout the wide channels associated with a slow trapping-type of exchange involving the less-accessible cavities? This situation is discussed by Sherry in his symposium lecture.

F. Wolf: As I indicated, the observed *D* (differential diffusion coefficient) is not only valid for zeolite A but also for zeolite X. The exchange experiments were carried out stepwise in dependence on the degree of cation exchange (about 20–25 single measurements, involving the whole degree of ion exchange). The slow process was observed. In this case, *D* decreased by a factor of about 100; the same results were indicated in this symposium by the contributions of Barrer and other authors.

R. A. Munson (U. S. Bureau of Mines, College Park, Md. 20740): You mentioned the importance of a relaxation process in the diffusion results with Type X zeolite. What is the relaxation process involved, and how does it affect your results?

F. Wolf: A relaxation process in the case of cation exchange of Zeolite X accompanies the diffusion process and must be evaluated by mathematical methods, which we will publish. The influence of relaxation is eliminated in the results given in my paper.

A. Dyer (University of Salford, Salford, Lancs., England): Were the experiments concerned with the kinetics of alkane adsorption carried out in the absence of water?

F. Wolf: The experiments were carried out in the absence of water.

L. V. C. Rees (Imperial College, London): In the kinetics of ion exchange measurements, can you tell me what length of time was required before the fractional attainment of equilibrium reached approximately 100%?

F. Wolf: It differed according to the kind of cation and the degree of cation exchange. In most experiments, it was possible to achieve results in between 20 and 100 minutes.

59

Separation of 2,7-Dimethylnaphthalene from 2,6-Dimethylnaphthalene with Molecular Sieves

JOHN A. HEDGE

Sun Oil Co., Marcus Hook, Pa. 19061

Dimethylnaphthalene concentrate contains significant amounts of 2,6-dimethylnaphthalene bound in a binary eutectic with 2,7-dimethylnaphthalene. This eutectic cannot be broken by distillation or solvent crystallization. A practical method for separating this eutectic mixture of 2,7-dimethylnaphthalene and 2,6-dimethylnaphthalene has been achieved. Selective adsorption of 2,7-dimethylnaphthalene from a dimethylnaphthalene concentrate is obtained with sodium type Y molecular sieves. 2,6-Dimethylnaphthalene then can be crystallized from the unadsorbed raffinate fraction. Separation factors of 6 to 8 are obtained, indicating the high selectivity of these particular molecular sieves for this adsorption. Previous work in this area achieved a separation factor of 2.7. A continuous method has been developed for adsorption and desorption of 2,7-dimethylnaphthalene. Toluene has been selected as the optimum desorbent. This process makes 2,7-dimethylnaphthalene potentially available.

Molecular sieves have been used thus far to separate distinct classes of organic compounds. The separation of *n*-paraffins from branched paraffins with 5A molecular sieves is well known. Selective adsorption of aromatics from mixed streams with 10X, 13X, and type Y sieves is perhaps less widely known. The use of 10X molecular sieves to separate mixtures of aromatics has been disclosed (*1, 2, 3*). These patents disclose separations of mixtures of monocyclic aromatics (*1, 2*) and dicyclic aromatics (*3*). These were the first indications that molecular sieves could separate compounds within a single chemical class.

Discussion

We, at Sun Oil Co., are greatly interested in separating 2,6-dimethylnaphthalene from 2,7-dimethylnaphthalene (2,7-DMN). These 2 dimethylnaphthalene isomers are too close-boiling (262.0°C for 2,6-DMN and 262.3°C for 2,7-DMN) to allow separation *via* fractional distillation. Furthermore, both isomers are solid at room temperature (mp 111°–112°C for 2,6-DMN and 96°–97°C for 2,7-DMN). Only a small fraction of the naturally-occurring 2,6-DMN in a refinery stream can be crystallized before a binary eutectic containing 42% 2,6-DMN and 58% 2,7-DMN begins to precipitate.

Naturally, we were interested by one of Union's patent examples (3) which described use of 10X molecular sieve to separate a 2,6-DMN and 2,7-DMN binary eutectic (separation factor 2.7). We have enhanced this separation in 3 ways. First, by substitution of Na-Y sieve for 10X sieve, we have improved selectivity for 2,7-DMN adsorption. Second, a much sharper separation is obtained if a mixed refinery stream containing liquid DMN isomers is utilized instead of a pure binary mixture. Third, careful control of water content of the sieve allows improved selectivity.

Batch Separation Studies

We have evaluated several molecular sieves for the separation of 2,6-DMN from 2,7-DMN in the presence of liquid isomers present in a 257°–265°C heart-cut of an aromatic extract of catalytic gas oil. This heart-cut contains:

12.5%	2,6-DMN (free 2,6-DMN removed)
16.9%	2,7-DMN
43.1%	Other DMN's
6.3%	1- and 2-Ethylnaphthalenes
0.5%	Biphenyl
20.7%	Saturates and monocyclic aromatics

Prior to evaluation, each sieve was conditioned carefully in moist air (usually at 125°C) to control the water content of the sieve. Batch adsorptions were run then in which 10.0 grams of heart-cut, 2.5 grams of isooctane, and 5.0 grams of sieve were held at 100°C for 2 hours to ensure complete equilibration between the raffinate and the adsorbate. This 100°C temperature was chosen because runs at ambient temperature were considerably slower in reaching equilibrium. The unadsorbed (raffinate) fraction was then filtered off, and the cooled sieve was washed with room temperature isooctane to remove the remainder of the unadsorbed fraction. The adsorbate was removed with refluxing benzene. The results of these evaluations are shown in Table I. Sodium-Y sieve

Table I. Batch Runs, Selectivity of Various Molecular Sieves for Separation of 2,6-DMN and 2,7-DMN in Presence of 257°–265°C Heart-Cut Isomers, 100°C, 2 Hrs.

Type of Sieve	*Sieve Pretreatment Temperature in Air, °C*	*Sieve Capacity, Grams Hydrocarbon/ 100 Grams Sieve*	*Separation Factor, α*
Sodium form, type Y, powder (SK-40)	125	16.6	8.0
Sodium form, type Y, 20–40 mesh[a]	125	11.4	7.3
Sodium form, type Y, tablet	125	15.0	6.5
Ammonium form, type Y, powder	125	13.2	5.8
Rare earth exchanged, type Y, pellet (SK-500)	125	11.8	4.7
Sodium form, type Y, 1/16″ extrudate	125	13.6	4.4
Linde 13-X 14 × 30 mesh	125	9.2	2.8
Linde 10-X 1/16″ extrudate	100	6.6	2.4
Potassium form, type L (SK-45)	125	11.4	0.53[b]

[a] 1/16″ Extrudate ground to 20–40 mesh.
[b] 2,6-DMN preferentially adsorbed.

is the most selective sieve studied for adsorption of 2,7-DMN, as shown by separation factor, α, where

$$\alpha = \frac{\%\ \text{2,7-DMN adsorbed}/\%\ \text{2,7-DMN unadsorbed}}{\%\ \text{2,6-DMN adsorbed}/\%\ \text{2,6-DMN unadsorbed}}$$

Further examination of the data in Table I reveals several interesting points. Sieve geometry affects selectivity somewhat as shown by the slight differences between powdered, 20–40 mesh size, and tableted Na-Y sieve. Both basic sodium and ammonium-Y sieves and acidic rare earth-Y sieves

Table II. Batch Runs, Comparison of Purified Binary

Charge Stock	*Type of Sieve*
Purified eutectic	Sodium form, type Y, powder
Heart-cut mixture	Sodium form, type Y, powder
Purified eutectic	Sodium form, type Y, tablet
Heart-cut mixture	Sodium form, type Y, tablet

show selectivity for 2,7-DMN. Type L sieves adsorb 2,6-DMN in preference to 2,7-DMN, the reverse of all other sieves shown in Table I.

Intuitively, one would predict that separation of a binary mixture should be simpler than separation of 2 components present in a complex mixture. However, this was not the case in our work. As shown in Table II, a much higher separation factor is obtained with a mixed stream than with a binary eutectic. Despite the dilution of 2,7-DMN present in adsorbate by liquid isomers, the weight of 2,7-DMN freed per 100 grams of sieve drops only slightly. This enhanced separation is illustrated further by the equilibrium diagrams for the binary eutectic (Figure 1) and for the heart-cut mixture (Figure 2).

Control of water content of this molecular sieve is quite important. An optimum H_2O content of 12.0% for Na-Y sieve is shown in Table III. This water content was determined by weight loss on ignition at 1900°F. Sieves having lower water contents are less selective. This finding implies that water present in sieves is structural and does affect sieve properties. Sieves with high water contents up to 25 wt % have reduced capacity for dinuclear aromatics. Sieves wet with methanol have greatly reduced selectivity; conditioning in moist air at 125°C restores original selectivity.

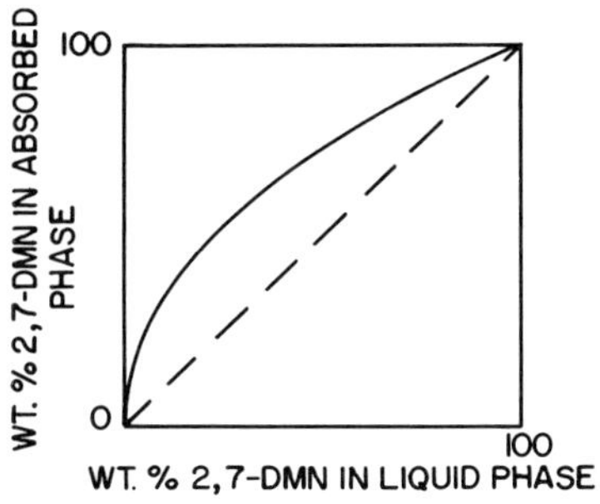

Figure 1. Separation of 2,6-DMN and 2,7-DMN binary eutectic with Na-Y sieve

Eutectic *vs.* 257°–265°C Heart-Cut Mixture, 100°C, 2 Hrs.

Adsorbed Phase			
% 2,6-DMN	*% 2,7-DMN*	*Separation Factor, α*	*Wt. 2, 7-DMN Freed/100 Grams Sieve*
23.7	76.3	3.1	5.8
4.4	38.0	8.0	5.2
28.8	71.2	2.8	4.8
4.8	37.0	6.8	4.4

Selection of Desorbent, Batch Competition Studies

To develop a cyclic separation process, our next step was to find a suitable desorbent. A suitable desorbent must allow selective adsorption and yet must desorb at a reasonable rate. Three aromatic desorbents were evaluated: benzene, toluene, and *o*-xylene.

Batch competition experiments with equal weights of desorbent and DMN heart-cut were run to determine the effect of desorbent on selectivity for adsorption of 2,7-DMN over 2,6-DMN. The relative strength with which desorbent or any other component of a mixture is adsorbed can be shown by β factor. β Factor is defined as the ratio of component x in the adsorbed phase over component x in the unadsorbed phase divided

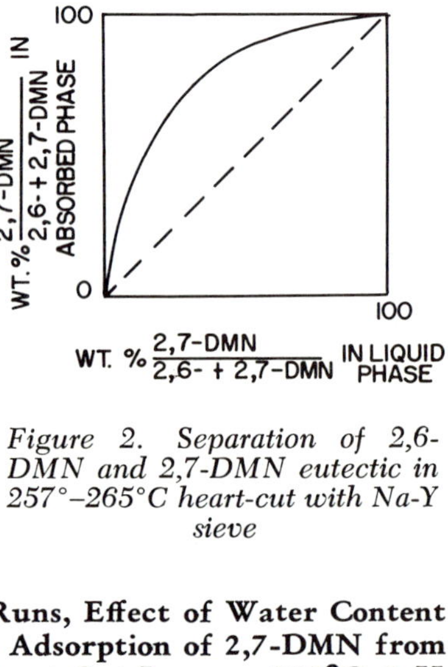

Figure 2. Separation of 2,6-DMN and 2,7-DMN eutectic in 257°–265°C heart-cut with Na-Y sieve

Table III. Batch Runs, Effect of Water Content on Molecular Sieve Selectivity in Adsorption of 2,7-DMN from 257°–265°C Heart-Cut Isomers, 100°C, 2 Hrs.

Type of Sieve	*Sieve Pretreatment Temperature in Air, °C*	*Wt % H_2O[a]*	*Sieve Capacity, G/100G Sieve*	*Separation Factor, α*
Sodium form, type Y, powder	125	12.00	16.6	8.0
Sodium form, type Y, powder	400	2.30	20.6	3.2
Sodium form, type Y, tablet	25	23.34	4.4	4.1
Sodium form, type Y, tablet	125	12.13	15.0	6.5
Sodium form, type Y, tablet	175	6.00	21.2	5.2
Sodium form, type Y, tablet	methanol washed	10.02	15.0	3.5

[a] Based on weight of conditioned sieve.

by wt 2,6-DMN adsorbed over wt 2,6-DMN unadsorbed. The β factor for 2,7-DMN is equal to α, as defined earlier.

$$\beta \text{ Factor for component } x = \frac{\text{Wt } x \text{ adsorbed/Wt } x \text{ unadsorbed}}{\text{Wt 2,6-DMN adsorbed/Wt 2,6-DMN unadsorbed}}$$

The results of our competition studies are shown in Table IV. The most selective adsorption of 2,7-DMN occurs in the presence of *o*-xylene, the most weakly adsorbed desorbent. The least selective adsorption occurs in the presence of benzene, the most strongly adsorbed desorbent. Other DMN isomers are adsorbed more strongly than 2,6-DMN. This may explain the higher separation factors obtained with the heart-cut mixture.

Table IV. Competition of Benzene, Toluene, and *o*-Xylene with Dimethylnaphthalenes for Adsorption on Molecular Sieve, β Factors Obtained from Batch Studies[a]

	Benzene	*Toluene*	*o-Xylene*
2,7-DMN	2.7	4.7	8.0
Other DMN's	2.0	2.2	2.3
Benzene	1.8	–	–
1,6-DMN	1.4	–	1.9
1- and 2-Ethylnaphthalenes	1.7	1.8	1.8
2,6-DMN	1.0	1.0	1.0
Toluene	–	0.9	–
o-Xylene	–	–	0.7
Mixed monocyclic aromatics	0.6	0.6	0.6

[a] β Factor 2,6-DMN chosen equal to 1.0.

These batch competition studies showed that benzene would cause the fastest desorption while *o*-xylene would allow the sharpest separation. Since we were faced with this dilemma, all 3 desorbents were carried into our column studies.

Column Separation Studies

The experimental apparatus used for our column studies consisted of a 3/4-inch i.d. by 36-inch glass column filled with Na-Y sieve containing about 12 wt % water. The DMN heart-cut containing eutectic 2,6-DMN and 2,7-DMN is pumped into the bottom of the column. At the end of this DMN charge, desorbent is pumped into the bottom of the column. The desorbent pushes out a 2,6-DMN enriched raffinate and desorbs the 2,7-DMN enriched adsorbate. When the last of the adsorbate is removed from the bottom portion of the sieve bed, the column is ready for a new cycle. The incoming DMN feed pushed out the remainder of the 2,7-DMN enriched desorbate. The effluent from the column was taken in small cuts for our initial studies. These cuts then were analyzed by gas chromatography.

The adsorbent column is held at a temperature just below the boiling point of the desorbent to speed diffusion into the sieve particles.

Our column runs showed (as predicted by the batch studies) that benzene is the most efficient stripping solvent. *o*-Xylene was quite slow to desorb 2,7-DMN and was dropped from consideration in any cyclic process.

Sharpness of separation is shown by plotting the breakthrough curve of 2,7-DMN into the raffinate. Breakthrough curves for these 3 desorbents are shown in Figure 3. Benzene allows a rapid breakthrough of 2,7-DMN, as predicted. Surprisingly, toluene is superior to *o*-xylene at the 8.0

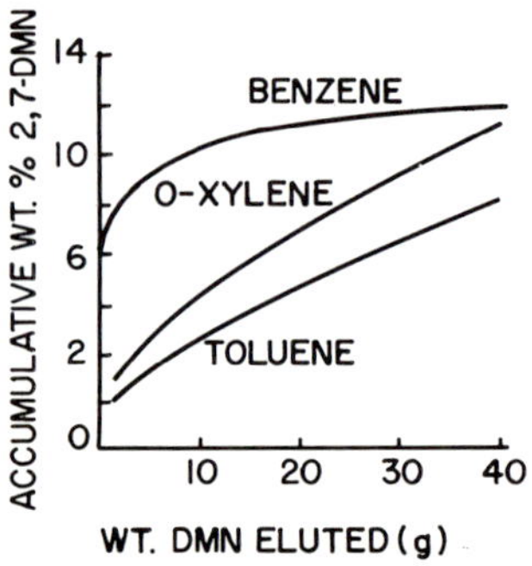

Figure 3. Breakthrough of 2,7-DMN into raffinate pump rate 8 ml/min

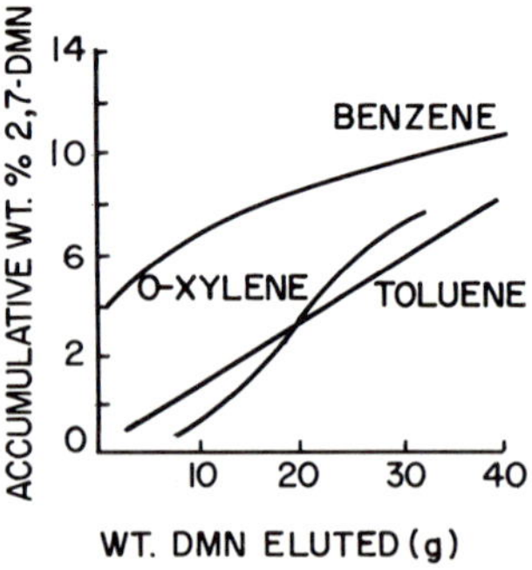

Figure 4. Breakthrough of 2,7-DMN into raffinate pump rate 4 ml/min

ml/min pump rate shown in Figure 3. At a slower 4.0 ml/min pump rate, the curves for toluene and *o*-xylene are superimposable (Figure 4). Presumably, toluene allows faster equilibration than *o*-xylene, and thus toluene is superior at higher flow rates. Therefore, our laboratory studies show that toluene is the best desorbent for our proposed separation process.

A separation process can be envisioned using 3 synchronized molecular sieve columns. At a given point in time, one column would be eluting 2,6-DMN enriched raffinate, the second column would be eluting poorly separated charge to be recycled, and the third column would be undergoing desorption of 2,7-DMN enriched adsorbate. A more efficient process could involve hardware used in the Molex process.

Literature Cited

(1) Eberly, P. E., Jr., Arey, W. F., Jr., U. S. Patent **3,126,425** (March 24, 1964).
(2) Fleck, R. N., Wight, C. G., U. S. Patent **3,114,782** (Dec. 17, 1963).
(3) *Ibid.*, **3,133,126** (May 12, 1964).

RECEIVED February 13, 1970.

Discussion

J. D. Sherman (Union Carbide, Tarrytown, N. Y. 10591): What error limits apply to the capacity data and separation factors in Table I? Have you measured rates of sorption on systems other than NaY with 12% H_2O?

J. A. Hedge: Error limits for capacities are ±1 gram hydrocarbon per 100 grams molecular sieve, and for separation factors are ±0.5. No systems other than NaY have been studied.

D. P. Roelofsen (Technological University, Delft, Netherlands): In your adsorption experiments, you tried to reach equilibrium by working at 100°C. In view of the earlier paper presented by Satterfield and Katzer which showed a very slow counterdiffusion of molecules such as 1-methylnaphthalene and cumene, I would imagine that the counterdiffusion of your 2,6- and 2,7-DMN would be even slower. This could well mean that you reach a pseudo-equilibrium in which the ratio 2,6:2,7 isomer is kinetically determined. In L sieve, one then would expect that 2,6-DMN, which is the more linear of the two, would be preferred by the one-dimensional channel system, in agreement with your data. With X and Y sieves, one can imagine that the nonlinear 2,7-DMN has just the right shape to diffuse through the tetrahedral channel system, which also agrees with your data.

J. A. Hedge: With Type X and Y sieves, equilibration was complete after two hours at 100°C. At room temperature, equilibration required 24 hours. We feel that selectivity for 2,7-DMN over 2,6-DMN is the result of the higher basicity of 2,7-DMN. With Type L sieve, we did not check for complete equilibration. This should be done.

G. C. Blytas (Shell Development Co., Oakland, Calif. 94623): In Table I, how do you explain the difference in capacity and selectivity of NaY, 1/16-inch extrudate, and of the same sample ground to 20–40-mesh? Is it an effect of the amount of binder present (which may change during grinding) or is it a kinetic effect (incomplete equilibration)? In your fixed-bed cyclic experiments, what is the duration of the entire cycle?

J. A. Hedge: The improved selectivity obtained by grinding 1/16-inch extrudate may result from removal of binder during grinding and sieving to 20–40-mesh material. Decrease in capacity may be owing to destruction of a portion of the crystalline structure during grinding. Cycle times for the fixed-bed column experiments ranged from 20 to 40 minutes, depending on pumping rate.

60

Diffusion in Granular Zeolites

D. P. TIMOFEEV

Institute of Physical Chemistry, Academy of Science of USSR,
31 Lenin Avenue, Moscow

Three modes of vapor diffusion in granular zeolites and some results are discussed. The first mode involves diffusion of adsorbate through crystals and along secondary pores. On the basis of steady-state diffusion experiments, it was concluded that diffusion within the zeolite crystals is not significant as a contribution to total diffusion of organic vapors through the granule. In the second and the third modes, diffusion to the central parts of a granule is limited to secondary pores. For these modes which differ in the ratio of diffusion rates in the crystals and in the secondary pores, approximate equations of kinetics of adsorption are derived, and their agreement with experimental data is discussed.

Synthetic zeolites often are employed as spherical or cylindrical pellets composing the aggregate of elementary crystals. Spaces between the crystals form the secondary porous structure with pore dimensions of approximately the same order as the dimensions of the crystals. Both the first (porosity of crystals) and the secondary porosity are significant to kinetics of adsorption. Diffusion coefficients in the crystals depend mainly on the type and ionic form of the zeolite and on the nature of the adsorbing substance (*2*). Dependence of the diffusion coefficients in secondary pores on these factors is relatively light; together, they are significantly dependent on the conditions of mass exchange with the ambient gas media. A quantitative description of diffusion in a granule is a rather difficult task, the total diffusion coefficient being a complex function of many variables. The qualitative side of the process is given below, with some quantitative ratios of use in the evaluation assessments.

Three modes of diffusion in a granule, schematically represented in Figure 1, are possible. The first mode (Figure 1a) involves diffusion of

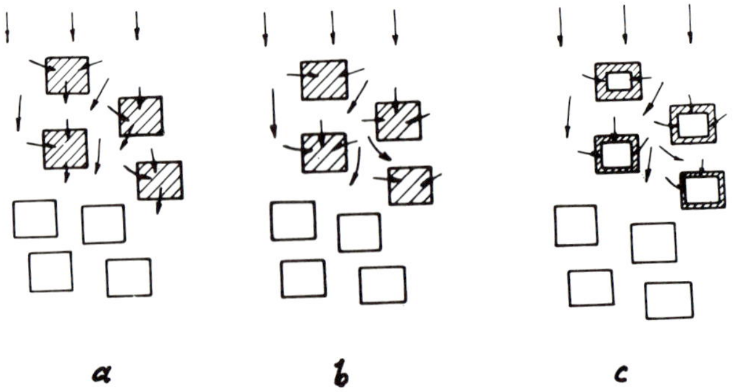

Figure 1. Three modes of diffusion of adsorbate in a granule

adsorbate molecules through gaps between the crystals and through the crystals. In the second and third modes, diffusion to central parts of a granule occurs only through secondary pores. According to the second mode, adsorbate molecules penetrate easily into the openings of adsorption cells and diffuse rapidly into the crystals (Figure 1b). The adsorption capacity of the crystals participating in the process is practically exhausted as the diffusion flow progresses into the depths of the granule. In the third mode (Figure 1c), diffusion into the crystals is slow and causes only partial filling.

In order to verify the diffusion through the crystals, we investigated the steady-state diffusion of a number of organic vapors in granules of zeolite CaA and NaX (*12, 13*). The experiments were carried out in the presence of carrier gas at atmospheric pressure. Granules of commercial zeolite samples were secured with hermetizing paste into the orifice of an aluminum membrane, and powdered crystallic zeolite was pressed into the orifice.

A pure carrier gas was streamed past one face of the membrane as a gas mixture containing adsorbate vapors was streamed over the other face. When a steady state was reached, the concentration on the granule's reverse side was determined, and the diffusion coefficient was found according to the equation

$$D = \frac{vLc_2}{F\ (c_1 - c_2)} \tag{1}$$

where c_1 and c_2 are the concentrations on the front and reverse side of the membrane ($c_2 \approx 0.1\ c_1$), respectively, v is velocity of carrier gas, L and F are the length and cross-sectional area of the sample.

Curves of the temperature dependence of the diffusion coefficient of ethyl alcohol and nitrogen in a granule of zeolite CaA-I, grain size

3 mm, are represented in Figure 2. Helium (curve 2) and nitrogen (curve 4) were used as carrier gas in the experiments with ethyl alcohol. The dotted line is computed for ethyl alcohol, as a nonadsorbing gas, on the basis of experimental data obtained for nitrogen according to the equation

$$D_{C_2H_5OH} = D_{N_2} \frac{\sigma_{1,2}\sqrt{\frac{1}{m_1} + \frac{1}{m_3}}}{\sigma_{1,3}\sqrt{\frac{1}{m_1} + \frac{1}{m_2}}} \tag{2}$$

where

$$\sigma_{1,2} = \frac{d_1 + d_2}{2}, \ \sigma_{1,3} = \frac{d_1 + d_3}{2}, \ m_1, m_2, m_3, d_1, d_2, \text{and } d_3$$

are the masses and gas–kinetic diameters of the molecules helium, nitrogen, and ethyl alcohol, respectively.

In the temperature range 50°–100°C, we may ignore adsorption of nitrogen, assuming it as a first approximation to be nonadsorbing.

Figure 2 shows that the diffusion coefficients of ethyl alcohol evaluated according to Equation 2 are well in agreement with the experimental values, implying that transport through the adsorption phase is of no conspicuous consequence. Figure 2 also shows the diffusion

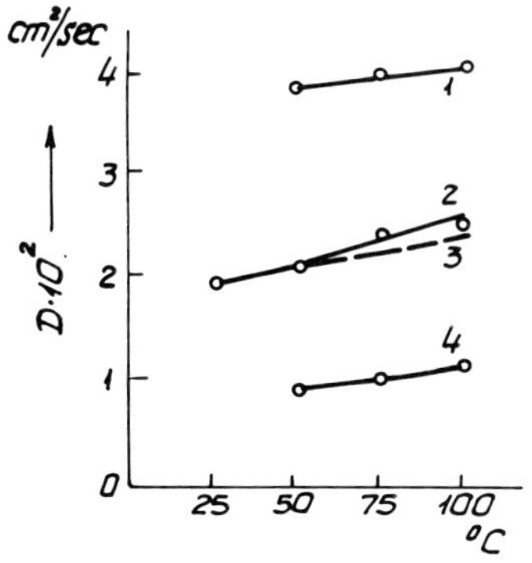

Figure 2. Effect of temperature on diffusion coefficient of nitrogen (1) and ethyl alcohol (2–4) in the granule of zeolite CaA-I

Carrier gas: 1,2,3 = helium, 4 = nitrogen
Curve 3 is calculated according to Equation 2

coefficients to be markedly higher in presence of helium as a carrier gas than in presence of nitrogen. The bulk of substance transport occurs in the secondary pores, since the change of carrier gas does not influence the transport of adsorbate in the zeolite crystal. The same conclusion follows from consideration of the temperature dependence of the diffusion coefficients. These coefficients increase with increasing temperature ($\sim T^{1,5}$) which is characteristic of gas-in-gas diffusion. Similar results are obtained with zeolite NaX (*13*).

F (Equation 1) is the full cross-sectional area of the sample. Taking account of the sample's porosity, ϵ, and of the tortuosity factor of the canals, k (*5*), we are able to find k using the equation

$$k = \sqrt{\frac{D_g \varepsilon}{D}} \tag{3}$$

where D_g is the literature value of the gas-in-gas diffusion coefficient.

Table I displays values of k for granulated zeolites CaA-I and CaA-II (*12*) and the pressed powdered zeolite NaX (*13*).

Table I. Values of *k*

Zeolite CaA-I			*Zeolite CaA-II*			*Pressed Powdered Zeolite*		
Substance	*Carrier Gas*	*k*	*Substance*	*Carrier Gas*	*k*	*Substance*	*Carrier Gas*	*k*
Nitrogen	Helium	2.6	Benzol	Nitrogen	2.2	Benzol	Helium	2.3
Ethyl alcohol	Helium	2.7	*n*-Hexane	Nitrogen	2.0	*n*-Hexane	Helium	2.2
Ethyl alcohol	Nitrogen	2.5				*n*-Hexane	Helium	2.2
n-Hexane	Nitrogen	2.5				Methyl alcohol	Helium	2.2
n-Heptane	Nitrogen	2.6						

Values of k for the same zeolite sample in experiments with various substances practically do not differ—*i.e.*, k is independent of the nature of substance. This fact indicates the absence of any considerable amount of substance transport through crystals in the investigated systems.

Let us consider the second mode of diffusion in granules of zeolite. Zeolites are characterized by high values of adsorption at small values of equilibrium pressure, the concentration in the adsorbed phase thus being much higher than in the gas phase. The second mode of diffusion presents a picture of layer-by-layer filling of the granule's adsorption capacity, running from the periphery to the center. In the limiting case (a rectangular isotherm), a sharp division will exist between the filled and the nonfilled parts of the granule. For homogeneous and isotropic granule porosity, the adsorption value is proportional to the volume of the filled part of the granule. In this case, the relative adsorption value may be expressed using the ratio of volumes. Consider a spherical granule of radius R, whose adsorption front penetration is x; the relative adsorption value will be cited (*14*)

$$\gamma = \frac{a}{a_0} = 3\alpha - 3\alpha^2 + \alpha^3 \tag{4}$$

where a is the adsorption value at time t, a_0 is the equilibrium value of adsorption, and

$$\alpha = \frac{x}{R}$$

We have similar case for a cylinder.

$$\gamma = 2\left(1 + \frac{1}{k_1}\right)\alpha - \left(1 + \frac{4}{k_1}\right)\alpha^2 + \frac{2}{k_1}\alpha^3 \tag{5}$$

where $k_1 = L/R$, R and L are radius and length of the cylinder. If $k = 2$, then Equation 5 is the same as Equation 4.

Since the movement of the adsorption front in a granule is relatively small, the adsorbate diffusion into the granule may be regarded as a pseudosteady process.

The adsorption front having progressed for distance x, the steady diffusion rate will be

$$\frac{dm}{dt} = 4\pi D \frac{R(R - x)c_0}{x} \tag{6}$$

where c_0 is the concentration on the external surface of the granule. The concentration of adsorbate in the nonfilled part of the granule is 0.

The adsorption front will travel a distance dx toward the center of the granule in a time dt at this particular distance from the center, the additionally filled volume being $4\pi\ (R - x)^2dx$. This volume is proportional to the adsorption increase dm, *i.e.*,

$$dm = 4\pi z\ (R - x)^2dx \tag{7}$$

where z is the capacity of volume unit of adsorbent.

From Equations 6 and 7, we find

$$\frac{dx}{dt} = \frac{DRc_0}{z(R - x)x} \tag{8}$$

By dividing variables and integrating, we find

$$3Rx^2 - 2x^3 = \mathrm{Kt} \quad (9)$$

where

$$\mathrm{K} = \frac{6DRc_0}{z} \quad (10)$$

Equation 9 assumes the form

$$x\,(3\alpha - 2\alpha^2) = \frac{\mathrm{K}t}{R^2} \quad (11)$$

and since

$$3\alpha - 2\alpha^2 \approx 3\alpha - 3\alpha^2 + \alpha^3 \quad (12)$$

we have

$$x\gamma = \frac{\mathrm{K}t}{R^2} \quad (13)$$

It follows from Equation 13 that

$$x \sim \frac{t}{\gamma} \quad (14)$$

and

$$\alpha \approx \frac{t}{\gamma t_\infty} \quad (15)$$

where t_∞ is the time in which the adsorption front is at the center of the granule.

After inserting value of α in Equation 4, we find

$$\gamma^2 = 3\frac{t}{t_\infty} - \frac{3}{\gamma}\left(\frac{t}{t_\infty}\right)^2 + \frac{1}{\gamma^2}\left(\frac{t}{t_\infty}\right)^3 \quad (16)$$

which approximates to

$$\gamma^2 = 3\frac{t}{t_\infty} - 3\left(\frac{t}{t_\infty}\right)^2 + \left(\frac{t}{t_\infty}\right)^3 \quad (17)$$

the order of precision being $\sim 10\%$.

Equation 17 enables us to find time t_∞ using only 1 point of the experimental curve, *e.g.*, $\gamma = 0.5$. According to Equation 17, adsorption value $\gamma = 0.5$ is reached in the relative time $t_{0.5}/t_\infty = 0.09$; thus, the time of complete filling of a granule (spherical) is $t_\infty = t_{0.5}/0.09$.

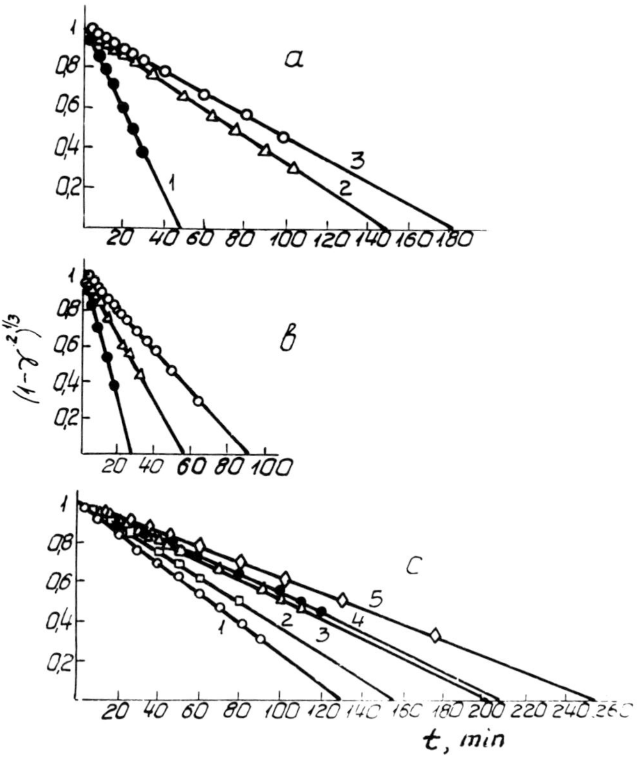

Figure 3. Experimental data plotted according to Equation 19; Zeolite CaA-II

a) n-*Hexane* (t, °C: *1 = 200°, 2 = 100°, 3 = 20°;* c *= 10 mg/l)*
b) Ethyl alcohol (t, °C: *1 = 200°, 2 = 100°, 3 = 20°;* c *= 20 mg/l)*
c) n-*Dodecane* (t, °C, c *mg/l: 1 = 195 and 7.1; 2 = 204 and 6.8; 3 = 111 and 9.8; 4 = 107 and 9.4; 5 = 110 and 8.9)*

If the diffusion coefficient is known, time t_∞ may be found according to the equation

$$t_\infty = \frac{zR^2}{6Dc_0} \tag{18}$$

which follows from Equation 13 after using the assumptions $\alpha = 1$ and $\gamma = 1$. The reverse is equally possible—*i.e.,* evaluation of D by an experimental value t_∞.

A similar mode of shell progressive combustion coke has been described by P. B. Weisz and R. D. Goodwin (*18*) and W. J. Blinow (*4*).

Equation 17 has been checked against experimental data in a number of investigations (*1, 16, 17*). Ref. *17* dealt with the kinetics of adsorption of vapors of *n*-hexane, *n*-octane, *n*-decane, *n*-dodecane, ethyl alcohol, propyl alcohol, and diethyl ether from a carrier gas flow (nitrogen) by zeolite CaA-II within the temperature range 20°–200°C. Measurements were carried out with single granules by the gravimetric method under dynamic conditions. The flow speed was chosen so that the kinetic was controlled by the rate of internal diffusion. Zeolite granules used in these experiments were in the form of cylinders whose lengths and diameters were equal.

Some results of experiments are shown in Figure 3 according to the linear form of Equation 17.

$$(1 - \gamma^2)^{1/3} = 1 - \frac{1}{t_\infty} \cdot t \qquad (19)$$

Figure 3 shows that experimental data are in satisfactory agreement with Equation 19. For values $(1 - \gamma^2)^{1/3} < 0.4$, corresponding to relative adsorption values $\gamma > 0.97$, experimental points were not plotted since within this range small increases of adsorption are reached in relatively large time intervals, which makes measurements unreliable. Similar results were obtained in investigations (*1, 16*) for zeolite NaX and CaA.

Values t_∞ obtained by plotting Equation 19 and the internal diffusion coefficients calculated by Equation 18 are represented in Table II. Column 7 of the table displays diffusion coefficients of substances in nitrogen (*6, 7, 10*).

Diffusion coefficients of hydrocarbons are less influenced by temperature than those of alcohols and diethyl ether, for which the dependence is close to that observed in a normal gas-in-gas diffusion. Equation 17 was derived for strong sorbable gases; thus, this equation could not be used for *n*-hexane isotherms in the higher temperature range, where the isotherm is almost linear.

Substituting $t_{0.5}$ in Equation 18 instead of t_∞, we obtain

$$D = 0.015 \frac{zR^2}{c_0 t_{0.5}} \qquad (20)$$

Similarly, for a linear adsorption isotherm

$$D = 0.308 \frac{zR^2}{\pi^2 c_0 t_{0.5}} = 0.031 \frac{zR^2}{c_0 t_{0.5}} \qquad (21)$$

The latter has a numerical ratio twice as large as the former. Calculation according to Equation 21 results for *n*-hexane at 200°C in $D = 0.012$ cm²/sec instead of the table value of 0.006 cm²/sec.

Thus, in spite of a satisfactory agreement of Equation 17 with experimental data for systems with weakly convex adsorption isotherms, the internal diffusion coefficients, evaluated for these cases according to Equation 17, are in fact below their actual value.

Table II. Values of t_∞ and D

Substance	T,°*C*	c_0 *Mg/L*	α_0 *Mg/ Gram*	t_∞, *Min.*	D • *10*³, *Cm*²/ *Sec.*	D_g[a] • *10*³, *Cm*²/ *Sec*	*No. of Granules*
n-Hexane	20	10.0	85.0	167	6.4	77.2	1
	100	10.0	71.1	131	6.7	113.6	1
	200	10.0	22.0	46	6.1	166.5	1
n-Octane	100	9.7	70.1	155	6.2	101.1	2
	112	9.7	66.6	150	6.2	106.4	2
	200	10.0	56.1	106	7.1	146.9	2
n-Decane	91	10.1	89.5	215	5.6	84.1	3
	121	10.7	77.7	150	6.0	99.6	4
	197	9.7	63.0	140	6.5	126.5	5
n-Dodecane	110	8.9	108.5	260	5.5	76.1	6
	195	7.1	48.2	130	6.2	106.1	6
	107	9.4	75.5	212	5.5	75.2	7
	111	9.8	76.9	205	5.5	76.3	7
	204	6.8	46.0	156	6.3	107.5	7
Ethyl alcohol	20	20.0	132.0	93	8.5	125.3	8
	100	20.0	110.0	56	11.7	191.2	8
	200	20.0	83.0	27	18.5	289.7	8
Propyl alcohol	100	15.7	115.0	117	7.8	159.8	9
	150	18.2	101.0	82	8.6	199.2	9
	200	12.7	80.0	66	11.2	242.2	9
Diethyl ether	150	10.0	84.8	93	10.8	174.4	10
	200	10.0	74.0	82	11.5	212.0	10
	250	10.0	59.8	43	17.5	252.8	10

[a] D_g = value of gas-in-gas diffusion coefficient.

The second mode of diffusion allows estimation of diffusion coefficients for different substances by using the value of a diffusion coefficient for a single substance, taken as a standard (*17*).

When the second mode of diffusion takes place, experimental data for $\gamma < 0.5$ can be expressed by straight lines passing through the origin by using coordinates γ and $\sqrt{t}$. In experiments with granular zeolites, frequently under vacuum conditions and sometimes for adsorption from carrier gases, these straight lines do not pass through the origin but cut off a part of the abscissa (Figure 4). [A part of the ordinate was cut off in some cases by the straight lines $\gamma - \sqrt{t}$. Causes for that will be discussed in Refs. *2, 3*.] Explanation for this lies in the fact that diffusion in crystals is slower than transport in secondary pores. This difference causes a shift of the process pattern toward the third mode of diffusion. A similar task was dealt with in studies of diffusion in polycrystalline metals (*8, 9, 11*). We are interested in another aspect. Thus, we obtain the approximate ratio between the adsorption value and the principal kinetic parameters for a sphere (a granule) of radius R, consisting of spherical particles of radius r ($r << R$) which are homogeneous with respect to their physical properties and size.

Let a granule be placed at $t = 0$ into media with constant concentration of c_0. Molecules diffuse into the granule in the secondary pores and hence into the crystals. Let us assume that the external surface of the crystals in contact with the gas phase is instantly saturated to the equilibrium value.

The distance of diffusion flow in the secondary pores is given by

$$x \sim \sqrt{D_2 t} \qquad (22)$$

where D_2 is the effective diffusion coefficient $\left(= \frac{D}{\partial a/\partial c}\right)$.

The distance of diffusion in the crystal for the same space of time is

$$y \sim \sqrt{D_1 t} \qquad (23)$$

where D_1 is diffusion coefficient in the crystal.

The relative depth of adsorbate penetration in the granule is

$$\alpha = \frac{x}{R} \sim \frac{1}{R}\sqrt{D_2 t} \qquad (24)$$

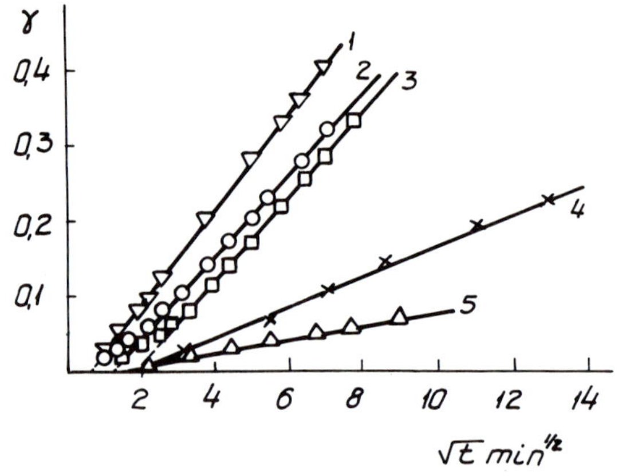

Figure 4. Crossing of abscissa by straight lines; sorption under vacuum conditions (1–3) and from a carrier gas flow (4–5)

1) n-*Heptane CaA-II*, p = *0.052 mm, 100°C*
2) *Diethyl ether Linde 5A*, p = *0.066 mm, 50°C*
3) n-*Heptane, CaA-II*, p = *0.034 mm, 20°C*
4) *Propyl alcohol, CaA-II*, c = *3 mg/l, 0°C*
5) *Diethyl ether, CaA-II*, c = *10 mg/l, 0°C*

and that in the crystal is

$$\beta = \frac{y}{r} \sim \frac{1}{r}\sqrt{D_1 t} \tag{25}$$

If $\frac{\beta}{\alpha} >> 1$ or $\frac{R}{r}\sqrt{\frac{D_1}{D_2}} >> 1$, then the kinetics of adsorption are determined by diffusion rate in the secondary porosity, which was considered for the second mode of diffusion.

If $\frac{R}{r}\sqrt{\frac{D_1}{D_2}} << 1$, then the diffusion flow in the secondary pores penetrates to a lower depth, the crystals being only partially filled.

Shallow penetration of adsorbate into crystals is

$$y = \sqrt{2D_1 t} \tag{26}$$

Since the dimensions of the cavities of synthetic zeolites are small, only accommodating a few molecules, the assumption of a layer-by-layer saturation of the crystal from surface to central parts is justified. The adsorption value in the surface layer whose thickness is y is

$$a \approx 4\pi r^2 y z_1 \tag{27}$$

where z_1 is the equilibrium value of adsorption per unit volume of the crystal.

If the whole crystal is saturated, then

$$\Upsilon_0 = \frac{4}{3}\pi r^3 z_1 \tag{28}$$

and the relative adsorption value is

$$\gamma_\beta = \frac{a}{a_0} \approx 3\beta \tag{29}$$

When the adsorbate diffuses into the granule, the crystals take part in the adsorption process in series, and the intensity of filling varies from crystal to crystal. If penetration depth is α', the general relative value of adsorption (for a granule) is given by

$$\gamma = 3\int_0^{\alpha'} \gamma_\beta\,(1 - \alpha)^2 d\alpha \tag{30}$$

Assuming $\alpha' = \frac{2D_2 t}{R}$, the approximate degree of filling of crystals near the granule surface will be

$$\gamma_\beta = 3\,\frac{R}{r}\sqrt{\frac{D_1}{D_2}}\cdot\alpha' = 3A\alpha' \tag{31}$$

whence

$$\gamma = 9A \int_0^{\alpha'} (\alpha' - \alpha)(1 - \alpha)^2 d\alpha \tag{32}$$

By integrating, we find

$$\gamma = 9A \left(0.5\alpha^2 - \frac{\alpha^3}{3} + \frac{\alpha^4}{12}\right) \tag{33}$$

or

$$\gamma = B\tau \left(1 - \frac{2\sqrt{2}}{3}\sqrt{\tau} + \frac{\tau}{3}\right) \tag{34}$$

where

$$B = 9\frac{R}{r}\sqrt{\frac{D_1}{D_2}} \quad \text{and} \quad \tau = \frac{D_2 t}{R^2} \tag{35}$$

It follows from Equation 34 that the adsorption rate is influenced by diffusion coefficients in the crystal and secondary porosity by the dimen-

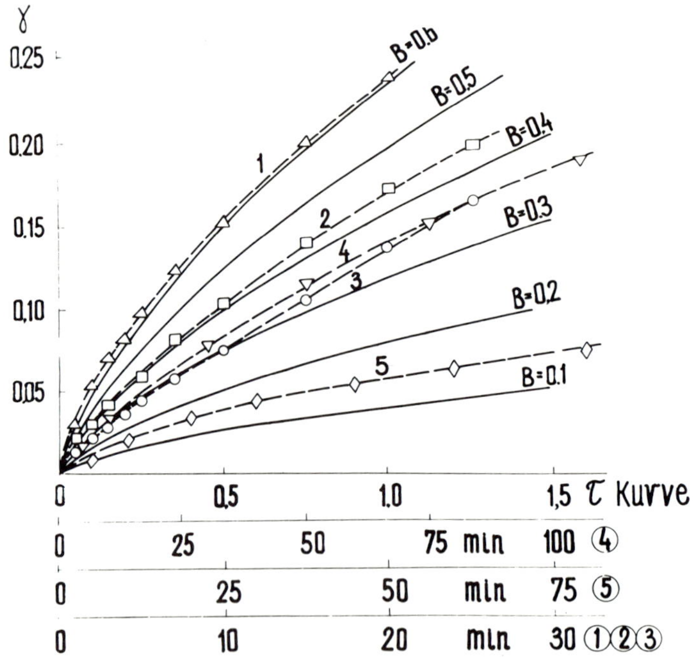

Figure 5. Curves γ — τ for various values of B *according to Equation 34*

Dotted lines represent experimental data
Curves 1–5 as in Figure 4

sions of crystals and granule. By virtue of assumptions 26 and 31, Equation 32 is valid up to $\gamma \approx 0.1$–0.2.

Figure 5 displays curves, plotted in accordance with Equation 34 for various values of B. It also shows experimental data obtained in evacuated systems and from adsorption from a carrier gas flow (nitrogen) (*15*). The experimental plots follow the same pattern as the computed ones.

According to Equation 35, the diffusion coefficient in crystals is given by

$$D_1 = \frac{B^2 r^2 \tau}{81 t} \qquad (36)$$

Equation 36 provides a basis for assessing the value D_1 from experimental data with granulated zeolites. Value B and ratio τ/t are found by $\gamma - t$ plot.

Assuming $r = 2.10^{-4}$ cm, we obtain the following values of D_1 for the data of Figure 5: *n*-heptane, 100°C– 1.5×10^{-13} cm²/sec, 20°C– 3.7×10^{-14}; propyl alcohol, 0°C– 1.6×10^{-14} cm²/sec; diethyl ether, 50°C– 6.9×10^{-14}, 0°C– 3.4×10^{-15} cm²/sec. These values are acceptable limits.

Literature Cited

(1) Alekseeva, N. I., Timofeev, D. P., Sharifova, E. M., *Zh. Fiz. Khim.* **1966,** 40, 238.
(2) Barrer, R. M., *Trans. Faraday Soc.* **1949,** 45, 358.
(3) Barrer, R. M., Ibitson, D. A., *Trans. Faraday Soc.* **1944,** 40, 206.
(4) Blinow, W. J., *Dokl. Akad. Nauk* **1946,** 52, 511.
(5) Carman, P. C., "Flow of Gases through Porous Media," Butterworths, London, 1956.
(6) Clarke, J. K., Ubbelohde, A. R., *J. Chem. Soc.* **1957,** 2050.
(7) Cummings, G. A., Ubbelohde, A. R., *J. Chem. Soc.* **1953,** 3751.
(8) Fischer, J. C., *J. Appl. Phys.* **1951,** 22, 74.
(9) Frischat, G. H., *Z. Angew. Phys.* **1967,** 22, 281.
(10) Fuller, E. N., Schetter, P. D., Giddings, S. C., *Ind. Eng. Chem.* **1966,** 5,
(11) Le Claire, A. D., *Phil. Mag.* **1951,** 42, 468.
(12) Ponomarev, A. S., Sharifova, E. M., Timofeev, D. P., *Dokl. Akad. Nauk* **1967,** 177, 395.
(13) Ponomarev, A. S., Timofeev, D. P., Sbornik "Tseolity, ikh sintez, svoistva i primenenie". Izd. "Nauka", 1965.
(14) Timofeev, D. P., *Zh. Fiz. Khim.* **1965,** 39, 2735.
(15) Timofeev, D. P., Tverdokhleb, N. A., Sbornik Trudov 3-go Soveshchaniya po Adsorbentam, 1969 (in print).
(16) Timofeev, D. P., Tverdokhleb, N. A., *Zh. Fiz. Khim.* **1966,** 40, 2351.
(17) Timofeev, D. P., Tverdokhleb, N. A., Sharifova, E. M., *Zh. Fiz. Khim.* **1968,** 42, 2899.
(18) Weisz, P. B., Goodwin, R. D., *J. Catalysis* **1963,** 2, 397.

RECEIVED January 30, 1970.

61

Some Perspectives on Zeolite Catalysis

PAUL B. VENUTO

Mobil Research and Development Corp., Paulsboro, N. J. 08066

The last 10 years have seen an extraordinary expansion in the catalytic applications of zeolites, particularly in the petroleum industry effort in catalytic cracking. Catalysis of a broad spectrum of "classical" organic reactions has been discovered for various base-exchanged forms of crystalline aluminosilicates, notably for the faujasite family. In this review, these reactions are classified, and certain mechanistic and phenomenological features are highlighted. Included are olefin-forming β-eliminations, olefin reactions, α-eliminations and reactions of one-carbon fragments, reactions involving electrophilic attack on aromatic rings, carbonyl condensations, a wide variety of molecular rearrangements, and oxidation–dehydrogenation type reactions (over transition metal-exchanged zeolites). Also considered are hydride transfer processes, reactions on zeolite external surfaces, intracrystalline aging pathways, and systems involving inhibition of reaction by adsorbed reactant.

The pattern of progress in a given scientific field resembles, in many respects, the advance of the tide on an irregular coast. The onrushing waters, while sometimes penetrating deeply inland, often leave large prominences of land intact and exposed. This typifies the history of catalysis over zeolites since the pioneering studies of Rabo *et al.* (*48*) and Weisz *et al.* (*70*) in the early 1960's. Considerable structural knowledge now exists; a great deal is known about some reaction patterns; a degree of understanding has been gained about the relationship of structure to reactivity in catalysis. Yet, significant and fundamental elements in the framework of understanding still are lacking. Many questions remain unanswered.

A certain chemical reaction may occur in concentrated sulfuric acid, or in the presence of aluminum halides, or over amorphous silica–alumina. The same reaction may proceed—with differences—over a zeolite catalyst.

But why is the reaction pattern changed over the zeolite? Is it the result of a specific site geometric or electronic configuration, or site energy distribution, or site spatial separation that is peculiar to a zeolite? Is it actually the consequence of a metal function that happens to be in a zeolitic environment? Does it merely reflect reduced degrees of freedom for the reactant—the concentrating effect of adsorption on the vast internal surface area of the porous aluminosilicate? Are we seeing the distorting influence of some unusual pore-channel geometry superimposed on what otherwise would be a "normal" product distribution? Are there unsuspected variables influencing which molecules are occluded and which are not; which react and which do not; which move quickly through the pores and which are more strongly held?

We will review selected topographic features of organic catalysis over zeolites, examining more recent findings and some earlier work with the perspective of a somewhat different system of mechanistic and phenomenological classification.

Olefin-Forming β-Eliminations and Related Reactions

The formation of olefins generally results when a proton is lost from 1 of 2 adjacent carbon atoms in a ring or chain, and a nucleophile, X, is lost from the other. Such β-eliminations (Equation 1) are among the

$$X-\overset{\alpha}{C}-\overset{\beta}{C}-H \xrightarrow[\text{Zeolite}]{-HX} \overset{\alpha}{C}=\overset{\beta}{C} \qquad (1)$$

$$X = -OH, -OCOCH_3, -Cl, -Br, -I, -SH$$

most characteristic reactions observed over zeolite catalysts. These have been reviewed in detail earlier (*62*), and we will comment only briefly on some recent studies.

Anderson and Imai (*2*) have studied β-elimination involving a different leaving group, acetate anion. Both *n*- and *sec*-butyl acetates decomposed cleanly over faujasite, mordenite, and A-type zeolites to form mixed linear butenes as shown. Poisoning with product acetic acid occurred in some cases.

$$C_4H_9-O-\overset{O}{\overset{\|}{C}}-CH_3 \xrightarrow[\text{1 atm}]{190°-290°C} n\text{-}C_4H_8 + HO-\overset{O}{\overset{\|}{C}}-CH_3 \qquad (2)$$

Barrer and Kravitz (*5*) reported some conversion of isoamyl- and cyclohexyl chlorides over CaA with long contact times at temperatures as low as 20°C. They observed ready conversion of *tert*-butyl halides over CaA, MgA, and Na mordenite at 20°–45°C. Mixtures of C_8-, C_{12}-, and C_{16}-polyolefins were formed, with trimer most abundant. It appears that either the polymerization rate becomes sharply lower after the trimer stage, or that a competing depolymerization reaction occurs. Since dimer formed small amounts of trimer (in addition to tetramer) when exposed to CaA at 25°C, a β-scission reaction, *i.e.*, $C_{16} \rightarrow C_4 + C_{12}$, may have been operative at these low temperatures. Significantly, double bond isomerization in olefins has been reported at −80°C over HY (*62*).

Mochida and Yoneda (*40*) recently studied elimination reactions of 1,1,2-trichloroethane over zeolites as shown:

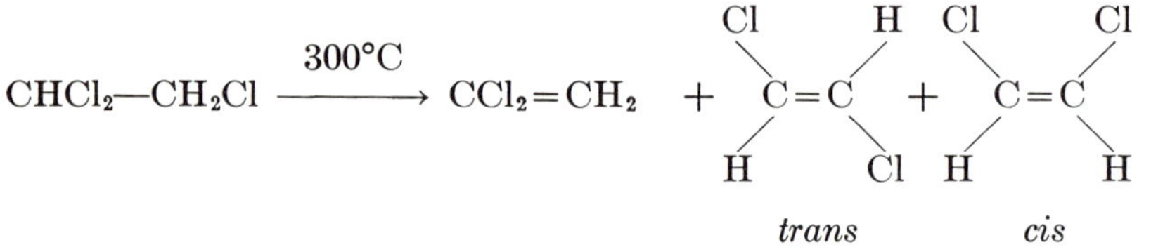

(3)

Low reactivity was noted for Linde 5A, and the relative yields of isomeric dichloroethylenes varied greatly with the catalyst. More acidic sieves (HY, MgX) generally showed lower 1,1/1,2 and trans/cis ratios, *i.e.*, cis > trans > 1,1; others (NaX, KX) gave unusually high (greater than equilibrium) trans/cis ratios.

Olefin Transformations

The fundamental carbonium ion–type reactions of olefins—including double bond and carbon skeleton isomerization, polymerization, isotopic exchange, and hydrogen transfer—have been reviewed earlier (*62*). The importance of a thorough understanding of the nature of olefin transformations over zeolite catalysts cannot be underestimated. Probably the most important and frequently recurring pattern is the transfer or redistribution of hydrogen that is observed with olefins over acidic crystalline aluminosilicate catalysts.

Venuto *et al.* (*60*) reported that at 60°C 1-hexene underwent isomerization and polymerization over REX—with formation of occluded intracrystalline polyolefins as large as $C_{30}H_{60}$—but that no aromatics were formed. Eberly (*16*), in studying the transformations of 1-hexene over HY in a high-temperature infrared cell, noted the adsorption of the olefin and the disappearance of C═C bond character at 93°C. At 150°C, hydrogen redistribution began, as evidenced by the appearance of a polyene-type band at 1600 cm^{-1}. At higher temperatures, 200°–260°C, the infrared band shifted to 1585–1580 cm^{-1}, a frequency characteristic of aromatic ring structures.

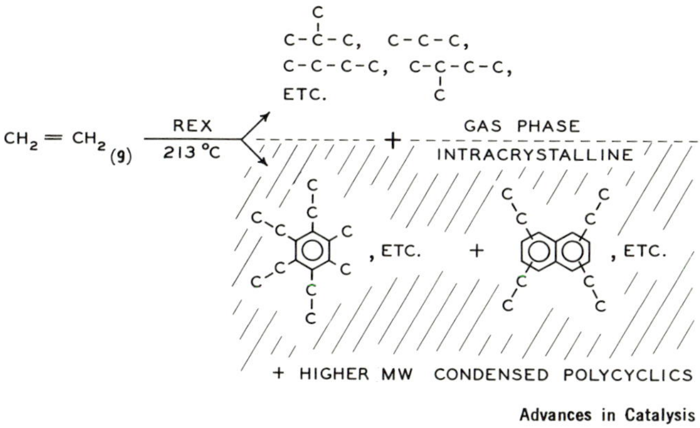

Figure 1. Schematic visualization of intracrystalline hydrogen transfer reactions of ethylene (62)

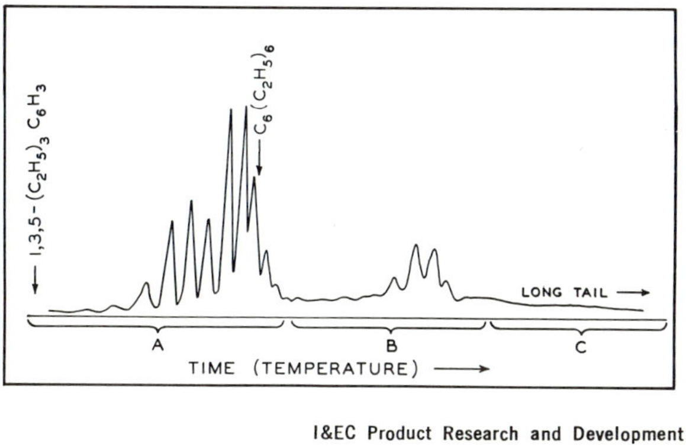

Figure 2. Gas chromatogram of complex mixture of products extracted from REX after intracrystalline reaction of ethylene at 213°C (58)

A similar pattern has been observed with C_2–C_5 monoolefins (*16, 60*). The reactions of the simplest olefin, ethylene, over REX at 213°C (Figures 1 and 2) typify the complexity of these transformations. After initial acid-catalyzed polymerization of ethylene to low molecular weight aliphatic polymer, a series of intermolecular hydrogen transfer and dehydrogenation–cyclization reactions accompanied by skeletal isomerization and β-scission must occur, leading to the observed products. As noted by Thomas and Barmby (*55*), such processes are similar to those occurring over silica–alumina and to conjunct polymerization, in which an olefin is converted by a strong acid into paraffins and hydrogen-deficient products. In concentrated sulfuric acid, such hydrogen-deficient products are cyclic allylic carbonium ions such as A or B, with B deriving from polyene C (*14*). Such cycloalkenyl cations undergo sig-

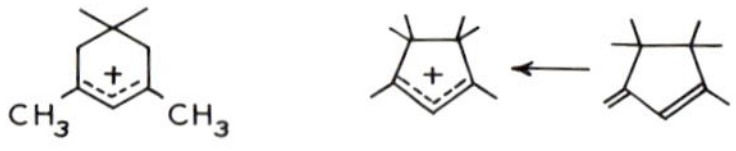

A B C

nificant rearrangement and (hydrogen–deuterium) exchange in 96% sulfuric acid. Somewhat similar hydrogen-transfer reactions also accompany ethylene polymerization at low temperatures over promoted aluminum chloride (*54*).

The cyclopropane ring is highly strained, and resembles the olefinic double bond in certain aspects; the bonds between carbons are "bent" somewhat like bananas (A), and 3 regions of high electron density lie outside the triangle of carbon nuclei (*24*). Habgood *et al.* (*6*) studied

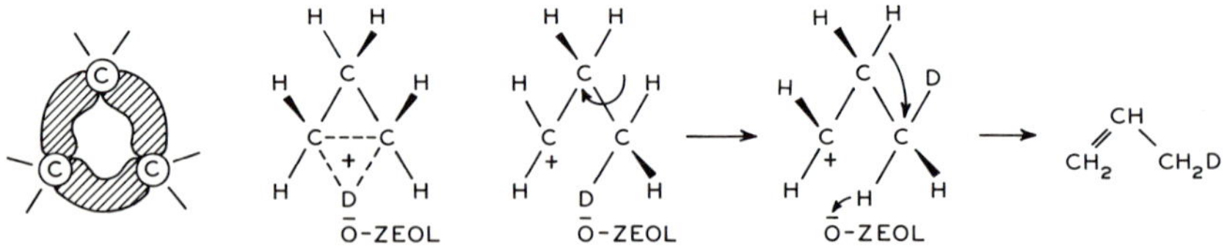

A **B** **C** **D** **E**

Scheme 1

the isomerization of cyclopropane over deuterated NaY zeolite at 200°–400°C. Reaction was promoted by small amounts of water, and a Bronsted mechanism was proposed in which both isotopic exchange and isomerization to propylene (E) proceeded *via* nonclassical carbonium ion (B). The isomerization step C → D involves an intramolecular hydrogen transfer.

Some Unusual Chemical Pathways and Reactions of One-Carbon Fragments

Many organic reactions occurring over crystalline aluminosilicates can be rationalized as proceeding *via* carbonium ions, or related organic cations. Scrutiny of the literature and examination of some recent work suggests that there may be exceptions. For instance, a free radical mechanism for isomerization of 1-butene over Cu-exchanged NaX has been proposed at low copper contents (*15*).

Less apparent is the mechanism of the reaction of α-methylnaphthalene and methanol over KX and RbX (*29*) shown in Scheme 2. Aging was extremely rapid, and the gases were mainly H_2, CO, and dimethyl

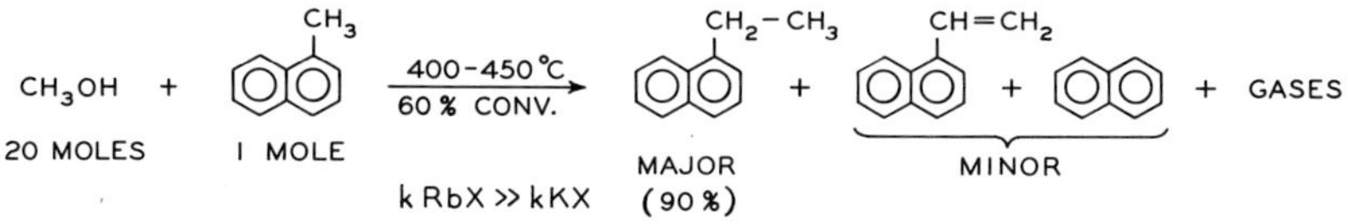

Scheme 2

ether, with traces of CO_2 and CH_4. An analogous product distribution was observed when toluene was substituted for α-methylnaphthalene. If the reaction proceeded *via* $CH_3OH \rightarrow HCHO + H_2$, α-vinylnaphthalene, and not α-ethylnaphthalene, would have been expected as major product. If at these high temperatures one considers the benzylic protons of

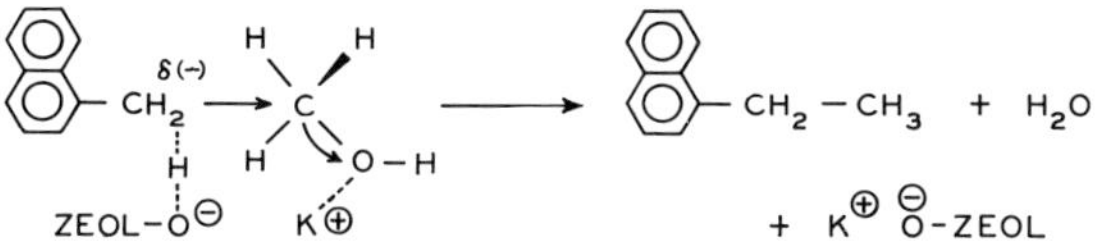

α-methylnaphthalene as weakly acidic and the zeolitic anionic lattice as weakly basic, a mechanism based on nucleophilic attack of a carbanion-like species on methanol (Equation 4) is not inconceivable.

Reactions of methanol (*37*), methyl halides (*4, 10*), methyl mercaptan (*62*), and benzyl mercaptan (*30*) over zeolites have been reviewed earlier (*62*) in terms of the possible intervention of carbenoid type intermediates or α-eliminations (Equation 5).

$$H—CH_2—OH \longrightarrow H_2O + [:CH_2] \quad (5)$$

Wu *et al.* (*74*) found that the thermal decomposition of intracrystalline tetramethylammonium cations (*A*) at 150°–400°C in Y-type faujasite, $(Na)_{28.4}(TTMA)_{17.7}(H)_{6.0}(AlO_2)_{52}(SiO_2)_{140}$, proceeds by a number of different pathways (Scheme 3). One path involves nucleophilic

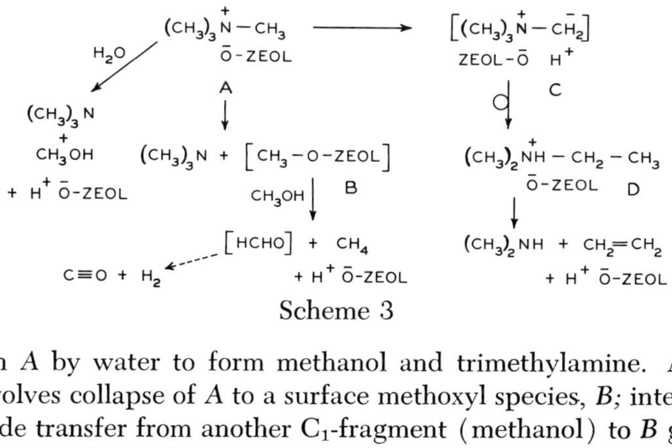

Scheme 3

attack on *A* by water to form methanol and trimethylamine. A second route involves collapse of *A* to a surface methoxyl species, *B;* intermolecular hydride transfer from another C_1-fragment (methanol) to *B* generates methane and, ultimately, CO and H_2 (methyl disproportionation reaction). Tetramethylammonium cation *A* may also react to form the ylide species, *C*. Stevens rearrangement (*73*) in *C* (intramolecular carbanionic attack) to form dimethylethylamine followed by reprotonation gives *D*. Ethylene is generated then by simple Hoffmann elimination. Thus, profound redistribution of hydrogen as well as C—C bond formation characterize this organic cation decomposition.

From the structural standpoint, the common feature of these one-carbon reactants, Y—CH_2—X (X = —OH, —SH, $-\overset{+}{N}\langle$) is the absence of a β-hydrogen atom allowing β-elimination and the formation of a C═C double bond; that is Y = H or aromatic carbon. From a reaction aspect, they all appear to generate olefins and, in some cases, undergo methyl disproportionation reactions. Whether or not the

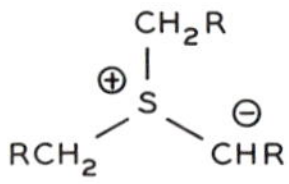

A

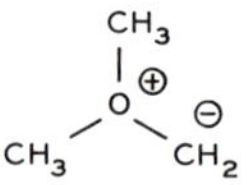

B

ylide–carbanion type mechanism of olefin formation shown for the nitrogen system in Scheme 3 can be extended to the sulfur system (*A*) or even conceivably to the oxygen system (*B*) is uncertain.

Electrophilic Aromatic Substitution and Related Reactions

Considerable work was done at an early stage in the area of electrophilic aromatic substitution reactions over zeolites, and a fairly coherent understanding has evolved (*62*). Basically, the process involves the attack of a positive species (electrophile, *E*) on an aromatic ring (ArH), formation of a new bond, and elimination of a proton, as shown in Scheme 4. The electrophile, *E*, may be generated from olefins or carbonyl compounds by protonation or related process; from alkyl or acyl halides, alcohols, and ethers by C—X or C—O heterolysis; from alkyl aromatics and paraffins by hydride abstraction or β-scission; from benzene, *via* actual ring fragmentation. A proton itself behaves as an electrophile, in contrast to hydrogen species generated in the presence of transition metals such as Pt (*20*) or Ni (*45*). Benzene bromination—presumably *via* electron-deficient Br-species—also has been reported (*33*) over iron-containing zeolites.

Alkylation. It is instructive to highlight briefly the key features of the earlier work on alkylation (*56, 58, 60, 61, 63, 64*) before discussing more recent work in this area. Based on an analysis of the structures of product alkylaromatics, patterns of substrate reactivity, and side reaction pathways, alkylations catalyzed by crystalline aluminosilicate catalysts

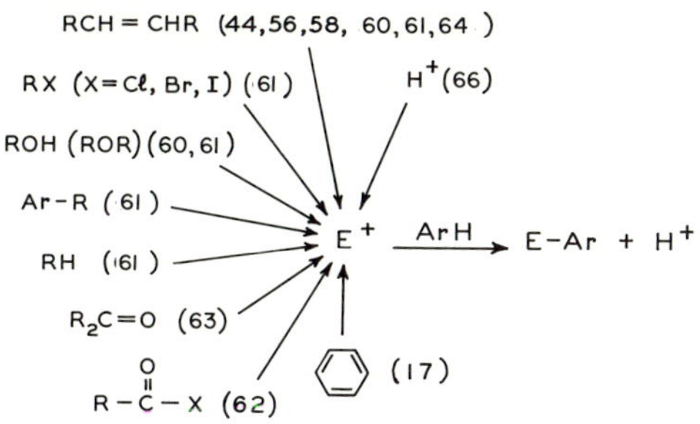

Scheme 4

generally proceed *via* carbonium ion type mechanisms; they show great similarity to the corresponding features commonly reported for electrophilic aromatic substitution in the presence of strong protonic acids such as concentrated H_2SO_4, liquid HF, or promoted Lewis acids. In general, *ortho,para*-orientation is observed, and in competitive situations, selectivity for attack on the more reactive (nucleophilic) aromatic nucleus is shown. A more quantitative expression of the ring orientation and substrate selectivity—at the interface between heterogeneous and homogeneous catalysis—has been found in obedience to the Brown selectivity

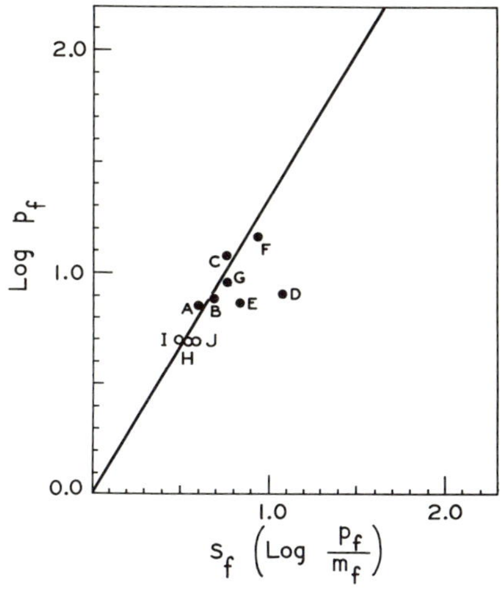

Figure 3. Plot of log p_f *and selectivity factor,* S_f

● *Data from competitive hydrogen deuterium exchange reactions of toluene and benzene over DHY* (66)
○ *Data from competitive ethylation reactions of toluene and benzene over REX* (56)

relationship (Figure 3) in zeolite-catalyzed ethylations and isotopic exchange.

Qualitative evidence (*56, 60, 62, 63, 64*) points toward operation of a Rideal-like mechanism for crystalline aluminosilicate-catalyzed alkylations. This is visualized for an olefin alkylation over a deuterated zeolite (*60*) in Scheme 5. Significantly, most modified faujasite-catalyzed alkyla-

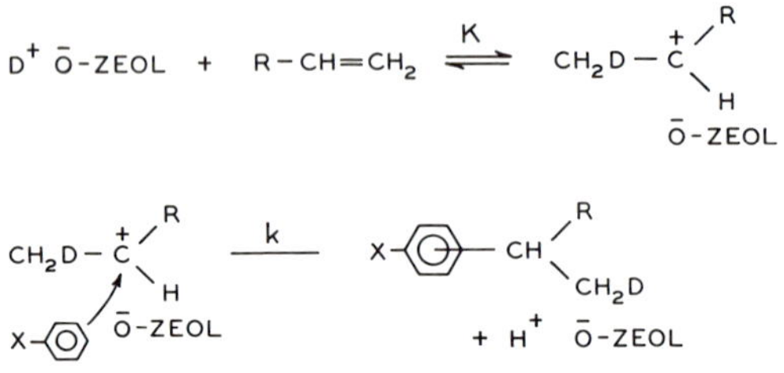

Scheme 5

tions proceed with greatest efficiency in the liquid phase. For similar reactants, they generally catalyze alkylation at lower temperatures than do amorphous silica–alumina-type catalysts. On the other hand, strong protonic acids such as HCl–$AlCl_3$ or 98% H_2SO_4 often show significant activity at lower temperatures than modified faujasites. Some possible explanations for the "temperature disguise" have been advanced (*60, 62*).

Pickert *et al.* (*44*) presented data on benzene–propylene alkylation over faujasite-type catalysts in which alkylation activity was enhanced as the temperature of calcination was increased. They observed maximum alkylation activity when all residual OH groups associated with catalytically-active sites were removed, and suggested that carbonium ion-like species were formed *via* polarization of reactant hydrocarbons by cation fields. Rabo *et al.* (*47*) conducted a similar study on the alkylation of toluene and propylene at 100°C over LaY faujasite. They observed that changing the temperature of activation had no significant effect on the alkylation activity—LaY, whether dehydrated at 550°C or activated near 700°, showed high activity for toluene propylation near room temperature. They suggested that either oxygen-deficient lattice sites (Lewis sites) created in LaY during activation at 700°C had activity for alkylation, or that some other unsuspected site configuration was responsible for this activity.

Haag (*25*) has examined the kinetics of toluene isopropylation and ethylation in the presence of REY catalyst. At 1 atm (total pressure) in the liquid phase, toluene alkylation with propylene is uncontrollably fast near 25°C, and decreases in rate with increasing temperature up to the toluene boiling point (110°C) because of decreasing partial pressure and solubility of propylene (some relative rates: $r_{100°C}/r_{110°C} = 3.1$). Even at 110°C, the rate is very fast. Under conditions where external gas–liquid and bulk liquid mass transfer problems are eliminated, the rate amounts to 3060 mmole of toluene reacted/hr/gram of catalyst. By comparison, rates of 60 times lower were reported by Rabo *et al.* (*47*) for the same isopropylation under similar conditions, suggesting the possible existence of external mass transfer problems. In agreement with expectations based on classical carbonium ion concepts, Haag observed that ethylation of toluene proceeds 10^{-3} times the rate of isopropylation.

Alkylation reactions, on solid ion exchange resins as heterogeneous Bronsted acids, obey Rideal kinetics (*25*), as shown in Equation 6; at low pressure (P olefin < 2 atm), where K Ol $<< 1$, the rate law simplifies to Equation 7.

$$r = k\,[\mathrm{Ar}]\,\frac{K\,[\mathrm{Ol}]}{1 + K\,[\mathrm{Ol}]} \qquad (6)$$

Ar = Aromatic
Ol = Olefin
K = Olefin adsorption constant

$$r = k'\,[\mathrm{Ar}]\,[\mathrm{Ol}] \qquad (7)$$

For toluene ethylation over REY catalyst, Haag (*25*) has established the rate law:

$$r = k'\,[\mathrm{Tol}]\,[\mathrm{Eth}] \text{ where } k' = kK \qquad (8)$$

k = second order rate constant
K = ethylene adsorption constant

At 80°C, the value of the pseudo second order rate constant, k', was 2.05×10^{-2} l^2 mole^{-1} hr^{-1} $(\text{gram cat})^{-1}$. These observed kinetics provide firm quantitative support for a Rideal mechanism, which requires fast, reversible, and noncompetitive adsorption of the olefin on the catalyst acid sites (*see* Scheme 5 above). The strictly first order dependency of toluene proves freedom from intracrystalline diffusion problems with respect to ethylene. Likewise, the product distribution, which contains only traces of disubstituted toluenes, shows absence of diffusion limitation for the product, ethyltoluene. Intracrystalline diffusion problems may, however, exist in the isopropylation experiments with extremely active catalysts.

Other Transformations of Alkylaromatics. An extensive amount of work has been done in the area of alkylaromatic isomerization–transalkylation, dealkylation, and related areas (*7, 8, 13, 32, 34, 35, 36, 41, 42, 60, 61, 62*). Ward (*67, 68, 69*) also has conducted a series of very thorough studies into the relationship of activity in catalytic reactions, such as cumene dealkylation and alkylaromatic isomerization, and structural properties in synthetic faujasites, using IR spectroscopy among other tools. In many of these reaction systems (*68, 69*), catalytic activity was related to Bronsted acidity.

Bolton *et al.* (*8*) proposed that isomerization of diethylbenzenes at 170°C over a modified Y-type faujasite occurs *via* a transalkylation mechanism involving diphenylethane type intermediates. A similar bimolecular mechanism was proposed for xylene isomerization (*32*). However, as Csicsery (*13*) observed, this reaction cannot proceed when easily-extractable α-hydrogen atoms are unavailable—*i.e.*, when the alkyl group is *tert*-butyl. In his studies of the isomerization of 1-methyl-2-ethylbenzene over Ca–NH_4–Y-type faujasite, Csicsery has shown further that intermolecular isomerization predominates at temperatures below 200°C, while at temperatures above 300°C, isomerization proceeds *via* 1,2-shifts.

Reactions Involving Condensations of Carbonyl Functions

Carbonyl condensation reactions occur with great facility over zeolite catalysts (*26, 62, 63*). From the organic structural standpoint, the common feature—and usually the primary event—is the attack of a nucleophile, *A*, on the positive end of the dipole of a C═O bond, forming a new C—A bond and a hydroxyl group. In Equation 9, the carbonyl

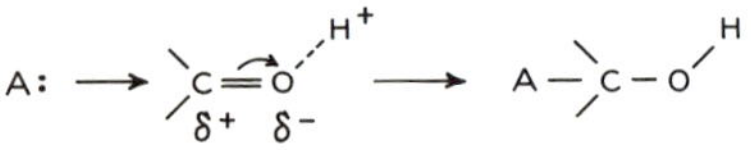

(9)

group is represented in conjugate acid form; other electron-deficient structures on the surface of crystalline aluminosilicates also could interact with the negative end of the dipole and assist in the shift of double bond electron density. The carbonyl function can reside in either aldehyde or ketone.

If the structural configuration shown in the left-hand member of Equation 10 exists in the primary condensation product, then a great

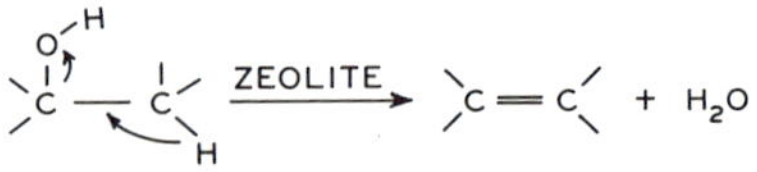

(10)

driving force for elimination of water exists in the presence of a zeolite surface. A further tendency toward formation of cyclic trimers, tetramers, etc., is observed under somewhat more severe conditions (*63*).

In the Cannizzaro reaction (*62*), which involves self-condensation of aldehydes that have no α-hydrogen atoms, the nucleophile is represented best as a hydride ion; in the Prins reaction (*26*), the nucleophile is an olefin; in the aldol condensation (*63*) and reaction of methyl acetate with formaldehyde (*62*), the nucleophile is a methylene group α- to a carbonyl function, which may be visualized conveniently as a carbanion; in the bisarylalkane synthesis (*63*), the nucleophile is an aromatic ring; in formation of heterocyclics (alkylpyridines) (*62*), the nucleophile is NH_3; in the formation of acetals and ketals (*62*), the nucleophile is an alcohol.

Other Rearrangements and Reactions

The Beckmann rearrangement of ketoximes to the corresponding amides (*31*), the Fischer indole cyclization, isomerization of epoxides to the corresponding aldehydes, ketones, or alcohols, hydration and ammonolysis of epoxides, oxygen–sulfur interchange, formation of diaryl–ureas and –thioureas from condensation of aniline and carbonyl sulfide, and olefin carbonylation occur over zeolite catalysts (*62*). The oxo reaction over rhodium and cobalt containing zeolites recently has been claimed (*22*).

Hydrogen Transfer Reactions

As emphasized earlier, the redistribution or transfer of hydrogen is one of the most dominant and recurrent reactions of olefins in the presence of acidic zeolite catalysts. The formation of hydrogen-rich paraffins and hydrogen-deficient aromatics is superimposed constantly on any reaction where a low molecular weight olefin—or olefin precursor—is either a reactant, a product, or an intermediate. Even more specifically, numerous examples of hydride-transfer processes during reactions over zeolite catalysts have been observed, and we will discuss some of these in detail.

The common feature in such reactions is the attack of a hydride ion on a relatively electron-deficient carbon atom. The hydride may migrate from another carbon atom in the same molecule, from a carbon atom on another molecule, or even from a Lewis-type site (—Al- - -H^-) on the catalyst. Whatever the source, the structural environment to which the hydride adds becomes more hydrogen-rich, and that from which it derives, more hydrogen-deficient. In Scheme 6, some of these reactions are tabulated.

Let us consider each of the reactions in Scheme 6: Reaction A involves intramolecular hydride transfer to the electron-deficient β-carbon atom following protonation of an epoxide, with the net effect being isomerization; the hydride-transfer to acylium ion in reaction B presumably occurs after initial attack of CO on adsorbed isopropyl carbonium ion in propylene carbonylation; in C and D, hydride transfer occurs to carbonium ions generated after initial attack (alkylation) by

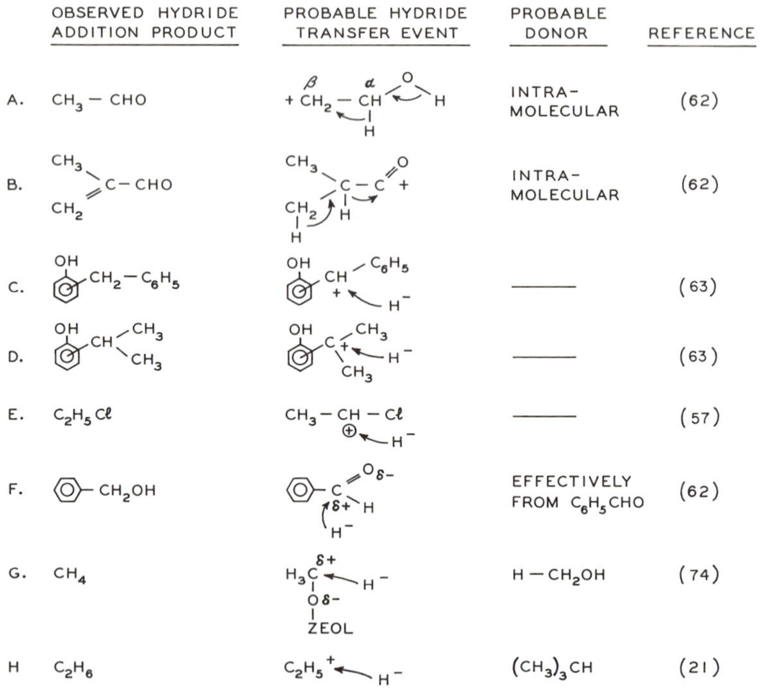

Scheme 6

phenol on benzaldehyde and acetone, respectively; in E, hydride transfer occurs to the carbonium ion derived from protonation of vinyl chloride, the primary product from dehydrohalogenation of 1,2-dichloroethane.

In F, benzyl alcohol is the hydrogen-rich product from hydride transfer to benzaldehyde in the Cannizzaro reaction; the hydrogen-deficient species is benzoic acid. Hydride transfer to the carbonium ion derived from benzyl alcohol also results in the formation of toluene, an even more "reduced" species. As shown in Equations 11 and 12, these Cannizzaro-like transformations are essentially redox type, at least in respect to hydrogen balance.

$$C_6H_5CHO + C_6H_5CHO \longrightarrow C_6H_5CH_2OH + C_6H_5COOH \quad (11)$$

$$C_6H_5CH_2OH + C_6H_5CHO \longrightarrow C_6H_5CH_3 + C_6H_5COOH \quad (12)$$

In G, the "methyl disproportionation" reaction observed in thermal decomposition of tetramethylammonium cations, the hydride receiver may be a surface methoxyl group, while the hydrogen-deficient ("oxidized") moiety is a formaldehyde-like species, and ultimately, $C{\equiv}O$ and H_2. Ethane was detected (H) as a minor product during the alkylation of isobutane with ethylene over REHX catalyst; assuming a classical carbonium ion chain reaction mechanism (*21*), ethane probably arose *via* hydride transfer from isobutane (or some other branched paraffin which forms a stable carbonium ion) to the carbonium ion generated by protonation of ethylene.

Oxidation, Dehydrogenation, and Related Reactions

Most crystalline aluminosilicates appear to have little intrinsic catalytic activity for hydrogenation–dehydrogenation type reactions, excluding hydrogen transfer reactions (*18, 19*). Minachev *et al.* (*38*) report slight catalytic activity of NdY for the hydrogenation of ethylene at 150°C; such activity may be an effect of the rare earth cations. When metallic or chalcogen functions are introduced into zeolites, the situation dramatically changes; one then must ask to what extent the observed reaction pattern really reflects the structural influence of the zeolite and to what extent it derives from the extraneously introduced function which happens to be supported on a zeolite.

The formation of α-methylstyrene in the reaction of cumene over Ba-faujasite at 550°–600°C (*49*) almost certainly occurs *via* thermal dehydrogenation, and the ethylene and acetylene present in the light gas stream probably arise from thermal (free radical) cracking. On the other hand in Ni-, Co-, Fe- or Cr-containing X- or A-type catalysts, moderate paraffin dehydrogenation activity—similar to that observed in nonzeolite catalyst systems—has been observed (*18, 19*). Zn-, Co-, and other transition metal zeolites also show activity (in the presence of NH_3) for the dehydrogenation of olefins to dienes, alkylaromatics to alkenylaromatics, and alkylaromatics to nitriles (*62*). Dehydrogenation of cyclohexene to benzene over Ni-X at 90°–140°C (*9*) and of simple alcohols to ketones or aldehydes (Equation 13) over Cr-containing X- and A-type zeolites at 300°–390°C (*43*) also have been reported.

(13)

The oxidation of intracrystalline ammonium cations has been reported (*3, 65*), but most oxidations over zeolites featuring molecular oxygen–hydrocarbon systems have featured transition metal zeolites. The oxidative dehydrogenation of ethylbenzene to styrene and selective oxidation of benzyl alcohol to benzaldehyde over MnY at 250°–370°C were reported earlier (*62*). Propylene has been oxidized to formaldehyde, CO_2, and minor quantities or acrolein and acetaldehyde over Cu(II)-exchanged Y-type zeolite (*39*).

Fripiat and coworkers have studied a number of transition metal zeolite-catalyzed oxidations recently (*50, 51, 52, 53*). In the liquid-phase oxidation of *n*-hexane (*52, 53*) at 160°C and pO_2 = 25 atm, yields of C_1–C_5 carboxylic acids, especially acetic acid, were greater relative to other products (alcohols, ketones, CO_2) in the presence of Ni(II)-, Co(II)-, and Mn(II)-X type zeolites than in their absence. In oxidations of *p*-xylene (*51*) in acetic acid (165°C, 32 atm total pressure), 90% conversion with 55% selectivity for formation of terephthalic acid was obtained with Co–Mn-X zeolite. The performance of the mixed cation-containing zeolite far surpassed that of either the Mn- or Co-exchanged zeolite alone. From magnetic susceptibility studies, a relationship between the high activity of the zeolites and promotion of electron-unpairing in the cations by the zeolite support was proposed. In propylene oxidation (*50*) (benzene, liquid phase, 150°C, 12 atm initial pO_2, 45 atm *p* total), low conversion (7.5%) but high selectivity for propylene oxide (up to 70%) was observed in the presence of Mo 13X. These authors concluded that the high selectivity of the Mo-zeolite—as contrasted to Mo-oxide suspensions, dissolved Mo-stearate, MoO_3/Al_2O_3—reflected a combination of the intrinsic catalytic activity of the crystalline aluminosilicate lattice and the activating effect of the lattice on the Mo-ions with which it was coordinated.

Shape-Selective Catalysis, Reactions on External Surface, and Special Effects

The principle of molecular shape-selective catalytic reaction was established by Weisz and associates in 1960 (*70*), and "molecular engineering" aspects have been reviewed recently by Chen and Weisz (*12*). The selective sorption of molecules differing only slightly in their critical dimensions has been applied in the development of industrial separation processes to remove straight chain hydrocarbons from a mixed hydrocarbon stream. In shape-selective catalysis, on the other hand, molecules of proper dimensions are continuously entering and leaving the intracrystalline cavities of the molecular sieve, thus allowing specific selectivities to occur. Reactant selectivity (Scheme 7A) occurs when 1 of 2

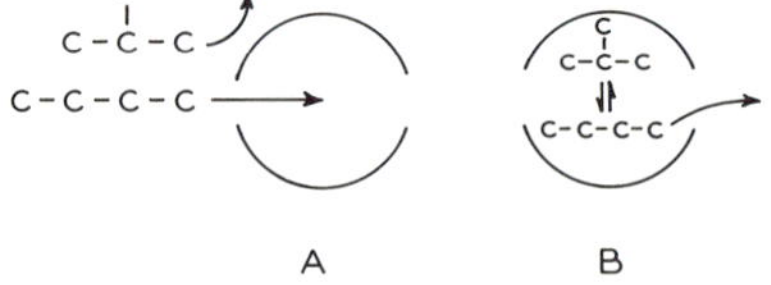

Scheme 7

classes of reactant molecules can pass through the pores. Product selectivity (Scheme 7B) occurs when, among all the product species formed within the pores, only those with the proper dimensions can diffuse out and appear as observed products. Formation of bulky products may occur within the cage, and these either may be drained off *via* equilibration to less bulky species, or eventually lead to pore blockage.

It has been concluded that, for most cases, catalysis over zeolites occurs within the intracrystalline voids. Strong supporting evidence for this was provided by Weisz (*71*), who compared the rates of dehydration of 2-butanol over Linde 10X and 5A zeolites at relatively high temperatures and low conversion. The rate constant per unit volume of 5A was 1/100–1/1000th that for 10X, a magnitude consistent with the ratio of available surface areas for the external area of 1–5μ-sized 5A crystals and for 10X, where the internal surface area was available to the alcohol. The strong driving force for occlusion within the intracrystalline zeolite voids is exemplified by the rapid adsorption kinetics and rectangular adsorption isotherms observed for molecules whose dimensions are not close to those of the entry pores.

Notwithstanding the above considerations, some catalytic role can be attributed to the external surface, particularly at high temperatures. Quite a few examples from the recent literature can be cited where reaction of olefins and olefin-precursors too large to pass through the 8-ring 4–5A entry pores of Linde 5A and chabazite has been observed (*62*). Similarly, formation of acetophenone tetramer over HY faujasite (with its 12-ring 8–10A entry pores) also must have occurred on the external surface (*63*). If we assume that the external surface area of a 1-μ cubic crystal (3–4 m^2/gram) of a typical zeolite is 1/200–1/300 of the total surface area, it becomes apparent that relatively few potential catalytic sites are available. Further, there is no reason to assume that these sites are similar in strength and structural environment to those within the narrow zeolite pore–channel system. The environment and concentration of reactant molecules is totally different, and there is freedom from possible intracrystalline diffusion restrictions. Differences in reactivity and selectivity thus may be expected to occur.

Katzer (*28*) observed that counterdiffusion of benzene and cumene within the pores of H-mordenite does not occur at low temperatures. However, H-mordenite shows activity for the alkylation of benzene with propylene to form cumene under the liquid phase conditions used for the diffusion studies, and he has suggested that reaction must occur on the external crystallite surface, or just within the pore mouth. In earlier studies on the isomerization of 2,3-dimethylbutene-1 at 0°–20°C over a deuterated Y-type faujasite (*62*), we observed that the extent of isomerization (2,3-dimethylbutene-2) was far greater than the extent of deuteration; only a fraction of the total deuterium on catalyst OD groups was exchanged. One possible explanation—assuming a protonic isomerization mechanism—is that, because of lowered intracrystalline diffusion rates

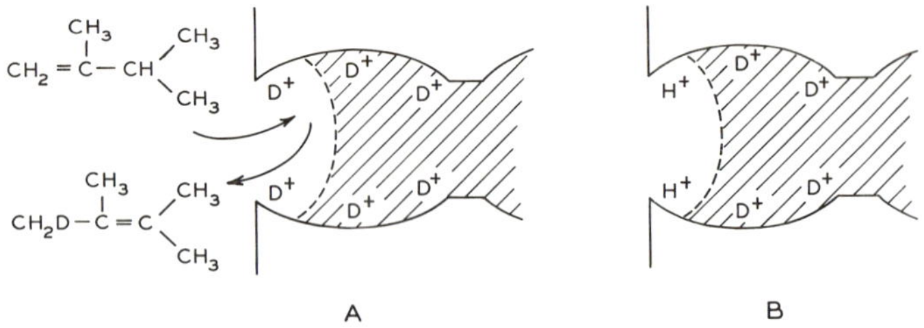

Scheme 8

under these conditions, only sites near the external surface were actually accessible (Scheme 8,A) (pore mouth catalysis), and that these were used repeatedly, with the majority of intracrystalline OD groups (shaded areas, Scheme 8 A, B) being effectively unavailable. A similar net effect would be observed if initial polymerization occurred, to form highly branched dimer or trimer that was adsorbed strongly near the pore mouth.

Apropos of the external–internal surface "problem," Thomas and Barmby (*55*), from calculations on published data, have suggested that the primary cracking of gas oil molecules (m.w. 175–252) occurs in the amorphous silica–alumina matrix or on the external surface of zeolite crystallites dispersed in the matrix in cracking catalysts such as Durabead 5. That is, the rate of diffusion of gas oil molecules through the zeolite pore–channel system may be too slow to account for the observed rates of cracking, and β-scission to form olefins takes place mainly outside of the molecular sieve crystals. The cracked fragments then diffuse into the zeolite cavities, and there undergo hydrogen transfer to paraffins and aromatics (and coke). Aditya, Chen, and Weisz (*1*) show that gas oil molecules can diffuse into the faujasite pores, and that the whole product distribution effect is fully explained by the large ratio of rate constants for H-transfer, relative to cracking, of the entire internal zeolite surface, in contrast to the conventionally small ratio for the amorphous catalyst.

Before leaving the subject of shape-selectivity and related phenomena, reference should be made to an unusual product distribution observed by Chen *et al.* (*11*) in the cracking of *n*-docosane ($C_{22}H_{46}$) over erionite at 250°–400°C. The carbon number distribution of the cracked products exhibited a trimodal pattern, with peaks at C_{11}, C_6, and $C_{3\text{-}4}$; products in the carbon number range of C_1–C_2, C_7–C_9, and $> C_{12}$ were missing. Selectivity for C_{10}–C_{12} products was not merely the result of "center cracking." Erionite contains cylindrical cavities about 15.1A long. These cages are bundled together with interconnecting 8-ring windows on the walls of the cylinders, and these windows provide for the passage of hydrocarbon molecules. The calculated path between such windows is about 15.4A, and the length of C_{10}–C_{12} *n*-paraffins (15.3–17.9A) nearly equals this distance. The authors suggest that C_{10}–C_{12} fragments—which

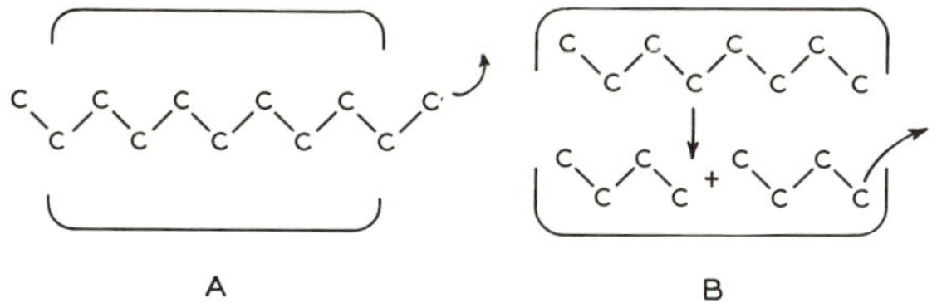

Scheme 9

could bridge across 2 windows—preferentially pass through the crystal unchanged (Scheme 9A), while those in the C_7–C_9 and $> C_{12}$ range undergo secondary reactions (Scheme 9B) and do not appear in the product. Thus, there appears to be the superimposition of a "cage" or "window" effect on the conventional cracking mechanism.

Systems Involving Inhibition of Reaction by Adsorbed Reactant

A priori, it is reasonable to assume that inhibition of catalytic reaction over a zeolite would occur whenever molecules from some source are strongly adsorbed at or near the entry pore mouths or on or near intracrystalline sites responsible for catalysis of the reaction in question. Let us consider the sorption–diffusion aspect briefly: Goldstein (*23*) observed inhibition in sorption of simple aliphatic carboxylic acids, nitriles, and nitro compounds in a series of zeolites containing 8-ring entry pores, where steric considerations definitely were not the determining factor. Katzer (*27*) has noted also that extremely small concentrations (*i.e.,* 0.4 mole %) of strongly adsorbed species (possibly radical ions from trace impurities in "high purity" 1-methylnaphthalene) greatly hinder the diffusion of 1-methylnaphthalene molecules, significantly lowering the observed diffusion coefficients in modified Y-type faujasite.

In early phenol alkylation studies (*61*), we noticed that alkylation of phenol with ethylene occurred at 204°C, a temperature much higher than that required for ethylation of the much less nucleophilic benzene nucleus (121°C) under similar conditions. Superficially, at least, this appears to violate the classical laws of electrophilic substitution (*24*). Closer examination of this system (*62, 64*), however, showed that phenol, at moderately low temperatures, was specifically adsorbed at sites active for alkylation, thus hindering adsorption of ethylene at these same sites, and preventing generation of the electrophile necessary for attack on the phenyl nucleus. That is, alkylation by a Rideal-type mechanism (*see* Scheme 5) cannot occur until temperatures high enough to desorb phenol from the active sites—and allow ethylene to compete for adsorption—are obtained. In such systems, alkylation can be facilitated by imposition of pressure (in the case of ethylene), or use of more polar or higher molecular weight alkylating agents. A similar "adsorption disguise" may explain the failure of propylene to alkylate naphthalene over a faujasite catalyst until a temperature of 205°C was reached (*61*). Also, the solvent effect observed in Beckmann rearrangement studies using REX faujasite (*31*)—decreased conversion with increased solvent polarity—probably stems from competitive adsorption of the polar solvent molecules at catalyst acid sites. The above solvent effect is exactly opposite to that observed in many homogeneous Beckmann rearrangement studies.

Such adsorption effects, are of course, well known in catalytic reactions over other heterogeneous systems. For a detailed discussion of the effect on catalytic cracking kinetics by molecules which do not compete, compete with near equal strength, and strongly compete with the reactant for adsorption on catalyst sites, the reader is referred to the classic work of Prater, Weisz, and associates (*46, 72*).

Aging In Zeolite Catalyst Systems

The term "aging" generally describes a loss in the activity—or selectivity—observed in a catalytic process after a certain period of reaction time. Aging may result from some change in the nature or number of catalyst sites, or in the accessibility of the sites to reactant molecules. Thus, such factors as formation of hydrogen-deficient organic residues ("coke"), selective adsorption of impurities from the charge [*i.e.*, adsorption of H_3PO_2 antioxidant from phenol (59)], or actual changes in the inorganic zeolitic structure will cause aging. The process may be gradual, while sometimes a sharp dramatic drop in activity occurs. The latter phenomenon, seen in the alkylation of benzene with decene over REX faujasite (*61*), represents the climax of a "masked" aging process which had been going on for some time, but which had not yet reached a critical stage.

It is interesting to compare structural features of the zeolites mordenite and faujasite in relation to aging. Mordenite, with noninterconnecting channels, can be visualized as a bundle of tubes with elliptical (6.96 × 5.81A) cross-section. If an obstruction develops within such a tube, access to catalytic sites can be decreased (Scheme 10A) or completely blocked (Scheme 10B), depending on its location. In faujasite,

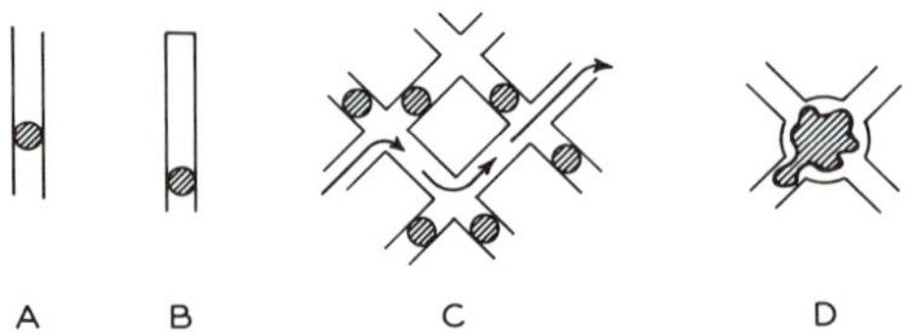

Scheme 10

however, each supercage opens *via* 8–9A diameter windows into 4 identical, tetrahedrally-distributed cavities. If 1 or even 2 of the 4 windows become obstructed, there is, in principle, the possibility of molecular diffusion through the faujasite crystal (Scheme 10C). (To state this in physiological terms, collateral circulation can develop around an embolus in the plexiform faujasite channel system!)

However, there are some disadvantages in the faujasite system that are not present in the mordenite system. The geometry of the mordenite pore-channel system dictates that no molecule with critical dimension larger than its 6.95 × 5.81A cross-section can be synthesized within its pores. Thus, conceivably, mild thermal treatment under vacuum or solvent extraction might remove the obstruction shown in Scheme 10A—provided its diffusivity is not too low. Within the faujasite supercage system, however, there is the potentiality for synthesis of molecules far too large to diffuse through the 8–9A 12-ring windows (Scheme 10D)—the phenomenon of "reverse molecular-size selectivity" (58). We thus have "the faujasite trap" (Figure 4), from which bulky molecules cannot be removed by simple physical means (vacuum, mild thermal or solvent desorption), but only by chemical processes (cracking, oxidation) that break C—C bonds. Details of the hydrogen transfer reactions of ethylene that lead to the genesis of bulky, hydrogen-deficient aromatics within the pores of faujasite—and aging—have been discussed earlier (*58, 60, 62*).

Conclusions, Analogies and Contrasts: The Over-all View

Considering zeolites as a group, the most striking feature that emerges is their versatility—the multiplicity of catalytic reactions for which they show activity. Certain other compositions have a broad spectrum of

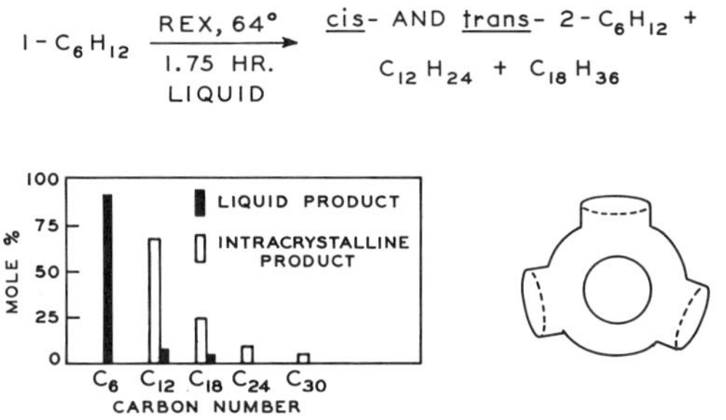

Figure 4. The "faujasite trap": steric occlusion visualized in an uncomplicated way in an olefin reaction at relatively low temperatures; hexene data from Ref. 60

catalytic activity, but they are useful only within narrow operating limits and may decompose under extreme conditions. A zeolite—unlike an enzyme, the remarkable structure and properties of which can persist over only a relatively narrow temperature and pH range—is thermally stable and catalytically active over a wide range of temperatures. It is capable of inducing double bond migration in a simple olefin at −80°C and yet can effect transformations at 600°C and higher—at the interface with thermal and free radical-type chemical reaction. Zeolites do not complex with reactant in the sense of $AlCl_3$, and unlike H_2SO_4, they do not corrode, sulfonate, or oxidize; great flexibility exists in mechanical handling of zeolites, and contact with reactants can be effected in diverse ways. In one sense, we can visualize "normal" or "classical" organic reaction patterns as occurring over zeolites, but with the superimposition of distortions (imposed by the peculiarities of zeolite structure) such as "cage" and "window" effects, and temperature, sorption, or diffusion "disguises."

Profound operations on the structure of organic molecules occur, ranging from delicate C=C isomerization to cracking activity powerful enough for fragmentation of the benzene ring. Carbon, oxygen, sulfur, nitrogen, and hydrogen atoms can be reshuffled; rings can be built and broken; HX can be added to molecules, and likewise, subtracted. Of prime importance are the transfer reactions of hydrogen from lattice oxygen to carbon, oxygen, and nitrogen as protons and *vice versa,* and from carbon to carbon as hydride. In some cases, the net process constitutes disproportionation, a dramatic redistribution of hydrogen between hydrogen-rich (paraffins) and hydrogen-deficient (often aromatic) species. A great many reactions appear to involve carbonium ion-type intermediates—*via* molecular interaction with the electron-deficient sites on the rigid, ordered, polar zeolite surface. Other intermediates also may occur: free-radicals and radical-ions have been observed in zeolite systems, and some appear to be stabilized by the lattice; there also is evidence to suggest the possible intervention of other "onium" ions, ylides, carbenes, and even carbanion-like species under certain circumstances. Remarkable deactivating effects occur, involving the intracrystalline synthesis and occlusion of molecules too large to escape—reminiscent of urea adduction polymerization. Evidence also suggests that the crystalline aluminosilicate in some way influences the catalytic activity of exchanged transition metal ions (the anionic lattice as a polymeric "ligand").

Many reactions over zeolites don't seem to require strong acidity. They are probably utilizing the polarity of the crystal lattice, and taking advantage of the tremendous reactant-concentrating effect of the vast, membrane-like internal surface area of the zeolite. Perhaps we even can look at zeolites as rugged, primitive "proto-enzymes." Of course, they are orders of magnitude below enzymes in their potentialities, but collectively they are also orders of magnitude more versatile, and have many more levels of discrimination than most other catalytic systems. In their ability to adsorb selectively one class of molecules and selectively exclude another, they are exhibiting a crude "lock and key" effect. In some cases, this selectivity is the result of size alone; in others, it is caused by polarity or some other physical property. Similarly, these same factors influence the rate at which molecules move through pores—and react. While most reactions occur within the pore–channel system, some may occur on the external surface.

The zeolite is rigid and ordered, and lacks conformational adaptability, in contrast to an enzyme, which can coil, uncoil, and twist around. Yet the zeolite can incorporate transition metal functions—these are of prime importance in enzyme catalysis—and it can effect redox reactions; reactions over zeolites can be inhibited by competitive adsorption of reactants, products, solvents, or poisons—a phenomenon observed in biological and some other inorganic heterogeneous catalytic systems; Rideal kinetics have been identified in some zeolite-catalyzed alkylations, a pattern which has its parallels in the enzyme field; a few cases of stereospecificity (such as *ortho*-alkylation effects, unusual olefin isomer ratios), where a transition state not otherwise attainable intervenes, may exist. What better group of catalysts than zeolites might there have been to activate the evolutionary process in the dark, fermenting Pre-Cambrian seas some 1,000,000,000 years ago?

Acknowledgment

Gratitude is expressed to L. A. Hamilton for encouragement and direction in the early stages of the author's studies on zeolites.

Literature Cited

(1) Aditya, S. K., Chen, N. Y., Weisz, P. B., private communication.
(2) Anderson, R. B., Imai, T., McMaster University, Hamilton, Ontario, private communication, 1969.
(3) Barrer, R. M., *Nature* **1949**, 164, 112.
(4) Barrer, R. M., Brook, D. W., *Trans. Faraday Soc.* **1953**, 49, 940.
(5) Barrer, R. M., Kravitz, S., "Molecular Sieves," p. 95, Society of the Chemical Industry (London) 1968.
(6) Bartley, B. H., Habgood, H. W., George, Z. M., *J. Phys. Chem.* **1968,** 72, 1689.
(7) Bolton, A. P., Lanewala, M. A., Pickert, P. E., *Natl. Meeting, ACS, 152nd, New York, 1966,* Preprints of paper, p. U22.
(8) Bolton, A. P., Lanewala, M. A., Pickert, P. E., *J. Org. Chem.* **1968,** 33, 1513.
(9) Borunova, N. V., *et al., Izv. Akad. Nauk SSR, Ser. Khim.* **1968,** 4, 773.

(10) Cannon, P., *J. Phys. Chem.* **1959,** 63, 160.
(11) Chen, N. Y., Lucki, S. J., Mower, E. B., *J. Catalysis* **1969,** 13, 329.
(12) Chen, N. Y., Weisz, P. B., *Chem. Eng. Progr. Symp. Ser.* **1967,** 63, 86.
(13) Csicsery, S. M., *J. Org. Chem.* **1969,** 34, 3338.
(14) Deno, N. C., Richey, H. G., Jr., Friedman, N., Hodge, J. D., Houser, J. J., Pittman, C. U., *J. Am. Chem. Soc.* **1963,** 85, 2991.
(15) Dimitrov, C., Leach, H. F., *J. Catalysis* **1969,** 14, 336.
(16) Eberly, P. E., Jr., *J. Phys. Chem.* **1967,** 71, 1717.
(17) Frilette, V. J., Rubin, M. K., *J. Catalysis* **1965,** 4, 310.
(18) Galich, P. N., *et al., Neftekhim., Akad. Nauk Turkm. SSR* **1963,** 63; *Chem. Abstr.* **1964,** 61, 14434.
(19) Galich, P. N., *et al., Neftekhim., Akad. Nauk Ukr. SSR, Inst. Khim. Vysokomolekul. Soedin.* **1964,** 13; *Chem. Abstr.* **1965,** 62, 8905.
(20) Garnett, J. L., Sollich, W. A., *J. Catalysis* **1963,** 2, 350.
(21) Garwood, W. E., Venuto, P. B., *J. Catalysis* **1968,** 11, 175.
(22) Gladrow, E. M., Mattox, W. J., U. S. Patent **3,352,924** (November 14, 1967).
(23) Goldstein, T. P., Joint Spring Symposium, Philadelphia and New York Catalysis Clubs, Princeton, N. J., 1965.
(24) Gould, E. S., "Mechanism and Structure in Organic Chemistry," p. 589, Holt, New York, 1959.
(25) Haag, W. O., Mobil Oil Corp., private communication.
(26) Jones, D. G., U. S. Patent **3,414,588** (December 3, 1968).
(27) Katzer, J. R., University of Delaware, private communication, 1969.
(28) Katzer, J. R., Thesis, Massachusetts Institute of Technology, Cambridge, Mass., 1969.
(29) Konoval'chikov, O. D., Galich, P. N., Gutyrya, V. S., Lugovskaya, G. P., *Kinetics Catalysis* **1968,** 9, 1146.
(30) Landis, P. S., U. S. Patent **3,254,131** (May 31, 1966).
(31) Landis, P. S., Venuto, P. B., *J. Catalysis* **1966,** 6, 245.
(32) Lanewala, M. A., Bolton, A. P., *Natl. Meeting, North American Catalysis Society, 1st, Atlantic City,* **1969.**
(33) Levina, S. A., Malashevich, L. N., Ermolenko, N. F., *Dokl. Akad. Nauk Belorussk. SSR* **1966,** 10, 236.
(34) Matsumoto, H., Morita, Y., *Kogyo Kagaku Zasshi* **1968,** 71, 1496.
(35) *Ibid.,* **1967,** 70, 1674.
(36) Matsumoto, H., Yasui, K., Morita, Y., *J. Catalysis* **1968,** 12, 84.
(37) Mattox, W. J., U. S. Patent **3,036,134** (1962).
(38) Minachev, Kh. M., *et al., Neftekhimiya* **1968,** 8, 37.
(39) Mochida, I., Hayata, S., Kato, A., Seiyama, T., *J. Catalysis* **1969,** 13, 314.
(40) Mochida, I., Yoneda, Y., *J. Org. Chem.* **1968,** 33, 2161.
(41) Morita, Y., Matsumoto, H., *Kogyo Kagaku Zasshi* **1967,** 70, 1363.
(42) Morita, Y., Takayasu, M., Matsumoto, H., *Kogyo Kagaku Zasshi* **1968,** 71, 1492.
(43) Pansevich-Kolyada, L. V., Ermolenko, N. F., *Russ. J. Phys. Chem.* **1966,** 40, 1282.
(44) Pickert, P. E., Bolton, A. P., Lanewala, M. A., *A.I.Ch.E. Meeting, 59th, Columbus, Ohio,* 1966.
(45) Pope, C. G., Kemball, C., *Trans. Faraday Soc.* **1969,** 65, 619.
(46) Prater, C. D., Lago, R. M., *Advan. Catalysis* **1956,** 8, 293.
(47) Rabo, J. A., Angell, C. L., Schomaker, V., *Proc. Intern. Congr. Catalysis, 4th, Moscow, 1968,* **1968,** 3, 966.
(48) Rabo, J. A., Pickert, P. E., Stamires, D. N., Boyle, J. E., *Proc. Intern. Congr. Catalysis, 2nd, Paris, 1960,* **1961,** 2, 2055.
(49) Richardson, J. T., *J. Catalysis* **1967,** 9, 182.
(50) Rouchaud, J., Fripiat, J., *Bull. Soc. Chem. Fr.* **1969,** 1, 78.
(51) Rouchaud, J., Mulkay, P., Fripiat, J., *Bull. Soc. Chim. Belg.* **1968,** 77, 537.

(52) Rouchaud, J., Sondegam, L., Fripiat, J. J., *Bull. Soc. Chim. Belg.* **1968,** 77, 505.
(53) Rouchaud, J., Sondegam, L., Fripiat, J. J., *Bull. Soc. Chim. Fr.* **1968**, 11, 4387.
(54) Thomas, C. A., "Anhydrous Aluminum Chloride in Organic Chemistry," p. 798, Reinhold, New York, 1941.
(55) Thomas, C. L., Barmby, D. S., *J. Catalysis* **1968,** 12, 341.
(56) Venuto, P. B., *J. Org. Chem.* **1967,** 32, 1272.
(57) Venuto, P. B., Givens, E. N., Hamilton, L. A., Landis, P. S., *J. Catalysis* **1966,** 5, 253.
(58) Venuto, P. B., Hamilton, L. A., *Ind. Eng. Chem. Prod. Res. Develop.* **1967,** 6, 190.
(59) Venuto, P. B., Hamilton, L. A., Landis, P. S., *Natl. Meeting, ACS, 151st, Pittsburgh, Pa.,* March, 1966.
(60) Venuto, P. B., Hamilton, L. A., Landis, P. S., *J. Catalysis* **1966,** 5, 484.
(61) Venuto, P. B., Hamilton, L. A., Landis, P. S., Wise, J. J., *J. Catalysis* **1966,** 5, 81.
(62) Venuto, P. B., Landis, P. S., *Advan. Catalysis* **1968**, 18, 259.
(63) Venuto, P. B., Landis, P. S., *J. Catalysis* **1966,** 5, 237.
(64) Venuto, P. B., Wu, E. L., *J. Catalysis* **1969,** 15, 205.
(65) Venuto, P. B., Wu, E. L., Cattanach, J., *Anal. Chem.* **1966,** 38, 1266.
(66) Venuto, P. B., Wu, E. L., Cattanach, J., "Molecular Sieves," p. 117, Society of the Chemical Industry, London, 1968.
(67) Ward, J. W., *J. Catalysis* **1968,** 10, 34.
(68) *Ibid.,* **1968,** 11, 259.
(69) *Ibid.,* **1969,** 14, 365.
(70) Weisz, P. B., Frilette, V. J., *J. Phys. Chem.* **1960**, 64, 382.
(71) Weisz, P. B., Frilette, V. J., Maatman, R. W., Mower, E. B., *J. Catalysis* **1962,** 1, 307.
(72) Weisz, P. B., Prater, C. D., *Advan. Catalysis* **1954,** 6, 143.
(73) Wittig, G., Krauss, D., *Ann.* **1964,** 679, 34.
(74) Wu, E. L., Kuhl, G. H., Whyte, T. E., Jr., Venuto, P. B., *Intern. Conf. Mol. Sieves, 2nd, Worcester, Mass.,* 1970.

RECEIVED February 4, 1970.

Discussion

W. K. Hall (Gulf Research & Development Co., Pittsburgh, Pa. 15230): Could not the evidence cited for pore mouth reaction really result from the formation of polymer layer which acts as a proton donor? Such a situation exists with silica–alumina catalyst.

P. B. Venuto: Such proton-donation from adsorbed polymer layer as you have suggested cannot be eliminated, and such mechanisms very probably may operate in certain zeolitic systems. Since, however, some deuterium incorporation from catalyst OD groups to hydrocarbon did occur, and since the reaction was effected at low temperatures in the liquid phase, a diffusion-limited pore mouth type catalysis may be more likely.

62

Structural Aspects of Catalysis With Zeolites; Cracking of Cumene and Hexane

J. A. RABO and M. L. POUTSMA

Union Carbide Research Institute, Tarrytown Technical Center, Tarrytown, N. Y. 10591

Catalytic reactions of hydrocarbons over zeolites are reviewed. The historical development of various mechanistic proposals, particularly of the carbonium ion type, is traced. In spite of numerous catalytic, spectroscopic, and structural studies which have been reported concerning the possible roles of Bronsted acid, Lewis acid, and cationic sites, it still is not possible to formulate a comprehensive mechanistic picture. New activity and product data for cumene cracking and isotope redistribution in deuterated benzenes over Ca- and La-exchanged Y zeolites is presented. Cracking of the isomeric hexanes over alkali metal–exchanged Y and L zeolites has been studied. This cracking is clearly radical rather than carbonium-ion in nature but certain distinct differences from thermal cracking are described.

The literature related to various aspects of catalysis with zeolites has grown within the past 10 years with a propulsion such as follows only major discoveries in science and technology. The published work has been collected in several review articles giving account of several hundred publications and patents. Unlike many other cases in catalysis, the origin of the chain of events which resulted in the discovery of zeolite catalysts lies in inorganic chemistry and crystallography rather than in catalysis itself. Consequently, a good deal of the scientific recognition is owed to early physicochemical studies, the complex structural work on several zeolite minerals, and to the discovery of new synthetic zeolites, particularly those varieties which have pores large enough to admit most of the hydrocarbons that occur in petroleum distillates.

There are bibliographic reviews on catalysis with zeolites (*62, 94, 100*) as well as on zeolites in general (*7, 19, 76, 85*) covering the literature up to about 1966. In the limited space of this paper, we shall give only a selective review on the progress in zeolite catalysis with type-X and -Y zeolites, with emphasis on the relationship between zeolite structure and catalytic properties. In addition, we shall describe here several catalytic experiments of our own.

The intriguing structural property of zeolites which permits them to admit or reject adsorbates based on molecular size was first reported by McBain (*59*), who also proposed the term "molecular sieve." His findings were confirmed and greatly extended by Barrer, who later prepared a synthetic zeolite for the first time. Inspired by the potential application of molecular sieves in the separation of gases and liquids, Union Carbide Corp. launched the first large-scale industrial research, conducted by Milton, Breck, and associates, for the synthesis and characterization of new zeolite varieties. This work was a natural extension of the laboratory's long-standing interest in synthetic crystals: sapphire, diamond, ruby, etc. With the discovery of X and Y zeolites (*20, 61*) with a pore size of about 10 A, large enough to admit not only gas molecules but also most hydrocarbons in petroleum distillates, the interest was promptly extended from adsorption studies to catalysis.

Three zeolite features were considered particularly attractive for catalysis studies: first, the intracrystalline pores and cavities offered a better-defined surface than ever before in catalysis; second, the exchangeable zeolite cations introduced a new chemical and physical variable on a more or less constant surface; and finally, the uniformity of the pores linking the large cavities within the crystal suggested that with a feed containing both small and large molecules, the catalytic effect will be restricted to those which are admitted through the pores, similar to the molecular sieve effect found in adsorption. However, this review will demonstrate that the intracrystalline surface of zeolites is not as well defined by x-ray as desired; that the catalytic phenomenon is likely centered on sites which may be called "defect sites" (and consequently tend to elude characterization by the methods of crystallography); that the molecular sieving effect may be an advantage as well as a restriction; and that out of the great variety of available metal cations, only a few have found commercial application in catalysis so far: alkaline earth and rare earth cations.

The major practical and scientific result of zeolite catalysis is a great improvement in several major petroleum technologies and a new and growing interest of scientists of various disciplines in catalysis. The result of the latter is a growing tendency to characterize the structural details of catalysts using all available physical and chemical methods, as a welcome complement to kinetic studies.

Review of Structural Aspects of Zeolite Catalysis

Carbonium Ion Type Catalysis. The first reports on carboniogenic catalysis—*i.e.*, activities apparently caused by the formation of carbonium ion–like intermediates—with zeolites were presented in 1960 by 2 groups working independently. At the 2nd International Congress on Catalysis, the research group at Union Carbide Corporation reported (*75*) that Pt- or Pd-loaded synthetic zeolites of type X and Y can be made to show excellent carboniogenic catalytic activities, depending on the selection of the Si/Al ratio and the charge of zeolitic cation introduced by ion exchange into the sodium-containing, as-synthesized materials. While the noble metal–loaded NaX and NaY showed only negligible activity for the isomerization of *n*-hexane, the Ca-exchanged X and Y zeolites demonstrated substantial activity, with the higher Si/Al ratio Y displaying much higher activity than the X zeolite. An isomerization activity comparable or even superior to the Ca-exchanged catalysts was reported with NH_4-exchanged X and Y, following heat treatment to decompose the NH_4 cation, products of which are often referred to as decationized or decationated zeolites. Here again, decationized Y was much superior to decationized X, and the latter lost its crystal structure on heat treatment in the decationization process. With the Y zeolite, an exchange of Ca for Na up to 50% caused only a small increase, but a more complete exchange caused a sudden rise in catalytic activity. In contrast, even small (10–20%) decationization gave rise to substantial catalytic activity.

In the same year, Weisz and Frilette at Socony Mobil Oil Co. reported (*110*) that NaX is more active for the cracking of large normal paraffins than the conventional silica–alumina cracking catalysts, and that the activity of NaX is further increased by replacing sodium with calcium ions. The product obtained with NaX was free of branched products while the product obtained with Ca-exchanged X contained branched hydrocarbons, similar to the product obtained with silica–alumina type catalysts. Several examples of the molecular sieve effect were demonstrated with CaA zeolite catalyst, using mixed feeds containing both smaller and larger molecules than the pore in the zeolite crystal.

These and the shortly following reports indicated that (1) the locus of catalytic activity is in the intracrystalline "zeolitic surface," as demonstrated by the molecular sieve phenomena (*75, 110, 111*); (2) bi- and poly-valent cations induce high carboniogenic activity in X and Y zeolites (*33, 75*); (3) the decationized Y obtained through heat treatment of NH_4-exchanged Y shows activity superior to cation-exchanged Y zeolites (*75*); (4) comparable forms of Y are more active and in certain forms more stable than X (*75*); (5) in spite of the lack of carboniogenic activity, the NaX cracks large *n*-paraffins more extensively than the standard silica–alumina catalysts (*110*); (6) loading of small amounts of noble metal by cation exchange greatly enhances the activity (*75*). It was suggested that the electrostatic shielding of surface cations (cations coordinated to framework oxide ions in the large intracrystalline cavities, Figure 1) of a valence higher than one is ineffective, resulting in strong electrostatic fields in the large cavities near these cations and near the AlO_4^- sites not associated with cation (*75*), and that adsorbed molecules are exposed to unusual coulomb fields (*110*). It was shown later (*70*) that with cation-exchanged X and Y zeolites, the carboniogenic activity not only increases with the larger Si/Al ratio but that it increases with the electrostatic potential of the cation—*i.e.*, increases with higher valence and smaller size. The activity was attributed to the strong accessible electrostatic field near surface cations and, in essence, to polarization of the substrate molecules to form carbonium ion–like reaction intermediates.

The existence of "surface" bivalent cations at lattice positions with 3-fold coordination in the large cavities (Site 2) exposed to adsorbed molecules was shown from a structure analysis of a Ca-exchanged faujasite single crystal (*12, 30*). The assignment of Ca ions to various lattice positions in mineral faujasite helped to explain by comparison the observation that with Y zeolite the catalytic activity emerges at approximately 50% Ca exchange, because of the preference of Ca ions for the fully

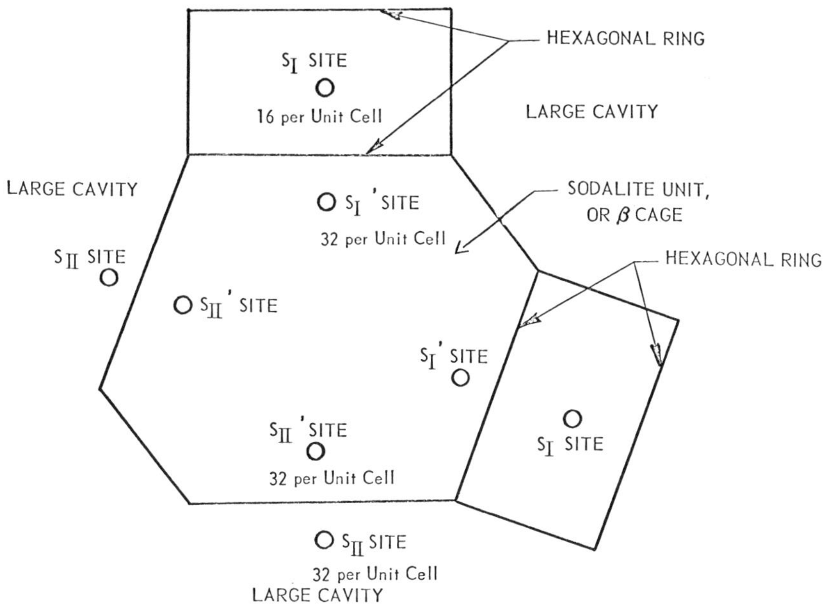

Figure 1. Cation positions in type X, Y zeolites

coordinated Site 1 in the hexagonal double ring. The unit cell contains 16 Site 1 and 32 Site 2, and with a fully exchanged Y of Si/Al = 2.5, there are ~27 Ca ions. The electrostatic field near surface cations has been estimated, by considering the contribution of the whole crystal lattice, to exceed 1 V/A in the large cavities at a distance of 3 A from the center of a bivalent cation (*28, 70*). Higher coulombic fields are expected in Y than in X zeolites (*70*). The direct coulombic interaction between surface cations and adsorbed molecules was demonstrated with adsorbed CO, CO_2 (*4, 6, 73*), etc., all giving rise to specific band shifts in the infrared spectrum, changing with the valence and size of cation, consistent with the suggested electrostatic effects near surface cations.

The electron affinity of surface cations was demonstrated by the formation and the unusual stability of Ni^+ ions in Y (*73*). The collective electron affinity of surface alkali metal cations, and thus of the cation-containing large cavity itself, also was demonstrated by the unusual stability of the electron captured from alkali metal vapor giving rise to Na_4^{3+} and Na_6^{5+} centers (*73*).

The simple idea involved in the "electrostatic theory" was very effective in predicting the catalytic activity of various cation-exchanged X and Y catalysts. However, it failed to explain quantitatively the difference in activity between cation-exchanged X and Y, alkaline earth cation X being less active than expected. It did not explain the cause for the similar behavior between cation-exchanged and decationized zeolites, and it did not offer satisfactory chemical evidence for the suggested reaction mechanism.

The presence of Bronsted acid sites in Ca-exchanged X was first reported by Norton (*64*) based on acid indicator tests, and the existence of various hydroxyl groups was recognized by infrared spectroscopy both in cation-exchanged (*5, 24, 26, 106*) and in decationized zeolites X and Y (*5, 45, 75, 87, 96, 104, 108, 112*). It was suggested that the small band at 3745 cm^{-1} is either owing to hydroxyls terminating the zeolite crystal (*96*) or, more likely, since they show no interaction with adsorbed molecules, to some form of occluded silica residue (*5*). The band at 3652–3636 cm^{-1} is medium size in bi- or multivalent cation Y, very large in decationized Y, shows a small but consistent shift to lower frequency with the stronger cation, and the lowest frequency in decationized Y. This band shifts in the presence of various adsorbed molecules, showing a shift of about 300 cm^{-1} with adsorbed benzene (*5*). The hydroxyl associated with this band in decationized Y protonates NH_3, pyridine, piperidine, and quinoline at room temperature (*16, 25, 45, 95, 104, 108*) but, according to one report, the hydroxyl with comparable frequency in bivalent cation Y protonates pyridine only at 85°–150°C (*26*). This band was assigned to hydrogen attached to the type of framework oxygen which serves as a bridge between the sodalite cages, O_1 (*74*). The large broad band at ~3540 cm^{-1} in decationized Y was assigned (*67*) on the basis of x-ray evidence to hydrogen attached to framework oxygen in the hexagonal double ring. This hydroxyl protonates piperidine but fails to protonate the weaker base pyridine at room temperature. The band with varying frequency around 3500–3600 cm^{-1} in bi- or multivalent cation Y shows significant shift with changing cation and usually does not indicate interaction with molecules adsorbed in the large cavities. Consequently, it is assigned to hydroxyls attached to cations in the sodalite cage (*74*).

Hirschler (*43*) and Planck (*71*) proposed that hydroxyl protons are the locus of carboniogenic activity. Hirschler also proposed that cations influence the geometry and the acidity of protonic sites, and Plank suggested that protons are introduced through hydrolysis of the cation:

$$Re^{3+} + H_2O \rightleftarrows [ReOH]^{2+} + H^+$$

Ward suggested (*106, 107*) that the amount of hydroxyl introduced either through hydrolysis during cation exchange or by ionization of water upon activation depends on the ionic strength of the cation; consequently, if activity depends on hydroxyl concentration, it will parallel the cation strength.

Information on the amount of "surface cation" in various cation-exchanged Y zeolites was obtained by Rabo, Angell, *et al.* (*73*) by using a combined method of CO sorption and infrared spectroscopy. They found that the surface cation content of bivalent cation-exchanged Y depends not only on activation temperature but also on the rate of heat-up; "flash activation" of NiY resulted in several times more surface Ni^{2+} cations than activation by slow heat-up. Fewer surface cations were found with bivalent transition metal Y than with alkaline earth Y (*73*). Barry *et al.* (*8, 9*) used spin tagging (Mn^{2+}) and concluded that certain transition metals display strong covalent character. They prefer sites of specific symmetry rather than sites of higher coordination number—*i.e.*, Zn^{2+} prefers tetrahedral coordination in the sodalite cage to the nearly octahedral Site 1. He also detects with cations of higher valence a tendency to occupy positions in the sodalite cage. Such a tendency for rare earth cations was shown by Bennett *et al.* (*13, 14*), Smith *et al.* (*84*), Olson *et al.* (*68*), and later by Rabo *et al.* (*74*). In addition, Smith *et al.* (*84*) and Bennett *et al.* (*15*) reported that the cation distribution in La-exchanged X and Y is temperature-dependent and that between room temperature and 700° about half of the cations change their position in a reversible manner.

The information obtained by x-ray crystallography shows considerable differences in the position of La^{3+} cation between microcrystalline

Type Y and single crystal faujasite (*13, 14, 84*). Since crystallographic studies are conducted with very small amounts of material, especially in the case of the single-crystal method, it is likely that the effective control of the environment for traces of water and other impurities becomes very critical and may affect the position of the cations.

The critical nature of the method of dehydration can be illustrated best by the effect of changing activation conditions on Ca-exchanged Y (*3*). It appears that, on activation at 550°C in air, a small fraction (~15%) of surface calcium ions react with water

$$Ca^{2+} + H_2O \rightleftarrows [Ca^{2+} - OH^-] + H^+$$

$$H^+ + O^{2-} \rightleftarrows OH^- \sim 3640 \text{ cm}^{-1} \text{ infrared band}$$

and that the CaOH groups occupy positions in the sodalite cage as previously reported by Olson (*66*). However, on treatment with steam at 600°C, the reaction with water goes to completion, resulting in the elimination of all surface calcium ions and in a sharp rise in the 3640 cm^{-1} hydroxyl band. The small structural change induced on activation at 550°C can be observed by x-ray methods only with difficulty and with limited accuracy, while the effect of this small structural change on catalytic activity is already very large.

Further evidence for the need of protons for carboniogenic activity is given by Matsumoto *et al.* (*58*), who have shown that NaY, which is inactive for cumene cracking, can be made into an active catalyst by the addition of HCl. This effect of HCl is reversible. In contrast, silica gains no activity on HCl addition, and γ-alumina is activated irreversibly by HCl. Similar observations were made by Kolesnikov *et al.* (*51*), who found that in the propylation of benzene, a treatment with propyl chloride promotes the activity of NaY and of type X zeolites.

While the addition of HCl enhanced catalytic activity, it was shown by Benesi (*11*) that on removal of hydroxyls through activation at progressively higher temperatures the catalytic activity first drops and then vanishes for the cracking of toluene, *n*-butane, and *n*-pentane; and that the activity is restored upon readdition of water to the catalyst. Similar deactivation upon dehydroxylation of rare earth–exchanged Y and Ca-exchanged Y was observed by Csicsery (*41*) and Hopkins (*44*), respectively. Ward (*104*) suggested that the activity found by Benesi with various activation temperatures corresponds well with the Bronsted acid concentration. Hopkins (*44*), Richardson (*80*), and Lunsford (*57*) pointed out, however, that according to the data reported earlier by Ward (*102, 106*) and Uytterhoeven (*96*), at the activation temperature (~600°C) which yielded maximum activity, a significant fraction of the hydroxyls must have been removed already. Similar, or even more extensive, dehydroxylation is expected on activation at 600° according to Cattanach *et al.* (*25*) and Bolton (*16*), respectively. Pickert *et al.* found that in the alkylation of benzene by propylene, the La(46%)-NH_4(44%)-Y shows highest activity following activation at 700°C (*69*). Rabo *et al.* found that activation of La-exchanged Y at a high temperature which resulted in the removal of the 3640 cm^{-1} band below the observable level, hence reducing the hydroxyl content below 1 OH/100 Al (*74*), led to only partial loss of activity for cracking of *n*-butane. Hopkins (*44*) also pointed out that the activity of NH_4-exchanged Y activated at 375°C, at which temperature all or nearly all NH_4^+ ions are decomposed and the hydroxyl content is near maximum (*16, 25, 31*) is small compared with the one activated at 600°C. Hopkins suggests that "only the strongest Bronsted sites are effective catalytically, and that the presence of Lewis acid sites is either necessary or has a synergistic effect on the activity."

A different type of observation was reported by Tung *et al.* (*93*). They compared the conversion of *n*-hexane on decationized Y with a series of samples ranging in the degree of NH_4^+ exchange—before decationization—from 15 to 95%. The conversion of *n*-hexane was observed at stepwise increased temperatures from 200° to 450°C. The catalytic activity increased with increasing degrees of NH_4-exchange as demonstrated by the higher conversions obtained with these zeolites at low temperatures. Reaching the higher temperatures in this stepwise testing procedure, the activity of the zeolites with lower NH_4 exchange showed increasing conversions, while those with high degrees of exchange showed peak activity at an intermediate temperature and a decline in activity at high temperatures, resulting in higher conversions at high temperatures with lower NH_4 exchange. It was reported also that, in contrast to the findings of other investigators who used cation-exchanged zeolites (*63*), coke burn-off by air with the spent decationized Y catalyst failed to restore the catalytic activity. According to Tung *et al.*, the locus of activity is a Lewis acid or surface electric field, and the deactivation found with decationized Y of high degree of exchange is caused by shifts of lattice oxygen to more stable positions. They also suggest that the shift of oxide ions at Lewis acid sites effects the desorption of the reaction product, and that product desorption (hence oxide jumps) controls the rate of reaction.

It is conceivable, however, that deactivation of Tung's more active catalysts at lower temperatures is caused by coke formation which has accumulated on the more active catalyst at lower temperatures. The lack of return of activity upon regeneration by air remains to be explained. The jump of framework oxygen as the rate-controlling step in catalysis is not likely in view of the large apparent activation energy of dehydroxylation reported by Venuto (*25*), $>$ 70–100 kcal/mole, a substantial fraction of which has to contribute to the activation energy of oxygen jump. While the jump of oxygen at the suggested high rate is not consistent with the high activation energy of dehydroxylation, occasional shifts of oxides are more probable. Considering that framework oxygen in the zeolite consists of 4 different species (*67*), it must be expected that on dehydroxylation the crystal will favor the removal of one of the possible 4, which may or may not be the one preferentially associated with hydrogen. In the latter case, the primary oxygen-deficient site created on dehydroxylation is not preferred by the crystal; hence, it may be subject to change through oxygen jump, with resulting changes in adjoining atoms, including hydroxyl groups, and possibly in catalytic activity. Similar structural change may result if a hydrogen jump, which requires only 6.9 kcal/mole activation energy (*38*), preceeds dehydroxylation. The 6.9 kcal/mole activation energy was derived from NMR measurement with a NH_4 (~70) exchanged Y zeolite which was thermally deammoniated at low temperature to minimize dehydroxylation.

Isakov *et al.* (*48*) estimated the fraction of O_6 rings unoccupied by surface cation in Y zeolites, using the change of heat of adsorption with loading as a measure for surface cations and consequently as an indirect measure for single O_6 rings unoccupied by surface cations (Site 2). They suggest a parallel relationship between catalytic activity and the fraction of O_6 rings unoccupied by cations. A similar conclusion was drawn by Minachev *et al.* (*63*).

Richardson studied the reactions of cumene with a variety of zeolites over a wide temperature range (*80*) and found that dealkylation to benzene proceeds at high rates with all alkaline earth Y zeolites. The activity of decationized Y was superior to all alkaline earth Y catalysts, while alkali metal Y zeolites produced no benzene at all. All cation-exchanged Y catalysts showed a sharp decline in benzene formation with time, presumably because of coke deposition. Coincidental with the decline in benzene formation, α-methylstyrene was observed in the product at 550°C. The production of α-methylstyrene was largely unaffected by coking, and it was produced even when benzene formation was eliminated, presumably because of deposition of coke on the catalyst. Similar experiments conducted with decationized Y showed also a sharp decline in benzene formation with time; however, no α-methylstyrene was observed after benzene formation ceased owing to coke deposits. The reaction with cumene at 550°C in the presence of increasing amounts of quinoline resulted in a gradual decline in benzene formation and in a coincidental increase in α-methylstyrene formation, similar to the effect of coke. Richardson suggests that quinoline prefers strongly acidic hydroxyls to Lewis acids (surface cations, trigonal Al), and that decline in benzene formation results from the attachment of coke or quinoline at the protonic sites. Since decationized Y deactivated by coke produced no α-methylstyrene, he concluded that dehydrogenation activity is centered on zeolite cations, that the activity parallels their coulombic strength, and that dehydrogenation is not catalyzed by Lewis acid sites found in decationized Y.

In Richardson's experiment, the amount of quinoline needed to eliminate most of the activity for benzene formation with Ag^{+}-exchanged Y is about 1×10^{20}/gram. For comparison, the amount of Al and the corresponding highest possible concentration of OH groups at Si/Al = 2.5 is 2.5×10^{21}/gram. If hydroxyls are introduced by reduction of only the surface Ag^{+} ions, and if some dehydroxylation occurs at the reaction temperature, the amount of quinoline used for full poisoning of benzene formation corresponds well with the expected number of surface hydroxyls. Accepting Richardson's conclusion that zeolite cations are the centers for dehydrogenation activity, the lack of effect of quinoline on dehydrogenation can be taken as evidence that at 550°C the quinoline is only weakly adsorbed on zeolite cations. Previously, Turkevich *et al.* (*95*) and Boreskova *et al.* (*18*) described similar poisoning experiments. According to the data shown by Turkevich, the amount of quinoline needed for the full poisoning of decationized X corresponds with the order of the estimated decationized sites (50–100%), while according to Boreskova *et al.*, the amount of quinoline needed to eliminate activity—presumably cumene → benzene + C_3—is several times higher than the amount of Al ions, and consequently higher than the highest known hydroxyl concentration in decationized zeolites.

Ward recently reported (*109*) *in situ* infrared spectroscopic observations of decationized Y during cumene cracking and observed that in a continuous flow of cumene and helium mixture at 250°, the hydroxyl band near 3640 cm^{-1} declines to a small fraction of its original intensity, while the band at 3540 cm^{-1} is unaffected. On raising the temperature to 365°C, a small decrease in the band at 3540 cm^{-1} was observed also. Ward suggests that the hydroxyls corresponding to 3640 cm^{-1} band interact readily with cumene by proton transfer at lower temperatures, but at higher temperatures the less active hydroxyl groups also become sufficiently acidic or activated to interact with cumene and thus serve as active centers. The spectra shown by Ward indicate an almost quantitative removal of the 3640 cm^{-1} hydroxyls following long exposure to cumene.

The complete elimination of the 3640 cm^{-1} band may give some support to the idea that most of the hydroxyls represented in this band act as catalytic centers; however, Ward points out that changes in hydroxyl group intensities could be attributed to the formation of coke or other surface species. Such species could be formed through secondary reactions with the propylene product (*17, 32, 40, 56*). Upon gradual elimination of hydroxyls, no frequency shift of the residual band at 3640 cm^{-1} was reported by Ward. Such shift could be taken as evidence that the original band at 3640 cm^{-1} is a composite band representing a variety of hydroxyls of slightly differing frequency and different activity. With the most active hydroxyls reacting first and being eliminated, it may be expected that their removal would cause a change in frequency of the residual band.

Since most catalytic reactions with zeolites are conducted at high temperatures, it was natural to investigate the infrared spectrum of zeolite hydroxyls at high temperatures. Fripiat *et al.* (*34, 35*) observed that changes in the apparent intensities of hydroxyl bands in micas and clays are temperature-dependent. These changes were explained on the basis of a progressive delocalization of the protons, and an activation energy of ~4.3 kcal/mole was calculated for the process. Similar results have been reported for hydroxyl groups of both X and Y zeolites by Uytterhoeven *et al.* (*97*) and by Ward (*105*). Ward suggested that about one-third of the hydroxyl hydrogen becomes mobile or delocalized at 450°C and that these hydrogens may be responsible for the superactivity of zeolites. However, Cant *et al.* (*23*) repeated the high-temperature infrared measurements and concluded that between room temperature and 600°C the behavior of the hydroxyl bands of silica–alumina, alumina, and decationized Y zeolite is indistinguishable from that of catalytically inert silica.

Richardson (*80*) estimated the number of active sites for cumene cracking in a variety of Y zeolites, based on kinetic measurements. He ascribed the small value found for the preexponential factor entirely to the number of active sites, by ignoring all other components, such as entropy change on chemisorption, etc. He obtained values for active sites which were less by 3 to 5 orders of magnitude than the amount of aluminum ions, and hence the largest amount of potential hydroxyl sites. In order to explain the small value obtained, he suggested that the band at 3640 cm^{-1} is a composite band representing hydroxyls of a wide range of acidity and catalytic activity. He suggested that only a very small fraction—the most acidic—contributes to catalytic activity.

The nature of the oxygen-deficient Lewis acid sites obtained upon dehydroxylation of decationized zeolites was studied by several investigators, and it was suggested that it is somewhat similar to comparable sites in silica–alumina. Stamires *et al.* (*86*) further characterized this site by capturing electrons dislocated *via* gamma irradiation near aluminum Lewis acids, and they were able to identify it by the hyperfine interaction of the captured electron with the ^{27}Al nucleus. They suggested that the site is an aluminum Lewis acid due to "lack of shielding by an oxygen atom." Lunsford (*57*) adsorbed NO radicals on various oxides and zeolites and found crystal field interactions of $\Delta = 0.21$ and 0.60 eV for NaY and decationized Y, respectively. He also observed that the shape and the magnetic parameters of the NO spectra obtained with decationzed Y were unaffected by the temperature of dehydroxylation, but the number of spins, *i.e.*, the number of chemisorbed NO molecules, was strongly influenced by the method of activation. The highest spin concentration of 1.1×10^{19} spin/gram was observed upon activation between 600°–700°, coinciding reasonably closely with the highest catalytic activity (*11, 41, 44, 67*). The broad signal observed was identified on the basis of hyperfine interaction with ^{27}Al nucleus and he suggested that NO may be bonded directly to the aluminum ion. In contrast to decationized Y, the NO adsorbed on NaY showed no interaction with aluminum ions.

Chemisorption of molecules on decationized Y zeolites, presumably on aluminum Lewis acid sites, forming either radical ions (perylene) (*78*) or charge transfer complexes (*56*) has been reported previously; however, the observed spin concentrations were much smaller than the one observed by Lunsford. Lunsford observed the parallel between the concentration of NO chemisorbed on aluminum Lewis acids and the catalytic activity, and suggested that Lewis acid sites increase the activity of hydroxyls through an induction effect intervening through the zeolite lattice. Accordingly, on activation from 300° to 600°, the number of catalytic centers is limited by the concentration of Lewis acid sites, while above 600°C the decreasing concentration of hydroxyls adjacent to Lewis acids becomes limiting (*57*). However, as an alternative to Lunsford's interpretation of the NO adsorption maximum at 600° to 700°, it can be argued that between 300° and 700° the NO sorption rises because more oxygen-deficient sites are formed at the higher temperature for NO sorption, and that above 700° a gradual structural degradation sets in.

Richardson (*79, 80*) conducted kinetic experiments on cumene cracking with NaY and with various cation-exchanged Y zeolites (Li, K, Mg) loaded with 2 wt % copper ions by cation exchange. He found great differences in reaction rates between CuKY and CuMgY. The activity of CuMgY was significantly more than expected by the linear combination of the activities of CuKY and MgY. He found a similar rise in the quantity of radical ions obtained from adsorbed anthracene, perylene, pyrene, and naphthalene when he combined Cu^{2+} with bivalent cations instead of alkali metal cations in Y zeolite. He attributed the increase in catalytic activity and the amount of radical ion formation owing to a synergistic effect between Cu^{2+} and Mg^{2+} ions to polarization of the oxide ions linked to copper ions by the alkaline earth cations, with a resulting increase in electron affinity of the copper ions. As a direct evidence for cation–cation interaction, he showed a change in the magnetic parameters of the Cu^{2+}

ions, changing with zeolite cations of higher electron affinity: K < Li < Mg. Richardson seems to assume that the distribution of Cu^{2+} ions, and particularly the fraction of surface Cu^{2+} ions, is identical in the various cation-exchanged zeolites. However, it is conceivable because of the preference of bivalent cations for sites in the hexagonal ring (Site 1) or in the sodalite cage (Site 1′, 2′) that in combination with alkali metal ions very few Cu^{2+} ions are at surface site (Site 2). However, in combination with alkaline-earth cations, all competing for the more highly coordinated sites (Site 1, 1′, 2′), a significant fraction of the Cu^{2+} ions may emerge at the surface site (Site 2), where they can be in direct contact with adsorbed molecules, and consequently, may give rise to the "unexpected" increase in certain properties.

The molecular sieve effect in catalysis with zeolites was well demonstrated with small-pore zeolites (~5 A) in the early publications (*33, 75, 110, 111*). Recently, Thomas *et al.* (*89*) suggested that in the cracking of gas oil the large molecules do not enter the intracrystalline large cavities of the zeolite crystal, cracking on the external surface of the zeolite crystals, while the smaller cracked molecules enter secondary reactions mainly through hydride shift in the intracrystalline large cavities. The substantial increase in gasoline yields and the reduction of olefin product with zeolite-containing cracking catalysts was ascribed to a combination of factors including those mentioned above.

Noncarbonium Ion Type Catalysis. Richardson (*80*) reported that cation-exchanged Y zeolites show considerable activity for the dehydrogenation of cumene to α-methylstyrene and that the activity parallels the cation strength to some extent.

$$\text{K,Na} = 0 \ll \text{Li} < \text{Ba} \sim \text{Sr} \sim \text{Ca} \sim \text{Mg}.$$

Agudo *et al.* (*1*) found that the rate of oxidation of hexanes at 200°–350°C is significantly enhanced by NaX. Replacement of Na ions by Ca or Mn^{2+} ions in NaX decreased activity for the oxidation of all hexane isomers, and the sequence of activity was NaX > MnX > CaX, while the cracking activity varied in the reverse order: CaX > MnX > NaX. It was proposed that nonheterogeneous processes are initiated at the zeolite surface, and that the superior oxidation activity of NaX is attributed to its greater ability to initiate reactions by H-atom abstraction and to sustain free-radical reactions, whereas CaX and MnX favor carbonium-ion reactions.

It will be shown later herein that in the cracking of the isomeric hexanes with alkali metal Y, the products indicate a noncarbonium ion process and that the activity increases in the order: thermal cracking < NaY < KY < SiO_2–Al_2O_3. More enhanced differences in the same order are observed for cracking of *n*-butane, with the activity of NaY being barely distinguishable from that of thermal cracking, in contrast to its significant activity demonstrated for cracking of large paraffins (*110*). It appears then that for small paraffins NaY is much less active, but for large paraffins it is more active than the silica–alumina gel type catalyst.

Catalytic activity of cation-exchanged zeolites containing transition metal ions was reported by Rouchaud *et al.* (*82*), who found that the oxidation of *n*-hexane to acetic acid is catalyzed by Mn^{2+}-exchanged Y at 160° and 25 atm. As another example of the effect of transition metal ions, Kruerke (*52*) found that acetylene trimerized to benzene on transition metal–exchanged Y at near room temperature. The activity changes with the transition metal: Ca and Na $= O < Mn^{2+} << Co^{2+} = Ni^{2+} >> Cu^{2+} = Zn^{2+} = O$. This activity pattern is consistent with a model according to which at least 2 acetylenes have to be directly coordinated with the transition metal cation at the same time; consequently, transition metal ions with 2 filled + 2 half-filled *d* orbitals are most active. Significantly, trimerization activity only appeared at and beyond 40% ion exchange which indicated the need for cations at the surface position, in direct contact with the acetylene. Kruerke (*52*) also found that Y zeolite containing Cu^{+} ions catalyzes the Diels-Alder addition of acetylene and butadiene to form 1,4-cyclohexadiene. Dimitrov *et al.* (*29*) found that Cu^{2+} cations in zeolite X have a profound effect on the activation energy, rate, and stereoselective nature of the *n*-butene isomerization process.

Active Sites in Zeolites. Based on the literature reviewed here, the main characteristics of carbonium ion type zeolite catalysts emerge in the following manner.

The recent studies on the relationship between activation temperature and "carbonium ion" type catalytic activity of both decationized and cation exchanged zeolites show that at and above the temperature required for the removal of all observable hydroxyls with vibrational frequencies between 3700–3500 cm^{-1} the activity sharply declines. The lowest concentration of acidic lattice hydroxyl required for carbonium ion activity seems to depend on the reaction involved. For example, dehydroxylation of La-exchanged Y to a level at which hydroxyl content was unobservable by currently-used infrared techniques led to total loss of activity to crack *n*-butane, but only partial loss of activity to crack cumene (*vide infra*) and to alkylate toluene with propylene (*74*). The activity and hydroxyl content lost on dehydroxylation can be restored upon subsequent treatment with water (*11*). Furthermore, alkali metal zeolites, which have little or no carbonium ion type activity can be made to show strong activity by the addition of a proton source, such as alkyl chlorides (*51, 58*). The similarity of the products obtained with the decationized Y, bi-, and multivalent cation-exchanged zeolites X and Y, and their parallel behavior in deuterium exchange with benzene (*101*) (*see* section on Cumene Cracking) are all consistent with a mechanism involving Bronsted acid. Of the several types of hydroxyls identified in zeolites, the one which gives rise to an infrared band near 3640 cm^{-1} appears to be the most accessible and probably most acidic, and it is generally regarded as the source of activity involving proton sites. Other hydroxyls which are less accessible and/or less acidic also may contribute to similar activity to a lesser extent.

It was suggested that "superactivity" of zeolites is caused by the high concentration of hydroxyls, exceeding that of silica–alumina catalysts by more than an order of magnitude (*39*). However, with decationized Y, it appears that upon activation at 550°–600°, which is required for highest activity for several carbonium ion-type reactions, a significant fraction of the hydroxyls are removed. This is indicated by the amount of chemisorbed pyridinium ion (*104*), the reionexchangeability (*16*), the amount of deuterium exchange (*96, 101*), and the weight loss of acid-exchanged zeolites (*16*) and of NH_4-exchanged Y (*16, 25*).

It is unfortunate that in the published catalytic studies with heat-treated NH_4-exchanged Y the catalyst substrate seems to be insufficiently characterized, and the assignment of structural details, which are most important for catalytic properties, are based on assumptions, usually drawn from comparisons with "similar" zeolite samples. Furthermore, the catalytic properties of zeolites which are more fully characterized usually are not reported in the literature. Consequently, comparison between catalysts prepared by different investigators is very difficult and risky since the catalytically important details depend on several chemical and structural characteristics, all of which usually are not described. As an example of this problem, the dehydroxylation of heat-treated NH_4^+-exchanged Y depends on the temperature, atmosphere, and time employed on activation, as well as on the Si/Al ratio, the degree of NH_4^+ exchange (*16*), and the residual cations (*108*).

On removal of hydroxyl groups, oxygen-deficient sites are formed which have been identified as aluminum sites with one oxygen missing: aluminum–Lewis-acid sites. It seems significant that with decationized Y the highest hydroxyl content does not coincide with highest catalytic activity, suggesting that some property introduced on dehydroxylation enhances catalytic activity, in spite of the decrease in hydroxyl content. Since in NH_4-exchanged Y the hydroxyls are formed upon evolution of NH_3, the effect of residual NH_4 ions or a poisoning of active sites by re-adsorbed NH_3 has to be considered, following activation at low temperatures. On isothermal heat treatment in air at 375°, the residual N content of decationized Y is unobservable (*16*), and upon heat treatment in helium at 550° only about 5–6% of the original NH_4 ions are not decomposed (*101*). According to the experiments using quinoline as poison (*18, 80, 95*), the poisoning at 6 quinoline per 100 hydrogen level is negligible. Therefore, the effect of residual NH_3 is likely to be insignificant.

In order to explain that the highest activity of decationized Y coincides with partial dehydroxylation and with the formation of Al–Lewis acid sites formed at the expense of hydroxyl groups, it was suggested that strong cations or aluminum Lewis acid sites—in the respective type of zeolites—intervene through the lattice to affect each other (*80*) as well as the structural hydroxyls, with a resulting increase in acidity of the latter (*57*). The effect of cation content on hydroxyls in various cation-exchanged zeolites and in decationized Y is demonstrated in the changes both in the size and the frequency of the ~3640 cm^{-1} band. The change to lower frequency is inversely related to the activity in protonating pyridine (*26*), and probably to hydroxyl acidity, suggesting that the frequency of the 3640 cm^{-1} band may be an indication for changes in the hydroxyl's intrinsic catalytic activity. However, the latter relationship has not been established by the comparison of various zeolites, *i.e.*, bivalent cation Y and decationized Y, at a comparable hydroxyl level.

Major changes in catalytic activity are observed not only with changing cation but also on gradual dehydroxylation of the decationized Y. Therefore, if higher acidity and catalytic activity are associated with a shift of the 3640 cm^{-1} band to lower frequency, then such a shift may be expected on dehydroxylation as well. However, no change in frequency of the 3640 cm^{-1} band was reported or can be observed in the published infrared spectra at increasing degrees of dehydroxylation; hence, there is no direct evidence for changes in the nature of this hydroxyl group with increasing number of aluminum Lewis acid sites.

Some evidence was shown for cation–cation interactions, and for an apparent "synergistic" catalytic effect with Cu–Mg cation combination in Y zeolite; however, the experimental data supporting the suggested interpretation are inconclusive, and an alternative explanation without invoking synergistic effects is also possible, based on simple structural consideration. Similarly, with gradually dehydroxylated decationized Y, the parallel between the quantity of chemisorbed NO and catalytic activity is not as unexpected as suggested, and it may be explained by alternative reasoning without linking the 2 phenomena. Furthermore, the lack of change in the spectrum of chemisorbed NO following increasing activation temperatures fails to give a direct evidence for lattice interactions. Hence, in the absence of substantiating evidence for lattice interactions between Lewis acid sites and hydroxyls, it appears that the catalytic mechanism of dehydroxylated Y may be explained either by the suggested interaction of the sites intervening through the lattice or by their interaction intervening through the reacting molecules, and possibly by other structural factors not considered so far.

The experimental data and the interpretation regarding the number of active sites are rather conflicting. The very small values derived from kinetic experiments for active sites, about 10^{-3}–10^{-6} per framework aluminum, cannot serve even as an approximation for the true number of active sites, as long as all important components of the preexponential factor are not evaluated and considered. The interesting suggestion developed in order to rationalize the small number derived for active sites, that the hydroxyl band at 3640 cm^{-1} represents a variety of hydroxyls of greatly varying acidity, is also unsubstantiated. The lack of shift of the 3640 cm^{-1} band on dehydroxylation, or on gradual elimination by coke formation in catalytic reaction, and particularly the quantitative aspect of the poisoning experiments with nitrogen bases all seem to weaken this argument. In fact, the poisoning experiments (*18, 80, 95*) indicate a gradual decline in activity with increasing amounts of base, and the amount of base applied for full poisoning is between 0.5 to 3 per hydroxyl. In weighing this evidence one may consider, however, that the interaction between strong nitrogen bases and strongly acidic hydroxyls may be thermodynamically greatly favored, and it may come to completion even with hydroxyls of weaker acidity without displaying an easily observable selectivity. Since the base strength of hydrocarbons is much weaker than those of the N-bases used for poisoning, their selectivity toward changes in hydroxyl acidity may be more profound.

Significant changes in the intrinsic activity of catalytic sites should result in corresponding changes in the activation energy for any particular process, and consequently changes in the observed activation energies should indicate any qualitative change in the catalytic sites introduced either by cation exchange, decationization, or dehydroxylation. Unfortunately, the published data are too few (*60*) and are often inconsistent to the extreme (*18, 81, 91*), and thus are inadequate for any structural confirmation.

It appears that in most carbonium ion–type catalysis, the primary role of zeolite cations is in the generation of hydroxyls through hydrolysis, and in addition, they may further enhance the activity by a cooperative effect, intervening through the lattice or through the reactant. The role of cations in noncarbonium ion–type reactions—dehydrogenation, oxidation, trimerization of acetylene to benzene, and possibly radical-type cracking—appears to be direct, through interaction between reactant and surface cations. The need for direct contact between reactant and cation is demonstrated both in the dehydrogenation of cumene and in the trimerization of acetylene.

Cumene Cracking

Catalytic dealkylation of cumene is a standard test (*72*) for characterizing acidic cracking catalysts. Over silica–alumina, the products are largely benzene and propylene (*72*), although minor amounts of propane, C_4, and C_6 products have been reported (*90*). Early reports of the facile dealkylation of cumene by cation-exchanged X and Y zeolites at 325°–500° have been reviewed (*62, 100*), and several more recent studies have appeared (*18, 58, 80, 81, 91, 104, 106, 109*). Although this reaction is clearly of the Friedel-Crafts type and generally is rationalized by proton attack at an aromatic carbon atom with displacement of the side-chain as a carbonium ion (*100*), the details of mechanism are the subject of continuing study *via* kinetic (*80, 81*) and spectroscopic (*109*) approaches. Comparatively less attention has been paid to a full identification of the products. Benzene and propylene are reported (*18, 58, 104, 106*) to be the only products under a variety of conditions, and Ward (*104, 106*) specifically states the absence of propane over various members of the Y series at 260° and relatively high conversions (although it must be noted that analyses were performed only after some deactivation of the catalyst had occurred) (*109*). However, Frilette *et al.* (*33*) reported a deficiency of propylene compared with benzene and the presence of "miscellaneous aliphatics" from a flow study over Ca-exchanged Y at 470°, and Richardson (*80*) reported a number of aromatics besides benzene from runs above 500° over alkaline earth–exchanged faujasites. We report herein a fairly complete activity and product study for cumene cracking at relatively high conversions over Ca-exchanged and La-exchanged Y under flow conditions at 325° as a function of catalyst activation temperature and of time on-stream (*80, 81*). Several of the catalysts were examined also for their ability to scramble hydrogen isotopes in a feed composed of normal and perdeuteriobenzene, this exchange being a simple model for electrophilic aromatic substitution; a related study has been described by Venuto *et al.* (*101*) for decationized Y zeolites.

In Table I are recorded results obtained from the following sequence of operations. A 3:1 mixture of benzene and perdeuteriobenzene, diluted with nitrogen, was passed over the activated catalyst at 325° for 2 hours in an apparatus similar to that previously described (*74*). The total effluent was collected and analyzed by low-voltage mass spectroscopy for deuterium distribution; the catalyst was flushed with nitrogen; finally, a mixture of cumene and nitrogen was fed for 1 hour before glpc examination of the effluent. As expected, NaY showed negligible activity for either deuterium scrambling or dealkylation. In contrast, Ca-exchanged Y, activated at 550°, gave a benzene–deuterium distribution approaching statistical scrambling (hence requiring a large number of individual ex-

Table I. Cracking of Cumene and Redistribution

Catalyst	*Activation*	*Benzene Hydrogen Distribution*[b] d_0	d_1	d_2	d_3	d_4	d_5	d_6
NaY	550°, 16 hr	76.4	0.0	0.0	0.0	0.0	0.4	23.1
CaY[d]	550°, 16 hr	29.9	38.3	22.9	7.5	1.4	0.1	0.0
CaY[d,e]	710°, 2 hr	73.5	1.7	0.1	0.0	0.1	2.3	22.3
LaY[f]	550°, 16 hr	22.4	43.2	22.7	9.5	2.0	0.2	0.0
LaY[f]	700°, 3 hr	43.7	23.8	10.5	5.6	5.3	6.1	5.0
LaY[e,f]	750°, 16 hr	63.4	10.1	1.6	0.7	2.0	6.3	15.9
Statistical Distribution		17.8	35.6	29.7	13.2	3.3	0.4	0.02

[a] 325°; 0.0248 moles/hr 3:1 C_6H_6:C_6D_6 plus 0.39 moles/hr N_2 fed 2 hr; pure N_2 fed 2 hr; then 0.0157 moles/hr cumene plus 0.39 moles/hr N_2 fed 1 hr; 10 grams catalyst.
[b] Initial feed—d_0:d_1:d_2:d_3:d_4:d_5:d_6 = 75.8:0.0:0.0:0.0:0.0:0.4:23.8.
[c] After 1 hr on cumene feed based on benzene: cumene ratio in effluent.

Table II. Cracking of

Time on Stream, Hr.	Relative Molar Amounts C_2[d]	C_3H_6	C_3H_8	C_4[e]	C_5	C_6[f]
						CaY
1.00	—	34.2	35.7	27.8	8.9	7.0
2.75	—	81.4[g]		9.8	1.6	1.9
4.75	—	61.4	17.6	6.3	1.2	1.2
6.50	—	89.3[g]		4.9	0.7	0.7
8.25[h]	—	73.3	14.5	4.4	0.8	0.8
						LaY
1.00	0.4	[i]	53.5	7.6[j]	4.4	5.4
2.75	1.3	2.2	40.6	12.7	4.3	3.6
4.75	1.6	21.9	42.1	9.2	4.4	2.6
6.50	1.1	55.2	28.7	2.9	1.1	1.3
8.25[k]	0.9	63.4	27.3	1.4	0.5	0.7

[a] 325°; 0.0157 moles/hr cumene plus 0.39 moles/hr N_2; 10 grams catalyst activated at 550° in vacuum for 16 hr.
[b] Normalized to benzene = 100.
[c] Does not include small amounts of materials less volatile than cumene; total collected effluent from CaY showed di-*i*-propylbenzenes (14% of the residual cumene) and other materials (13% of residual cumene).
[d] Largely ethylene.
[e] Largely paraffinic with isobutane predominant.

changes) and showed high activity to destroy cumene; however, the dealkylation product contained as much propane as propylene. La-exchanged Y, activated at 550°, was even more active in both tests, and the C_3 fraction was largely propane. Activation of both catalysts at 700° led to dehydroxylation, as judged from infrared spectra (*74*). This treatment removed essentially all activity in both tests for the Ca-exchanged Y.

of Benzene Hydrogens Over Y Zeolites[a]

Cumene "Conversion"[c]	C_3H_8/C_3H_6
< 0.1	
>97.	1.0
2.	
>98.	>20.0
60.	< 0.05
20.	< 0.02

[d] CaY: Si/Al = 2.43; Ca(II) exchange = 83%; equivalent cation/Al = 0.99.
[e] X-ray analysis showed no significant loss of crystallinity after activation and reaction.
[f] LaY; Si/Al = 2.46; La(III) exchange = 86%; equivalent cation/Al = 1.03.

Cumene Over Y Zeolites[a]

in Effluent[b,c]

C_6H_6	*Toluene*	*Ethylbenzene*	*Cumene*
100	—	—	4.1
100	1.4	5.8	31.0
100	1.6	4.2	41.9
100	1.5	3.2	42.7
100	1.3	3.6	59.7
100	19.8	8.6	0.02
100	9.2	5.4	1.3
100	5.8	4.6	6.6
100	3.3	2.7	26.5
100	2.6	2.6	33.8

[f] Other than benzene.
[g] Propylene plus propane.
[h] 5.5% of total feed had been converted to nonvolatile carbonaceous residue on the catalyst at this point.
[i] Too small to measure.
[j] Minimum value.
[k] 10.4% of total feed had been converted to residue.

However, for La-exchanged Y, dehydroxylation, even at 750°, gave only partial loss of activity in both tests; however, no propane was produced now, even at 60% conversion.

Additional results for cumene cracking over Ca- and La-exchanged Y, each activated at 550°, are listed in Table II. The following concurrent changes occurred for La-exchanged Y with increasing time on-stream. Conversion gradually decreased as visible darkening of the catalyst occurred; carbon analyses of spent catalysts after varying periods on-stream showed that residue formation was most serious at the beginning of a run and tapered off as activity dropped. The C_3 fraction changed from being propane-rich to being propylene-rich. The amount of C_2, C_4, C_5, C_6, toluene, and ethylbenzene byproducts became less important in comparison to the C_3's and benzene. The ratio of total C_3's:benzene, initially only ~0.5, approached the simple 1:1 stoichiometric ratio. Results for Ca-exchanged Y were generally parallel. These results show how complex a product mixture is produced at high levels of cracking, presumably *via* secondary reactions particularly of propylene—olefin-destroying reactions over a variety of zeolites are well known (*32, 56, 65, 98*)—and show that gradual deactivation of the catalysts by residue formation leads more nearly to a "clean" mixture of propylene and benzene.

The most striking product result is the extensive formation of propane over very active catalysts. Venuto *et al.* (*99*) reported analogously that dealkylation of *tert*-butylbenzene over rare earth–exchanged X zeolite at 260° gave isobutane as the major gaseous product. Such paraffin formation is presumably the result of hydride transfer reactions to the carbonium ions formed by initial electrophilic cleavage of the alkylbenzene (*100*) or by protonation of the olefin. Reasonable hydride donors are cumene and propylene; the resultant hydrogen-deficient species are then precursors of residue formation (*32, 89*). Paraffin formation by treatment of alkylbenzenes with aluminum halides in the presence of cyclohexane or decalin has been known for 30 years (*47*), and there is ample evidence for hydride transfer between carbonium ions and hydrocarbons (*10, 22, 27, 53*).

Cracking of Hexanes Over Alkali Metal–Exchanged Zeolites

Two mechanisms are generally recognized for cracking of paraffins. Noncatalytic thermal cracking (*ca.* 550°–700°) proceeds by a free-radical route of moderate chain length (*77*) to give a characteristic and predictable mixture of products. Radical mechanisms also have been postulated for cracking over "inert" solids such as silica gel, which lead to some rate enhancement over the vapor phase but give typical "thermal" products (*36*). However, typical catalytic cracking over acidic materials such as silica–alumina gives greater enhancement in rate and a distinctly different product distribution, and is generally accepted as proceeding through carbonium ion intermediates (*36, 88*).

Under radical conditions, the paraffin is activated by hydrogen abstraction to produce a mixture of alkyl radicals corresponding to relative reactivities of tert:sec:prim hydrogens of *ca.* 10:3:1 at 600° (*77*). These radicals undergo β-scission to produce an olefin and a smaller radical, preferably the more stable one if a competition is possible (*50*). This new radical ($R\cdot$) has 2 major reaction paths available: further β-scission (rate $\propto k_{\beta}^{R\cdot}$) or hydrogen abstraction from the paraffin substrate (S) to produce an alkane, RH, and a new chain-carrying radical (rate $\propto k_H^{R\cdot}$ [S]). At the usual temperatures and [S] employed in thermal cracking, $k_{\beta}^{R\cdot}/(k_H^{R\cdot}[S]) >> 1$ for all radicals, $R\cdot$, which can cleave to form a tert, sec, or prim radical. However, when hydrogen atom or methyl radical must be lost, β-scission and hydrogen abstraction become competitive in rate. Thus, the product mix from a radical chain is characterized by formation of specific olefins with double-bond placement predicted by β-scission without any rearrangement of the intermediate radicals, only small paraffins, a relatively large proportion of C_1 and C_2 products, and by the absence of skeletal isomerization of substrate or products.

In carbonium ion cracking, β-scission of carbonium ions formed by hydride abstraction from the paraffin is now the mode of C–C bond cleavage. However, the isomerization reactions available to carbonium ions by hydrogen and alkyl group migrations, which are more rapid than those in radicals, lead to conversion of *prim*-carbonium ions to more stable ions before cleavage. The resulting product mix is characterized by formation of highly isomerized olefins, paraffins $> C_3$, low yields of C_1 and C_2 compared to C_3 and C_4 products, and by extensive skeletal isomerization of substrate and products.

Although many studies (*12, 44, 60, 74, 89*) have appeared concerning the cracking of paraffins over "acidic" zeolites which follow the "carbonium ion" pattern, less attention has been paid to the parent alkali metal–exchanged zeolites. The cracking of *n*-paraffins over NaX and Ca-exchanged X zeolites has been studied (*33*), and an activity order Ca-exchanged X > NaX > silica–alumina was reported. Whereas Ca-exchanged X gave a typical "carbonium ion" product mix from *n*-hexane, NaX gave a product comparatively much richer in C_1 and C_2 products, poorer in C_4 and C_5 paraffins, and free from skeletal isomerization (although double bond migration had occurred in the olefins). Although no direct experimental comparison of activity or products was made with thermal cracking of *n*-hexane, a radical mechanism was assigned therefore to the cracking over NaX. Similar results were reported (*92*) in the Y zeolite series where cracking of *n*-hexane at 450° was somewhat more rapid over NaY than over silica–alumina and gave a "radical" product mix; introduction of Ca ions led to sharply increased activity and formation of a "carbonium ion" product beyond 60% ion exchange.

We wish to report a study of the cracking of *n*-hexane, 2-methylpentane, 3-methylpentane, and 2,3-dimethylbutane over K-exchanged Y, NaY, and Na,K-exchanged L zeolites at 500° and 1 atm at low conversion levels (LHSV ~0.3), as well as thermal cracking in a quartz wool–packed

Table III. Cracking of Hexanes

Moles/100 Moles

Substrate	*n-Hexane*				
Catalyst	*Thermal*	*KY*	*Na,KL*	*NaY*	*Aerocat*
Conversion, %	*2*	*12–15*	*6*	*6*	*16*
Hydrogen	2.8	3.1	6.9	5.0	45.9
Methane	50.3	23.5	29.0	27.2	14.5
Ethylene	56.5	26.0	30.6	29.0	15.2
Ethane	37.7	37.4	40.7	38.3	9.7
Propylene	55.7	58.5	53.9	55.9	46.4
Propane	10.7	38.2	31.8	37.1	72.2
1-Butene	31.3	12.0	9.6	9.8	5.0
2-Butene		19.6	23.7	21.9	10.6[b]
Isobutene					
Butanes					11.6[c]
1-Pentene	7.3	1.9	1.5	1.5	3.8[d]
2-Pentene		4.7	6.1	1.5	1.1[e]
2-Methyl-1-butene					0.3
3-Methyl-1-butene					0.1
2-Methyl-2-butene				1.7	0.7
Pentanes					

[a] 500°C; 10 grams catalyst activated 550° in vacuum; hexane feed 6.14 ml/hr.
[b] Could contain *n*-butane.
[c] Isobutane.
[d] Could contain isopentane.

reactor under identical conditions. Conversion and product data for samples activated at 550° are shown in Table III.

The thermal cracking data for the 4 substrates can be fit with the Rice radical mechanism (*77*) if one assumes that essentially all ethyl radicals abstract hydrogen rather than cleave (note the low yield of hydrogen compared with ethane), but that *n*-propyl and isopropyl radicals have $k_\beta^{R\cdot} \simeq k_H^{R\cdot}$ [S] (note the propane product) at the conditions studied. This is reasonable since 500° is lower than the temperatures usually employed for thermal cracking, and β-scission has a higher activation energy than hydrogen abstraction (*50*). There is, of course, not enough data to fix uniquely all the parameters necessary to describe the decomposition of each substrate, but an averaged value of 10–15:2.5–3:1 for the relative reactivity of tert:sec:prim hydrogens reproduces the data quite well. With respect to products, note for example the accuracy of the predictions that *n*-hexane should give 1-butene and 1-pentene free from 2-isomers, that 2-methylpentane should give isobutene free from linear butenes and 3-methyl-1-butene free from the other isopentenes, that 3-methylpentane should give 2-methyl-1-butene as the exclusive C_5

Over Alkali Metal Zeolites[a]

Decomposed

2-Methylpentane			3-Methylpentane			2,3-Dimethylbutane		
Thermal 2	*KY 9.5*	*Na,KL 5.5*	*Thermal 2.5*	*KY 11*	*Na,KL 5*	*Thermal 2*	*KY 11*	*Na,KL 5*
24.4	4.3	7.0	5.6	3.9	3.7	14.5	3.8	5.5
27.9	21.7	25.0	63.5	57.3	58.0	78.5	76.8	80.7
11.5	6.5	7.5	23.7	23.0	20.9	0.5	0.4	0.4
42.4	43.5	43.3	44.2	44.3	42.8	0.1	0.5	0.3
57.8	44.2	43.4	20.5	15.7	13.4	44.6	37.8	34.3
19.2	36.4	31.5		0.5	0.7	21.5	33.2	26.2
			7.6	11.2	11.7			
			34.2	34.8	34.4	0.5	0.9	1.1
40.5	40.1	40.7						
0.2[c]	1.6[c]	1.7[c]						
1.0	2.3	2.8		[f]	[g]			
8.5	8.4	9.1	13.3	6.4	6.8			
			20.9	[f]	[g]	1.0	17.2	18.3
4.5	1.6	1.0		1.0	1.2	0.7	2.4	3.2
	1.7	2.6		19.5	19.8	60.9	41.3	44.2
						1.2	0.2	0.6

[e] Could contain *n*-pentane.
[f] Sum = 8.3.
[g] Sum = 9.9.

olefin, and that 2,3-dimethylbutane should give C_3 and C_5 but neither C_2 nor C_4 olefins.

In contrast, cracking of *n*-hexane over amorphous silica–alumina (Aerocat) gave modestly increased activity and a typical "carbonium ion" product mix containing more hydrogen, less methane and C_2's, more propane, and isomerized higher paraffins (isobutane).

Cracking over K-exchanged Y also led to modest increases in activity (of the same order of magnitude for *n*-hexane as did silica–alumina) compared with thermal cracking, but product mixes different from both thermal cracking and silica–alumina cracking were observed. However, the product still can be correlated with a radical mechanism if 2 modifications caused by the zeolite are introduced. First, we note that there were no significant changes in C_4 and C_5 yields if olefins with a given carbon skeleton were summed; for example, *n*-hexane gives 31.6 moles/100 moles of 1- plus 2-butene compared with 31.3 moles/100 moles of 1-butene thermally with no isobutene in either case. Therefore, we suspected that double bond shifts (with resultant approach to equilibrium within a given carbon skeleton) might be occurring as a secondary reaction subsequent to rather than as an inherent part of cracking. This was experimentally demonstrated for the 2-butenes from *n*-hexane. Feeding individual *n*-butenes over K-exchanged Y at 500° gave extensive double-bond isomerization but negligible cracking. Also, a series of cracking runs with progressively shorter contact times (N_2 dilution) gave 1-:2-butene ratios of 0.60, 0.62, 1.80, and 2.32 at 12.7, 5.9, 1.8, and 0.7% *n*-hexane conversion. Hence, we conclude that 1-butene is the primary C_4 cracking product as in thermal cracking, but it is partially isomerized by further contact with the catalyst.

Secondly, having considered the C_4 and C_5 products, we note that the differences in C_1–C_3 products between thermal and K-exchanged Y-catalyzed cracking can be rationalized effectively by introducing a second effect of the catalyst; namely, for any given radical, $R\cdot$, the ratio $k_\beta^{R\cdot}/(k_H^{R\cdot}[S])$ is smaller over K-exchanged Y than thermally. All radicals which can cleave to form new radicals as stable as ethyl still do so, but the propyl radicals for which β-scission and hydrogen abstraction were closely balanced thermally now give comparatively more propane and less ethylene plus methane from *n*-propyl radical and more propane and less propylene plus hydrogen from isopropyl radical. The operation of this effect can be illustrated for each substrate. 3-Methylpentane gives the closest correspondence between thermal and catalytic results, and examination of the Rice scheme (*77*) shows no C_3 radicals. *n*-Hexane gives increased propane at the expense of methane and ethylene with no change in ethane and propylene, in accord with less cleavage of the *n*-propyl radical. 2,3-Dimethylbutane, in contrast, has isopropyl radical as a chain carrier and the expected increase in propane at the expense of propylene and hydrogen characterizes the observed products. Finally, 2-methylpentane, whose decomposition involves both *n*- and isopropyl radicals, shows both effects.

Compared with materials such as alumina (*21, 42*), silica–alumina (*42, 55*), chromia–alumina (*2*), or silicotungstic acid (*21*), K-exchanged Y is relatively inactive for catalyzing double bond shifts in the *n*-butenes, a temperature of 400° being required to achieve $> 5\%$ conversion at a nominal contact time of 4 sec; however, since an even higher temperature is required to achieve significant paraffin cracking, isomerization is rapid at cracking conditions. The mechanism of such double-bond shifts has been considered recently in some detail (*42*). We have observed that NaY is considerably more active than KY for this reaction (*see also 29, 37, 83*), in contrast to paraffin cracking.

The nature of the effect of the zeolite on $k_\beta^{R\cdot}/(k_H^{R\cdot}[S])$ has not been clarified. Modification of radical behavior by solvents has been observed for certain radicals (*46, 103*) but such effects on carbon radicals are minimal even at ambient temperatures. Carbon radicals associated with silica and silica–alumina surfaces have been observed spectroscopically at very low temperatures, but the strength of the association seems too weak to cause a significant effect on relative reactivities, *i.e.*, $k_\beta^{R\cdot}/k_H^{R\cdot}$, at 500° (*49*). However, another contributor for altering the β-scission:abstraction ratio may be [S] rather than $k_\beta^{R\cdot}/k_H^{R\cdot}$. An effectively higher "concentration" of substrate within the catalyst pores than in the vapor phase at the same applied pressure would explain the observed effect.

Although we point out the *ca.* 5-fold increase in rate of cracking over K-exchanged Y compared with thermal cracking, attempting to rationalize this small a rate effect seems premature at this time. Whereas we can consider only the chain steps when predicting products because initiation steps contribute only slightly to the products, we must consider the nature of the initiation and termination steps (*77*) when concerned with rate phenomena. There is still controversy as to whether initiation and termination are homogeneous or heterogeneous in the purely thermal reaction (*54, 102*). Initiation of radical chain reactions by the surface of NaX has been postulated by Agudo *et al.* (*1*) for the oxidation of hexanes.

The product mixtures over NaY and Na,K-exchanged L zeolites were indistinguishable within experimental error from those over K-exchanged Y, whereas the activities were only one-half that of K-exchanged Y.

Similar experiments also were carried out for cracking of *n*-butane at 550°; under identical conditions, conversions over Aerocat (silica–alumina), K-exchanged Y, NaY, and an empty tube were 40, 12, 3, and 1.5%, respectively.

Literature Cited

(1) Agudo, A. L., Badcock, F. R., Stone, F. S., *Intern. Congr. Catalysis, 4th, Moscow, 1968*, Preprint of Paper No. 59.
(2) Amenomiya, Y., Cvetanovic, R. J., *Can. J. Chem.* **1962,** 40, 2130.
(3) Angell, C. L., Bennett, J. M., Rabo, J. A., private communication.
(4) Angell, C. L., Howell, M. V., *Can. J. Chem.* **1969,** 47, 3831.
(5) Angell, C. L., Schaffer, P. C., *J. Phys. Chem.* **1965,** 69, 3463.
(6) *Ibid.,* **1966,** 70, 1413.
(7) Barrer, R. M., *Endeavour* **1964,** 23, 122.
(8) Barry, T. I., Lay, L. A., *J. Phys. Chem. Solids* **1966,** 27, 1821.
(9) *Ibid.,* **1968,** 29, 1395.
(10) Bartlett, P. D., Condon, F. E., Schneider, A., *J. Am. Chem. Soc.* **1944,** 66, 1531.
(11) Benesi, H. A., *J. Catalysis* **1967,** 8, 368.
(12) Bennett, J. M., Smith, J. V., *Mater. Res. Bull.* **1968,** 3, 633.
(13) *Ibid.,* **1968,** 3, 865.
(14) *Ibid.,* **1969,** 4, 7.
(15) Bennett, J. M., Smith, J. W., Angell, C. L., *Mater. Res. Bull.* **1969,** 4, 77.
(16) Bolton, A. P., Lanewala, M. A., *J. Catalysis* **1970,** in press.
(17) Bolton, A. P., Lanewala, M. A., Pickert, P. E., *J. Org. Chem.* **1968,** 33, 1513.
(18) Boreskova, E. G., Topchieva, K. V., Piguzova, L. I., *Kinetics Catalysis* **1964,** 5, 792.

(19) Breck, D. W., *J. Chem. Educ.* **1964,** 41, 678.
(20) Breck, D. W., U. S. Patent **3,130,007** (1964).
(21) Brouwer, D. M., *J. Catalysis* **1962,** 1, 22.
(22) Brouwer, D. M., Oelderik, J. M., *ACS Natl. Meeting, San Francisco, April, 1968,* Preprints of Papers, Div. Petrol. Chem., 184.
(23) Cant, N. W., Hall, W. K., *Trans. Faraday Soc.* **1968,** 64, 1093.
(24) Carter, J. L., Lucchesi, P. J., Yates, D. J. C., *J. Phys. Chem.* **1964,** 68, 1385.
(25) Cattanach, J. E., Wu, L., Venuto, P. B., *J. Catalysis* **1968,** 11, 342.
(26) Christner, L. G., Liengme, B. V., Hall, W. K., *Trans. Faraday Soc.* **1968,** 64, 1679.
(27) Condon, F. E., "Catalysis," Vol. VI, P. H. Emmett, Ed., Reinhold, New York, 1958.
(28) Dempsey, E., *Soc. Chem. Ind. (London)* **1968,** 293.
(29) Dimitrov, C., Leach, H. F., *J. Catalysis* **1969,** 14, 336.
(30) Dodge, R. P., Schomaker, V., Union Carbide Research Institute, private communication, 1961.
(31) Dollish, F. R., Hall, W. K., *J. Phys. Chem.* **1967,** 71, 1005.
(32) Eberly, P. E., Jr., *J. Phys. Chem.* **1967,** 71, 1717.
(33) Frilette, V. J., Weisz, P. B., Golden, R. L., *J. Catalysis* **1962,** 1, 301.
(34) Fripiat, J. J., Rouxhet, P., Jacobs, H., *Am. Mineralogist* **1965,** 50, 1937.
(35) Fripiat, J. J., Toussaint, T., *J. Phys. Chem.* **1963,** 67, 30.
(36) Greensfelder, B. S., Voge, H. H., Good, G. M., *Ind. Eng. Chem.* **1949,** 41, 2573.
(37) Gryaznova, Z. V., Glotova, Z. F., Balandin, A. A., Tsitsishvili, G. V., Andronokashvili, T. G., Krupennikova, A. Y., *Bull. Acad. Sci. USSR Div. Chem. Sci.* **1968,** 1231.
(38) Hall, W. K., private communication.
(39) Hansford, R. C., Ward, J. W., *J. Catalysis* **1969,** 13, 316.
(40) Hattori, H., Shiba, T., *J. Catalysis* **1968,** 12, 111.
(41) Hickson, D. A., Csicsery, S. M., *J. Catalysis* **1968,** 10, 27.
(42) Hightower, J. W., Hall, W. K., *J. Phys. Chem.* **1967,** 71, 1014.
(43) Hirschler, A. E., *J. Catalysis* **1963,** 2, 428.
(44) Hopkins, P. D., *J. Catalysis* **1968,** 12, 325.
(45) Hughes, T. R., White, H. M., *J. Phys. Chem.* **1967,** 71, 2192.
(46) Huyser, E. S., "Advances in Free-Radical Chemistry," Vol. I, G. H. Williams, Ed., Logos, London, 1965.
(47) Ipatieff, V. N., Pines, H., *J. Am. Chem. Soc.* **1937,** 59, 56.
(48) Isakova, Ya. I., Kylachko-Gurvich, A. L., Khudiev, A. T., Minachev, Kh. M., Rubinstein, A. M., *Intern. Congr. Catalysis, 4th, Moscow, 1968,* Paper No. 56.
(49) Kasansky, Y. B., Pariisky, G. B., *Proc. Intern. Congr. Catalysis, 3rd, Amsterdam, 1965,* Vol. I, p. 367.
(50) Kerr, J. A., Lloyd, A. C., *Quart. Rev. (London)* **1968,** 22, 549.
(51) Kolesnikov, I. M., Panchenkov, G. M., Tret'yakova, V. A., *Russ. J. Phys. Chem.* **1967,** 41, 586.
(52) Kruerke, U., Union Carbide Research Institute, Tarrytown, N. Y., private communication.
(53) Kramer, G. M., *J. Am. Chem. Soc.* **1969,** 91, 4819.
(54) Laidler, K. J., Wojciechowski, B. W., *Proc. Roy. Soc.* **1961,** A250, 91, 103.
(55) Leftin, H. P., Hermana, E., *Proc. Intern. Congr. Catalysis, 3rd, Amsterdam, 1946,* Vol. II, p. 1064.
(56) Liengme, B. V., Hall, W. K., *Trans. Faraday Soc.* **1966,** 62, 3229.
(57) Lunsford, J. H., *J. Phys. Chem.* **1968,** 72, 4163.
(58) Matsumoto, H., Yasui, K., Morita, Y., *J. Catalysis* **1968,** 12, 84.

(59) McBain, J. W., "The Sorption of Gases and Vapors by Solids," Ch. 5, George Rutledge & Sons, Ltd., London, 1932.
(60) Miale, J. N., Chen, N. Y., Weisz, P. B., *J. Catalysis* **1966,** 6, 278.
(61) Milton, R. M., U. S. Patent **2,882,244** (1959).
(62) Minachev, Kl. M., Garanin, V. I., Isakov, Ya. I., *Russ. Chem. Rev.* **1966,** 35, 904.
(63) Minachev, Kl. M., Isakov, Ya. I., *Izv. Akad. Nauk SSSR, Ser. Khim.* **1968,** 943.
(64) Norton, C. J., *Proc. Intern. Congr. Catalysis, 2nd, Paris, 1960,* 2073.
(65) Norton, C. J., *Ind. Eng. Chem. Proc. Design Develop.* **1964,** 3, 230.
(66) Olson, D. H., *J. Phys. Chem.* **1968,** 72, 1400.
(67) Olson, D. H., Dempsey, E., *J. Catalysis* **1969,** 13, 221.
(68) Olson, D. H., Kokotailo, G. T., Charnell, J. F., *Nature* **1967,** 215, 270.
(69) Pickert, P. E., Bolton, A. P., Lanewala, M. A., *Chem. Eng. Progr. Symp. Ser.* **1967,** 53, 50.
(70) Pickert, P. E., Rabo, J. A., Dempsey, E., Schomaker, V., *Proc. Intern. Congr. Catalysis, 3rd, Amsterdam, 1964,* 714.
(71) Planck, C. J., *Proc. Intern. Congr. Catalysis, 3rd, Amsterdam, 1964,* 727.
(72) Prater, C. D., Lago, R. M., *Advan. Catalysis* **1956,** 8, 293.
(73) Rabo, J. A., Angell, C. L., Kasai, P. H., Schomaker, V., *Discussions Faraday Soc.* **1966,** 328.
(74) Rabo, J. A., Angell, C. L., Schomaker, V., *Intern. Congr. Catalysis, 4th, Moscow, 1968,* Preprint No. 54.
(75) Rabo, J. A., Pickert, P. E., Stamires, D. N., Boyle, J. E., *Proc. Intern. Congr. Catalysis, 2nd, Paris, 1960,* 2055.
(76) Ralek, M., Grubner, O., Jiru, P., "Molekularsiebe," VEB Deutscher Verlag der Wissenschaften, Berlin, 1968.
(77) Rice, R. O., Rice, K. K., *J. Am. Chem. Soc.* **1943,** 65, 590.
(78) Richardson, J. T., *J. Catalysis* **1967,** 9, 50.
(79) *Ibid.,* **1967,** 9, 178.
(80) *Ibid.,* **1967,** 9, 182.
(81) Romanovskii, B. V., Skhoang, K. S., Topchieva, K. V., Piguzova, L. I., *Kinetics Catalysis* **1966,** 7, 739.
(82) Rouchaud, J., Sondegam, L., Fripiat, J. J., *Bull. Soc. Chim. France* **1968,** 11, 4387.
(83) Sharf, V. Z., Freidlin, L. K., Samokhvalov, G. I., Neimark, I. E., German, E. N., Piontkovskaya, M. A., Nekrosov, A. S., Kurtii, V. N., *Bull. Acad. Sci. USSR Div. Chem. Sci.* **1968,** 780.
(84) Smith, J. V., Bennett, J. M., Flanigen, E. M., *Nature* **1967,** 215, 270.
(85) Spandov, S. P., Egorova, E. N., "Chemistry of Zeolites," Science Publ., Leningrad, 1968.
(86) Stamires, D., Turkevich, J., *J. Am. Chem. Soc.* **1964,** 86, 757.
(87) Szymanski, H. A., Stamires, D. N., Lynch, G. R., *J. Opt. Soc. Am.* **1960,** 50, 1323.
(88) Thomas, C. L., *Ind. Eng. Chem.* **1949,** 41, 2564.
(89) Thomas, C. L., Barmby, D. S., *J. Catalysis* **1968,** 12, 341.
(90) Thomas, C. L., Hoekstra, J., Pinkston, J. T., *J. Am. Chem. Soc.* **1944,** 66, 1964.
(91) Topchieva, K. V., Romanovsky, B. V., Piguzova, L. I., Thoang, H. S., Bizreh, Y. W., *Intern. Congr. Catalysis, 4th, Moscow, 1968,* Preprint of Papers No. 57.
(92) Tung, S. E., McIninch, E., *J. Catalysis* **1968,** 10, 166.
(93) *Ibid.,* **1968,** 10, 175.
(94) Turkevich, J., *Catalysis Rev.* **1967,** 1, 1.
(95) Turkevich, J., Nozaki, F., Stamires, D., *Proc. Intern. Congr. Catalysis, 3rd, Amsterdam, 1964,* **1965,** 1, 586.

(96) Uytterhoeven, J. B., Christener, L. G., Hall, W. K., *J. Phys. Chem.* **1965,** 69, 2117.
(97) Uytterhoeven, J. B., Schoonheidt, R., Fripiat, J. J., *Intern. Symp. Reaction Mechanisms Inorg. Solids, Aberdeen, 1966.*
(98) Venuto, P. B., Hamilton, L. A., Landis, P. S., *J. Catalysis* **1966,** 5, 484.
(99) Venuto, P. B., Hamilton, L. A., Landis, P. S., Wise, J. J., *J. Catalysis* **1966,** 5, 81.
(100) Venuto, P. B., Landis, P., *Advan. Catalysis* **1968,** 18, 259.
(101) Venuto, P. B., Wu, E. L., Cattanach, J., *Soc. Chem. Ind. (London)* **1968,** 117.
(102) Voevodskii, V. V., *Kinetics Catalysis* **1964,** 5, 533.
(103) Walling, C., Wagner, P. J., *J. Am. Chem. Soc.* **1964,** 86, 3368.
(104) Ward, J. W., *J. Catalysis* **1967,** 9, 225.
(105) *Ibid.,* **1967,** 9, 396.
(106) *Ibid.,* **1968,** 10, 34.
(107) *Ibid.,* **1968,** 11, 238.
(108) *Ibid.,* **1968,** 11, 251.
(109) *Ibid.,* **1968,** 11, 259.
(110) Weisz, P. B., Frilette, V. J., *J. Phys. Chem.* **1960,** 64, 382.
(111) Weisz, P. B., Frilette, V. J., Maatman, R. W., Mower, E. B., *J. Catalysis* **1962,** 1, 307.
(112) Zhdanov, I. P., Kiselev, A. V., Lygin, V. I., Titova, T. I., *Dokl. Akad. Nauk SSSR* **1963,** 150, 584.

RECEIVED February 18, 1970.

Discussion

Jack H. Lunsford (Texas A & M University, College Station, Tex. 77843): I would like to clarify one point concerning the discussion of our work on the change in NO spin concentration as a function of activation temperature for decationized Y zeolites. Our interpretation has been that the number of NO molecules at aluminum sites increases following activation between 300° and 700° because of the formation of oxygen-deficient sites, and that the number of paramagnetic species decreases above 700° because these defect sites begin to anneal out. This is not an alternate to our interpretation, as Rabo suggested.

J. A. Rabo: Thank you.

Alfred E. Hirschler (Sun Oil Co., Marcus Hook, Pa. 19061): I would like to point out two significant omissions in Rabo's review dealing with Bronsted sites in zeolites. The first substantive evidence for the presence of Bronsted sites in cationic zeolites was our 1963 paper (*J. Catalysis* **1963,** 2, 428), which showed that arylmethanol indicators generate carbonium ions on contact with CaX and other cationic zeolites. The cited work of Norton did not establish the presence of Bronsted acidity for two reasons. He used H_0 indicators which give acid colors with Lewis as well as Bronsted acids, and his zeolites contained 20% of clay binder which itself possesses considerable acidity.

Secondly, I believe our 1963 paper was the first to suggest, albeit briefly, what is now believed to be the origin of the Bronsted acidity—*i.e.*, the cation field can polarize adjacent hydroxyl groups or water molecules adsorbed on the cations such that they become proton donors. We explicitly suggested that Bronsted acid strength should increase with increasing charge and decreasing size of the cation. As supporting evidence we cited Rabo's 1960 paper wherein the catalytic activity of cationic zeolites was found to depend on the electrostatic field of the cation and our own acidity data on cationic zeolites and cation-exchanged silica–alumina. At the 1964 Catalysis Congress in Amsterdam (*Proceedings*, Vol. 5, p. 726), we pointed out that the hexane isomerization data of Pickert and Rabo for a series of cationic zeolites were a linear function of the loss on ignition, which is the amount of potential Bronsted sites derived from hydroxyl groups and adsorbed water. We also pointed out that hydrogen zeolites had stronger Bronsted acidity than the cationic zeolites and that therefore alkaline earth cations partially poisoned the acidity of hydrogen zeolites just as in amorphous silica–alumina. We reported here also that the cumene cracking data of the hydrogen and cationic zeolites paralleled their strong Bronsted acidity as measured by H_R acidity titration. We therefore suggested that the active centers are Bronsted sites; the electrostatic field of the cations does not directly polarize the hydrocarbon molecules but exerts its influence *via* the Bronsted acidity which is influenced by the cation field. Later workers have provided much more detailed experimental evidence for this viewpoint as well as more explicit mechanistic interpretations and have often overlooked our prior suggestion.

J. A. Rabo: You must have overlooked p. 289 and Ref. *43* in our review.

W. H. Flank (Houdry Laboratories, Marcus Hook, Pa. 19061): In connection with the problem of relating the hydroxyl content of the calcined NH_4-exchanged synthetic faujasites to the behavior of these materials, it might be noted that there are amounts that are possibly quite significant catalytically, that are not going to be detected by infrared or some of the other techniques that have been used to look for them, and these hydroxyl groups do persist to very high temperatures. Kerr and more recently some people at Linde, Bolton and Lanewala, reported that they have observed a dehydroxylation step in thermogravimetric curves that they have obtained. We have also noted that such steps exist and that they increase in size with increasing degree of exchange and with decreasing SiO_2/Al_2O_3 ratio. We have further noted in agreement with Bolton and

Lanewala that the temperature at which this step occurs decreases with increasing degree of exchange and with decreasing SiO_2/Al_2O_3 ratio. But in almost all cases, we have repeatedly found that there are significant amounts of residual hydroxyl that remain after this step in the thermogravimetric curve. These resistant groups may be some of the ones that you are looking for. But I don't think that their concentration need be very high.

J. Rabo: Your comment is well taken. However, the chemical composition of the product released in the referred TGA step should be established before any assumption is made regarding its origin or its relationship to catalytic properties.

63

Catalytic Activity and the Nature of Active Centers in Zeolites

JOHN TURKEVICH and YOSHIO ONO

Princeton University, Princeton, N. J.

Catalyst preparations with controlled number and nature of acid sites were used to study the cracking of cumene, 2,3-dimethylbutane, and isomerization of xylenes. In all these cases, the Bronsted acid with its surface OH group was responsible for the catalytic activity. The cracking of 2,3-dimethylbutane required a small number of Lewis acid sites to initiate the reaction, presumably caused by abstraction of a hydride ion from the saturated hydrocarbon. The cracking of 2,3-dimethylbutenes was studied also. The hydrogen–deuterium equilibration is catalyzed by the Lewis acid sites. Introduction of palladium into the catalyst and the use of hydrogen as the gas carrier affects the cumene cracking favorably, but represses the branched chain hydrocarbon cracking.

The catalytic properties of the zeolites are of unusual interest. Technically, they are the most important catalyst used in the petroleum industry. Scientifically, they are specimens "par excellence" for studying the nature of active centers on alumina silica catalysts and for determining the mode of activation of the various molecules whose reactions they accelerate. They have the advantage over alumina silica gel catalysts in that they are crystalline. The main features of their structural arrangement can be delineated from x-ray investigation. This permits the identification and characterization of a limited number of discrete sites of catalytic activity. On the other hand, the structure of the alumina silica gel catalyst can only be surmised from general theoretical considerations. The x-ray and electron diffraction patterns are too broad to give any useful structural interpretation. Furthermore, indirect evidence indicates a broad continuous spectrum of active sites reflecting the amorphous state. This complicates characterization of the activation process. Another complicating feature of the gel type catalysts is the variable pore structure, making the preparations irreproducible and the catalytic activity variable.

We present some recent results obtained in the Princeton University Chemistry Department on the nature of the active center in catalysts derived from sodium type faujasite catalyst (*17, 18, 23, 25, 26, 27, 28, 29, 30, 31*). A general survey of the structural characteristics has been given by one of us and by others (*1, 4, 9, 25, 33*).

We shall review briefly those structural results which will be relevant to the determination of the nature of the active site. The formula for sodium faujasite is $Na_{56}(AlO_2)_{56}(SiO_2)_{136} \cdot nH_2O$. The water can be removed by heat treatment without destroying the crystal structure. The number of sodium ions is equal to the number of aluminum atoms. In zeolites, this number is at most equal to the number of silicon atoms. The aluminum atoms are never in adjoining oxygen tetrahedra but are separated by at least 1 silicon-oxygen tetrahedron. The sodium ion can be ion exchanged by monovalent, divalent, and certain trivalent ions. In this ion exchange, certain precautions must be observed. The direct replacement of the sodium ion by hydrogen ion using acid treatment may result in the destruction of the zeolite crystal structure, particularly when the silicon–aluminum ratio is close to 1. This is avoided by replacing sodium by ammonium ion and then decomposing the ammonium ion.

The washing of the zeolite must be carried out carefully to avoid hydrolysis and replacement of the cation by the hydrogen ion. This is particularly true of the divalent and trivalent cations. Another complication may arise in the case of divalent and trivalent cations, in that on dehydration of a hydroxy salt cation may be formed such as $[M(OH)]^+$, $[M(III)(OH)_2]^+$, $[M(III)(OH)]^{2+}$. Since the catalytic activity of the acid form of the zeolite is very high, the hydrolysis processes may mask the action of the multivalent ions and be ascribed to them. Sodium and hydrogen nuclear magnetic studies indicate that the sodium ions are hydrated by 6 water molecules in an octahedral array and are free to move in the zeolite cavity when the zeolite is hydrated. On dehydration, when there are less than 6 water molecules around the sodium ion, it becomes localized near 1 of the oxygen ions next to the aluminum.

In the faujasite type zeolite, the silica-alumina tetrahedra are linked through sharing of oxygen into 2 primary cages: the hexagonal prism and the truncated octahedron (sodalite). These primary cages combine to form a superstructure of cages by having 4 of the hexagonal faces of the sodalite linked in a tetrahedral manner *via* "virtual" hexagonal prisms to give the faujasite structure with the important secondary cage, the faujasite cage. This is the cage wherein the catalytic activity takes place.

Its 4 ports have a diameter 8–9A, permitting both straight and branched hydrocarbons to enter the pore. The port holes of the sodalite cage and the hexagonal prisms are too small for the hydrocarbon to enter, but they do play an important role in the activation process. The number of acid sites can be controlled by the depth of the replacement of sodium by the ammonium ion. There are 3 different sites for the sodium ion, 16 in the interior of the hexagonal prisms, 32 at the hexagonal faces in the sodalite cages, and 8 in the large cage next to the square oxygen faces (*20, 21*).

The kinetics of cracking of cumene was determined by the pulse technique introduced by Emmett, Kokes, and Tobin (*8*), modified by Turkevich and coworkers to determine the number of active sites and the specific activity (*30*). The number of sites active for cumene cracking was equal to the number of sodium ions replaced by hydrogen ions (*via* the ammonium route) until about half of the sodium ions had been replaced. From that, the number of active sites remained constant. Apparently, half of the acid sites produced by replacement of the sodium ion by ammonium ion and subsequent heating were not available for catalytic activity and must be located in either the sodalite cage or the hexagonal prism but not in the faujasite supercage. Furthermore, the technique of alternating pulses of reaction (cumene) and poison (quinoline) permitted the determination of activity per active site. This increased with the number of sites even when the sites are not available to the reactant, indicating an interaction between the sites *via* transport of hydrogen ion or electrons, both of which could pass through the small port holes. Furthermore, measurement of the temperature coefficient of the rate of cracking of cumene permitted the determination of both the activation energy of the cracking reaction and the entropy of activation. The activation energy decreased with increase in number of sites because of the interaction of sites. On the other hand, the entropy of activation was negative, indicating that in the transition state the adsorbed molecules are ordered. Furthermore, with increase in number of sites, the entropy of activation decreases, indicating that the greater the number of sites, the greater the required ordering. In the catalytic cracking of cumene, a large number of cumene molecules adsorbed on a large number of interacting sites, with 1 of these adsorbed molecules undergoing the reaction. The process does not involve 1 cumene molecule reacting with 1 site. It is the crystalline structure of the zeolite, which permits interaction between a large number of acid sites, that makes the zeolite catalyst superior to the amorphous alumina silica gel catalyst.

The next step in the determination of the mechanism of the catalytic action of the acid zeolite catalyst is elucidating the nature of the active site—is it a Bronsted acid, Lewis acid, or base site? The Bronsted acid site is a proton adsorbed on an oxygen atom next to an aluminum atom. The Lewis acid sites are produced from 2 Bronsted sites by elimination of water with 2 protons and an oxygen from the aluminum–silicon bond. The dehydration process was quantitatively studied by Turkevich and Ciborowski (*27*), using a stream of helium gas for dehydration and a thermal conductivity for determination of the amount of water or ammonia produced at any desired temperature. All the water was removed from the sodium faujasite by 250°C and no further water was evolved even when the sodium faujasite was heated to 900°C. This was taken as proof that in a sodium zeolite there are no free hydroxyl groups nor any hydrogen bonds holding the structure together. These findings are consistent with the structure of the faujasite as determined by x-ray crystallography. The water evolved in the dehydration of the sodium faujasite was the water in the cavities. When the same procedure was applied to ammonium faujasite, the same amount of water was produced below 250°C as from the pure sodium faujasite. The ammonia evolution is completed by heat treatment at 400°C. The product at this stage is essentially a Bronsted acid. Further heat treatment in the temperature range of 450°–600°C produces further dehydration, which converts the Bronsted acids into a Lewis acid and a Bronsted base site with removal of 2 hydrogen atoms and an oxygen between an aluminum and silicon atom in the faujasite skeleton. Thus, the conditions have been established for the preparation of the acid zeolite in the Bronsted form—helium flow gas at a temperature up to 400°C—and the Lewis acid Bronsted base form by dehydration in a flow of helium at a temperature at least of 550°C. A mixture of both types of acid centers is produced in the temperature range of 450°–550°C. We are thus in a position to establish which type of acidity is responsible for the various types of catalytic activity of the zeolites.

Other workers have studied the thermal decomposition of sodium and ammonium zeolites both before and after the Turkevich and Ciborowski studies.

The results were at best semiquantitative, since they were carried out in a static gas flow or in a vacuum where the rate of removal of gaseous products, water, and ammonia was undoubtedly diffusion controlled. The analysis was made using moist litmus paper or infrared spectral bands, both methods qualitative or semiquantitative. No quantitative relationship was established between the amount of ammonia evolved in the first stage and the water evolved in the process of dehydroxylation in the second stage (*2, 3, 5, 6, 7, 10, 11, 12, 13, 14, 15, 16, 19, 22, 24, 32, 34, 35, 36*).

Experimental

The catalyst was prepared by 70-hour treatment of 100 grams of Linde sodium Y zeolite with 1700 ml of 22.1 wt % of ammonium nitrate. The slurry was occasionally stirred. After filtration, the solid was washed several times with distilled water and dried at room temperature. The amount of sodium replaced by ammonium ion was 52% by analysis of the filtrate for sodium using the magnesium uranyl acetate reagent. This compares favorably with the result obtained by Turkevich and Ciborowski. The palladium catalyst was prepared by contacting a sample of sodium Y zeolite with $Pd(NH_3)_4Cl_2$ to give Pd 3.5%, Na 96.4% Y zeolite and similarly treating a sample of NH_4^+ 54%, Na 46% Y zeolite to give Pd 2.6%, NH_4^+ 54%, Na 46% Y zeolite.

Experiments studying the change in retention time of branched hydrocarbons showed that there was no change in pore size, and consequently crystalline structure, on the heat pretreatment in the temperature studied.

A conventional pulse catalytic microreactor was used with 15–65 mg of the catalyst for the cumene runs and 65 mg for the 2,3-dimethylbutane runs. The catalyst was held between 2 small plugs of borosilicate glass wool in a 5-mm ID diameter borosilicate reactor. In some experiments, the catalyst was diluted with 96% silica porous glass powder. The helium gas was purified by passage through alumina kept at liquid nitrogen temperature. The reaction temperature was measured by a thermocouple located adjacent to the reactor. The catalyst was pretreated at the desired temperature for 16 hours in a stream of helium. The products were analyzed with a dioctyl phthalate gas chromatography column at 110°C.

Cumene Cracking

The cumene cracking reaction was studied at 325°C. The simple kinetic equation

$$k = \frac{F}{W} \ln \frac{1}{1 - x}$$

(where F is the flow rate of the carrier gas, W the weight of the catalyst, k the reaction constant, and x the initial fraction converted) could not be used because the active catalysts (those prepared by heat treatment below 500°C) always gave conversion above the equilibrium conversion. This was caused by the separation of the products from each other and from the cumene in the reactor. There is a sharp drop in the activity of the catalyst on pretreatment of the catalyst at temperatures above 500°C. The activation energy of the reaction in the temperature range of 285°–325°C for an active catalyst was estimated to be 24 kcal. The quinoline titration of the samples obtained by heat treatment at various temperatures showed a similar decrease in the number of active centers with increase in the temperature of pretreatment (Figure 1), and this is similar to the curve obtained by dehydration of the Bronsted acids into

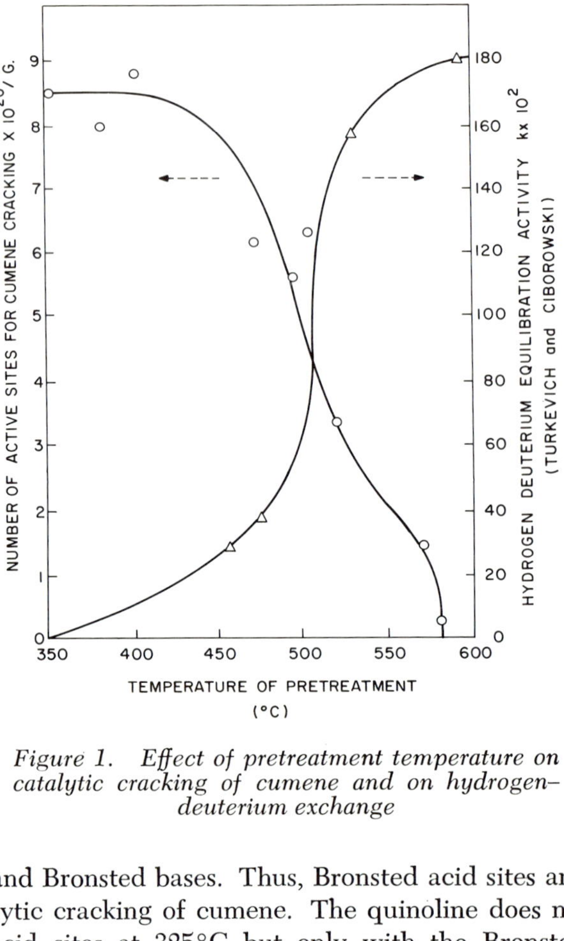

Figure 1. Effect of pretreatment temperature on catalytic cracking of cumene and on hydrogen–deuterium exchange

Lewis acid and Bronsted bases. Thus, Bronsted acid sites are responsible for the catalytic cracking of cumene. The quinoline does not react with the Lewis acid sites at 325°C but only with the Bronsted acid sites. Furthermore, for samples whose pretreatment temperature is equal or less than 400°C, the number of active sites is 8.3×10^{20}. This value is to be compared with 9.9×10^{20}/gram calculated from stoichiometry, assuming 25 wt % of water and ammonia as determined by Turkevich and Ciborowski.

The role of palladium in zeolite was investigated. The Na 100% zeolite has no activity for cumene cracking at 450°C. On the other hand, Pd 35% Na 96.4% Y zeolite cracks cumene in a stream of hydrogen with rate constants of 16.7 ml/min/gram at 400°C and 63.8 ml/min/gram at 443°C. The chromatographic column did not distinguish between propylene and propane. The palladium undoubtedly is reduced to the metal, producing protons on the surface which act as Bronsted acid. The introduction of decationated sites enhances the favorable effect of palladium, the rate constant at 325°C for Pd 2.6% NH_4 54% Na 43% Y zeolite being 1350 after the catalyst is placed in a stream of hydrogen overnight at 440°C. The palladium can be visualized as activating the hydrogen to remove carbonaceous material which forms on the surface at these high temperatures. However, it is also possible that the palladium heterolytically dissociates the hydrogen molecule into H^- and H^+, and these convert any small number of Lewis acid–Bronsted base on the surface into aluminum hydride and Bronsted acid sites. Experiments carried out at higher temperatures confirm this point of view. With palladium on decationated zeolite, cracking of cumene can take place when catalyst is treated at much higher temperatures than without palladium—*e.g.*, the rate constant (at 325°C) for the catalyst treated at 490°C is 1850, at 540°C is 1190, at 454°C is 1070, at 556°C is 440, and at 600°C is 300. At these temperatures, the Bronsted acids would be converted to Lewis acid–Bronsted base sites. These would be ineffective for cumene cracking and would favor carbon formation.

Cracking of 2,3-Dimethylbutane

The cracking of this branched chain paraffin was carried out at 400°–450°C. The products of the reaction were not simply propylene and propane, as would be expected from simple scission, but 38.4 mole % C_3H_8, 24.8% C_3H_6, 13.8% C_2H_4, 3.7% C_2H_6, 4.2% C_4H_{10}, 1.3% and 0.8% of a nonidentifiable C_4 hydrocarbon. The conversion was proportional to the weight of the catalyst used and inversely proportional to the flow rate of the carrier gas. An examination of the degree of decationization on the reaction showed that the 12% ammonia exchange for sodium did not produce active centers, but from that point the activity of the catalyst was proportional to the degree of decationation. The activation energy is 27 kcal/mole. The effect of the temperature of pretreatment on the activity of the catalyst shows a maximum of activity after pretreatment at 450°C and then a marked drop in the 500°–530°C region (Figure 2). These results suggest that Lewis acid sites are necessary in addition to Bronsted acid sites to crack branched chain hydrocarbons. These Lewis acid sites abstract a hydride ion, producing a carbonium ion which then undergoes various carbonium ion reactions with the help of Bronsted acid and base sites. The necessity of a certain number of Lewis sites for the cracking of paraffins finds further support in the results obtained with palladium. Replacement of 3.4% of the sodium ions by palladium does not enhance the cracking of 2,3-dimethylbutane. The introduction of palladium in the decationated Y zeolite (Pd 2.6%, NH_4 54%, Na 43.4%) and treatment with hydrogen at 360° or 495°C overnight produced a catalyst which showed no activity for

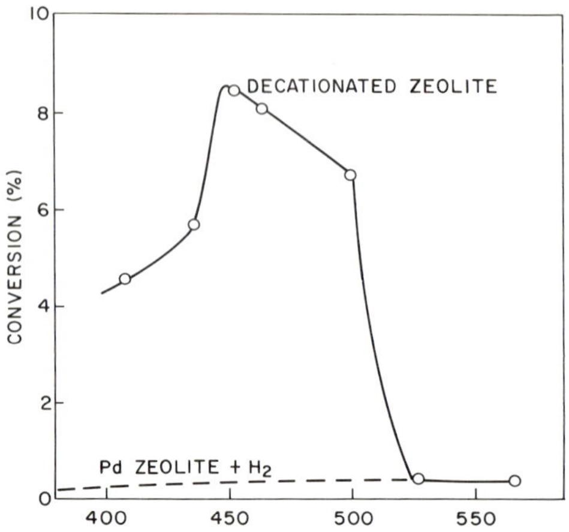

Figure 2. Effect of pretreatment temperature on catalytic cracking of 2,3-dimethylbutane, using a decationated zeolite and a palladium-decationated zeolite

the cracking of 2,3-dimethylbutane at 400°C, while it cracked cumene readily at 325°C. The function of palladium and probably other metal hydrogenating additions in supported oxide systems is to catalyze the heterolytic splitting of hydrogen gas into negative hydride ion and a proton. The negative hydride ion serves as a hydrogen bridge between the aluminum and silicon atoms of the Lewis acid site, and the proton neutralizes the Bronsted base site to make a Bronsted acid. This acid may be all the stronger because of the close proximity of the Al–H–Si bond. Thus, by using palladium and a decationated zeolite, the cracking of industrially valuable branched chain hydrocarbons is minimized while the "clipping off" of alkyl side chains from aromatics can be carried out readily even at high temperatures.

Cracking of Branched Chain Olefins

The cracking of the 2,3-dimethyl-2-butene and 2,3-dimethyl-1-butene showed that the rates of the various reactions they undergo were independent of the position of the double bond or the temperature of pretreatment of the catalyst. Both Bronsted and Lewis acids were effective. At 200°C, the double bond migration reaction was dominant and a significant carbon skeleton isomerization to 3,3-dimethyl-1-butene was observed. A slight amount of cracking was noted. At 300°C, the 2,3-dimethylbutenes had isomerized extensively into the 3,3-dimethylbutene, and this compound underwent cracking to produce 13% C_5's and 26%

C_4's. At 400°C, an appreciable amount of 3,3-dimethylbutene still persists, the C_4's and C_5's remain at the same level as at 300°C, but the yield of propylene increases to 10% of the product. These results indicate that the cracking of an olefin first involves a skeleton isomerization to an olefin whose subsequent cracking reactions are not explained readily by a simple carbonium ion mechanism.

Xylene Isomerization

The conversion of *ortho-* to *meta-* and *para-*xylene was carried out on a series of decationated catalysts which were subjected to thermal treatment at various temperatures. 10-μl pulses of *o*-xylene were used, the catalyst amount was 300–350 mg, and the flow rate of the helium carrier gas was 50–100 ml/min. The products were analyzed on 7.8-benzoquinoline on Chromosorb W gas-chromatographic column. The *m*-xylene predominated over the para isomer. A small toluene production seemed to parallel that of isomerization. The conversion of xylene ×

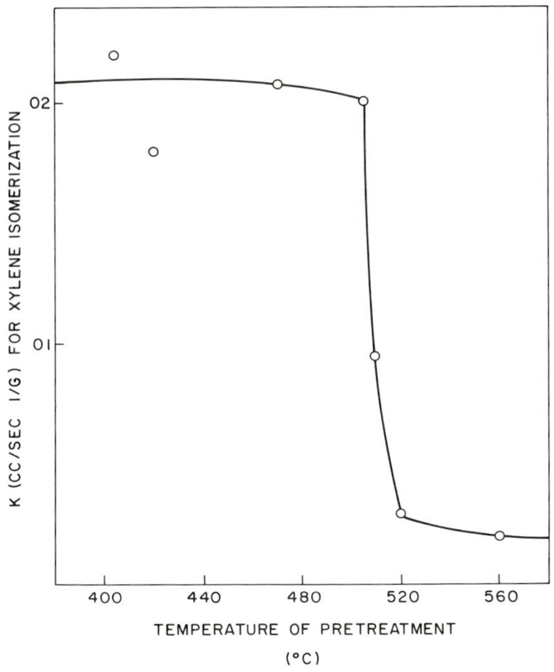

Figure 3. Effect of pretreatment temperature on catalytic isomerization of ortho-*xylene*

(ratio of *meta*- and *para*-xylenes to total xylenes) followed first-order kinetics

$$k = \frac{F}{W} \ln \left(\frac{x_{eq}}{x_{eq} - x} \right)$$

where x_{eq} was the equilibrium value of 0.782. The plot of the logarithm term against W/F_o gave a straight line whose slope gave the first-order reaction constant, k. A plot of k against the temperature of the pretreatment of the catalyst showed a marked drop in the curve at 500°–520°C, again indicating that Bronsted acid sites are responsible for xylene isomerization (Figure 3) (*15, 36*).

Not all reactions take place on the Bronsted acid sites. The hydrogen–deuterium equilibration takes place on the Lewis acid sites (Figure 1).

The characterization of these Lewis acid sites by magnetic resonance techniques has been extensively pursued in the laboratory.

Literature Cited

(1) Barrer, R. M., *Endeavour* **1964,** 23, 122.
(2) Benesi, H. A., *J. Catalysis* **1967,** 8, 368.
(3) Bertsch, L., Hapgood, H. W., *J. Phys. Chem.* **1963,** 67, 1621.
(4) Breck, D. W., *J. Chem. Educ.* **1964,** 41, 678.
(5) Carter, J. L., Lucchesi, P. J., Yates, D. J. C., *J. Phys. Chem.* **1961,** 68, 1385.
(6) Cattanach, J., Wu, E. L., Venuto, P. B., *J. Catalysis* **1968,** 11, 342.
(7) Eberly, P. E., *J. Phys. Chem.* **1968,** 72, 1042.
(8) Emmett, P. H., Kokes, R. V., Tobin, H. P., *J. Am. Chem. Soc.* **1955,** 77, 5860.
(9) Fischer, L. F., Meier, W. M., *Fortschr. Mineral.* **1965,** 42, 50–86.
(10) Hughes, T. R., White, H. M., *J. Phys. Chem.* **1967,** 71, 2192.
(11) Kerr, G. T., *J. Catalysis* **1969,** 15, 200.
(12) Kerr, G. T., *J. Phys. Chem.* **1967,** 71, 4155.
(13) *Ibid.*, **1969,** 73, 2780.
(14) Kerr, G. T., Shipman, G. F., *J. Phys. Chem.* **1968,** 72, 3071.
(15) Hansford, R. C., Ward, J. W., *J. Catalysis* **1969,** 13, 316.
(16) Liengme, B. V., Hall, W. K., *Trans. Faraday Soc.* **1966,** 62, 3229.
(17) Mackey, J., Thomas, W. H., Turkevich, J., *Actes Congr. Intern. Catalyse, 2nd, Paris, 1960,* **1961,** 1815.
(18) Nicula, A., Stamires, D., Turkevich, J., *J. Chem. Phys.* **1965,** 42, 3684.
(19) Peri, J. B., *Actes Congr. Intern. Catalyse, 2nd, Paris, 1960,* **1961,** 1, 1333.
(20) Pickert, P. E., Rabo, J. A., Dempsey, E., Schomaker, V., *Proc. Intern. Congr. Catalysis, 3rd, Amsterdam, 1964,* **1965,** 714.
(21) Rabo, J. A., Angell, C. L., Kasai, P. H., Shomaker, V., *Disc. Faraday Soc.* **1966,** 41, 328.
(22) Smith, J. V., Bennett, J. M., Flanigen, E. M., *Nature* **1967,** 215, 241.
(23) Stamires, D. N., Turkevich, J., *J. Am. Chem. Soc.* **1964,** 86, 749, 757.
(24) Szymanski, H. A., Stamires, D. N., Lynch, G. R., *J. Apt. Soc. Am.* **1960,** 50, 1323.
(25) Turkevich, J., *Catalysis Rev.* **1967,** 1, 1–35.

(26) Turkevich, J., Preprints of Papers from USA for Japan USA Seminar on Catalytic Science, A-3-1–A-3-27, Tokyo and Kyoto, Japan, May 6, 1968.
(27) Turkevich, J., Ciborowski, S., *J. Phys. Chem.* **1967**, 71, 3208.
(28) Turkevich, J., Ichikawa, A., Ikawa, T., 140th Meeting, ACS, Chicago, 1961.
(29) Turkevich, J., Murakami, Y., Nozaki, F., Ciborowski, S., *Chem. Eng. Progr.* **1967**, 63, 75.
(30) Turkevich, J., Nozaki, F., Stamires, D. N., *Proc. Intern. Congr. Catalysis, 3rd, Amsterdam, 1964,* **1965**, 586.
(31) Turkevich, J., Ono, Y., *Bull. Polytech. Inst. Jassy (Roumania),* in press.
(32) Uytterhoeven, J. B., Christner, L. G., Hall, W. K., *J. Phys. Chem.* **1965**, 69, 2117.
(33) Venuto, S. B., Landis, P. S., *Advan. Catalysis* **1968**, 18, 259.
(34) Ward, J. W., *J. Catalysis* **1967**, 9, 225.
(35) *Ibid.*, **1968**, 10, 34.
(36) *Ibid.*, **1969**, 13, 321.

RECEIVED January 30, 1970.

Discussion

R. C. Pink (Queens University, Belfast, Northern Ireland): Turkevich has said that only the D–H exchange reaction can, with certainty, be attributed to the Lewis acid activity. The cyclopropane isomerization reaction, however, seems to respond to both the Bronsted and the Lewis activity. NH_4^+ Y zeolites activated at different temperatures show two temperature regions of activity, one corresponding closely to the Bronsted activity of the catalyst, the other at a much higher temperature ($\sim$650°) corresponding to the temperature at which the electron donor–acceptor properties are at a maximum.

J. Turkevich: I was in error in the statement that hydrogen–deuterium exchange reaction is the only reaction that takes place with the help of acid sites, for we have found that butene-1 to butene-2 transformation is catalyzed by both Bronsted and Lewis acid sites. The Bronsted sites, however, give a marked stereospecificity in producing nonequilibrium mixtures of cis and trans, while the Lewis acid sites give an equilibrium mixture of the geometric isomers.

M. S. Goldstein (American Cyanamid, Stamford, Conn. 06902): I think that the results from your pulse method and our continuous flow method for poisoning with quinoline (Goldstein, M. S., Morgan, T. R., *J. Catalysis* **1970**, 16, 232) are in general agreement. However, we found a correspondence between quinoline adsorption and quinoline poisoning. We also found that NaY adsorbed quinoline as well as HY. We interpreted the quinoline poisoning as caused by supercage blockage. Would you agree that our results indicate that quinoline is not a specific poison for Bronsted sites?

J. Turkevich: No, we are convinced more than ever from the quantitative results that we have published that quinoline in the poisoning experiments blocks specific sites in supercages. Total adsorption of quinoline can take place beyond this specific poisoning and, being nonspecific to sites, has no relevance to catalysis.

P. B. Venuto (Mobil Oil Corp., Paulsboro, N. J. 08034): In your second slide, you showed a plot of catalytic activity *vs.* number of catalytic sites for an ammonium-exchanged Y system. In this slide, catalytic activity did not increase in a simple linear relationship; rather, it increased greatly at the higher site concentrations. Do you attribute this to the appearance of sites with higher energy (different type sites) or to the "collective" action of sites of similar types?

J. Turkevich: The slide was from an earlier published work presented at the Amsterdam International Congress on Catalysis. It indicates first that the catalytic activity per site is not constant as one increases the number of sites, but more dramatically that it increases with increases in site number even when these new sites are not available to the substrate (cumene) or poison quinoline. This we interpret to be owing to interaction and migration of protons (and/or electrons) among the various sites.

64

Acidic and Exchange Properties of X and Y Zeolites. Correlation with Catalytic Activity for Some of Them

R. BEAUMONT, D. BARTHOMEUF, and Y. TRAMBOUZE

Institut de Recherches sur la Catalyse, 39 Boulevard du 11 Novembre 1918, 69-Villeurbanne, France and Faculté des Sciences de Lyon, France

The acidic properties of X and Y zeolites containing various amounts of Na^{+}, Ca^{2+}, and La^{3+} ions have been measured. The exchange of one sodium ion by H^{+} provides a fraction a *of acid site which characterizes the type of the zeolite (X or Y). The number of acid sites provided when calcium or lanthanum ions are exchanged depends mainly on the valency of the ion. The cracking of isooctane (used to study catalytic properties of zeolites) is enhanced considerably by hydrogen. Then a particular kind of sites has hydrogenative properties. Either acidic or catalytic properties are lowered by ions located near or in the supercage, and they increase when inner sites are exchanged.*

Properties of zeolites depend strongly on the nature and content of the cation. A number of investigators have shown that location of cations is important to consider. It depends on several factors, such as the valency of the ion and the pretreatment temperature. Therefore, it is rather difficult to compare different catalysts if they do not contain exactly the same number of metallic ions and have not been heat-treated in the same way.

For a given metallic ion, neither the acidity nor the catalytic activity are directly proportional to cation content, and the relations between these factors are rather complex (*1, 2, 7, 8, 9, 11–19*).

In order to have a better understanding of the correlations between different properties, we will first systematically study the variation of acid properties with the degree of exchange of Na^+ ions by protons, describing the results concerning only the whole acidity, Bronsted and/or Lewis, without trying to distinguish between the 2 forms. Different strengths of acidity will be pointed out when increasing the degree of exchange.

Results obtained lead to extension of the study to the case of polyvalent ions in order to verify the hypothesis of partial neutralization of acidity by bi- or trivalent ions.

Cracking of isooctane is used to check catalytic activity of progressively exchanged Na-zeolites and to try to correlate acid and catalytic properties.

Experimental

Materials and Sample Preparation. The starting materials are commercial Union Carbide zeolites NaX and NaY, in powder form. The chemical compositions are $Na_{86}(AlO_2)_{86}(SiO_2)_{106} \cdot x\ H_2O$ and $Na_{56}(AlO_2)_{56}(SiO_2)_{136} \cdot x\ H_2O$, respectively.

The sodium–ammonium forms ($Na\text{-}NH_4\text{-}X$ and $Na\text{-}NH_4\text{-}Y$) were prepared by repeated exchange of the sodium zeolite with ammonium acetate solutions at room temperature. The sodium–hydrogen forms (Na-H-X and Na-H-Y) were obtained by heating the ammonium zeolite at 380°C for 15 hours.

Several cationic forms were prepared from Na-Y. Nearly 95% of Na ions were exchanged by K and Ca ions with chloride solutions. These zeolites named K-Y and Ca-Y were progressively exchanged by NH_4 ions and transformed in K-H-Y and Ca-H-Y forms. La ions were introduced in a zeolite NH_4-Y, in which the residual sodium content is only 1%, by repeated exchange with $LaCl_3$ solutions. The samples were heated in dry air at 380° and 550°C for 15 hours.

X-ray diffraction studies showed that the structure of Na-X is destroyed if more than 50% of Na ions are exchanged by NH_4 ions. For all other catalysts, x-ray diffraction measurements show the samples to be highly crystalline.

Since zeolites differ from each other in the content and atomic weight of the cations, all results (acidity and catalytic activity) are expressed for a unit cell.

Technique. ACIDITY. The number of acid sites was determined by titration with *n*-butylamine. Hammett and arylmethanol indicators chosen according to Drushel and Sommers' studies (*4*) were used to obtain the catalyst acidity in terms of acid strength (*7*). Numerical results obtained by this method are similar to those obtained by other methods, for example, adsorption (*1, 14, 15*).

CATALYTIC ACTIVITY. The cracking of isooctane was studied. For catalytic activity measurements, the samples were heated in flowing helium or hydrogen from room temperature to 465°C for 15 hours. Then the gas saturated with isooctane passed through the catalyst in the microreactor and was analyzed by gas chromatography.

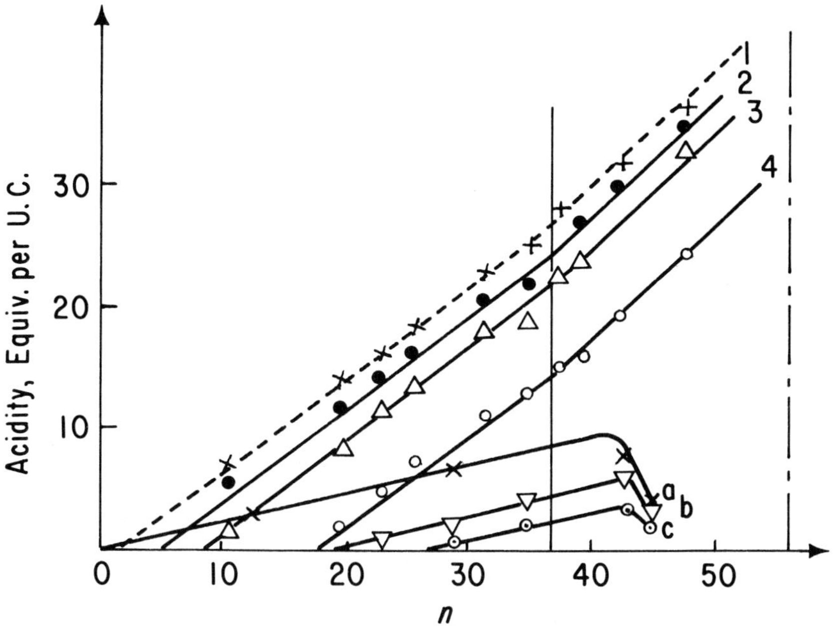

Figure 1. Acidity as a function of n *Na*$^+$ *ions exchanged per U.C.*

NaHY	NaHX	
1		*Hammett indicator (3.10^{-4}% H_2SO_4)*
2	a	*Arylmethanol (50% H_2SO_4)*
3	b	*Arylmethanol (77% H_2SO_4)*
4	c	*Arylmethanol (88% H_2SO_4)*

Results

Acidity. ZEOLITES NA-H-X AND NA-H-Y. Figure 1 shows the increase of acidity (at least equal in strength to the acid strength noted) *vs.* the number of sodium ions exchanged per unit cell.

Curves concerning Na-H-X are meaningless when the degree of exchange is greater than 50%; the crystalline structure is destroyed.

For zeolites X and Y, 2 important facts may be deduced from Figure 1. For a given type of zeolite X and Y, the slopes of the straight lines are similar, regardless of indicator (Hammett or arylmethanol) and strength of acidity. Secondly, the acidity A is a linear function of the number N of exchanged ions $A = aN + b$.

Numerical values of a can be determined from the straight lines of Figure 1 within an error of $\pm$ 4%. (The coefficient b will not be discussed here.) The slope a gives the number of equivalents of acidity provided by the exchange of one equivalent of Na^+ ion.

For NaY, when less than 37–38 atoms have been exchanged, $a = 0.78$. This cannot be related to a restriction imposed by the large molecules of butylamine and indicators, since results obtained by NH_3 adsorption give yet smaller values for a (*1*). Furthermore, the value of a only depends on the valencies of the cations, not on their size—*i.e.*, not on the size of the pore. The value of $a = 1$ (more than 37–38 atoms have been exchanged) will be discussed further.

The slope for zeolite X is only 0.22. The difference between the 2 slopes (0.22 and 0.78) can be explained only in terms of chemical structure.

Figure 1 and Table I show that for zeolites X and Y the number of exchanged Na^+ ions itself is not important, but in order to get strong acid sites the ratio n_{Na}/n_o % must be the same.

This result emphasizes the location of sites in hydrated zeolites X and Y. Sites S_I (16 per unit cell) are in the hexagonal prism between sodalite cages. Sites S_{II} (32 per unit cell) are located on the free six-membered rings of oxygen of the sodalite cage; sites S_{III} (Y : 8 and X : 48 per unit cell) are located on the walls of the large cavity (supercage).

In dehydrated forms, $S_{I'}$ and $S_{II'}$ lie inside the sodalite cage and face S_I and S_{II} across the relevant hexagonal windows. They may be occupied at the expense of S_{III} sites (*5*).

In hydrated forms in solution, Sherry (*12*) has shown that S_{III} sites are exchanged first, then sites S_{II} and S_I. The results described here lead us to a similar conclusion (Figure 1 and Table I).

The titration of the acidity in dehydrated catalysts by butylamine, which probably only goes through larger cavities, depends on several factors, particularly the interactions in the supercage. The acidity of one acid site which has been exchanged may be partially lowered by the neutralization by a Na^+ ion coming from another position, since under heating ions have the opportunity to migrate and then to lose their mobility by dehydration (*6, 10, 14, 16*). The greater the number of cations in the supercage, the greater will be this neutralization effect. The acidity of X zeolites which contain more ions in (or near) the supercage than Y zeolites will be much lowered. Then a (X) is smaller than a (Y).

Table I. Effect of $n_{Na}/n_{o(X)}$ on Acid Strength

Acid Strength, % H_2SO_4	*Zeolite Na–H–X* n_{Na}	$\frac{n_{Na}}{n_{0(X)}}$ %[a]	*Na–H–Y* n_{Na}	$\frac{n_{Na}}{n_{0(Y)}}$ %
50	0	0	4.5	7.7
77	19	22	8	14.5
88	27	31.5	17	30

[a] n_{Na} = number of Na^+ ions exchanged; n_0 = number of Na^+ ions in starting zeolites: X = $n_{0(X)}$ = 86; $n_{0(Y)}$ = 56.

Table II. *a* Values at 50% Acid Strength

Zeolite	*Slope* a	$\frac{a_0}{a}$	*Cation Valency*
Na-H-Y	0.78	1	1
K-H-Y	0.81 a_0	1	1
Ca-H-Y	0.40	2	2
La-H-Y (550°)	0.51	3/2	3
La-H-Y (300°)	0.25	3	3

The steadiness of *a* values (0.78 and 0.22) within a large degree of exchange means that as long as there are Na^+ ions in (or near) a supercage the partial neutralization goes on.

Beyond the exchange of 37–38 atoms in Na-H-Y, slope *a* becomes 1. This means that the last-exchanged sites, probably located in S_I and $S_{I'}$ positions, may induce in the supercage an acidity which depends on the mobility of OH group protons and ability of the adsorbate to delocalize H atoms (*16, 19*). As there is no more Na^+ ion in such a position to lower the efficiency of acid sites, the titrated number of acid sites becomes equal to the number of exchanged Na^+ ions. These considerations on the values of $a_{Na(X)}$ and $a_{Na(Y)}$ have led to some conclusions on the importance of some discrepancy between X and Y zeolites. They explain why X zeolites, which contain more Al atoms than Y zeolites, are less acid. Studies of *a* values for Y zeolites containing various cations will give important results.

ZEOLITES K-H-Y–Ca-H-Y–La-H-Y. The acidity of these 3 series of catalysts is studied in the same way as for the NaY. Acid strength can be determined when samples are not colored. Only results concerning 50% H_2SO_4 are described here (Table II). These acidities (equivalents per unit cell) are plotted (Figure 2) *vs.* the number of equivalents of exchanged cation per unit cell. Except for one case, all the samples were heated at 550°C. La-H-Y, heated at 300°C after its preparation, gives curve 5.

Sodium and potassium, which are monovalent ions, lead to very similar slopes. K^+ ions are located in positions rather similar to the Na^+ ions, and the results for *a* are not surprising. $a_o = 0.8$ is the average value of *a* for monovalent ions. Ca^{2+} ions are located in S_I and S_{II} sites, S_I being the most difficult to exchange. Results have to be compared for degree of exchange implying S_I sites to be filled. The slope $a_{(Ca)}$ is half of *a* for monovalent ions. Since calcium is bivalent, the ratio of slopes $a_o/a_{(Ca)}$ equals the ratio of valencies of Ca^{2+}/Na^+. This means that 2 equivalents of Ca^{2+} ions (one Ca^{2+} ion) provide the same number of acid sites as 1 equivalent or ion of Na^+. This hypothesis has been assumed by other authors (*11*), according to the equilibrium for Ca on the surface,

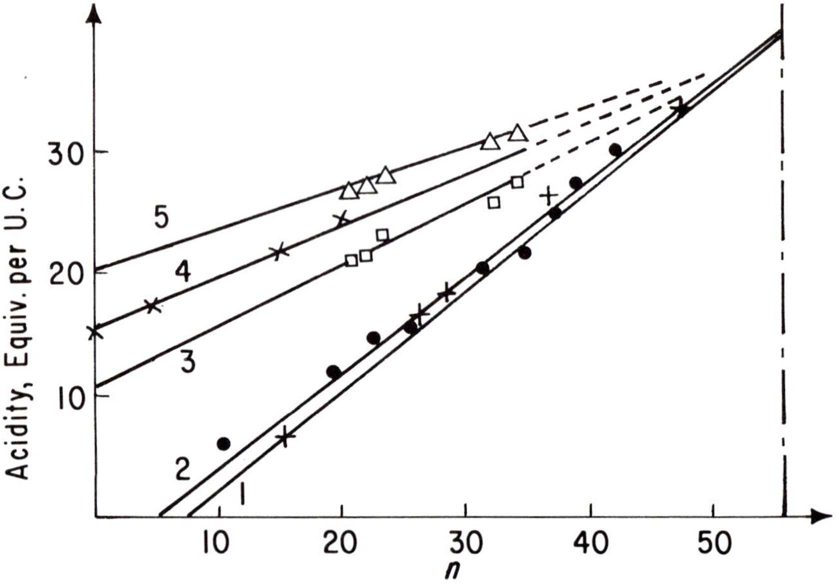

Figure 2. Acidity (strength ≥ 50% H_2SO_4) as a function of n *equivalents of cations exchanged per U.C.*

1	(+)	*KY*
2	(·)	*NaY*
3	(□)	*LaY (550°C)*
4	(×)	*CaY*
5	(△)	*LaY (300°C)*

$Ca^{2+}(OH_2) \rightleftharpoons Ca(OH)^+ + H^+$. One Ca atom on the surface replaces 2 equivalents of sodium, but neutralizes only one acid site.

The results obtained here confirm the hypothesis and verify it numerically for the degrees of exchange studied and for all acid strengths, since the slopes are independent of the indicator used. If the La-H-Y samples are heated at 300°C, the ratio a_o/a_{La-300}, calculated from the slopes, becomes equal to the valency of La^{3+} ion (3 equivalents) and gives the same number of acid sites as the exchange of one Na^+ (equivalent or ion). Infrared and EPR studies show (*3, 17, 18*) that heating at 300°C leads to an equilibrium on the surface

$$La(OH_2)_2^{3+} \rightleftharpoons La(OH)_2^{+} + 2\ H^+$$

The exchange of one equivalent of La^{3+} ion provides $a_o/3$ acid site, and the slope of a_{La-300} is one-third of a_o.

Furthermore, results obtained after heating at 550°C are in good agreement with the postulated dehydroxylation of La-H-Y zeolite under heating (*3*). According to this idea, increasing the temperature beyond 300°–400°C eliminates one proton from the surface; then the slope

a_{La-550} shows that one cationic site becomes $La(OH)^{2+}$ under heating, and the exchange of one La^{3+} ion (3 equivalents) provides 2 a_o equivalents of acidity.

$$La(OH)_2{}^{+} + 2H^{+} \rightarrow La(OH)^{2+} + H^{+} + H_2O$$

For La zeolite, the slopes a_{La-300} and a_{La-550} verify the assumptions based on results obtained by different experimental methods (infrared analysis and EPR).

In conclusion, for all the samples studied, values of a and their ratio explain the acid properties of Me-H-Y zeolites, completing results already known. In particular, the constant value of a within a large degree of exchange and for several acid strengths verifies the postulated acid properties of zeolites containing bi- and trivalent cations when varying exchange or acid strength. The method of determining the slope a can be applied to other polyvalent ions and rapidly provides easily understandable results.

The good agreement between slopes, ratios, and valencies shows that the slope a depends mainly on valency and that other factors such as the ion size are of minor importance.

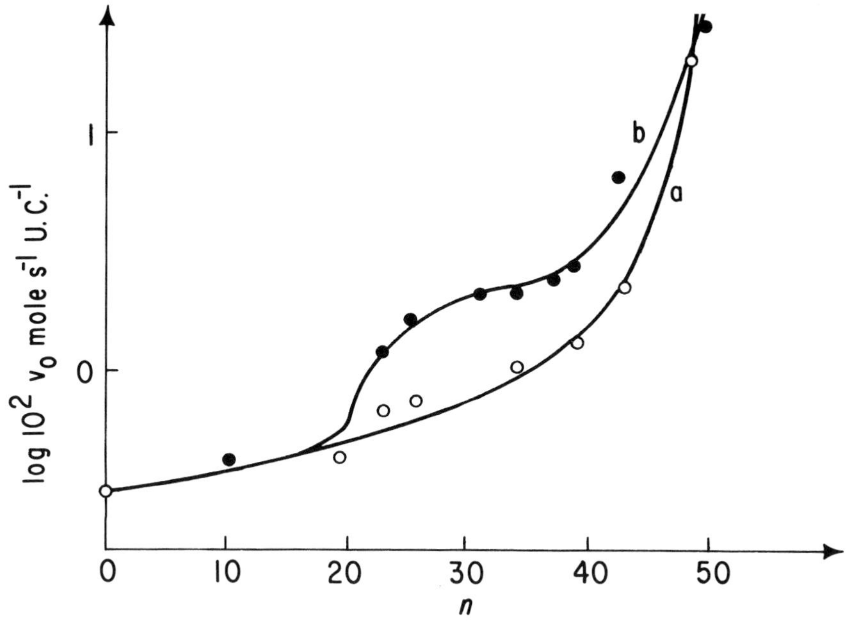

Figure 3. Catalytic activity of NaHY (v_o = initial rate for isobutene production) as a function of n *Na^+ ions exchanged per U.C.*

a = Helium stream
b = Hydrogen stream

Catalytic Activity. The catalytic properties of NaHY zeolites were studied. Results are shown in Figure 3. A semilogarithmic scale is chosen to show more clearly the variation of activity for an exchange including 17–37 atoms.

The increase of catalytic activity with ion exchange has already been shown. Figure 3 indicates that results depend on experimental factors. Curve *a* deals with helium as carrier gas, resembling published curves obtained with the same gas (*15, 19*). On the other hand, curve *b* (hydrogen stream) is similar to results of benzene alkylation with propylene (*8*). In this work, the authors studied catalytic activity with the exchange of Na^+ for Ca^{2+} ions. S_I sites are occupied either by Na^+ or by Ca^{2+} ions, and no effect can be seen of the influence of the exchange on activity of the last sites. Accordingly, curve *a* (Figure 3), limited to an exchange from 0 to 28 ions, is exactly similar to curves obtained for benzene alkylation (*8*).

Figure 3 shows that up to the exchange of nearly 17 Na^+ ions there is no difference attributable to the nature of the gas. Beyond this point, a rapid increase in activity is measured with hydrogen but not with helium. A second and important increase starts with the exchange of 37–38 Na ions. At the same time, activity in helium rapidly becomes greater. The higher effectiveness of the last-exchanged sites has been described (*15, 19*). Results and conclusions of the study of acidity described in the first part of this paper show that up to 37–38 ions exchanged, properties in the supercage depend on Na^+ ions in (or near) the supercage. Beyond this value, slope *a* increases, and the acidity only depends on the number of last-exchanged sites. Then properties induced in the supercage by unoccupied last-exchanged sites greatly differ in nature from properties in the supercage before their exchange.

Up to 37–38 atoms, catalytic properties are lowered by cations in (or near) the supercage. After 37–38 atoms exchanged, catalytic activity depends on the new properties in the supercage owing to replacement of sodium ion in inner sites, whatever be the nature of the gas.

In hydrogen atmosphere, the first increase in activity for the exchange of 17 ions may be explained partially according to the views of Rubinstein *et al.* (*8*). They demonstrated that catalytic activity varies, for this degree of exchange, with the number of unoccupied S_{II} sites. Nevertheless, curve *b* of Figure 3 shows some other properties of zeolites (*2*). The specific effect of hydrogen is detected for all the products obtained in isooctane cracking (propene, isobutene, isobutane, 2-butene). Nevertheless, the rise in activity (from He to H_2 for a given degree of exchange) varies from 9 for isobutane to 3 and 2 for propene and isobutene, respectively. The enhancement is the most important for isobutane. Several experiments showed that isobutane comes from isobutene hydrogenation. High activity for isobutane production is linked to hydrogenating properties of the zeolites employed. The increase in production of other hydrocarbons may be explained as follows: as in bifunctional catalysts, some sites of employed zeolite are active in hydrogenation reactions, forming heavy products of the isooctane cracking. This prevents coke deposition and increases the whole catalytic activity. Isobutene, which can be hydrogenated easily, gives more isobutane than in absence of hydrogen. This explanation is corroborated by the color of the zeolite after the cracking reaction: samples are grey in H_2 and black in He stream.

The effect of hydrogenating sites is only important when the exchange increases from 17 to 37–38 atoms. After that, the influence of S_I sites prevails and hydrogenating properties decrease.

Accordingly, for successive exchanges of sites, hydrogenating properties may come from a fraction of S_{II} sites.

Conclusions

The acidity of progressively exchanged zeolites has been studied, providing the number of acid sites (a) obtained when one equivalent of metallic ion is removed.

It may be considered that a characterizes the acidic efficiency of one exchanged site and is independent of the acid strength but depends on 3 main factors:

For a given cation (Na^+) and when the degree of exchange is not too high, a varies with the type of zeolite X and Y. This effect has to be connected to the number of ions which are located in (or near) the supercage.

For a given cation (Na^+) and a given type of zeolite (Y), a depends on the location of the ions. This factor is linked to the degree of exchange; *i.e.*, $a < 1$ when there are ions in (or near) the supercage and $a = 1$ when these ions are all removed and therefore the inner sites are exchanged.

For a given type of zeolite, a depends mainly on the valency of the cation; other factors such as ion size are of minor importance.

Catalytic activity is influenced also by ions in (or near) the supercage. As long as these ions are not exchanged, the cracking of isooctane is not important. In Na-H-Y zeolites, only the exchange of the last ions noticeably increases it.

Acid and catalytic studies enable us to assume that cations located near the walls of the supercage reduce the ability and change the character of molecule adsorption on the acid and active sites. After the whole exchange of these cations, the removal of ions from the inner sites leads to another kind of adsorption of the molecules.

Furthermore, hydrogenating properties are detected for intermediate degrees of exchange. Some exchanged sites in the supercage can hydrogenate olefins.

Literature Cited

(1) Bandiera, J., Ben Taarit, Y., Naccache, C., *Bull. Soc. Chim. France* **1969,** 3419.

(2) Beaumont, R., Barthomeuf, D., *Compt. Rend. Acad. Sci. Paris,* **1969,** C, 269, 617.

(3) Ben Taarit, Y., Bandiera, J., Mathieu, M. V., Naccache, C., *J. Chim. Phys.* **1970,** 67, 37.

(4) Drushel, H. V., Sommers, A. L., ACS, New York Meeting, 1966.

(5) Eulenberger, G. R., Shoemaker, D. P., Keil, J. G., *J. Phys. Chem.* **1967,** 71, 1812.

(6) Freeman, D. C., Stamires, D. N., *J. Chem. Phys.* **1961,** 35, 799.

(7) Hirschler, A. E., *J. Catalysis* **1963,** 5, 428.

(8) Isakov, Ya. I., Klyachko-Gurvich, A. L., Khudiev, A. T., Minachev, Kh., Rubinstein, A. M., *Intern. Congr. Catalysis, 4th, Moscow* **1968,** paper 56.

(9) Olson, D. H., Dempsey, E. J., *J. Catalysis* **1969,** 13, 221.

(10) Olson, D. H., Kokotailo, G. T., Charnell, J. F., *Nature* **1967,** 215, 270.

(11) Rabo, J. A., Pickert, P. E., Stamires, D. N., Boyle, J. E., *Proc. Intern. Congr. Catalysis, 2nd, Paris* **1960,** 2, 2055.

(12) Sherry, H. S., *J. Phys. Chem.* **1966,** 70, 1158.

(13) Turkevich, J., *Catalysis Rev.* **1967,** 1, 1.

(14) Turkevich, J., Murakami, Y., Nozaki, F., Ciborowski, S., *Chem. Eng. Progr. Symp. Ser.* **1967,** 63, 75.

(15) Turkevich, J., Nozaki, F., Stamires, D. N., *Proc. Intern. Congr. Catalysis, 3rd, Amsterdam* **1965,** 1, 586.

(16) Uytterhoeven, J. B., Jacobs, P., Makay, K., Schoonheydt, R., *J. Phys. Chem.* **1968,** 72, 1768.

(17) Venuto, P. B., Hamilton, L. A., Landis, P. S., *J. Catalysis* **1966,** 5, 484.

(18) Ward, J. W., *J. Catalysis* **1969,** 13, 321.

(19) Ward, J. W., Hansford, R. C., *J. Catalysis* **1969,** 13, 364.

RECEIVED February 4, 1970.

65

Acidity, Accessibility, and Stability of Hydroxyl Groups in Y-Zeolites

L. MOSCOU

Koninklijke Zwavelzuurfabrieken v/h Ketjen N.V., P.O. Box 15-C, Amsterdam, The Netherlands

Determination of acidic OH groups in Y-zeolites by $LiAlH_4$ reaction and Karl Fischer titration indicates that after heating to 200°–300°C, REY zeolite contains only OH groups in α-cages equivalent to one OH for each rare earth ion introduced. Further heating causes dehydroxylation, which is accompanied by a new formation of Bronsted sites in α-cages, up to a maximum of half the amount present at 250°C. In NH_4Y zeolite, deammoniation causes OH group formation in both α- and β-cages. Fully exchanged NH_4Y zeolite contains 10^{21} OH groups in α-cages per gram of zeolite. HY zeolite behavior compares well with the NH_4Y zeolite properties. The data support the assumption that OH groups with 3640 cm^{-1} infrared frequency are the accessible OH groups in α-cages.

Catalytic properties of crystalline aluminosilicates generally are correlated with acidic species in the zeolite framework. Many reports deal with the nature and location of this acidity, which has been extensively investigated with infrared spectroscopy before and after reaction of the zeolite with basic species.

The limitation of infrared studies on zeolites is the strong adsorption of water, giving rise to broad overlapping bands in the infrared spectrum. For this reason, only those zeolites can be studied which are highly dehydrated, either by heating to 200°C and above or by evacuation.

Prior work (*3*) has indicated that a combination of 3 analytical techniques—loss of weight on ignition (LOI), Karl Fischer titration (KF), and reaction with $LiAlH_4$—results in quantitative data on the amount of acidic OH groups in the zeolite, even in nonheated wet samples. This method discriminates between acidic and non-acidic OH

Table I. Measurement of Water and Hydroxyl Groups in Y-Zeolites

		Measured With		
Group	*Position*	*LOI*	*KF*	*$LiAlH_4$*
H_2O	Between crystallites	X	X	X
H_2O	α-Cages	X	X	X
H_2O	β-Cages	X		
Acidic OH	α-Cages	X		X
Nonacidic OH	α-Cages	X		
Acidic + nonacidic OH	β-Cages	X		

Table II. Chemical Analysis of Zeolites (Dry Basis)

		% Na_2O	*% RE_2O_3*	*% NH_3*
NaY zeolite sample	A	13.7		
REY zeolite sample	B	3.53	16.1	
REY zeolite sample	C	5.90	12.9	
REY zeolite sample	D	9.1	7.2	
NH_4Y zeolite sample	E	5.95		4.02
NH_4Y zeolite sample	F	0.54		7.7
HY zeolite sample	G	5.90		

groups and also between groups present in the supercages (α-cages) of the crystal and groups present in sodalite cages (β-cages) and hexagonal prisms.

The present paper deals with the application of these techniques to study the formation, accessibility, and stability of acidic hydroxyl groups in Y-type zeolites. Comparison is made between NaY, REY, NH_4Y, and HY zeolites.

Experimental

The analytical techniques used are described fully in the earlier paper (*3*). A summary is given in Table I, showing the various methods and the groups determined by them.

The LOI is defined as the weight loss of the heat-treated sample on its subsequent calcination at 1000°C for 1 hr. Thus, it gives the sum of residual water and OH groups which can be dehydroxylated at 1000°C.

For NH_4-containing zeolites, the LOI is corrected for the weight of NH_3 in the sample. The KF titration method is based on the reaction of water with iodine and SO_2; a possible slow reaction of hydroxyl groups with the reagent is excluded by extrapolating the titration curve to zero time, with constant excess of reagent (*4*). The analysis with $LiAlH_4$ is based on its reaction with water and acidic hydroxyl groups, both under evolution of hydrogen gas.

Acidity of Hydroxyl Groups. Table I shows that the amount of acidic OH groups (Bronsted sites, OH^+) in α-cages is obtained by taking the difference between $LiAlH_4$ and KF titrations. Earlier work indicated that even weakly acidic silanol groups in amorphous silica react quantitatively with $LiAlH_4$ (*1*). Each hydrogen ion with an acid strength $\geqq$ silanol acidity is expected to react with the reagent. Basic hydroxyl groups like those present in rare earth hydroxydes are inert to $LiAlH_4$.

Accessibility of Acidic Hydroxyl Groups. The difference between $LiAlH_4$ and KF titration data gives the acidic OH content in α-cages only because neither the $LiAlH_4$ nor the iodine molecule can enter the β-cage.

Stability of Acidic Hydroxyl Groups. The stability of acidic OH groups is investigated by titrating zeolites after various heat treatments. Care was taken that during handling of the heat-treated zeolite only negligible amounts of water could be readsorbed.

Zeolite Preparation. NaY zeolite, obtained from Union Carbide, Linde Division, was purified from free sodium silicate by repeated washing with water until the molar ratio of the zeolite was $Na_2O:Al_2O_3:SiO_2 = 1.03:1.00:4.90$. The REY zeolites were obtained from the purified NaY zeolite by exchange of sodium ions with rare earth ions in $RECl_3$ solution. NH_4Y zeolites were obtained by treating the purified NaY zeolite with NH_4Cl solution (50% exchange). More exchange steps were needed to remove 95% of the Na^+ ions. HY zeolite was obtained from the purified NaY zeolite by treating a zeolite–water suspension with a weakly acidic ion-exchange resin in the proton form at 20°C (*2*). Table II shows the chemical composition of the zeolites obtained. All zeolites were highly crystalline in x-ray analysis.

Results

Analysis data of 4 samples of zeolites are summarized in Table III, where NaY, REY, NH_4Y, and HY zeolites are compared. For each zeolite, 3 sets of data are given: the loss on ignition, the difference between LOI and $LiAlH_4 = H_2O + OH$ in β-cages (and nonacidic OH in α-cages), and the difference between $LiAlH_4$ and KF = acidic OH in α-cages. Nonacidic OH in α-cages cannot be distinguished separately and hence are not discussed further. Heat pretreatment of zeolites was carried out in 2 ways—*i.e.*, heating for a few hours at a constant temperature and heating at a constant rate of 10°C/min until a given temperature is reached.

X-ray diffraction analysis of the various heat-treated samples indicates that framework collapse occurs above the following temperatures: NaY, 800°C; REY, 900°C; NH_4Y, 800°C; HY, 800°C. The data in Table III show that in NaY zeolite no acidic OH groups are present

Table III. Analysis of Zeolites,

Pretreatment Temp., °C	*NaY, Sample A* LOI	$H_2O + OH$ (β)	OH^+ (α)	*REY, Sample B* LOI	$H_2O + OH$ (β)	OH^+ (α)
At 120				15.3	5.8	0.0
At 160	7.9	3.4	0.0	8.0	3.9	1.0
Till 200	10.3	2.9	0.0	19.4	4.2	0.0
Till 250						
Till 300	4.1	2.1	0.0	4.8	2.5	0.9
Till 350				4.5	3.4	0.3
Till 400	2.0	1.1	0.0	2.9	1.8	0.5
Till 500	1.6	1.1	0.1	2.4	1.7	0.5
Till 600	1.3	0.9	0.0	1.9	1.3	0.5
Till 700	1.0	0.6	0.0	1.6	1.1	0.4
Till 800				1.4	1.1	0.1
Till 900				0.6	0.4	0.0

[a] All data in % H_2O d.b.

in α-cages. The maximal acidic OH–α content found for the 3 REY zeolites under investigation is given in Table IV, comparing the experimentally found values with the values calculated from the RE content of the zeolite, assuming that each RE ion gives one acidic OH group (5). Table IV indicates that each rare earth ion in the zeolite forms approximately one acidic hydroxyl group in α-cages.

In Figure 1, the difference between LOI and $LiAlH_4$ is plotted against the LOI for the 4 types of zeolites under investigation. It shows that the curves for the REY and NaY zeolites above 4.5% LOI are nearly identical. At lower LOI values, the REY curve is identical to the NH_4Y and HY curves.

The thermal stability of acidic OH groups in α-cages is shown in Figure 2. The curve for REY zeolite shows 2 characteristics—*viz.*, a significant minimum in the curve at temperatures between 350° and 500°C, and the height of the second part of the curve is half that of the first part of the curve. All 3 REY zeolite samples (B, C, and D) showed these characteristics.

The results from the NH_4Y and HY zeolite analysis are given in Tables III and V and in Figures 1 and 2.

Figure 1 indicates that for both NH_4Y and HY zeolites, the content of water and OH groups in β-cages is significantly higher than for NaY and REY zeolites, which indicates that in contrast to NaY and REY, the NH_4Y and HY zeolites do contain OH groups in the β-cages. This agrees well with the data in Tables III and V, showing that NH_4Y sample E exposes a maximum of acidic OH groups in α-cages after a 250°C treat-

Dried at Different Temperatures[a]

NH_4Y, Sample E			*HY, Sample G*		
LOI	*H_2O + OH* (β)	*OH^+* (α)	*LOI*	*H_2O + OH* (β)	*OH^+* (α)
10.4	5.2	0.8	11.8	7.9	0.0
15.8	5.9	0.0	21.9	6.4	0.0
6.0	3.9	1.1			
			6.4	5.1	0.4
3.9	2.8	0.6	3.6	2.5	0.7
3.6	2.4	0.6	3.4	2.6	0.6
2.8	1.9	0.4	1.6	0.9	0.4
1.9	1.3	0.1	1.3	0.7	0.4
1.6	1.0	0.2	1.3	0.8	0.2

Table IV. Comparison of Measured and Calculated OH (α) Densities in RE-Y Zeolites

	Measured		*Calculated*	
Sample	*Max. OH (α) content, % H_2O*	*Number of OH (α) per gram zeolite*	*% RE_2O_3 in zeolite*	*Number of OH per gram zeolite*
REY sample B	1.04	6.9×10^{20}	16.1	6.4×10^{20}
REY sample C	0.82	5.5×10^{20}	12.9	5.2×10^{20}
REY sample D	0.58	3.9×10^{20}	7.2	2.9×10^{20}

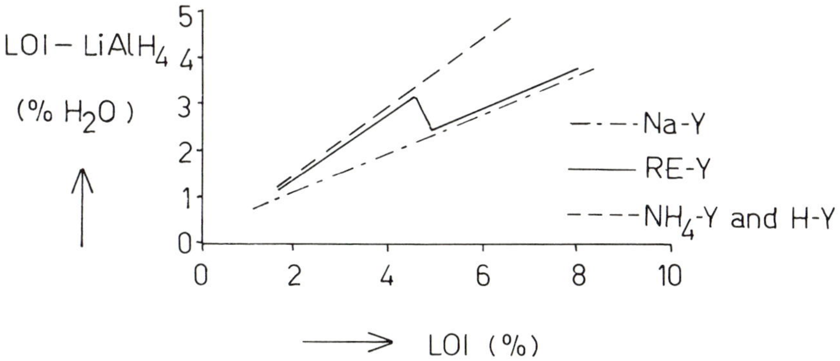

Figure 1. Relation between the difference of LOI and $LiAlH_4$ (H_2O + OH in β-cages and nonacidic OH in α-cages) and loss on ignition for Y-type zeolites

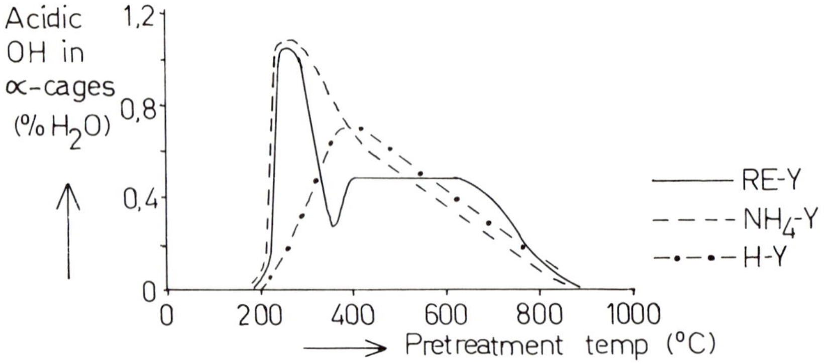

Figure 2. Thermal stability of acidic OH groups in α-cages of Y-type zeolites

ment which is exactly half the amount of OH groups, as calculated from the NH_4 content of the zeolite. For the HY zeolite, 1/3 of the total amount of OH groups is present in α-cages. The temperature regions where formation and dehydroxylation of OH groups in α-cages of HY and NH_4Y zeolites occur are clear from Figure 2.

Discussion

In NaY zeolite, no acidic hydroxyl groups are found in α-cages; the presence of OH groups in β-cages cannot be determined quantitatively, but seems very unlikely from infrared studies (*10*).

The maximum in the acidic OH content in α-cages of REY zeolites is present after heating the zeolite to 200°–300°C (Figure 2), while under these conditions OH groups in β-cages are probably absent, as can be inferred from the identity of the NaY and REY curves in Figure 1. The absolute quantity of OH (α) groups is in good agreement with the

Table V. Comparison of Measured and Calculated

	Measured		
Sample	*Max. OH (α) content, % H_2O*	*Number of OH (α) per gram zeolite*	
NH_4Y sample E	1.06	7.1×10^{20}	$\Delta = 2.5 \times 10^{20}$
NH_4Y sample F	1.44	9.6×10^{20}	
HY sample G	0.69	4.6×10^{20}	

proposed formation of acid sites by Plank (*5*)

$$RE^{3+} (H_2O) \longrightarrow [RE(OH)]^{2+} + H^+$$

where each rare earth ion forms one Bronsted site (Table IV).

At higher temperatures (300°–400°C), dehydroxylation occurs; the distinct minimum in the REY curve of Figure 2 indicates that at the same moment new hydroxyls are formed by another dissociation of water molecules, up to a maximum which has half the value of the low-temperature maximum. Perhaps this can be explained by the dehydroxylation mechanism where 50% of the OH groups lose their oxygen atom during dehydroxylation and consequently cannot be restored by a newly formed proton. It is likely that the increase in the REY curve in Figure 1 during drying from 5 to 4% LOI (which occurs between 300° and 400°C under the conditions used) is caused by the formation of OH groups in the β-cage system.

The observed formation and dehydroxylation of Bronsted sites in REY zeolite can be summarized as follows: between 200° and 300°C each RE ion dissociates one water molecule with the formation of one Bronsted site in α-cages, followed by a dehydroxylation between 300° and 400°C. This is accompanied by a new dissociation of water molecules, giving rise to $\frac{1}{2}$ Bronsted site in α-cages for each RE ion and to the formation of OH groups in β-cages.

This is in reasonable agreement with Ward's conclusion that for every 3 exchange sites a maximum of 2 Bronsted sites are formed in REY zeolite (*8*). The thermal stability curve for acidic OH (α) groups in REY zeolites is in remarkable agreement with its alkylation activity curve as given by Venuto (*7*). The analysis of NH_4Y zeolites shows that exchange of Na^+ by NH_4^+ primarily results in equal amounts of OH groups in α- and β-cages (Table V, Sample E).

Exchange of the more difficultly exchangeable Na^+ ions mainly leads to inaccessible OH groups in β-cages (Table V, Δ between samples E and F), which is in agreement with Ward's conclusions (*10*). This agree-

OH (α) Densities in NH_4Y and HY Zeolites

% NH_4 in zeolite	*Calculated: Total number of OH (α) per gram zeolite*		*Ratio of Measured OH (α) and Total Number of OH*
4.02	14.0×10^{20}	$\Delta = 13 \times 10^{20}$	1 : 2.0
			1 : 5.2
7.7	27.0×10^{20}		1 : 2.7
	14.0×10^{20}		1 : 3.0

ment supports the assumption that OH groups with 3640 cm^{-1} infrared frequency are the accessible OH groups in α-cages of the zeolite (*6, 9, 11*). The behavior of HY zeolite resembles closely that of NH_4Y zeolite; however, it is surprising that HY zeolite does not show accessible OH groups after low-temperature drying of the samples.

Acknowledgment

The author is grateful to M. Lakeman and S. Lub for excellent assistance in the experimental work, and to P. G. Menon for many valuable comments and advice on the manuscript. Thanks are due to J. I. de Jong for his continuous stimulating interest in this work.

Literature Cited

(1) Moscou, L., unpublished results, 1967.
(2) Moscou, L., Dutch Patent Application **6713340.**
(3) Moscou, L., Lakeman, M., *J. Catalysis* **1970**, 16, 173.
(4) Noll, W., Damm, K., Fauss, R., *Kolloid-Z.* **1960**, 169, 18.
(5) Plank, C. J., *Proc. Intern. Congr. Catalysis, 3rd,* **1965,** 1, 727.
(6) Uytterhoeven, J. B., Jacobs, P., Makay, K., Schoonheydt, R., *J. Phys. Chem.* **1968,** 72, 1768.
(7) Venuto, P. B., Hamilton, L. A., Landis, P. S., Wise, J. J., *J. Catalysis* **1966,** 5, 81.
(8) Ward, J. W., *J. Catalysis* **1969,** 13, 321.
(9) Ward, J. W., *J. Phys. Chem.* **1969,** 73, 2086.
(10) Ward, J. W., Hansford, R. C., *J. Catalysis* **1969,** 13, 364.
(11) White, J. L., Jelli, A. N., André, J. M., Fripiat, J. J., *Trans. Faraday Soc.* **1967,** 63, 461.

RECEIVED January 21, 1970.

Discussion

J. Rabo (Union Carbide Research Institute, Tarrytown, N. Y. 10591): Your conclusions generally agree with the results we reported at the last International Congress on Catalysis in Moscow. What rare earth cations did you use, and what was the total cation equivalent to aluminum ratio of your catalyst preparations?

L. Moscou: We used a commercially available rare earth chloride solution in which the main rare earth elements are lanthanum and cerium. The total cation equivalent to aluminum ratio was between 1.00 and 1.05.

D. A. Hickson (Chevron Research Co., Richmond, Calif. 94802): If, as shown in Figure 2, acidic hydroxyl groups in acid-exchanged Y zeolite appear only on heating above 200°C, where are these groups located in the zeolite structure below 200°C?

L. Moscou: In our opinion, the H–Y curve in Figure 2 indicates that at lower temperatures the acidic hydroxyl groups are located in β cages and that they move into α cages at ~200°C. These data support the concept of proton mobility in zeolites.

F. W. Kirsch (Sun Oil Co., Marcus Hook, Pa. 19061): Can you tell me the size range of samples used for Karl Fischer and lithium aluminum hydride titrations?

L. Moscou: The size of samples depends strongly on the water content and thus on pretreatment temperature of the sample. In practice, it varies between 0.1 and 1.0 gram.

P. Chu (Mobil Research & Development Corp., Paulsboro, N. J. 08066): I am very interested in the way you prepared the H–Y sample. Why did weakly acidic organic exchange resins have to be used? What degree of exchange can you achieve without damage to the zeolite structure? How stable thermally is the final H–Y sample?

L. Moscou: Generally, weakly acidic ion exchange resins have been used in order to prevent loss of zeolite crystallinity. This method enables one to exchange at least 75% of the sodium in Y-type zeolites. Thermal stability of these H–Na–Y zeolites is between 700° and 800°C.

66

Development of Acidity in the Sodium-Y Zeolites

EDUARDO A. LOMBARDO,[1] G. A. SILL, and W. KEITH HALL[2]

Mellon Institute, Carnegie–Mellon University, Pittsburgh, Pa. 15213

The isomerization of n-*butenes was used as a test reaction to follow the development of catalytic activity in Na-Y zeolite. The acidity was varied by the substitution of some of the* Na^+ *(0.3 to 5.7%) with* Ca^{2+} *and by creating a* Na^+ *deficiency (up to 0.94%); both series of catalysts were studied with and without water added as cocatalyst. A pure Na-Y zeolite containing no decationated sites was found catalytically inactive for this reaction. In contrast with silica–alumina catalysts, carbonaceous residues did not appear to play a role in the formation of the catalytic sites as long as* H_2O *was used as cocatalyst.*

Although there is general agreement regarding the very low catalytic activity of Group IA X or Y zeolites for carbonium ion reactions, considerable controversy exists concerning the effects of vacancies, impurities (mainly divalent cations), and cocatalysts on this property. One group of investigators reported no correlation between cation deficiency and catalytic activity (*7*). Several others (*1, 5, 6, 7, 13*) reported a marked increase on addition of proton donors to Na- and Ca-X or Y zeolites. Another (*17*) reported only a small increase for Ca-Y zeolite, but no variation was found (*19*) for pure Na-Y zeolite. Two firmly established conclusions derived from IR spectroscopy are that pure Na-X or Y zeolites, either partially or totally dehydrated, show no evidence of structural hydroxyl groups (*4, 18, 19, 20, 21, 22*), and Bronsted acidity

[1] Present address: Facultad de Ingenieria Quimica, Universidad Nacional del Litoral, Santa Fe, Argentina.
[2] To whom correspondence should be addressed; Gulf Research and Development Co., P.O. Box 2038, Pittsburgh, Pa. 15230.

develops in divalent (Me^{2+}) X and Y zeolites during dehydration (*2, 4, 14, 15, 18, 20, 21, 22*) through Reaction I.

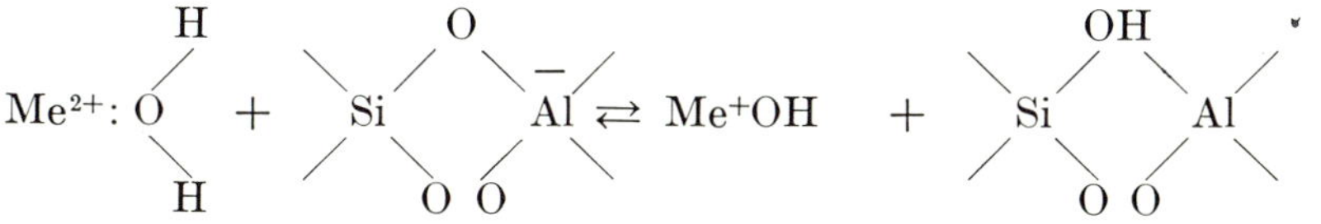

(1)

This chemistry suggests a role for H_2O as cocatalyst with the monovalent zeolites if the equilibrium is moved to the right in the presence of a base. In the present work, the effects of small deliberate replacements of Na^+ by Ca^{2+}, of cation deficiency, and of H_2O on the catalytic properties of Na-Y zeolite were studied. Also investigated was the possibility that carbonaceous residues form the catalytic sites, as was reported for the isomerization of *n*-butenes over silica–alumina catalysts (*3, 8, 9*). The isomerization of the *n*-butenes provided a useful tool for these studies because it follows first-order kinetics (*10*) and proceeds over Na-Y zeolite *via* the *sec*-butylcarbonium ion (*11, 12*).

Experimental

The static reactor, which was connected to a conventional BET system, and the purification of the reactants have been described elsewhere (*10*).

Catalyst Preparation. Na-Y zeolite was supplied by Linde (Lot No. 1280–133). The chemical composition provided by the supplier was:

Component	Al_2O_3	SiO_2	Na_2O	CaO	MgO	$(TiO_2+Fe_2O_3)$
%	22.4	63.2	12.6	0.7	0.2	0.1

Accordingly, the total cationic charge ($Na^+ + 2Mg^{2+} + 2Ca^{2+}$) was just about balanced by the total negative charge on the lattice; the ratio was $0.449/0.440 = 1.02$. The impurity alkaline earth ions had a profound effect on the catalytic activity for butene isomerization and, therefore, the concentration of these was reduced greatly by successive exchanges with purified NaAc solution. The resulting catalyst was washed thoroughly with slightly alkaline distilled water (pH = 10), to avoid decationation. Aliquots of this "parent catalyst" (I) containing only 0.02% Ca^{2+} underwent 2 kinds of treatments: Catalysts II to V were back exchanged to increase the Ca^{2+} content from 0.02 to 0.4% (Table I), and Catalysts VI to X were made by washing aliquots of the parent material with increasing amounts of distilled deionized water. In this way, some Na^+ was removed, creating varying degrees of decationation (Table II). The amount of Na^+ extracted was determined in the washing water, and the degree of decationation was calculated therefrom. No Ca, Mg, or Fe was detected by atomic absorption spectroscopy in the washing water. There was no significant variation in the level of minor impurities (B, Ba, Cr, Cu, Fe, Mg, Mn, Ti) between the parent material and the final catalyst.

Procedure. All the catalysts were pretreated with O_2 at 500°C and evacuated overnight at this temperature before each run. The experiments were all made using the same amount of reactant (55 cc STP) and catalyst. The required amount of H_2O was vacuum transferred into the reactor before adding the reactant. The carbonaceous material which would not desorb on overnight evacuation at room temperature was determined by combustion to CO_2 [*see also* Ref. *12*]. Product analysis was by GLC.

Results

The effect of increasing amounts of water, added as cocatalyst, is shown in Figure 1 for the catalyst with highest Ca^{2+} content (V). The reaction rate increased sharply with small water additions and then leveled off at about 10 H_2O/Ca^{2+} ($2H_2O$/cage). Similar behavior was observed with the other 4 catalysts of this series. Separate experiments showed that an increasing fraction of this H_2O was in the gas phase with increasing addition under reaction conditions.

The over-all rate constants ($k_{21} + k_{31}$) for 1-butene disappearance for Catalysts I to V are given in Table I. The rate constants are defined by

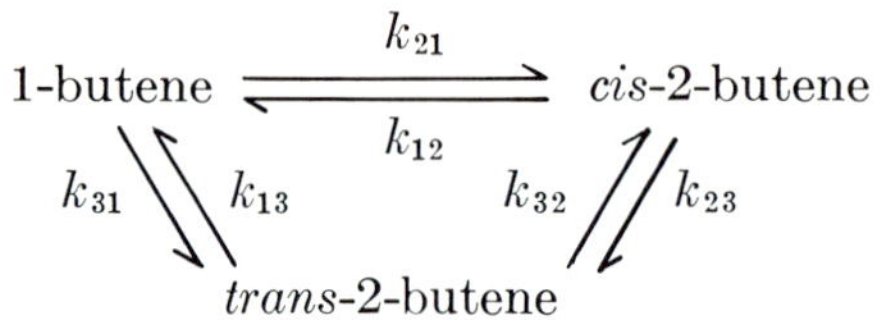

When the rate constants corresponding to $2H_2O$/cage were plotted against Ca^{2+} concentration, a straight line passing through the origin (zero rate at zero Ca^{2+}) resulted. In each case, the addition of this amount of H_2O, which was sufficient to maximize the rate, increased the rate by a factor of about 40.

The catalytic activity is correlated with increasing extent of decationation in Table II. When the rates for the dry catalyst were plotted *vs.* cation deficiency, a straight line passing through the origin again resulted. When the rates for the wet catalyst were plotted, another straight line was obtained, but with an intercept corresponding to the rate of Catalyst I in Table I (6×10^{-2} min^{-1} $gram^{-1}$).

With all these catalysts, a small portion of the reactant was retained by the catalyst which could not be removed by an overnight evacuation

Table I. Variation of Catalytic Activity with Ca^{2+} Content of Na-Y Zeolite[a]

Catalyst No.	*Calcium Analysis, %*	*Number of Na^+ Replaced by Ca^{2+}, %*	*Rate of Isomerization of 1-Butene $(k_{21} + k_{31}) \times 10^2$ Min^{-1} G^{-1}* Dry Catalyst	$2H_2O$/Cage	*Ads. Gases + Residues, n-Butene/Cage[b]*
I	0.02	0.3	0.01	6.0	0.08
II	0.07	1.0	0.55	21.9	0.10
III	0.1	1.4	–	20.0	–
IV	0.2	2.9	–	38.4	–
V	0.4	5.7	2.20	84.0	0.40

[a] Static reactor, 55 cc (STP) of reactant, 76 mg of catalyst (dry basis), reaction temperature 260°C.
[b] Residues determined after runs using water as cocatalyst when the *n*-butene mixture was close to equilibrium.

Table II. Variation of Catalytic Activity with Decationation of Na-Y Zeolite[a]

Catalyst No.	*Extent of Decationation Number of Na^+ Removed, %*	*Rate of Isomerization of 1-Butene $(k_{21} + k_{31}) \times 10^2$ Min^{-1} G^{-1}* Dry Catalyst	$2H_2O$/Cage	*Adsorbed Gases + Residues, n-Butenes/Cage[b]*
VI	0.15	0.45	–	0.15
VII	0.35	0.70	6.7	0.60
VIII	0.54	1.00	8.9	–
IX	0.79	1.8	10.4	0.40
X	0.94	2.0	10.4	0.15
X	0.94	–	11.9	–

[a] Same as for Table I.
[b] Residues were determined after using dry catalyst when the *n*-butene mixture was close to equilibrium.

at room temperature. Using radioactive tracers and Catalyst I, it was shown that this consisted of 2 fractions, 1 corresponding to exchangeable adsorbed butenes, and the other to residues or tightly held (polymerized?) molecules (*12*). The total amount of this material is reported in the last columns of Tables I and II. The amount of gas retained by the solid increased with increasing Ca^{2+} content (Table II).

Discussion

The results showed that divalent cations, present at impurity levels (as normally found in the commercial Linde zeolites) markedly increased the catalytic activity of Na-Y zeolite. If these ions were all in the hexagonal prism positions, as commonly supposed for these low replacement levels, this effect would not be expected. Relatively much larger increases per Ca^{2+} added were observed after the S_I sites were filled. Note that these effects appeared with the dry catalyst; it is possible that the added H_2O played a role in drawing more cations from these positions. Slight decationation (or cation deficiency) produced a similar effect. The intercepts of the rate plots showed that the activity for acid-catalyzed reactions would be negligibly low for an impurity-free, nondecationated, Na-Y zeolite. These facts support the idea that the catalytic activity is associated with Bronsted sites formed by interaction of H_2O with divalent cations or by decationation through hydrolysis. If the *in situ* interaction of Na^+ with water contributes to the catalytic activity, this contribution is neglible for all practical purposes. This point is clarified by the data shown in Figure 1. The tangent to the experimental curve (solid line) corresponds to 1 H_2O/Ca^{2+}; the catalytic activity would follow this curve if every H_2O added could be adsorbed on Ca^{2+}. The dashed line, shown for comparison, corresponds to 2 H_2O/Ca^{2+}. The rate does not reach its maximum saturation value, however, until 10 H_2O/Ca^{2+} have been added. This is probably because with larger additions, a major fraction of the H_2O remains in the gas phase. Movement of divalent ions from the S_I sites, effected by increasing solvation of the zeolite surface, could contribute also. This analysis, together with the fact that the reaction rate levels off much before 1 H_2O/Na^+ is present in the system, supports the idea that the mechanism represented in Equation 1 is responsible for the catalytic activity, *viz.*, that the Bronsted sites are on bridging oxygens between silica and alumina tetrahedra and that these may be formed either by hydrolysis of divalent cations or by creating a cation deficiency. It is, of course, a moot question whether more of these sites are formed adjacent to Ca^{2+} in the presence of H_2O or this molecule played a related role as cocatalyst. In either case, the picture agrees well not only with the available spectroscopic data, but also with the kinetic results of Gourisetti *et al.* (*6*) for alcohol dehydration. Careful analysis of the present rate data suggests that Bronsted sites formed by introduction of divalent ions are more active than those formed by decationation. This may be explained by the polarizing effect of the divalent cations, as discussed by Richardson (*16*). There is, however, one other possibility not ruled out by the present data. The Bronsted sites created by decationation or by Equation 1 could have been partially dehydroxylated during pretreatment. The effect of H_2O on the rate could then correspond to the interaction of H_2O molecules with these dehydroxylated sites. In view of the present state of knowledge concerning the thermal stability of these materials, this explanation is deemed both unlikely and unnecessary.

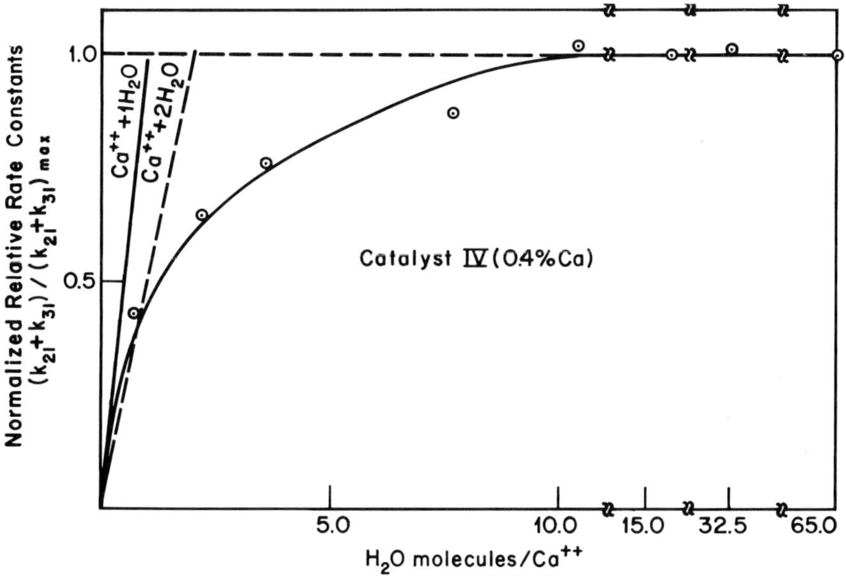

Figure 1. Variation of catalytic activity with amount of cocatalyst

The results presented herein afford an explanation for some of the discrepancies found in the literature (*1, 5, 7, 13, 19, 23*) where usually the chemical composition of the zeolite has not been specified carefully —*e.g.*, Habgood (*7*) noted that the activity changed from one lot of zeolite to another and reported that the activity of the Na zeolites increased on addition of water; Ward (*19*), on the other hand, found that a pure Na-Y zeolite was not activated by H_2O. Therefore, to define clearly a Group IA X or Y zeolite catalyst, it is very important to specify the kind and amount of impurities present.

The carbonaceous material which was retained by the catalysts after evacuation was held in 2 forms: 1 could be recovered as butene molecules (*12*) by exchange with isotopically labelled 1-butene; the other could be removed only by combustion to CO_2. With silica–alumina catalysts, the latter (residues) is thought to form the active sites for carbonium ion activity (*3, 8, 9*). The present results showed that the activity correlated with the degree of decationation or the Ca^{2+} content of the catalyst. Moreover, the amount of residue retained by the catalyst (nonexchangeable) was about 2 orders of magnitude smaller than the number of decationated sites of the sample (*12*). It seems probable, therefore, that residues do not play an important role in the development of catalytic activity of these materials in the presence of H_2O; however, residues may supply the necessary protons for catalytic activity in its absence.

Acknowledgment

This work was sponsored by the Gulf Research & Development Co. as part of the research program of the Fellowship on Petroleum. One of us (EAL) is indebted to the Organization of American States for the award of a Fellowship.

Literature Cited

(1) Bartley, B. H., Habgood, H. W., George, Z. M., *J. Phys. Chem.* **1968**, 72, 1689.
(2) Christner, L. G., Liengme, B. V., Hall, W. K., *Trans. Faraday Soc.* **1968**, 64, 1679.
(3) Clark, A., *Proc. Intern. Congr. Catalysis, 4th*, **1968**, Paper No. 75.
(4) Eberly, P. E., *J. Phys. Chem.* **1968**, 72, 1042.
(5) Frilette, V. J., Munns, G. W., *J. Catalysis* **1965**, 4, 504.
(6) Gourisetti, B., Cosyns, J., Leprince, P., *Bull. Soc. Chim. France* **1966**, 3, 1085.
(7) Habgood, H. W., George, Z. M., "London Conference on Molecular Sieves," **1968**, 130.
(8) Hightower, J. W., Hall, W. K., *Chem. Eng. Progr. Symp. Ser.* **1967**, 63 (73), 122.
(9) Hightower, J. W., Hall, W. K., *J. Am. Chem. Soc.* **1967**, 89, 778.
(10) Lombardo, E. A., Hall, W. K., *A.I.Ch.E. J.*, in press.
(11) Lombardo, E. A., Hall, W. K., in preparation.
(12) Lombardo, E. A., Sill, G. A., Hall, W. K., in preparation.
(13) Matsumato, H., Yasul, K., Morita, Y., *J. Catalysis* **1968**, 12, 84.
(14) Olson, D. H., *J. Phys. Chem.* **1968**, 72, 1400.
(15) Pickert, P. E., Rabo, J. A., Dempsey, E., Schomaker, V., *Proc. Intern. Congr. Catalysis, 3rd*, **1965**, 1, 714.
(16) Richardson, J. T., *J. Catalysis* **1967**, 9, 182.
(17) Topchieva, K. F., Romanovski, B. V., Piguzova, L. I., Thoang, H., Bizreh, Y. W., *Proc. Intern. Congr. Catalysis, 4th*, **1968**, Paper No. 57.
(18) Uytterhoeven, J. B., Schoonheydt, R. S., Liengme, B. V., Hall, W. K., *J. Catalysis* **1969**, 13, 425.
(19) Ward, J., *J. Catalysis* **1968**, 11, 238.
(20) *Ibid.*, **1968**, 10, 34.
(21) Ward, J., *J. Phys. Chem.* **1968**, 72, 1042.
(22) *Ibid.*, **1968**, 72, 4211.
(23) Watanake, I., Habgood, H. H., *J. Phys. Chem.* **1968**, 72, 3066.

RECEIVED January 12, 1970.

Discussion

Gourisetti Balamalliah (Indian Institute of Petroleum, Dehradum, India): I would like to draw the authors' attention to the fact that the relationship between the number of water molecules and the number of calcium ions for maximum catalytic activity of CaX and CaY molecular sieves in the dehydration of *tert*-butyl alcohol was first reported by Gourisetti *et al., Compt. Rend.* **1964,** 258 (18), 4547–9. It was obtained by drawing tangents to rate curves as shown in the present paper.

E. Lombardo: We reported in our paper under Ref. *6* a later work of Gourisetti, B., *et al.* that includes the data given in the short communication referred to by Gourisetti.

67

Catalytic Activity of Alkaline Earth Hydrogen Y Zeolites

ROWLAND C. HANSFORD and JOHN W. WARD

Union Oil Company of California, Union Research Center, Brea, Calif. 92621

Studies on ammonium Y zeolite back-exchanged with calcium and magnesium are reported. Calcium content above 55% of the exchange capacity leads to a sharp drop in concentration of the 3650 cm^{-1} hydroxyl groups and in catalytic activity. Exchange levels above 55% involve replacement of ammonium in accessible positions by calcium, decreasing accessible acid site concentration. In the sodium system, calcium promotes the activity. In both cases, the acid site concentration is influenced by calcium but in different ways. Similar results are obtained with magnesium. At lower exchange levels, the 3650 cm^{-1} hydroxyl group concentration is constant but the activity changes with the cation. These differences in activity may be ascribed to variations in the polarizing effects of the 2 cations on the site strength.

Divalent and ammonium forms of Y zeolite possess catalytic activities for many acid-catalyzed reactions (*11, 12, 16, 19*). Although it is generally accepted that acidic hydroxyl groups are responsible for the activity of the calcined ammonium form (*4, 6, 19, 24*), the reasons for the activity of the cation forms have been less well elucidated (*6, 7, 9, 15, 19, 20, 25*). Since the catalytic activity of the divalent cation forms usually is exhibited when sufficient cations have been introduced so as to occupy accessible positions in the lattice, it has been suggested that various properties of the cations may be responsible for the introduction of active sites. Properties of the cations such as their electrostatic field and surface diffusion have been suggested (*12, 15*). Alternatively, it has been suggested that the divalent cations can introduce Bronsted acidity by dissociation of water (*3, 5, 6, 18, 19, 20*), thus

$$Ca(OH_2)^{2+} + {}^{-}O\text{-Zeolite} \rightarrow CaOH^{+} + H\text{-}O\text{-Zeolite}$$

Both the OH groups attached to the cation and to the zeolite lattice could act as active sites, although recent evidence indicates that the $CaOH^{+}$ groups are nonacidic (*18*, *22*). The H-O-zeolite groups are the same as those found in hydrogen Y zeolite. Introduction of calcium or other divalent cations into sodium Y zeolite to an extent greater than necessary to fill the inaccessible exchange sites introduces $CaOH^{+}$ and H-O-zeolite groups in accessible positions (*18*, *22*). At the same time, catalytic activity is introduced (*7*, *15*) after about 55% exchange, which corresponds to the introduction of about 16–18 calcium ions per unit cell. The first 17 calcium ions enter the S_I and S_I' positions (inaccessible), and the remainder enter the S_{II} positions (accessible) (*1*, *2*). Similarly, the first 16–18 calcium ions introduced into ammonium Y zeolite occupy inaccessible positions and the remainder accessible positions (*23*). Hence, the influence of the calcium content on the catalytic activity of ammonium Y zeolite should give an indication of the role of the cation as an active site. Furthermore, it should also show if the Bronsted acid sites are modified by the cations (*5*, *13*).

Experimental

Materials. Ammonium Y zeolite was prepared by conventional ion exchange of sodium Y zeolite ($SiO_2/Al_2O_3 = 4.9$) to a sodium content of 1.1%. The alkaline earth ammonium Y zeolites were prepared by ion exchange of the parent ammonium Y zeolite (surface area 965 m^2 $gram^{-1}$ after heating in flowing helium at 400°C) with suitable solutions of calcium and magnesium salts. The degrees of exchange and surface areas (obtained by nitrogen adsorption at $P/P_o \approx 0.02$) are listed in Table I. The surface area and x-ray diffraction measurements show all samples to be highly crystalline.

Table I. Analysis of Zeolite Samples

Calcium Hydrogen Y		*Magnesium Hydrogen Y*	
% Exchanged to M^{2+} Form	*Surface Area M^2 $Gram^{-1}$*	*% Exchanged to M^{2+} Form*	*Surface Area M^2 $Gram^{-1}$*
92	842	79	891
74	860	63	863
65	890	52	861
56	900	50	868
51	908	45	884
37	924	35	948
18	920		

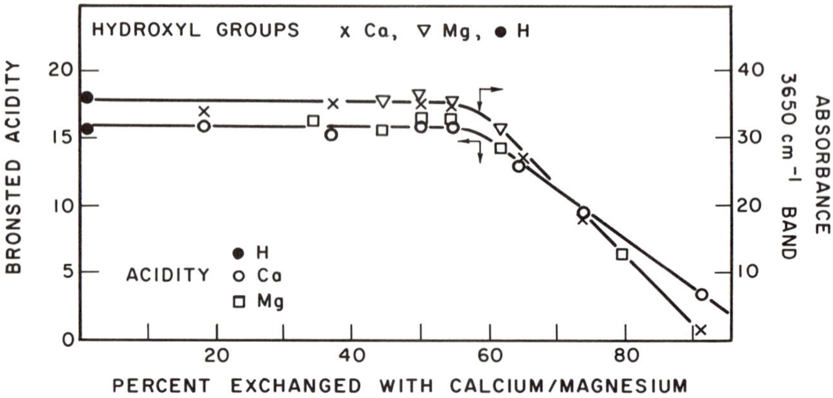

Figure 1. Dependence of acidity and 3650 cm⁻¹ OH groups on cation content

Experimental Techniques. Infrared and catalyst activity measurements were made as described elsewhere (*24*). Surface areas also were measured after carrying out the xylene isomerization reaction to confirm that the structure had not been destroyed.

Results

Infrared Spectra of Structural Hydroxyl Groups. Spectra of hydroxyl groups on alkaline earth Y and magnesium hydrogen Y zeolites have been discussed in detail elsewhere (*18, 20, 21, 22, 23*). After calcination at 450°C, absorption bands are observed at 3740 and 3650 cm^{-1}. Depending upon the degree of exchange, a band may be observed also at 3550 cm^{-1}. Since the 3550 cm^{-1} band probably represents hydroxyl groups inside the hexagonal prisms of the structure (*10*) and the 3740 cm^{-1} silanol-type groups are inert, they are unimportant to this discussion of the catalytic properties of alkaline earth hydrogen Y zeolites. However, by analogy with other systems, the 3650 cm^{-1} hydroxyl groups are probably important catalytic centers. The variation of the intensity, expressed as absorbance/sample mass, of the 3650 cm^{-1} band as a function of the degree of ion exchange is plotted in Figure 1.

Acidity Studies. Adsorption of pyridine on Bronsted acid sites results in the formation of pyridinium ions with a characteristic absorption band near 1540 cm^{-1}. Adsorption on Lewis acid sites produces a band at about 1451 cm^{-1}, while interaction with the exchangeable cations produces bands at 1444 and 1448 cm^{-1} for the calcium and magnesium forms, respectively. Typical spectra of pyridine adsorbed on magnesium hydrogen Y zeolite have been reported previously (*22*). Spectra of calcium hydrogen Y zeolite are similar except for the frequency of the pyridine–cation absorption band. The pyridine interacts with the 3650 cm^{-1} hydroxyl groups, causing their elimination, but does not appear to interact with the 3550 cm^{-1} hydroxyl groups.

For the samples studied, no pyridine adsorbed on Lewis acid sites was detected, indicating that the zeolite had not been partially dehydroxylated to form such sites. The band reflecting pyridine–cation interaction was detected only after about 16 alkaline earth ions had been introduced into the unit cell. It grew steadily in intensity as more divalent ions were introduced into the structure. In Figure 1, the intensity of the pyridinium ion band, expressed as absorbance/sample mass, is plotted as a function of the per cent exchange by divalent ion. The concentration of acid sites is a function of the degree of exchange.

Measurements of acid site strengths with H_R indicators showed that the Bronsted acid sites were stronger than 70% sulfuric acid. No variation with the degree of cation exchange was observed.

Activity Measurements. The rate constants at 260°C are plotted as a function of per cent exchange in Figure 2, which shows that they are a complex function of the calcium and magnesium contents of the zeo-

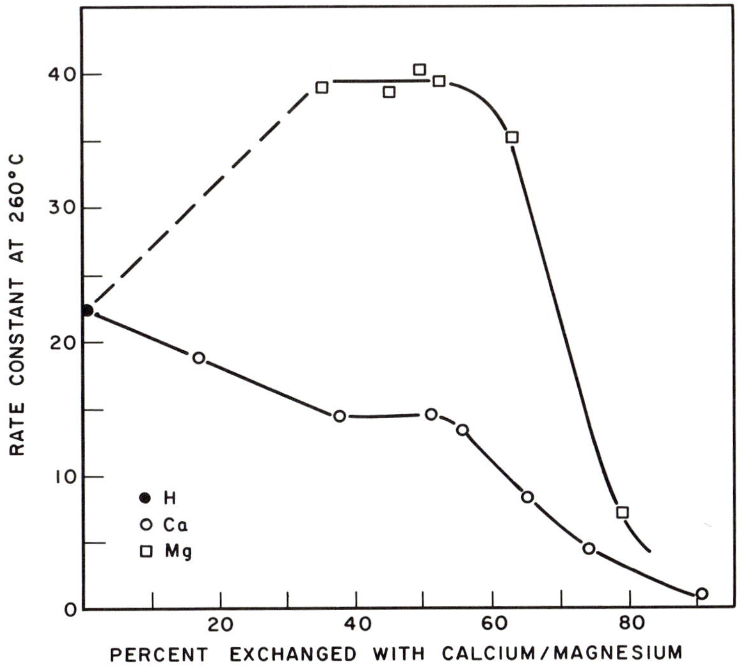

Figure 2. Dependence of rate constant on cation content

lites. The activation energy for the *o*-xylene isomerization was calculated to be 20 ± 1 kcal/mole. This value is the same as that found for the sodium hydrogen Y system and for a number of cation Y zeolites.

Discussion

For calcium zeolites, x-ray diffraction and infrared studies together have shown that the first 16 or 18 calcium ions introduced into the unit cell of the sodium form are located in the hexagonal prisms and sodalite portions of the structure (*1, 2, 23*). Addition of further calcium ions results in the occupation of sites in the supercages. The changes in intensity of the hydroxyl infrared bands and the onset of the cation–pyridine interactions indicate that the same preferential location of calcium ions in calcium hydrogen Y zeolite occurs (*23*).

Thus, the 3550 cm^{-1} band has been eliminated and the cation–pyridine band is observed after incorporation of about 16–18 calcium ions into ammonium Y zeolite. The data of Figure 1 show that the 3650 cm^{-1} hydroxyl group concentrations remain constant until about 16–18 calcium ions (55% exchange) have been introduced. The hydroxyl groups then progressively decrease in intensity as the concentration of calcium ions is increased. These data indicate that after introduction of about 16 calcium ions, calcium ions replace ammonium ions located in the supercages, probably near the S_{II} sites, thus reducing the number of hydroxyl groups on deammination. The 3650 cm^{-1} hydroxyl groups are almost certainly the same type as those existing on the hydrogen Y zeolite.

Chemisorption of pyridine results in the disappearance of these groups and the formation of pyridinium ions. The concentration of pyridinium ions, and hence accessible Bronsted acid sites, follows a similar relationship to that of the hydroxyl group concentrations. Thus, the acid site concentration remains constant until about 16–18 calcium ions have been introduced. The acidity concentration then decreases rapidly as the calcium ions are exchanged for ammonium ions in accessible positions. At the same degree of exchange, the cation–pyridine band near 1444 cm^{-1} is observed first, confirming the appearance of calcium ions in accessible positions in the structure.

These observations can be compared with the acidity of the sodium hydrogen Y zeolites (*24*). In that system, the acidity remains approximately constant until 16–18 sodium ions have been introduced and then decreases rapidly as more sodium ions are added.

The data of Figure 2 indicate a complex relationship between the catalytic activity and degree of calcium exchange. Topchieva *et al.* (*14*) studied CaHY with calcium contents of 20, 40, 62, and 90% of the exchange capacity and concluded that the activity was a complex function of the cation content. Their data follow the trend of Figure 2. However, the change in activity can be rationalized in terms of the physical properties of the zeolite. At about 55% exchange, corresponding to the calcium concentration at which cations are expected to start ion-exchanging into accessible positions, the catalytic activity starts to decrease rapidly from a plateau value of the rate constant. By comparison of Figures 1 and 2, it is seen that the concentration of hydroxyl groups and Bronsted acid sites and the catalytic activity follow the same pattern, that is, above 55% exchange, they all decrease in a similar manner with increasing calcium content. From these observations, it can be concluded that the calcium ions, in contrast to their effect in the CaNaY system, deactivate the ammonium Y system, probably by reducing the concentration of accessible acidic hydroxyl groups. The process can be represented by the following schemes:

For NaY $$2NaY + Ca^{2+} + H_2O \rightarrow CaOHY + HY + 2Na^+$$

For NH_4Y $$2NH_4Y \rightarrow 2HY + 2NH_3$$

$$2NH_4Y + Ca^{2+} + H_2O \rightarrow CaOHY + HY + 2NH_4^+$$

An equilibrium involving the divalent ion, $Ca(OH_2)^{2+} \rightleftharpoons CaOH^+ + H^+$, probably exists.

Hence, whereas in the case of sodium Y up to one acid site is introduced for each calcium ion, with ammonium Y at least one acid site is eliminated for each calcium ion. The results suggest that the calcium ions themselves or hydroxyl groups attached to cations are not active sites, but both in the sodium and ammonium systems, the acidic hydroxyl groups represented by the 3650 cm^{-1} band are the primary active centers. These sites probably are introduced because of the polarizing effects of the cation fields (*3, 5, 8, 12, 13, 19, 20*); thus, $Ca(OH_2)^{2+} \rightleftharpoons CaOH^+ + H^+$. Similar results were obtained with magnesium in place of calcium.

A second phenomenon is illustrated by the data of Figure 2. Whereas the concentration of the 3650 cm^{-1} hydroxyl groups and of acid sites remains constant up to about 55% exchange, the catalytic activity drops when calcium ions, insufficient to fill the inaccessible sites, are added. With magnesium ions, the catalytic activity actually increases in comparison with hydrogen Y zeolite, although the acid site concentration is the same. It is probable that the differences in activity result from different polarization effects exerted on the acidic hydroxyl groups by the fields of the different cations. Such effects were first proposed by Hirschler (*5*) and Richardson (*13*). The stronger polarizing effect of the magnesium compared with the calcium would be expected to produce stronger acid sites which might be more active catalytic centers.

Acknowledgment

The authors express their appreciation to S. J. Boardman and L. L. Sharar for their excellent technical assistance.

Literature Cited

(1) Angell, C. L., Schaffer, P. C., *J. Phys. Chem.* **1966,** 70, 2420.
(2) Bennett, J. M., Smith, J. V., *Mater. Res. Bull.* **1968,** 3, 633.
(3) Christner, L. G., Liengme, B. V., Hall, W. K., *Trans. Faraday Soc.* **1968,** 64, 1679.
(4) Hall, W. K., *Chem. Eng. Progr. Symp. Ser.* **1967,** 63 (73), 68.
(5) Hirschler, A. E., *J. Catalysis* **1963,** 2, 428.
(6) Hopkins, P. D., *J. Catalysis* **1968,** 12, 325.
(7) Klyachko-Gurvich, A. L., Khudiev, A. T., Rubinshtein, A. M., *Dokl. Akad. Nauk SSSR* **1967,** 177, 1379.
(8) Lunsford, J. H., *J. Phys. Chem.* **1968,** 72, 4163.
(9) Matsumoto, Y., Morita, Y., *Bull. Japan. Petrol. Inst.* **1969,** 11, 40.
(10) Olson, D. H., Dempsey, E., *J. Catalysis* **1969,** 13, 221.
(11) Pickert, P. E., Bolton, A. P., Lanewala, M. A., *Chem. Eng.* **1968** (July 29), 133.
(12) Pickert, P. E., Rabo, J. A., Dempsey, E., Schomaker, V., *Proc. Intern. Congr. Catalysis, 3rd, Amsterdam, 1964,* **1965,** p. 714.
(13) Richardson, J. T., *J. Catalysis* **1967,** 9, 182.
(14) Topchieva, K. V., Romanovsky, B. V., Piguzova, L. I., Thoang, H. S., Bizreh, Y. W., *Intern. Congr. Catalysis, 4th, Moscow,* **1968,** Preprint 57.
(15) Tung, S. E., McInnich, E., *J. Catalysis* **1968,** 10, 1966.
(16) Turkevich, J., *Catalysis Rev.* **1967,** 1, 1.
(17) Turkevich, J., Nozaki, F., Stamires, D. N., *Proc. Intern. Congr. Catalysis, 3rd, Amsterdam, 1964,* **1965,** p. 586.
(18) Uytterhoeven, J. B., Schoonheydt, R., Liengme, B. V., Hall, W. K., *J. Catalysis* **1969,** 13, 425.
(19) Venuto, P. B., Landis, P. S., *Advan. Catalysis* **1968,** 18, 259.
(20) Ward, J. W., *J. Catalysis* **1968,** 10, 34.
(21) *Ibid.*, **1968,** 11, 251.
(22) Ward, J. W., *J. Phys. Chem.* **1968,** 72, 4211.
(23) *Ibid.*, in press.
(24) Ward, J. W., Hansford, R. C., *J. Catalysis* **1969,** 13, 364.
(25) Zhavoronkov, M. V., Rosolovskaya, E. N., Topchieva, K. V., *Kinetics Catalysis* **1969,** 10, 186.

RECEIVED February 4, 1970.

Discussion

J. B. Uytterhoeven (University Leuven, 3030 Heverlee, Belgium): Your samples were pretreated at 460°C. The samples used in the catalytic experiment were probably calcined in a deeper bed than the infrared samples. Is it possible that this deep-bed calcination produced, by dehydroxylation, a given amount of a kind of ultra-stable material? The increasing amount of Mg^{2+} ions could favor this dehydroxylation. This could be another explanation for the increasing catalytic activity. Some of the ultra-stable faujasites contain hydroxyls vibrating around 3600 cm^{-1}. Such a band is also observed in Ca^{2+} and Mg^{2+} zeolites, although, in that case, this band is usually attributed to a $Ca(OH)^{+}$ species.

J. W. Ward: It was previously shown (Ward, *J. Catalysis* **1968,** 11, 251, "Nature of Active Sites on Zeolites" VI) that alkaline earth ions stabilized the hydrogen–Y zeolite. Examination of the infrared spectra, unit cell constant, ion exchange capacity, thermal analysis data, and other properties indicates that this material is not the so-called ultra-stable zeolite.

Furthermore, although the absorption frequency of the band assigned to CaOH groups is similar to that of bands observed in "ultra-stable" zeolite, substitution of other alkaline earth cations—*e.g.*, Mg, Sr, Ba—results in a frequency shift of the absorption band under discussion, the shift varying regularly with the cation size and electrostatic properties. This strongly suggests association of the hydroxyl group with the cation (Ward, J. W., *J. Phys. Chem.* **1968,** 72, 4211).

68

Acidic and Oxidizing Properties of Rare Earth Exchanged Y Zeolites

YOUNES BEN TAARIT, MICHEL-VITAL MATHIEU, and CLAUDE NACCACHE

Institut de Recherches sur la Catalyse, C.N.R.S., 39, Boulevard du Onze Novembre 1918, 69-Villeurbanne, France

Infrared techniques are used to study the nature of the OH groups present on rare earth forms of Y zeolite. The infrared spectra of pyridine adsorbed on REY, after calcination at a series of temperatures, show that Bronsted and Lewis acidities are present. ESR techniques are used to measure the number of anthracene radical ions which can be formed on the REY zeolite. Electron transfer occurs at the tricoordinated aluminum atoms in the case of LaY zeolite, while for $Ce^{4+}Y$*, the centers have been identified as* Ce^{4+} *ions.*

Synthetic Y type zeolites have been studied widely in recent years. Many techniques were used—particularly infrared spectroscopy—to determine the nature of active sites. At first, only the HX and HY forms were concerned, and the acidic nature of the hydrogen held on these catalysts was demonstrated by several infrared investigations (*10, 20*). The origin of the structural hydroxyl groups was a subject of general agreement. On the contrary, the attribution of some (OH) bands, such as the band at 3550 cm^{-1}, remains under discussion (*6, 9, 18, 21, 26*). More recently, the M(II)- and M(III)-exchanged X and Y forms came into study. Similar hydroxyl groups were observed and, under certain conditions, some OH groups were recognized as bound to the cation (*4, 13, 25*). The origin of these hydroxyl groups has been attributed unanimously to the ionization of water molecules by the strong electrostatic fields in the cation neighborhood, as stated by Rabo (*13*). However, the nature and role of hydroxyl groups is not clearly demonstrated. Recently, Ward (*24*) concluded an exhaustive study of the nature of active sites for cumene cracking in this way: "In conclusion, a study of various cation-exchanged X zeolites has shown that the catalytically active forms are Bronsted acids and contain structural hydroxyl groups. . . . However, there is no apparent relationship between the concentration of Bronsted acids sites and catalytic activity. It is possible that the acid site strength is important."

It seemed of some interest to test the ability of a series of REY zeolites to ionize polynuclear aromatics since the oxidizing properties of zeolites were pointed out (*8, 16*), but the nature of the electron acceptor site is still under discussion. Hall *et al.* (*5*), studying dehydroxylated HY zeolites, presumed it to be molecular oxygen trapped in an anion vacancy, while Hirschler (*7*) asserted that the protons may be the oxidative centers. In a previous work, as stated by Turkevich *et al.* (*16*), we concluded that the active sites are Lewis centers, while the chemisorbed oxygen increases their electron affinity (*27*). In a recent work, Richardson (*14*) related spin concentration to the electron affinity of the cation, presuming that the electron transfer took place from the anthracene to the cupric ion, but he could not observe any variation of the Cu^{2+} peak intensities.

The slightest change in the pretreatment conditions may have a great influence on the oxidative properties of the sample, making a simultaneous infrared study necessary.

Experimental

Materials. All the samples are issued from the same Linde NaY starting zeolite. The NH_4CeY, CeNaY, NaLaY, and NaCuY forms were obtained by conventional ion-exchange technique. In the case of the NH_4CeY, the ammonium form is obtained first and then exchanged with $Ce(NO_3)_3$ solution to the desired level.

The NaCeY, NaLaY, and NaCuY are exchanged with nitrate solutions; the degree of ion exchange is determined by flame spectrometry analysis of the residual sodium. The results are set in Table I. X-ray examination showed all the samples to be highly crystalline.

The pyridine is dried over a 5-A molecular sieve. The polynuclear aromatic solutions are dried and degassed by the freeze–pump–thaw technique.

Techniques. Infrared measurements were conducted in a previously described cell (*11*), allowing the sample thin wafers to be heated under vacuum or equilibrated with pyridine vapor. The samples were com-

Table I. Degree of Ion Exchange

Sample	*Sodium Ion*	*Ce Ion*	*La Ion*	*NH_4 Ion*	*Cupric Ion*
NaCeY	38	6	0	0	0
NaNH$_4$CeY	23	6	0	15	0
NaCuY	16	0	0	0	20
NaLaY	14	0	14	0	0

pressed under 6000 kg/cm^2 into 18-mm-diameter disks. These disks contain 12–16 mg of solid/cm^2. Spectra were recorded after cooling the sample at room temperature on a Perkin-Elmer Model 125 spectrophotometer using a spectral slit width of 3 cm^{-1} between 4000 and 1200 cm^{-1}.

The EPR studies are carried out in a special cell allowing the adsorption of aromatics and screening the catalyst from the action of moisture or oxygen. The spectra are recorded continuously before and after anthracene adsorption. After equilibrium, the spin concentration is measured by comparing the radical cation EPR spectrum with that of calibrated DPPH solution.

The samples either are dehydrated directly under vacuum or calcined in oxygen and evacuated.

Results

Samples which have not been pretreated with O_2 show no radical ion formation; with the same pretreatment conditions, NH_4Y forms have the same behavior. These results seem somewhat disconcerting, since the oxygen-pretreated NH_4Y zeolite was oxidative enough to ionize anthracene even at low exchange levels. Such strange results may be explained by the formation of a coke deposit. The carbon may either poison the sites or merely prevent the anthracene from coming close enough to the active centers.

When Ce(III)Y zeolites are calcined under oxygen at 400°–500°C, they turn pale yellow, probably by Ce^{3+} oxidation to Ce^{4+}. Adsorption of anthracene on these samples gives rise to a strong EPR signal. The same spectrum was observed by Rooney and Pink on silica–alumina (*15*) and interpreted as arising from an anthracene positive radical.

Results for NaCe(IV)Y, NH_4Ce(IV)Y, and $NaCu^{2+}Y$ are given in Table II.

In the same way, the Ce(III)Y samples which have been heated under vacuum are inactive. Further oxidation gives them the same properties as if they were oxygen-pretreated, in contrast to the NH_4Y form.

The electron-acceptor property of CeY zeolites may be attributed to Ce^{4+} ions. These results could be compared to Richardson's works on CuY zeolites (*14*). $La^{3+}Y$ is more stable than $Ce^{3+}Y$, and La^{3+} remains in the three-valency state when treated under oxygen. Figure 1 gives the spin concentration *vs.* the activation temperature for LaY. The number of

Table II. Anthracene Positive Radicals Concentration

Sample	*NaCeY*	*NH_4CeY*	*NaCuY*
Calcination temperature	500	500	500
(A.$^+$)/gram	2.10^{19}	16.10^{19}	6.10^{18}

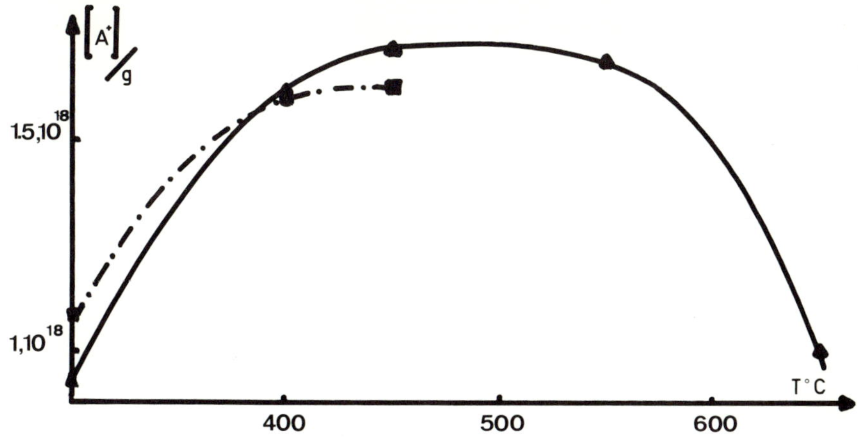

Figure 1. Concentration of anthracene positive ions formed on LaY as a function of activation temperature; -·-·- = rehydrated

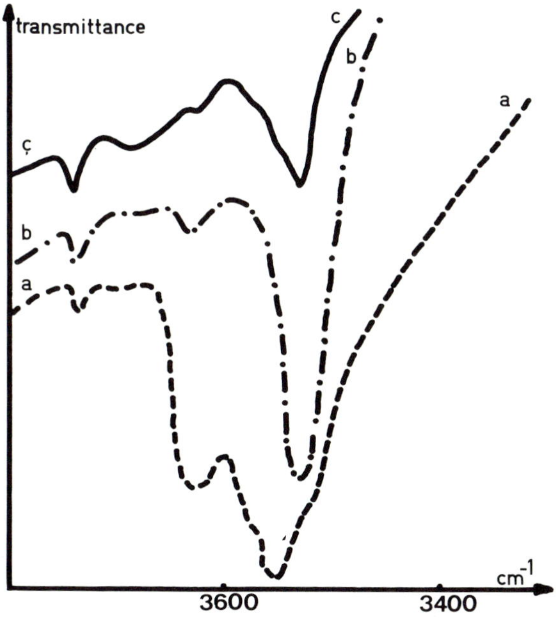

Figure 2. Spectra of the stretching hydroxyls of LaY

(a) Evacuated at 200°C
(b) Evacuated at 450°C
(c) Evacuated at 550°C

active sites increases with increasing temperature until 400°C and then remains constant. The slight decrease observed after 600°C may be attributed to a structure collapse or at least to a probable decrease in the diameter of the cavities. Rehydration results in a large decrease of the spin concentration.

Infrared Measurements. OH GROUPS. Infrared spectra in the hydroxyl stretching region are shown in Figure 2 for LaY zeolite.

Physically adsorbed water is removed by desorption of the sample at 200°C, as may be concluded from the disappearance of the 1640 cm^{-1} band. In such conditions, (OH) bands at 3745, 3640, and 3530 cm^{-1} with a shoulder at 3550 cm^{-1} are observed. Further dehydration reduced these bands' intensities, except for the one at 3745 cm^{-1}. Evacuation at 500°C affects the 3550 cm^{-1} and 3640 cm^{-1} bands, while the 3530 cm^{-1} band still

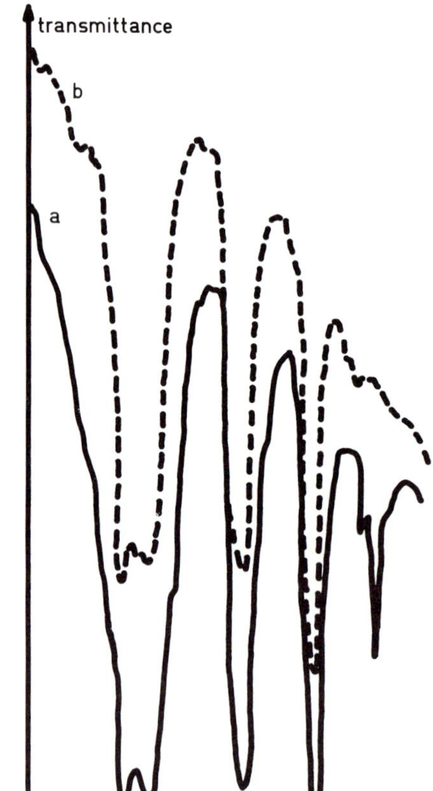

Figure 3. Spectra of pyridine adsorbed on LaY activated at 450°C

(a) Evacuated at 150°C
(b) Evacuated at 250°C

remains. These spectra are quite similar to those reported by Rabo *et al.* (*13*) and Ward (*25*).

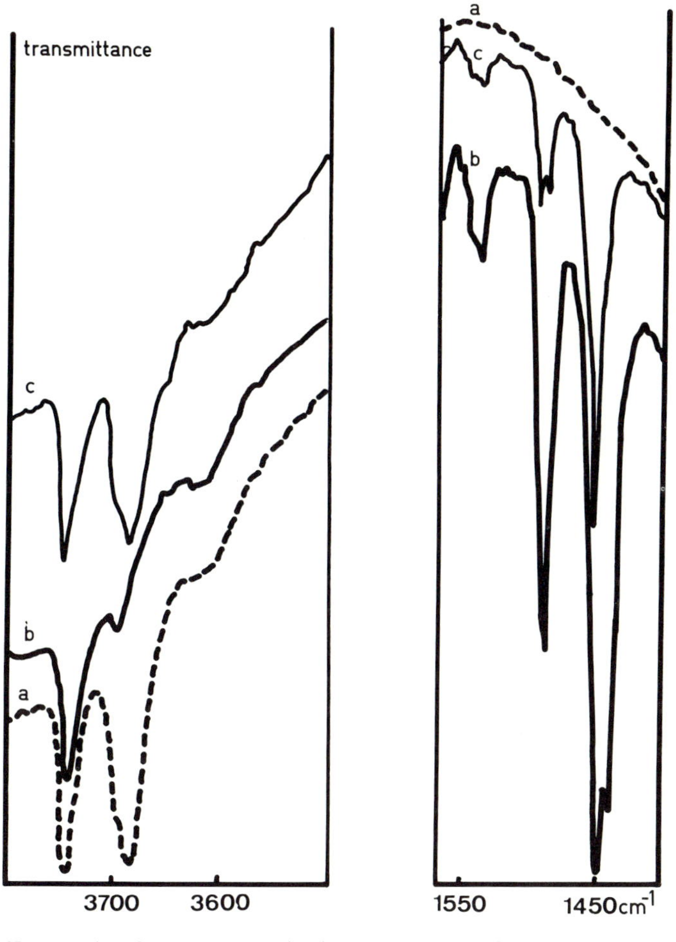

Figure 4. Spectra of Ce(IV)NH_4Y activated at 450°C under water vapor pressure, then evacuated at 450°C

(a) Initial
(b) Pyridine evacuated at 150°C
(c) Further evacuation at 250°C

In similar dehydration conditions, the 3640 cm^{-1} band is more stable for the NH_4CeY samples. These results may be related to Ward's conclusions about NH_4MgY zeolites (*23*); it is probable that cerium ions stabilize the HY form zeolite.

Calcination of CeY samples under 20 torr of water vapor at 450°C produces slight modifications in the infrared spectra of the hydroxyl groups: the 3640 and 3550 cm^{-1} are removed at 450°C. Simultaneously, a band at 3680 cm^{-1} is developed.

PYRIDINE ADSORPTION. Pyridine adsorption has been used to determine the nature of acid sites (*2*, *12*). The spectra are recorded after equilibrating the sample with an excess of pyridine vapor and further evacuation at various temperatures. Typical spectra are shown in Figures 3, 4, and 5. Table III gives the frequencies of the observed bands between 1700 and 1400 cm^{-1}.

After pyridine adsorption, the band at 3550 cm^{-1} is broadened while that at 3640 and 3680 cm^{-1} disappear. The 3680 and 3640 cm^{-1} bands are restored by evacuation at 250° and 350°C, respectively. More interesting results are obtained for the NaCeY samples; when pyridine is desorbed

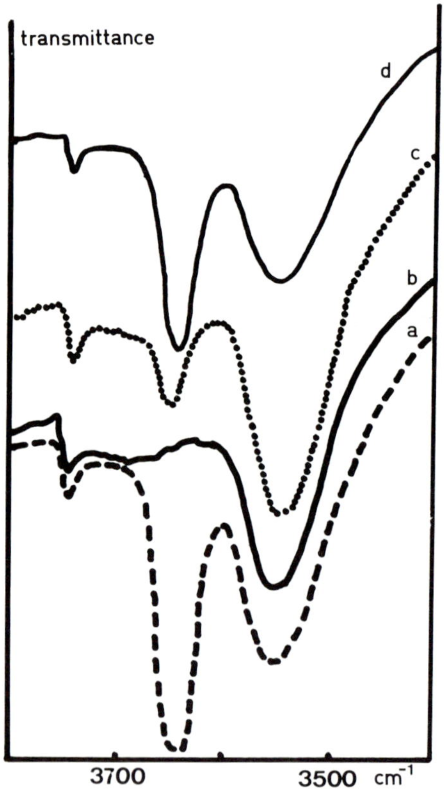

Figure 5. OH spectra of Ce(III)Y activated at 450°C

(a) Initial
(b) Pyridine evacuated at 150°C
(c) At 250°C
(d) At 350°C

Table III. Infrared Bands of Adsorbed Pyridine

	8a–8b	*19b* *Bronsted (B)*	*19a (B + L)*	*19b* *Lewis (L)*
Ce(III)Y (450°C)	1635–1625 1608–1600	1541	1489	1443
Ce(III)Y (600°C)	1625–1608 1600	1541	1489	1451–1443
Ce(IV)Y (450°C)	1630–1593	1544	1490	1444
Ce(IV)Y (600°C)	1620–1597	1538	1483	1451–1444
Ce(IV)NH_4Y (450°C)	1630	1540	1488	1451–1444
Ce(IV)NH_4Y (600°C)	1620–1592	1538	1489	1453–1444
LaY (200°C)	1630–1610 1590	1540	1489	1440
LaY (450°C)	1630–1610 1590	1540	1489	1451–1442

at 250°C, the 19b and 19a vibrations are split into 2 components (1550 and 1430 for 19b; 1495 and 1488 for 19a). Further evacuation at 350°C produces a greater decrease for the low-frequency components than for the high-frequency ones. The last remaining component for 19a vibration is 12 cm^{-1} shifted to the higher frequencies.

Discussion

The 3745 cm^{-1} band may be attributed either to Si–OH groups inherent to siliceous impurities or silanol groups which terminate the zeolite lattice as stated by different authors (*1, 3, 17*), while the 3640 and 3550 cm^{-1} bands were assigned to acidic OH groups. The disagreement in the assignment of the 3680–90 cm^{-1} bands (*1, 3, 19, 22*) seems to be owing to the different origin of the studied zeolites or to diversity of thermal treatment. Nevertheless, in the present work, it seems inconsistent to assign the 3680 cm^{-1} band to water molecules associated with the cation since no band is observed at 3400 and at 1640 cm^{-1}. Moreover, the conditions of the formation of these OH and their thermal stability must be related rather to the formation of Al–OH groups by an

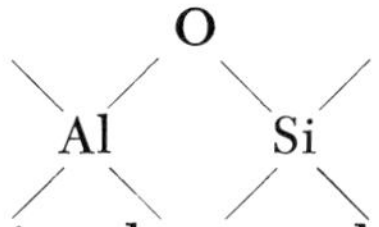

bridge hydrolysis at 460°C. The band at 3530 cm^{-1} is assigned as stated by Ward (*25*) and Rabo (*13*) to RE–OH groups. This band has no acidic property. On the contrary, the acidic nature of the 3640 and 3680 cm^{-1} bands is definitely established since they disappear when pyridine is adsorbed and, simultaneously, characteristic infrared bands of PyH^+ ions are observed in the spectral region 1700–1400 cm^{-1}. Furthermore, the evacuation of pyridine points out the difference in the acid strength of the 2 hydroxyl groups, since the latter is restored by evacuation at 250°C while the former is restored only after evacuation at 350°C. On the other hand, LaY and Ce(III)Y seem to be rather protonic acid solids after evacuation at 200°C, though adsorption of pyridine on such solids gives rise to a band at 1445 cm^{-1} which may be assigned as suggested by Ward (*22*) and Hall (*4*) to pyridine coordinated by the cation. On the other hand, the samples calcined at 500°C contain more Lewis acid sites than Bronsted acid sites. Though characteristic PyH^+ bands are still present in the spectra, the appearance of a marked shoulder at 1451 cm^{-1} shows the existence of true Lewis acid sites. These Lewis sites can only be created by dehydroxylation at the expense of Bronsted acid hydroxyl groups. Such phenomena had already been observed by Ward (*22*). The original result obtained for NaCe(IV)Y samples is more difficult to explain; the 19b vibration splitting may be interpreted as an evidence of the existence of different acid site strengths. The splitting of the 19a vibration, very insensitive to the bonding type of the N atom, is much more difficult to explain. However, it seems that there are 2 types of silanol acid groups—one near a sodium ion, the other near a cerium ion. The former type would give rise after pyridine equilibration to the low-frequency band. Since this band is removed at 250°C, the strength of such sites is rather weak. The latter type would provide a new species: a Π bonding is established between the pyridine and the empty *d* orbitals of the ceric ion. By activation at 350°C, these species would give rise to a charge-transfer complex $C_5H_5NH^{2+}$. The perturbation of the Π electron is strong enough to account for such an important shift to higher frequencies of the 19a vibration.

The EPR spectra of anthracene adsorbed on REY are characteristic of the anthracene positive radical. Our results provide strong evidence of the oxidizing properties of these solids, as pointed out by Hirschler (*8*). Anthracene ionization is inhibited by coke deposit on the sample during its activation under vacuum. No radical ion generation is observed by adsorption of anthracene or perylene on NH_4Y evacuated at 500°C even if the sample is oxygen-treated later. An opposite behavior is observed for Ce(III)Y form: an evacuated sample is inactive but further activation under oxygen produces a pale yellow solid, and a strong EPR signal is observed after adsorption of anthracene. The direct role of the cation in the electron transfer, as has been shown for CuY (*14*), is once more demonstrated. The electron acceptor sites would be the Ce^{4+} ions, which have powerful oxidizing properties; results for the NH_4Ce(III)Y form corroborate these hypotheses since no anthracene oxidation occurs on such solids after activation under vacuum at 500°C, though infrared evidence of the existence of Lewis acid sites is given by the presence of the band at 1451 cm^{-1} on the spectra after pyridine adsorption. The lack of activity is attributed once more to coke poisoning. In a similar manner, the hydrocarbon ionization needs the solid to be activated under oxygen and hence Ce^{4+} to be formed. LaY zeolite seems to behave in a similar way as NH_4Y. The dependence of the spin concentration on activation temperature is similar. Moreover, the numbers of radical ions are in the same magnitude range and the rehydration of the solid gives rise to a large decrease of spin numbers; on the other hand, the La^{3+} has no oxidizing properties. It seemed reasonable to postulate that the La^{3+} ions are not involved in the electron transfer. The similar behavior of LaY and NH_4Y makes us think that the active sites are the Lewis acid sites as in the case of the dehydroxylated NH_4Y. This seems consistent since a band at 1451 cm^{-1} appears when pyridine is adsorbed on activated LaY samples, giving strong evidence of the presence of such Lewis sites at the surface of this solid.

In conclusion, structural hydroxyl groups are observed on the surface of REY. Their formation is attributed to water ionzation by strong fields near the cation. This hydrolysis gives rise to RE–OH and a proton; the latter reacts with a surface oxygen in the way stated by several investigators:

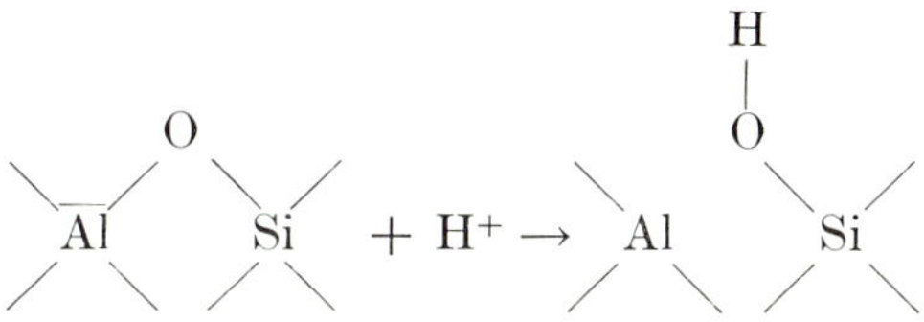

The acidic properties of some hydroxyl groups have been demonstrated by the existence of a 1540 cm^{-1} band in all the spectra of adsorbed pyridine on solids activated at low temperature. Further dehydrated samples behave either as Bronsted or Lewis acid solids; the latter are created by dehydroxylation of the zeolite, which occurs in two different ways: First

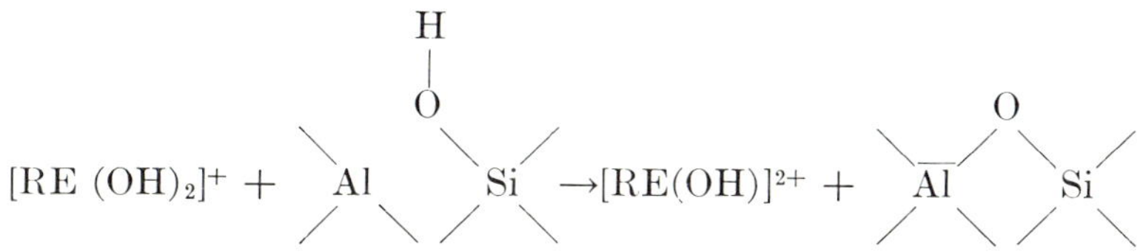

or second

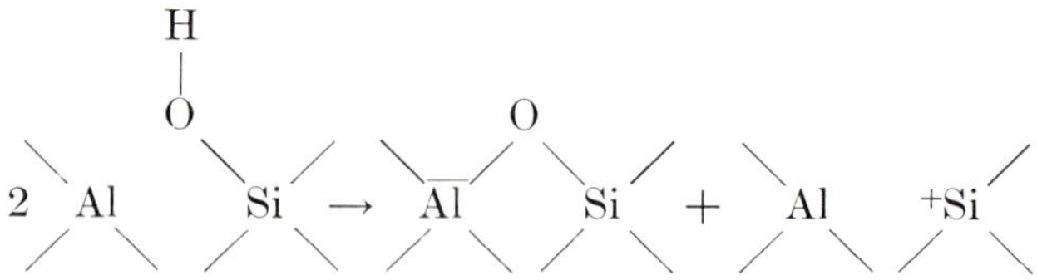

The anion vacancy, $RE(OH)^{2+}$, or RE^{3+} may all coordinate pyridine. Simultaneously, NaY zeolite exchanged with high-electron-affinity cations such as Cu^{2+} and Ce^{4+} acquires important oxidizing properties. The electron is transfered from the aromatic hydrocarbon to the ceric or cupric ion, whereas LaY zeolite behaves like a dehydroxylated NH_4Y zeolite. It seems that the NaCe(IV)Y form also provides species such as PyH^{2+}.

Literature Cited

(1) Angell, C. L., Schaffer, P. C., *J. Phys. Chem.* **1965,** 69, 3436.
(2) Basila, M. R., Kantner, T. R., *J. Phys. Chem.* **1966,** 70, 1681.
(3) Carter, J. L., Lucchesi, P. J., Yates, D. J. C., *J. Phys. Chem.* **1964,** 68, 1385.
(4) Christner, L. G., Liengme, D. V., Hall, W. K., *Trans. Faraday Soc.* **1968,** 64, 1679.
(5) Dollish, F. R., Hall, W. K., *J. Phys. Chem.* **1967,** 71, 1005.
(6) Habgood, H. W., *J. Phys. Chem.* **1965,** 69, 1764.
(7) Hirschler, A. E., *J. Catalysis* **1966,** 5, 196.
(8) Hirschler, A. E., Neikam, W. C., Barmby, D. S., James, R. L., *J. Catalysis* **1965,** 4, 628.
(9) Hugues, T. R., White, H. M., *J. Phys. Chem.* **1967,** 71, 2192.
(10) Liengme, B. V., Hall, W. K., *Trans. Faraday Soc.* **1966,** 62, 3229.
(11) Mathieu, M. V., Pichat, P., "La Catalyse au Laboratoire et dans l'Industrie," p. 320, Editions Masson, Paris, 1967.
(12) Parry, E. P., *J. Catalysis* **1962,** 2, 371.
(13) Rabo, J. A., Angell, C. L., Kasai, P. H., Schoemaker, V., *Discussions Faraday Soc.* **1966,** 41, 328.
(14) Richardson, J. T., *J. Catalysis* **1967,** 9, 172.
(15) Rooney, A. J., Pink, R. C., *Proc. Chem. Soc.* **1961,** 70.
(16) Stammires, D., Turkevich, J., *J. Am. Chem. Soc.* **1964,** 86, 749.
(17) Uytterhoeven, J. B., Christner, L. G., Hall, W. K., *J. Phys. Chem.* **1965,** 69, 2117.
(18) Uytterhoeven, J. B., Jacobs, P., Makay, K., Shoonheydt, R., *J. Phys. Chem.* **1968,** 72, 1768.
(19) Uytterhoeven, J. B., Shoonheydt, R., Liengme, B. V., Hall, W. K., *J. Catalysis* **1969,** 13, 425.
(20) Ward, J. W., *J. Catalysis* **1967,** 9, 225.
(21) *Ibid.*, **1967,** 9, 396.
(22) *Ibid.*, **1968,** 10, 34.
(23) *Ibid.*, **1968,** 11, 251.
(24) *Ibid.*, **1969,** 14, 365.
(25) Ward, J. W., *J. Phys. Chem.* **1968,** 72, 4211.
(26) White, J. L., Jelli, A. N., Andre, J. A., Frippiat, J. J., *Trans. Faraday Soc.* **1967,** 63, 461.
(27) Ben Taarit, Y., Naccache, C., Imelik, B., *J. Chim. Phys.* **1970,** 67, 389.

RECEIVED March 10, 1970.

Discussion

B. D. McNicol (Koninklijke/Shell Laboratorium, Amsterdam, Netherlands): With reference to Figure 1 of your paper regarding the spin concentration of positive anthracene ions, you state that "rehydration results in a large decrease in spin concentration." This is certainly not apparent in the figure. Also, were no measurements made on rehydrated samples which had been activated above 550°C? This would shed some light on the reason for the decrease in spin concentration observed above 550°C. In the absence of x-ray or other evidence, this decrease may not be due to structure collapse.

C. Naccache: Figure 1 shows that rehydration results in a large decrease of spin concentration. For example, as can be determined in this figure (dotted line corresponds to sample activated at 450° and then rehydrated), LaY zeolite dehydrated at 450° gives about 1.5×10^{18} positive radical ions, while after rehydration the number of radical ions is only 1.1×10^{18}. The decrease in electron-acceptor sites resulting from rehydration of the zeolite is demonstrated.

In fact, the decrease of spin concentration observed above 550°C could be better explained by migration of cations in S_I position when the cation is the active site (Ce^{4+} and Cu^{2+}, for example). In the case of LaY, we have no other explanation than that given in our paper. I would like to know if you have another hypothesis which can better explain this decrease.

69

Infrared Spectroscopy and Cumene Cracking Studies on Single Component Rare Earth Forms of Synthetic Faujasite

P. E. EBERLY, JR., and C. N. KIMBERLIN, JR.

Esso Research Laboratories, Humble Oil & Refining Co.,
Baton Rouge, La. 70821

Infrared spectroscopy and cumene cracking studies were conducted on the Y, La, Ce, Pr, Sm, Eu, Gd, Dy, Er, Yb, and Th forms of synthetic faujasite. OH groups having a frequency between 3470 and 3520 cm^{-1} are characteristic of the rare earth faujasites. This frequency is specific to the rare earth ion and increases linearly with ionic radius. These groups are indirectly perturbed by pyridine, which causes their absorption frequency to shift 20-30 cm^{-1} to a higher value. Although Bronsted acidity increases slightly with ionic radius, it is more strongly affected by calcination conditions. Lower calcination temperatures produce more acidic and more active catalysts for cumene cracking. The La, Eu, and Yb forms have higher activity than the other rare earths. Bronsted acidity cannot explain all nuances of catalytic activity.

Rare earth forms of zeolites X and Y type faujasites possess superior catalytic properties for various reactions such as alkylation, isomerization, and cracking (*9, 12, 18*). Structural studies involving x-ray diffraction and CO chemisorption have been made to locate the positions of the rare earth (*11, 14, 16*). Hydroxyl groups and their relationship to surface acidity have been studied by infrared spectroscopy, utilizing the adsorption of pyridine and other basic molecules (*2, 6, 21, 22, 23*). Since much of the previous research has involved measurements on mixed rare earth faujasites, a need existed for a more systematic study of the individual rare earth zeolites, in regard to both structural and catalytic properties. The present investigation deals with the Y, La, Ce, Pr, Sm,

Eu, Gd, Dy, Er, Yb, and Th forms. Infrared spectral measurements were conducted *in situ* to characterize the OH groups and acidity. Catalytic activities were measured by cumene cracking.

The location of rare earth ions in faujasite type zeolites has been the subject of several investigations. Using x-ray diffraction techniques on a cerium exchanged crystal of natural faujasite, Olson, Kokotailo, and Charnell (*11*) found that the Ce^{3+} ions are inside the large adsorption cavity in the freshly exchanged sample. Upon dehydration, however, the ions enter the sodalite cage cavity, where they are no longer available for interaction with adsorbates. Rabo, Angell, and Schomaker (*14*) reached similar conclusions with LaY. No cation-specific interaction of the rare earth ions with adsorbed CO was observed, indicating the absence of ions in the large adsorption cavities. Smith, Bennett, and Flanigen (*16*) found by x-ray diffraction that all the La^{3+} ions occupy S_I' sites within the sodalite unit adjacent to the hexagonal prism. These ions are electrostatically shielded from one another by an OH group or oxygen anion. The latter authors report that the La ions show a gradual, reversible shift in position upon heating. At 725°C, a fourth of the cations shift from S_I' to the S_I bridge position inside the hexagonal prism, and another fourth into S_{II} positions inside the large adsorption cavity.

To characterize the OH structure, infrared spectral techniques have been used. In general, 3 types of hydroxyl groups are observed. Those having an absorption frequency of 3740–3750 cm^{-1} are unreactive and are believed to terminate the external faces of the faujasite crystal or to exist on some extraneous silica-containing phase. Groups absorbing at 3650-3630 cm^{-1} are accessible to hydrocarbon molecules and are similar to those found on HY (*6, 10, 19, 21*). Several authors attribute the carbonium ion type catalytic activity of faujasite to this particular group (*4, 21*). A low-frequency band near 3520 cm^{-1} has been reported for mixed rare earth, La, and Ce forms of synthetic faujasite. These groups do not interact with adsorbed molecules such as pyridine or piperidine (*21*). Rabo, Angell, and Schomaker (*14*) suggest that this group serves as a shield between the rare earth ions inside the sodalite unit.

Various reactions have been studied on mixed rare earth and the La and Ce forms. These include ethylation of benzene (*18*), propylation of toluene (*14*), *o*-xylene isomerization (*21*), butane cracking (*14*), cracking of *n*-hexane, *n*-heptane, and ethylbenzene (*8*), and isomerization and disproportionation of 1-methyl-2-ethylbenzene (*7*). Other reactions are summarized by Venuto and Landis (*18*). In several reports, an optimum calcination temperature for best catalytic performance has been demonstrated (*7, 8, 14, 18, 21*).

Experimental

Materials. The degree of exchange and surface area of the faujasites are given in Table I. The experimental surface areas were obtained by nitrogen adsorption techniques, using the Langmuir method.

Table I. Properties of Various Ion-exchanged Forms of Synthetic Faujasite

Form	*% Na^+ Ions Exchanged*	*Surface Area, M^2/g*	
		Exptl.	*Calcd.*
Na	0	839	854
H	89	952	952
Y	68	840	837
La	75	809	792
Ce	75	809	791
Pr	76	813	789
Sm	74	763	785
Eu	69	789	787
Gd	74	782	778
Dy	72	774	780
Er	68	776	777
Yb	66	757	774
Th	87	556	752

The hydrogen form was obtained by calcining the ammonium form at 427°C in air. The latter was prepared by ion-exchanging NaY with NH_4NO_3 solutions, using conventional techniques. Rare earth chlorides (99.9%) were obtained from the American Potash and Chemical Corp. to prepare the various rare earth forms from NaY. ThY was prepared using $Th(NO_3)_4$.

Taking the value of 952 M^2/gram for HY, one can calculate what the surface areas should be for the other samples if the crystal structure remained intact and all that changed was the weight of the unit cell. These results are given in the calculated column and, in general, compare quite favorably with the experimental values. ThY, however, has a surface area which is nearly 200 M^2/gram lower than predicted. This suggests that some loss in crystallinity may have occurred. X-ray diffraction lines have lower than expected intensities.

Infrared Spectral Measurements. Sample preparation techniques and apparatus have been described previously (*3, 4*). The spectra were recorded on disks of powdered material with a Cary-White Model 90 infrared spectrophotometer. Spectra were obtained from 4000 to 1200 cm^{-1} at a spectral slit width of 4 cm^{-1} and a scan speed of 3 cm^{-1}/sec. The infrared cell was so constructed as to permit calcinations to be done either under vacuum or in a stream of dry air. All spectra were recorded *in situ* at the high temperatures indicated in the tables and figures.

Cumene Cracking Experiments. The cumene cracking studies were done in a manner similar to that reported by Richardson (*15*). The faujasites were compressed into pills which were cracked, and a 28-35 mesh fraction was isolated. A 300-mg portion was first calcined at an elevated temperature. This generally involved a 16-hour treatment in air at 538°C. Other calcination conditions are as noted. The catalyst was then transferred under nitrogen to a nominal 1/4-inch stainless steel, 20-gage tubular reactor which was placed in a constant-temperature fluidized sandbath. After equilibration with helium carrier gas, the cumene was introduced to the reactor by permitting the carrier gas to

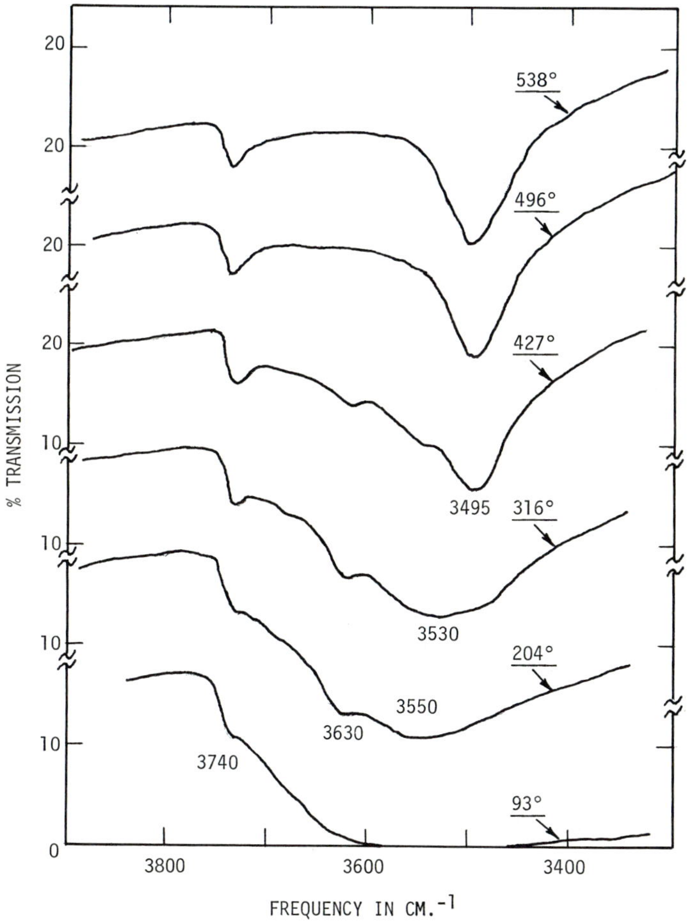

Figure 1. Infrared spectra of SmY under vacuum

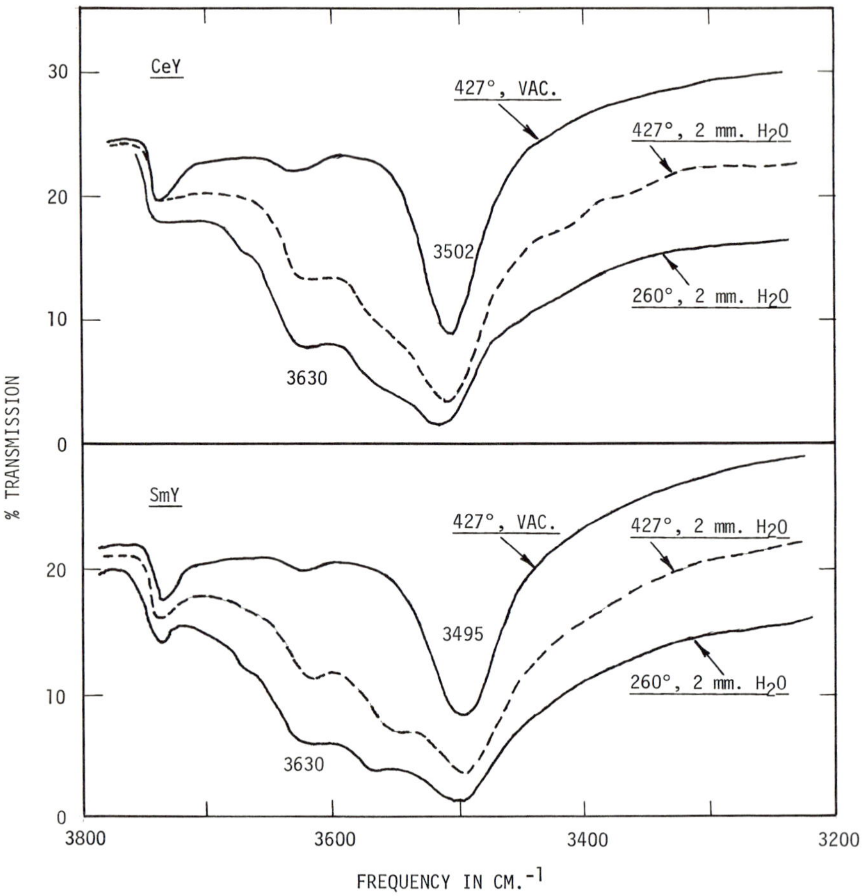

Figure 2. Infrared spectra in OH region of CeY and SmY

bubble through a cumene saturator maintained at 18°C. The flow rate was adjusted to provide a space velocity of 0.43 W/hr/W. Peroxides, which are known to affect cracking rate (*13*), were removed by passing the cumene feed through a bed of 13X molecular sieve.

Results

Infrared Spectral Studies. Characteristics of OH Groups. The vibrational properties of OH groups and water adsorbed on zeolites depend strongly on the amount of water present. Furthermore, for a given state of dehydration, the nature of the infrared bands in the OH region shows some dependence on temperature (*1*).

Spectra are given in Figure 1 for a sample of SmY. The disk was placed in the spectrometer cell, which was then evacuated. Spectra were recorded as the sample was heated from 93° to 538°C. The OH bands observed in these high-temperature spectra are broader than those measured previously at room temperature on other rare earth forms (*2, 14, 21*), indicating possibly a greater degree of proton mobility. The 3740 cm^{-1} band changes little in intensity. As the physically adsorbed water is removed, 2 bands begin to appear at 3630 and 3550 cm^{-1}. The latter shows a gradual decrease in frequency to 3495 cm^{-1} as the temperature is raised to 427°C and above. The 3630 cm^{-1} band is easily seen at intermediate temperatures but is not apparent in the spectra recorded at 496° and 538°C. The concentration of OH groups responsible for this band is too low for detection by infrared spectroscopy.

After dehydration, the OH groups can be reconstituted by exposure to water at elevated temperatures. This is illustrated by the spectra in Figure 2 for CeY and SmY. Except for a small change in frequency of the high-intensity OH band, the spectra are similar. After evacuation at 427°C, 2 mm of water pressure is placed on the sample and an immediate increase in OH band intensities is observed. This is particularly true for the 3630 and the 3500 cm^{-1} bands. Upon lowering to 260°C, a further increase in intensity occurs; however, some physically adsorbed water is also present at these conditions, as indicated by a band near 1635 cm^{-1} (not shown).

Spectra of the rare earth faujasites were recorded at 216°C under vacuum, after a previous evacuation at 427°C. Little difference was seen in the 3740 and 3630 cm^{-1} groups. The frequency of the predominant OH group, however, increases continually with ionic radius from 3470 cm^{-1} for YbY to 3522 cm^{-1} for LaY. This is shown by the open squares in Figure 3. The integrated absorbance (open circles), which is more difficult to estimate, exhibits no regular dependence with radius.

The OH structure of ThY is different from the rare earth forms. Groups at 3630 cm^{-1} are present, but the low-frequency groups between 3470-3520 cm^{-1} are not observed. The intensity of the 3740 cm^{-1} band is over 3 times as great. This fact and the lower than expected surface area suggest some loss of crystallinity.

PYRIDINE ADSORPTION. *Vacuum-Calcined Faujasites.* Pyridine adsorption was conducted at 260°C and 2 mm of pressure after previous evacuation of the solid at 427°C. Pyridinium ions formed by interaction with Bronsted sites have an infrared absorption band at 1538 cm^{-1}. Pyridine, coordinately bound either to Lewis sites or cations, is observed by bands in the 1440-1450 cm^{-1} region.

Pyridine interacts strongly with the OH groups at 3630 cm^{-1}, causing the band to disappear. However, some effect is also seen on the low-

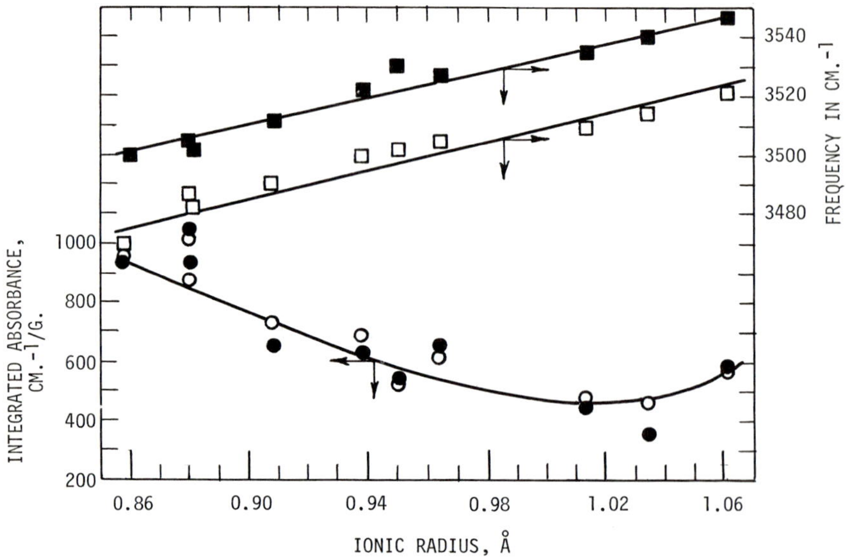

Figure 3. Effect of pyridine adsorption at 260°C on low-frequency OH band

Open points—vacuum before pyridine adsorption
Solid points—2 mm pyridine

frequency OH groups. This is illustrated in Figure 3. The top lines pertaining to the frequency of the band show that pyridine adsorption causes a shift in frequency of 20-30 cm^{-1} to higher values. Since most direct interactions, such as hydrogen bonding, cause a shift in the other direction, this phenomenon cannot be explained on this basis. Moreover, there is little change in the integrated absorbance of the band, as seen by the results on the bottom curve.

Acidities as expressed by the absorbances of the infrared bands are plotted in Figure 4. Data were recorded under 2 mm of pressure and after evacuation for 30 minutes to remove reversibly adsorbed species. The 3230 cm^{-1} band, like the 1538 cm^{-1} band, is caused by pyridinium ions and can be useful for estimating acidity when the latter band becomes too intense. Although some scatter exists, the Bronsted acidity tends to increase with ionic radius. Nearly 30-40% of this acidity is reversible. In the 1450 cm^{-1} region, 2 types of adsorbed pyridine can be distinguished, 1 species absorbing at 1452 cm^{-1} and the other at 1438 cm^{-1}. The latter band is more intense and removed easily by evacuation. The remaining 1452 cm^{-1} band arises from Lewis acidity. Its intensity is the same for all the rare earth faujasites.

Air-Calcined Faujasites. Since calcination conditions were shown to have a strong effect on catalytic activity, acidity measurements were

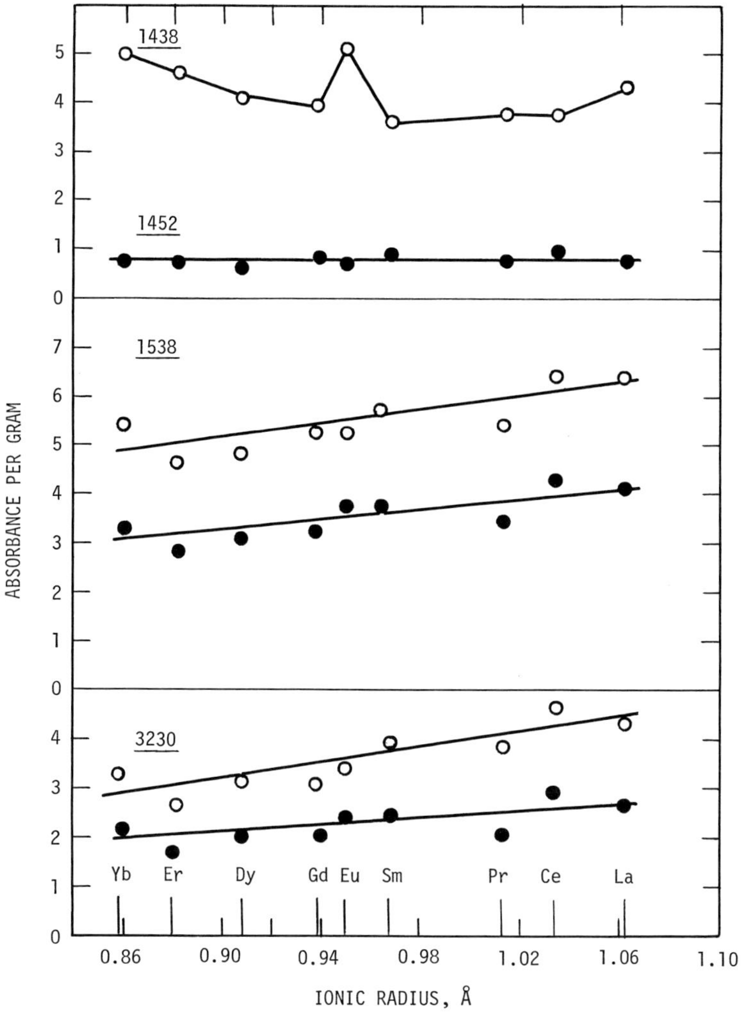

Figure 4. Absorbance of pyridine infrared bands on rare earth forms of zeolite Y at 260°C

Open points—2 mm pyridine
Solid points—30-min evacuation after pyridine adsorption

performed on samples calcined in air at various temperatures to obtain a more valid correlation with activity. After air calcination in the infrared spectrometer, the sample was evacuated briefly, and the temperature was lowered to 260°C for pyridine adsorption at the standard conditions. Results are shown in Table II for SmY. Data were recorded under 2 mm of pressure and after evacuation for 30 minutes, to distinguish between the reversible and irreversible acidity.

Table II. Infrared Absorbance[a] of Pyridine Bands on SmY after Calcination in Air

Calcination Temp., °C	Bronsted Acidity				Lewis Acidity	
	3230 Cm^{-1}		1538 Cm^{-1}		1438 Cm^{-1},	1452 Cm^{-1},
	Rev.	Irrev.	Rev.	Irrev.	Rev.	Irrev.
274	1.5	4.8	b	b	b	b
316	0.7	4.0	b	b	b	b
427	1.2	3.2	1.2	4.5	2.3	0.8
538	1.4	2.5	2.0	3.7	2.7	0.9
650	1.9	1.4	3.4	1.7	5.9	0.4

[a] Per gram of zeolite.
[b] Not measurable.

The total number of Bronsted sites is lowered by increasing calcination temperature. This results primarily from a sharp reduction in strong sites, as reflected by a decrease in the irreversible portion of acidity. The reversible acidity tends to increase. This is true for both the Bronsted and Lewis types.

Intensities of the pyridine bands on SmY could not be estimated in the low-frequency region because of interference from carbonate bands. Data are given in Table III which show that these bands disappear only at temperatures above 482°C.

Table III. Carbonate Bands on SmY after Calcination in Air

Calcination Temp., °C	Absorbance Per Gram	
	1510 Cm^{-1}	1440 Cm^{-1}
93	7.8	9.9
204	5.8	7.2
316	4.3	5.2
427	2.5	2.1
482	0.64	1.0
538	0	0

These species could result from impurities in the rare earth salt and may be of some importance catalytically.

Cumene Cracking Studies. Results are shown in Figure 5 for 3 samples of SmY calcined at different temperatures. The differences in activity are caused by changes in the preexponential factor rather than activation energy. The latter amounts to 20 kcal/mole. This is lower than the 28 kcal/mole reported by Richardson (*15*) for the alkali and alkaline earth forms but is the same as that observed by Topchieva *et al.* (*17*) for the yttrium and lanthanum faujasites. Conversion data for

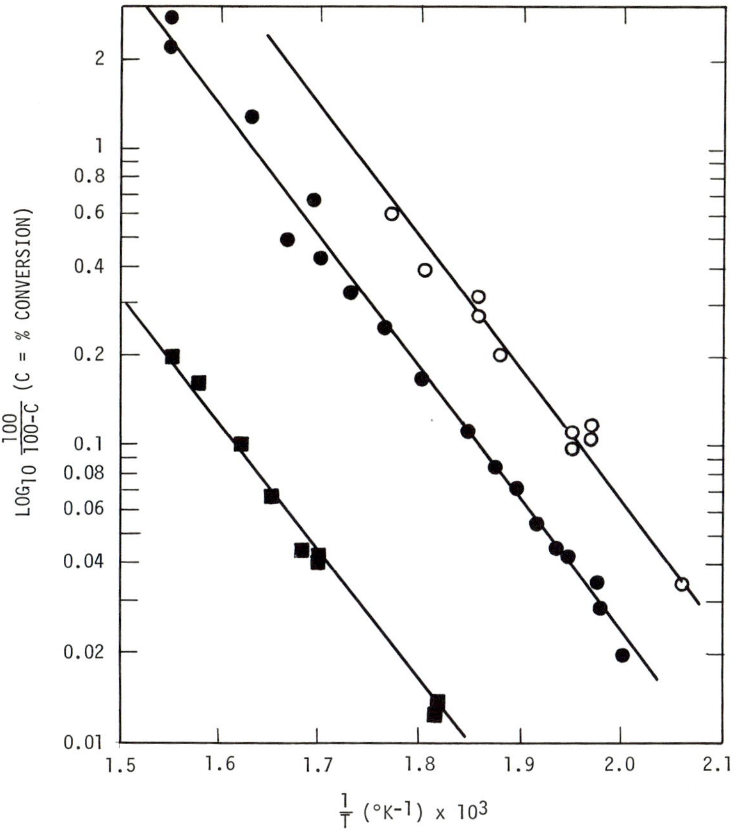

Figure 5. Effect of temperature on cumene cracking activity of Sm faujasites (0.43 W/hr/W)

○ *SmY calcined at 427°C*
● *SmY calcined at 538°C*
■ *SmY calcined at 650°C*

SmY calcined at other conditions are indicated by the solid points in Figure 6.

The various rare earth faujasites were calcined at 538°C and then tested for cumene cracking activity. Conversion levels are plotted in Figure 6. HY is superior to the rare earth forms. With the exception of La, Eu, and Yb, the rare earths generally give conversions in the 15-25% range.

Discussion

High-temperature spectra *in situ* permit characterization of the OH groups and other structural features as they present themselves to the reactant hydrocarbons. The groups near 3630 cm^{-1} have a particular importance since they are readily accessible to adsorbed molecules and are responsible for Bronsted acidity. The groups having a frequency between 3470-3520 cm^{-1} appear to be isolated from adsorbed molecules and are closely associated with the rare earth ion. The particular frequency at which they adsorb increases linearly with ionic radius. Pyridine adsorption causes this band to shift 20-30 cm^{-1} to higher values without any change in integrated absorbance. This indirect interaction can be explained if the group is inside the sodalite cage connected to a rare earth ion which, in turn, is connected to an accessible OH group (3630 cm^{-1}). Pyridine forms an ion with the latter which apparently causes an alternate strengthening and weakening of bonds along the —O—RE—O—H· · · linkage resulting in a strengthening of the bond in the inaccessible OH group and an increase in frequency.

Results suggest that the OH groups absorbing near 3630 cm^{-1} are primarily responsible for cracking activity. Hence, the decrease in their concentration caused by more severe calcination appears to account for the decrease in catalytic activity. The acidity or protonation ability of these groups, as measured by pyridine adsorption, would be expected to be the prime variable. Correlations of catalytic activity with Bronsted acidity have been reported previously (*20, 21, 24, 25*). Results from the

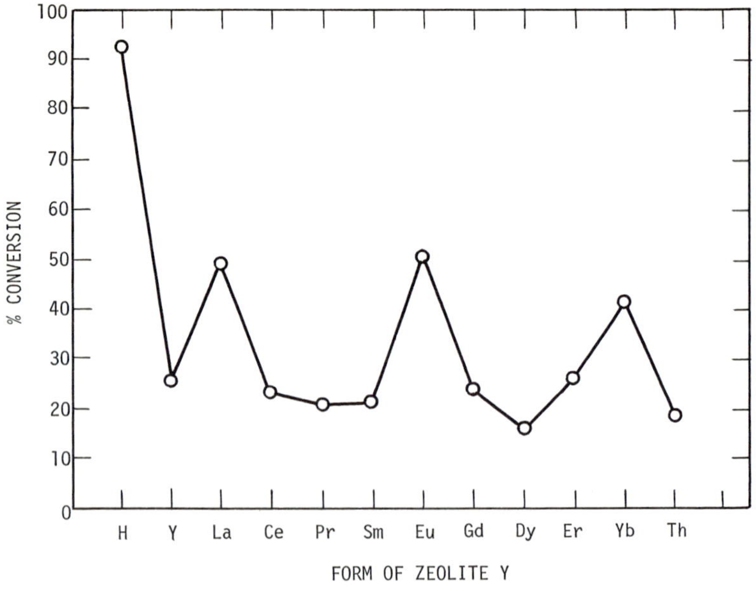

Figure 6. Cumene cracking on ion-exchanged forms of zeolite Y at 274°C and 0.43 W/hr/W

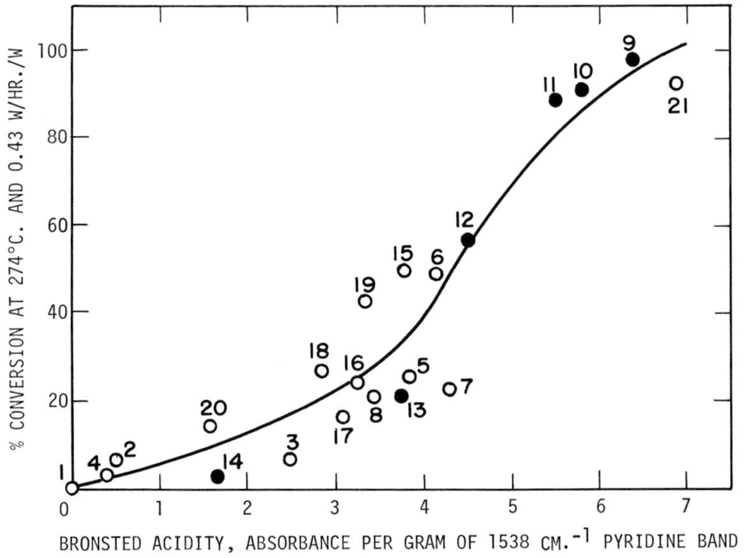

Figure 7. Cumene cracking activity vs. *irreversible Bronsted acidity*

		Calcd. Temp.
1.	n-Al_2O_3	538
2.	13% Al_2O_3—87% SiO_2	538
3.	CaY (75% exch.)	427
4.	CaY (75% exch.)	538
5.	YY	538
6.	LaY	538
7.	CeY	538
8.	PrY	538
9.	SmY	204
10.	SmY	274
11.	SmY	316
12.	SmY	427
13.	SmY	538
14.	SmY	650
15.	EuY	538
16.	GdY	538
17.	DyY	538
18.	ErY	538
19.	YbY	538
20.	ThY	538
21.	HY	538

present study are shown in Figure 7 for zeolitic and other catalysts. There is a general correlation, but it is obvious that exceptions exist. For a given rare earth faujasite such as SmY, the conversion *vs.* acidity follows a smooth S-type curve (solid points). However, if the rare earth ion is varied, the correlation with acidity is no longer convincing. For example, EuY gives about twice the conversion as CeY, even though the acidities are about the same. The results in Figure 6 show that among the rare earths, the La, Eu, and Yb forms have prominent activities which are not reflected by Bronsted acidity. There is no single property which immediately suggests itself that could account *a priori* for the activity pattern. ThY is probably less active because of a partial loss in structure. In any event, activity is not a simple function of acidity. Perhaps, as Ward (*21*) has suggested, activity can be caused by an interaction between Bronsted and Lewis sites.

Christner, Liengme, and Hall (*2*) observed that CO_2 addition increases the formation of pyridinium ions on faujasites. Frilette and Munns (*5*) found that CO_2 increases dehydration activity in NaY. Thus, the presence of carbonate ions on faujasites may have catalytic significance. Our results show that the activity of SmY decreases not only with decreasing Bronsted acidity but carbonate content as well.

Acknowledgment

The authors wish to express their sincere appreciation to Dorothy Webb for her excellent technical assistance in obtaining infrared spectra and measuring catalytic activity.

Literature Cited

(1) Cant, N. W., Hall, W. Keith, *Trans. Faraday Soc.* **1968,** 64, 1093.
(2) Christner, L. G., Liengme, B. V., Hall, W. K., *Trans. Faraday Soc.* **1968,** 64, 1679.
(3) Eberly, P. E., Jr., *J. Phys. Chem.* **1967,** 71, 1717.
(4) Eberly, P. E., Jr., *J. Phys. Chem.* **1968,** 72, 1042.
(5) Frilette, V. J., Munns, G. W., Jr., *J. Catalysis* **1965,** 4, 504.
(6) Hattori, H., Shiba, T., *J. Catalysis* **1968,** 12, 111.
(7) Hickson, D. A., Csicsery, S. M., *J. Catalysis* **1968,** 10, 27.
(8) Hopkins, P. Donald, *J. Catalysis* **1968,** 12, 325.
(9) Koffler, R. L., *Proc. Rare Earth Res. Conf., 7th,* **1968,** 2, 697.
(10) Olson, D. H., Dempsey, E., *J. Catalysis* **1969,** 13, 221.
(11) Olson, D. H., Kokotailo, G. T., Charnell, J. F., *J. Colloid Interface Sci.* **1968,** 28, 305.
(12) Plank, C. J., Rosinski, E. J., Hawthorne, W. P., *Ind. Eng. Chem., Prod. Res. Develop.* **1964,** 3, 165.
(13) Prater, C. D., Lago, R. M., *Advan. Catalysis* **1956,** 8, 293.
(14) Rabo, J. A., Angell, C. L., Schomaker, V., *Intern. Congr. Catalysis, 4th, Moscow,* **1968,** Preprint 54.
(15) Richardson, J. T., *J. Catalysis* **1967,** 9, 182.
(16) Smith, J. V., Bennett, J. M., Flanigen, E. M., *Nature* **1967,** 215, 241.
(17) Topchieva, K. V., Rosolovskaya, E. N., Zhavoronkov, M. N., Razanova, O. N., Parmenova, A. S., *Dokl. Akad. Nauk SSSR* **1969,** 185, 132.

(18) Venuto, P. B., Landis, P. S., *Advan. Catalysis* **1968,** 18, 259.
(19) Ward, J. W., *J. Catalysis* **1967,** 9, 225.
(20) Ward, J. W., *J. Catalysis* **1968,** 10, 34.
(21) *Ibid.,* **1969,** 13, 321.
(22) Ward, J. W., *J. Phys. Chem.* **1968,** 72, 4211.
(23) *Ibid.,* **1969,** 73, 2086.
(24) Ward, J. W., Hansford, R. C., *J. Catalysis* **1969,** 13, 154.
(25) *Ibid.,* **1969,** 13, 364.

RECEIVED November 13, 1969.

Discussion

W. H. Flank (Houdry Laboratories, Marcus Hook, Pa. 19061): I get the impression that we are viewing zeolitic catalysts as thermodynamically perfect crystals, and we view active sites in terms of stoichiometry rather than in catalytic amounts. I think we accept this over-simplification too readily. Catalytic sites should be given credit for being able to work hard. We must also consider that zeolitic catalysts under actual reaction conditions at elevated temperatures have probably undergone significant dynamic structural changes, especially with regard to protons. The problem is: what are we really looking at?

P. E. Eberly, Jr.: I share your concern about structural changes with temperature. Consequently, we have consistently given consideration to what conditions would be most desirable for measuring acidity. We have effected a compromise. At reaction temperatures much above 260°C, pyridine can undergo some decomposition, and this clouds the experimental results. We selected 260°C for pyridine adsorption. This is higher than that used by other investigators and hence should yield acidities which more nearly represent those at the cracking temperature of 274°C used in this study.

A. E. Hirschler (Sun Oil Co., Marcus Hook, Pa. 19061): I should like to make a comment relating to your statement that activity is not a simple function of acidity. This statement appears to contain the implicit assumption that all acid sites able to protonate pyridine would have the same catalytic activity. Actually, one might expect an intensity factor (acid strength) to be operative in acid catalysis as well as an extensive factor, and the apparent lack of correlation you observed may result from an influence of acid strength on catalytic activity. Pyridine adsorption data alone do not completely define Bronsted acidity. One may need also some measure of acid strength. Acidity measurement with H_R indicators, such as we and others have reported, may supply useful information relative to acid strength as well as number of acid sites.

P. E. Eberly, Jr.: I agree in general with your statements. As I indicated in my talk there is probably a distribution of strengths of acid sites responsible for protonating pyridine and creating the band at 1538 cm^{-1}. More data are needed. Possibly, the techniques you mention might be helpful. However, I have had some reservations in using the H_R indicators in view of their generally large molecular size and also in view of the measurements being conducted in the liquid phase. Some severe rate limitations may arise in measuring acidities by this technique. Secondly, protons in the hydroxyl groups are reported to become somewhat mobile as the temperature is raised. Consequently, measurements made at nominal temperatures might not truly reflect the acidity available at reaction conditions.

W. K. Hall (Gulf Research & Development Co., Pittsburgh, Pa. 15230): With the alkaline earth ions, the band formed on the adsorption of pyridine in the 1438 to 1450 cm^{-1} region shifts to higher frequency as the electric field from the ion increases. Did you observe such effects with the rare earth ions? If not, does this mean the rare earth ions are inaccessible to the gas?

P. E. Eberly, Jr.: No, we do not observe any shift in frequency of the 1438 cm^{-1} band with the nature of the rare earth ion. Thus, I believe the rare earth ions are inside the sodalite cage, not directly accessible to pyridine. Ward has suggested that the 1438 cm^{-1} band could arise from interaction of pyridine with the residual sodium ions in the supercage.

J. W. Ward (Union Oil Co., Brea, Calif. 92621): Since the exchanged zeolites used are ion-exchanged with the individual rare earth cations to about 75%, is it not possible that the absorption band at 1438 cm^{-1} in the spectrum of adsorbed pyridine is caused by interaction with the remaining sodium ions? What type of correlation does one obtain between the catalytic activity and the easily removed pyridinium ion? Is it not possible that the difference between activity and acidity result from different pretreatments of the sample used in each study?

P. E. Eberly, Jr.: Yes. This seems plausible. The band appears at 1438 cm^{-1} and does not change with the nature of the rare earth ion. The reversible Bronsted acidity does not correlate well with catalyst activity. Pretreatments were identical for the cracking studies in Figure 6 and the acidity measurements in Figure 7.

70

Isomerization of *n*-Butenes over Ion-Exchanged X Zeolites

N. E. CROSS, C. KEMBALL, and H. F. LEACH

Department of Chemistry, University of Edinburgh, West Mains Road, Edinburgh EH9 3JJ, Scotland

Several zeolites, prepared from 13X by replacing some of the sodium ions by other cations, have been used as catalysts for the isomerization of n-*butenes. Detailed studies, including the examination of the deuterium content of reactant and products after isomerization in the presence of deuterium or deuterium oxide, were carried out with zeolites containing varying amounts of* Ce^{3+} *or* Zn^{2+} *ions. Results with CeX zeolites (and the majority of the other zeolites examined) were indicative of a carbonium ion mechanism, but other work gave evidence for radical intermediates on NiX sieves. The behavior of the ZnX lay between these extremes and depended on the extent of exchange.*

The application of zeolites as catalysts for a variety of reactions of commercial significance has been reviewed by Venuto and Landis (*23*). In general, the multivalent and decationated forms of the zeolites exhibit a carbonium-ion type of catalytic activity, whereas the alkali metal forms show primarily a radical type of activity.

In the present work, the reaction selected to examine the effect of altering the cation on the catalytic properties of the zeolite was *n*-butene isomerization. This reaction has a highly selective character and does not require extreme conditions. The rate of 1-butene disappearance gives a convenient measure of activity, and the initial product ratio (*cis*-2-butene/*trans*-2-butene) can provide a guide to the nature of the reaction mechanism (*7*).

A synthetic near-faujasite molecular sieve of type X (*5*) was used as the parent zeolite, and the sodium cations were replaced by a range of other cations. Details of the structure and possible positions that the cations may occupy have been published (*1, 6, 18*). With some of the zeolites, a more detailed investigation was undertaken of the effect of the extent of cation exchange; with these zeolites, the butene isomerization was carried out also in the presence of deuterium and deuterium oxide.

Experimental

Apparatus. The apparatus and experimental procedure have been described (*9*). A static system was used with catalyst samples (normally 0.1 gram of hydrated material) in the bottom of the reaction vessel (volume 1×10^{-4} m^3). The reaction was followed by periodically sampling the gas phase above the catalyst and analyzing by GLC techniques.

Catalysts. The starting material for all the catalysts was Linde 13X in powder form, free of clay binder. The bulk sample had a silicon-to-aluminum ratio of 1.27 ± 0.02 (*16*). In all cases, the zeolites were prepared by ion exchange with salt solution (either chloride, sulfate, or nitrate) of the required cation at room temperature. The extent of exchange was estimated from either analysis of the residual salt solution or determination of the residual sodium in the zeolite sample. After preparation, the exchanged zeolites were stored over saturated calcium nitrate solution.

Reagents. The butenes (Matheson, CP grade) were purified further before use by repeated low-temperature distillation. Analysis by GLC showed that the individual butenes then contained only negligible amounts of lower molecular weight compounds. Deuterium (Matheson) was purified by diffusion through a palladium thimble. The deuterium oxide (99.7%) was obtained from I.C.I. Ltd. and was degassed before use.

Characterization of Catalysts. Comparative surface area measurements were made on a selection of the catalyst samples as described previously (*9*). X-ray diffraction traces were obtained for each zeolite both before and after use.

The residual water content of each catalyst after drying and storage over saturated calcium nitrate solution was determined by TG in air, using a Stanton Mass Flow balance (heating rate of $0.083°C\ sec^{-1}$).

Results

General Survey. The activity of a series of the ion-exchanged X-type zeolites was investigated by measuring the temperature range at which 1-butene isomerization occurred. A 0.1-gram sample of the hydrated catalyst was used in each case, and the catalyst outgassed at 420°C for 12 hours prior to use. A charge of approximately 1.2×10^{20} molecules of 1-butene was admitted to the reaction vessel at room temperature,

Table I. Catalytic Data of Ion-Exchanged X Zeolites

Catalyst	*% Na Exchanged*	*Approx. Temp. Range of Reaction, T,°C*	*Rate of 1-Butene Disappearance, % Min^{-1} (at T°C)*	*Product Ratio*[a]
NaX	0	200–300	0.17(250)	1.0
NiX	68	35–60	0.23(35)	0.5
CrX	35	50–100	0.77(64)	1.5
ZnX	70	50–100	0.63(50)	0.5
CeX	79	35–50	0.58(36)	1.5
LiX	85	190–300	0.18(190)	1.0
MgX	74	100–150	0.25(108)	1.0
CaX	59	150–200	0.08(157)	1.2
MnX	79	100–160	0.09(108)	0.3
CoX	52	100–150	0.35(126)	1.4
CuX	62	20–90	0.30(40)	0.95
LaX	88	0–40	2.30(16)	1.25
ThX	10	150–200	1.20(168)	1.0

[a] Initial value for *cis*-2-butene/*trans*-2-butene product ratio.

and the temperature increased slowly until isomerization occurred. The temperature was held at this value while the rate of 1-butene disappearance was measured and then raised to such a value that equilibrium between the butenes was rapidly established. In this manner, an approximate temperature range of activity and the initial *cis*-2-butene/*trans*-2-butene product ratio were determined for each catalyst. The results are listed in Table I. All the zeolites listed in this table displayed an x-ray diffraction pattern characteristic of the faujasite crystal structure, and surface areas were all in the range 870 ± 30 m^2 $gram^{-1}$ (using the point B method with nitrogen as adsorbate).

A more detailed investigation, including the effect of varying the amount of cation exchange, was undertaken for catalysts containing zinc and cerium cations as they were both active at low temperatures but exhibited different product selectivities. A similar study of the nickel-exchanged zeolite will be reported elsewhere (*8*).

ZnX and CeX. With both metals, 4 catalysts with different amounts of sodium cations replaced were prepared. The isomerization of 1-butene was studied over each catalyst at a range of temperatures, using catalyst samples weighing 0.1 gram after dehydration. The data from the TG experiments were used to calculate the amounts of hydrated catalysts required to give this weight of anhydrous material. In most cases, the course of the reaction at a given temperature followed the reversible first order equation

$$k_1at/(a - x_e) = \ln (a - x_e) - \ln (x - x_e)$$

Table II. Kinetic Data for *n*-Butene

		1-*Butene Data*			
Catalyst	*% Exch.*	*Rate*[a]	*Activn. Energy*[b]	*Log A*[c]	*Product Ratio*[d]
ZnX-I	70	1.14(73)	40	20.7	0.5
ZnX-II	50	1.53(170)	59	21.3	1.0
ZnX-III	40	1.68(178)	59	21.3	1.3
ZnX-IV	21	1.90(192)	95	25.2	1.2
CeX-I	77	1.17(41)	111	33.0	1.5
CeX-II	68	1.05(70)	88	27.8	1.4
CeX-III	36	1.52(175)	71	22.8	1.2
CeX-IV	15	0.94(177)	93	25.0	1.2

[a] Initial rate of 1-butene disappearance (% min^{-1} at T°C).
[b] Activation energy in kJ $mole^{-1}$.
[c] A in moles sec^{-1} m^{-2}.

where k_1 represents the first order rate constant of the forward reaction, a the initial concentration of 1-butene, x the 1-butene concentration at time t, and x_e the 1-butene concentration at equilibrium. The kinetic data, with Arrhenius parameters, are summarized in Table II. The rate of reaction increased with percentage cation exchange for both ZnX and CeX, as also did the activation energy and the pre-exponential factor for CeX. With ZnX, however, there was a decrease in the latter 2 quantities with increase in zinc content.

Experiments using a higher initial 1-butene pressure resulted in a decrease in percentage rate, but the absolute rate remained effectively constant—*e.g.*, with ZnX-II at 170°C, doubling the pressure changed the initial rate from 1.22×10^{18} to 1.14×10^{18} molecules $minute^{-1}$.

Experiments were carried out with both cerium and zinc catalysts in which the catalysts were reused after outgassing at reaction temperature. Invariably, a slow rate of reaction was observed in the repeat experiment, and an increased preference for *cis*-2-butene in the products was exhibited also. With CeX-I at 41° and 50°C, a 25% decrease in first order rate was noted, and the initial product ratio increased from 1.5 to 1.75. With ZnX-I at 50°C, the initial product ratio was 0.45 in the first experiment, but in the repeat experiment, which was much slower, the ratio was well over unity.

Reactions with Deuterium and Deuterium Oxide. The experiments were undertaken to establish if any exchange reactions (with incorporation of deuterium into the butenes) occurred during the 1-butene isomerization in the presence of deuterium or deuterium oxide. It was thought that the chance of obtaining exchanged products would be greater with deuterium if a radical mechanism were operating and, conversely, greater with deuterium oxide on catalysts favoring an ionic mechanism.

Isomerization over ZnX and CeX

cis-2-Butene Data

Rate[e]	Product Ratio[f]
0.50(94)	0.7
0.40(160)	0.7
0.58(172)	1.2
0.48(180)	1.1
~1.0(50)	1.0
0.46(80)	1.5
0.51(175)	1.1
0.51(190)	0.8

[d] Initial *cis*-2-butene/*trans*-2-butene product ratio.
[e] Initial rate of *cis*-2-butene disappearance (% min^{-1} at T°C).
[f] Initial 1-butene/*trans*-2-butene product ratio.

In the experiments with deuterium, a deuterium/1-butene ratio of 2 was employed. When deuterium oxide was used, the catalyst was allowed to pre-adsorb a small, controlled amount of D_2O (normally approximately one molecule per cage) prior to the admission of the 1-butene. These reaction systems were examined over the least and the highest exchanged forms of the zinc and the cerium zeolites.

The presence of deuterium did not influence the initial product ratio of any of the 4 catalysts examined and led to an increase in the rate of reaction only with ZnX-I—*e.g.*, twice the normal rate at 73°C. In contrast, the presence of deuterium oxide caused an increase in rate of reaction with all 4 catalysts: the factors were 4, 12, 25, and 6 for the catalysts ZnX-I, ZnX-IV, CeX-I, and CeX-IV, respectively, at the temperatures used for the reactions in Table III. The presence of deuterium oxide increased the initial product ratio at 50°C with ZnX-I from 0.5 to 1.0 but did not alter the ratio with the other catalysts.

At the end of the experiments, the reaction products were separated by chromatographic means and the individual species then analyzed mass spectrometrically to establish the extent of exchange. The experimental results are listed in Table III.

Discussion

The parent NaX has been shown to be active for 1-butene isomerization in the temperature range 200°–300°C. LiX has a similar activity, but with all the other zeolites listed in Table I, an enhancement of catalytic activity resulted from the replacement of sodium by other cations.

The initial product distribution resulting from 1-butene isomerization can be a useful guide to the type of reaction mechanism operative.

Table III. Product Distribution for Exchange

Catalytic System	T,°C	*% Conversion*	*Compound*[a]
ZnX-I with D_2	73	73	B1
			TB2
			CB2
ZnX-IV with D_2	190	42	B1
			TB2
			CB2
CeX-I with D_2	50	66	B1
			TB2
			CB2
CeX-IV with D_2	190	35	B1
			TB2
			CB2
ZnX-I with D_2O	50	66	B1
			TB2
			CB2
ZnX-IV with D_2O	170	78	B1
			TB2
			CB2
CeX-I with D_2O	25	58	B1
			TB2
			CB2
CeX-IV with D_2O	160	35	B1
			TB2
			CB2

[a] B1 ≡ 1-butene, TB2 ≡ *trans*-2-butene, and CB2 ≡ *cis*-2-butene.
[b] 0.4% of d_5 species was observed in the 1-butene.

Bond and Wells (*4*) observed cis/trans ratios < 1 for reactions over metal catalysts. Such results have been interpreted in terms of a radical mechanism in which the stability of the various conformations of the adsorbed butyl radical control the ratio. From energy differences, it can be calculated that the cis/trans ratio should be between 0.5 and 0.8 for the temperature range 0°–150°C. On the other hand, studies over acidic-type catalysts (*11, 14*) have suggested that if the reaction intermediate is a secondary butyl carbonium ion, then a cis/trans product ratio $\geqslant 1$ is to be expected with 1-butene as starting material, and a *trans*-1-butene

of Butenes with D_2 and D_2O over ZnX and CeX

Percentage of Isotopic Species				
d_0	d_1	d_2	d_3	d_4
		No exchange found		
94.9	5.1	0		
77.5	18.8	3.7	0	
79.9	20.1	0		
96.6	1.4	2.0	0	
96.7	1.5	1.8	0	
96.5	1.3	2.2	0	
99.8	0.2	0		
97.2	2.8	0		
98.7	1.3	0		
85.3	12.6	1.7	0.4	0
76.5	22.1	1.2	0.2	0
73.3	25.2	1.3	0.2	0
65.0	23.6	8.5	1.8	1.1
47.8	44.7	6.6	0.8	0.1
48.1	43.9	7.0	0.9	0.1
96.7	2.6	0.5	0.2	0
80.5	18.4	0.9	0.4	0
80.4	18.8	0.5	0.3	0
92.0	5.6	0.9	0.6	0.5[b]
46.0	51.2	2.4	0.4	0
47.2	50.3	2.3	0.2	0

product ratio from *cis*-2-butene isomerization $\geqslant 1$. Higher cis/trans product ratios have been interpreted in terms of allyl carbanion intermediates (*19*) where restricted rotation will occur. These differing product ratio values are not by themselves sufficient to define the operative mechanism.

The catalytic activity of zeolites containing monovalent ions has been interpreted in terms of mechanisms involving radical rather than ionic species (*12, 21*). The product ratio with NaX and LiX did not show preferential formation of *trans*-2-butene, as expected for a radical mechanism, but this may be a consequence of the need to use temperatures about 200°C.

The initial product ratios shown in Table I suggest that radical mechanisms could be operative with NiX, ZnX, MnX, and possibly CuX. Adsorption experiments using *cis*- and *trans*-2-butene with NiX (8) showed that both olefins were adsorbed to an equal extent, and thus the greater proportion of the trans-isomer in the initial products is not caused by preferential adsorption of the cis-isomer. With the other zeolites, an ionic type of activity is suggested, and with the exception of ThX (which was only 10% exchanged), there is a qualitative correlation between the activity and the valency of the ion. Monovalent NaX and LiX required temperatures $> 200°C$, divalent MgX, CaX, and CoX were active in the temperature range 100°–200°C, whereas the trivalent CeX, CrX, and LaX were all active at temperatures $< 100°C$.

CeX. The catalytic activity of the four CeX zeolites examined increased with increase in cerium content. The activity pattern suggests that very little cerium is located in surface sites in the lower exchanged forms. If full S_I site occupancy occurred initially, then all the cerium ions could be so located in CeX-III and CeX-IV, but the higher exchanged forms must contain cations in surface sites. Various workers (*10, 17, 20*) have reported that cerium ions migrate into S_I sites upon dehydration.

The initial product ratios are indicative of a carbonium ion type of activity for all the CeX zeolites, and acidic sites have previously been invoked for rare-earth exchanged zeolites. The active sites for isomerization are envisaged as arising through an essentially reversible transfer of a proton from a residual, strongly held water molecule, either solvating the cation or under the influence of its field, to oxygen of the associated AlO_4 tetrahedron. Venuto and coworkers (*22*) have postulated the following scheme:

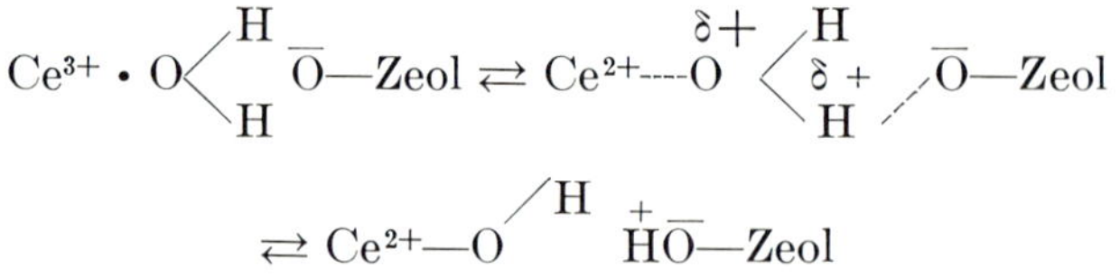

The absence of any effect of deuterium on the rate of reaction and the small extent of deuterium incorporation in the products with CeX suggest that the mechanism is ionic rather than radical. The small amounts of exchange noted could result from the very strong electrostatic fields associated with the Ce^{3+} ion causing some heterolytic splitting of the deuterium molecule. In contrast, the enhanced isomerization rates exhibited in the presence of D_2O provide added confirmation of the protonic activity of these catalysts, as also does the substantial amount of deuterium incorporation. As Table III indicates, the majority of exchange takes place with the products and is therefore assumed to result from the isomerization process.

The highly exchanged cerium catalysts exhibited an initial deceleratory period in the 1-butene isomerization. This may be caused by some strong sorption of the butene molecules which would be expected at the low reaction temperatures used for CeX-I and CeX-II and with the high fields associated with Ce^{3+} ions in surface sites. Such sorption might lead to a reduction in the number of sites for catalysis.

The behavior of CeX in the presence of D_2 and D_2O is in marked contrast to that of NiX. The latter is believed (*8*) to operate *via* a radical mechanism: when deuterium was present, appreciable hydrogenation to butane was observed and the rate of isomerization was markedly enhanced, whereas the rate of isomerization was not altered in the presence of D_2O. Again in contrast to CeX, with NiX there was extensive exchange with deuterium but virtually no deuterium incorporation into the butenes when D_2O was employed.

ZnX. Barry and Lay (*2, 3*), using ESR techniques, have reported that Zn^{2+} has a marked preference for S_{II} type sites rather than the relatively inaccessible S_I sites. The bulk of cations in the four ZnX zeolites used here are therefore likely to be in surface sites, although some migration to S_I sites is likely on dehydration. The increase in catalytic activity with increase in zinc content thus can be accounted for in qualitative terms.

The manner in which the initial product ratio decreases with increase in zinc content suggests that the character of the reaction is changing from ionic to radical, but it is difficult to envisage why an increase in zinc cations should cause this. The fact that ZnX-I when re-used gives an increased ratio of *cis*-2-butene/*trans*-2-butene could be attributed to hydrocarbon residues from the first experiment acting as ion-forming centers (*15*). Such residues could be responsible also for the observed reduction in rate of 1-butene disappearance.

The behavior of ZnX appears to exhibit some of the characteristics of the ionic CeX and some of the radical NiX. The enhancement of isomerization rate for ZnX-I in the presence of deuterium lends some support to the radical nature of this catalyst. Likewise, the lack of enhancement exhibited by ZnX-IV would suggest that the mechanism with this zeolite is ionic. As expected on this basis, more incorporation of D in the products formed on ZnX-IV is found with heavy water than with deuterium. The absence of exchanged products using ZnX-I in the presence of deuterium may be caused by the difficulty of dissociating deuterium on the catalyst at the low temperature used.

The enhancement of rate for both ZnX-I ($\times$ 4) and ZnX-IV ($\times$ 12) on the introduction of the proton donor D_2O strongly suggests that protons generally play an important role in the isomerization process over ZnX zeolites. This is supported by the fact that with ZnX-I the initial product ratio was raised to 1.0 in the presence of D_2O, and also that over both ZnX zeolites substantial amounts of deuterium were incorporated into the butenes.

Ward (*24*) has suggested that with ZnX the active centers are essentially Bronsted acid sites associated with residual hydroxyl groups. Partial hydration could promote movement of cations and result in the formation of additional such sites. The activity of the ZnX-IV zeolite can be explained satisfactorily in this manner, but with ZnX-I other factors must be involved. It is possible that with the higher exchanged zeolite, the increase in surface cations provides stronger adsorption sites for the butenes. If this were the case, then a slow rate of desorption of the products (relative to the surface reaction) could allow the 2-butene product ratio to approach that of thermodynamic equilibrium (*13*).

Acknowledgment

One of us (N.E.C.) acknowledges the financial support of the S.R.C.

Literature Cited

(1) Barrer, R. M., Bultitude, F. W., Sutherland, J. W., *Trans. Faraday Soc.* **1957,** 53, 1111.
(2) Barry, T. I., Lay, L. A., *J. Phys. Chem. Solids* **1966,** 27, 1821.
(3) Barry, T. I., Lay, L. A., *J. Phys. Chem. Solids* **1968,** 29, 1395.
(4) Bond, G. C., Wells, P. B., *Advan. Catalysis* **1964,** 15, 92.
(5) Breck, D. W., Eversole, W. G., Milton, R. M., *J. Am. Chem. Soc.* **1956,** 78, 2338.
(6) Broussard, L., Shoemaker, D. P., *J. Am. Chem. Soc.* **1960,** 82, 1041.
(7) Cross, N. E., Ph.D. Thesis, University of Edinburgh, 1969.
(8) Cross, N. E., Kemball, C., Leach, H. F., to be published.
(9) Dimitrov, Chr., Leach, H. F., *J. Catalysis* **1969,** 14, 336.
(10) Egerton, T. A., Ph.D. Thesis, University of Bristol, 1968.
(11) Foster, N. F., Cvetanovic, R. J., *J. Am. Chem. Soc.* **1960,** 82, 4274.
(12) Frilette, J., Weisz, P. B., Golden, R. L., *J. Catalysis* **1962,** 1, 301.
(13) Hightower, J. W., Gerberich, H. R., Hall, W. K., *J. Catalysis* **1967,** 7, 57.
(14) Hightower, J. W., Hall, W. K., *J. Catalysis* **1966,** 5, 99.
(15) Hightower, J. W., Hall, W. K., *J. Am. Chem. Soc.* **1967,** 89, 778.
(16) McCosh, R., Ph.D. Thesis, University of Edinburgh, 1968.
(17) Olson, J. A., Kokotailo, G. T., Charnell, J. F., *Nature* **1967,** 215, 270.
(18) Pickert, P. E., Rabo, J. A., Dempsey, E., Schomaker, V., *Proc. Intern. Congr. Catalysis, 3rd,* **1965,** 1, 714.
(19) Pines, H., Schaap, L. A., *Advan. Catalysis* **1960,** 12, 117.
(20) Rabo, J. A., Angell, C. L., Schomaker, V., *Intern. Congr. Catalysis, 4th,* **1968,** preprint 54.
(21) Rabo, J. A., Pickert, P. E., Stamires, D. N., Boyle, J. E., *Proc. Intern. Congr. Catalysis, 2nd,* **1960,** 2, 2055.
(22) Venuto, P. B., Hamilton, L. A., Landis, P. S., *J. Catalysis* **1966,** 5, 484.
(23) Venuto, P. B., Landis, P. S., *Advan. Catalysis* **1968,** 18, 259.
(24) Ward, J. W., *J. Catalysis* **1969,** 14, 365.

RECEIVED January 23, 1970.

Discussion

W. K. Hall (Gulf Research & Development Co., Pittsburgh, Pa. 15230): How sure are you that the Ni^{2+} and Zn^{2+} of catalysts containing these base-exchange cations were not partially reduced? Would not small amounts of these metals effect the same general pattern of results?

H. F. Leach: We do not think that reduction of Ni^{2+} and Zn^{2+} is very likely under the experimental conditions used. The observed color changes support this, but it is not possible to state categorically that there was no reduction to the metal.

When a NiX sieve was deliberately reduced to Ni metal (using hydrogen at 370°C), the same general pattern of a low cis/trans ratio was observed.

P. B. Venuto (Mobil Oil, Paulsboro, N. J. 08066): Even though your butene reactions occur at low temperatures—relative to thermal free radical reactions—with the presence of polyvalent transition metal ions, radical-type or electron unpairing-type reactions may certainly occur. In your reactions of butene over CeX at 190°C, did you notice any evidence of an intermolecular hydrogen (hydride) type reaction, as evidenced by light paraffins in the gas phase and hydrogen-deficient species (aromatics) within pores?

H. F. Leach: In none of our reactions did we detect the presence of light paraffins in the gas phase, and we have no evidence for aromatic species being present within the catalyst.

J. W. Ward (Union Oil Co. of California, Brea, Calif. 92621): Could you provide more information regarding the characterization of your catalysts, in particular, the Ni, Cu, Cr, and Th exchanged forms? Is not the interpretation of the data complicated by the presence of decomposition products?

H. F. Leach: As mentioned briefly in the paper, the catalysts were examined by x-ray diffraction, and for all the samples used there was no apparent loss of crystallinity relative to the parent NaX. The catalysts were also characterized by measurement of surface area; the value for NaX was 870 m^2gram^{-1} and for all the other zeolites lay within the range 840–900 m^2gram^{-1}. We considered that these results indicated that very little decomposition of the zeolite to amorphous silica or silica–alumina occurred.

71

Physical and Catalytic Properties of the Zeolite Mordenite

B. W. BURBIDGE,[1] I. M. KEEN, and M. K. EYLES

British Petroleum Co., Ltd., Research Centre, Sunbury-on-Thames, Middlesex, England

Extensive studies have shown that mordenite is a remarkably versatile catalytic material. Some of the properties which probably govern adsorption and diffusion in and catalytic activity of this zeolite are reviewed briefly. Reactions which can be achieved over mordenite-based catalysts are selective hydrocracking of normal and near-normal paraffins, hydrogenation, disproportionation, and isomerization. Examples are given which illustrate its potential in petroleum refining technology. These include the selective removal of normal paraffinic wax by a hydrocracking operation from high-boiling petroleum distillates to produce diesel, fuel, and lubricating oil components of improved flow properties at low temperatures.

Catalysts based on zeolites are finding increasing application throughout the chemical industry, in particular in the petroleum and petrochemical sectors. Their catalytic properties derive partly from their regular structure with its regular pore characteristics, partly from the large, highly polar surface which they present to reactant molecules, and partly from their robust nature.

At the first International Conference on Molecular Sieves, Mays and Pickert described many aspects of zeolite catalysts, referring mainly to X- and Y-types (*26*). Since then, more data have become available on the catalytic properties of mordenite. These indicated that mordenite-based catalysts, like those of zeolite-Y, participated in carbonium-ion

[1] Present address: Society Francaise des Petroles BP, Lavera Refinery, pres Martigues, Bouche de Rhone, France.

reactions. Our aim in writing this paper is to review the general physical and catalytic properties of mordenite-based catalysts, to make comparison with catalysts based on faujasite-type zeolites, and to describe a remarkable selectivity of precious-metal hydrogen mordenites.

Structure and General Properties

Mordenite is a naturally-occurring silica-rich zeolite (Si/Al $\sim$5) which also can be synthesized readily (*34*). Meier determined the structure of its fully hydrated sodium form (*28*). The mordenite framework is characterized by a micropore system composed essentially of parallel elliptical cylinders of maximum and minimum crystallographic free diameters of 7.0 and 5.7A, respectively, these main channels being interconnected by smaller side channels.

The mordenite framework has high thermal stability and is much more resistant to acids than that of zeolite Y. About 80 % wt of the (AlO_4) tetrahedra in sodium mordenite can be hydrolyzed out with strong acid without significant loss in crystallinity or contraction of the interplanar (*d*-) spacings (*17*). It is thought that the (AlO_4) tetrahedra are replaced by $(OH)_4$ tetrahedra of about the same size as the (SiO_4) ones (*27*). "Decationized" mordenites and Y-type zeolites each contain both Bronsted and Lewis acid sites (*5*). All the acid sites in normal and dealuminated mordenites can be detected wholly from the infrared spectra of adsorbed pyridine (*14, 15*). Compared with hydrogen-Y, hydrogen–mordenite has a maximum of about half the number of potentially exchangeable protons per unit surface area—*i.e.*, 4 to 5 microequivalents (H^+/m^2)—but contains some stronger acid sites (*5*).

Adsorption and Diffusion in Mordenites

The usefulness of mordenite-based catalysts will depend on the extent and character of adsorption and rates of diffusion of reactant molecules and products in the mordenite base.

Early adsorption studies indicated that the effective micropore openings in natural and synthetic mordenites were about 4A (*2*) and not *ca.* 7A as required by structural data. Later it was shown that this probably was caused by partial blocking of the main channels by calcium and/or sodium cations (*3, 22*). Sorption in the side channels is limited normally to molecules smaller than *n*-butane (4.9A) (*3*), while the main channels in hydrogen–mordenite are nearly filled by the larger cyclohexane (*22*), benzene, and neopentane molecules (*3*) of critical diameters 6.6, 6.7, and 7.1A, respectively. Recently, it has been claimed that on dealumination, the effective micropore diameter of hydrogen–mordenite is increased to about that of zeolite-Y (*13*), enabling sorption of cumene and 1,3,5-triethylbenzene of critical diameters *ca.* 7.6 and 8.5A, respectively (*33*). The resulting cumene adsorption capacity passes through a maximum for hydrogen–mordenite of *ca.* 19:1 molar silica/alumina ratio. Mordenites also adsorb isoprenoids—*e.g.*, 2,6,10,14-tetramethylhexadecane—from mixtures with cyclics such as steranes and triterpanes, in "denormalized" shale oils (*16*).

Diffusion of permanent gases in zeolites is an activated process, activation energy increasing with gas molecular weight to, for example, 15 kcal/mole for SF_6 in sodium–mordenite (*18*). Kr diffuses less rapidly in mordenites than in either 5A or Y-zeolites (*18*) but *n*-octane diffuses slightly more rapidly in mordenite than 5A-zeolite (*4, 19*).

The activation energy for diffusion of a given adsorbate in mordenite is influenced by the cation present, being low for Li^+ and high for NH_4^+. K^+ ions partly obstruct mordenite's main channels, creating energy barriers sensitive to molecular size—*e.g.*, hydrogen (2.4A diameter) diffusing 10^5 times as fast as Kr (3.9A) in K-mordenite at −78°C (*2*). Diffusion coefficients for C_1 to C_4, C_8, and C_{10} *n*-paraffins, toluene, and decalin in mordenites are much smaller than those of inert gases (*35*), while cyclo- and isoparaffins diffuse more slowly than *n*-paraffins of the same carbon number (*4, 36*). Mordenite's relatively high resistance to gaseous diffusion can be reduced significantly by extensive dealumination which reduces the framework charge. Thus, hydrogen–mordenites of from 25 to 90:1 silica/alumina ratios are claimed to adsorb selectively paraffins or cycloparaffins from mixtures with aromatics (*20*).

Catalytic Reactions of Low Molecular Weight Hydrocarbons

Hydrogen and multivalent-ion exchanged zeolites—*e.g.*, chabazite, faujasite, gmelinite, and mordenite—have rate constants for *n*-hexane cracking of over 10^4 times greater than commercial silica–alumina cracking catalyst, compared at the same temperature (*29*). The cracking activities of hydrogen–mordenites for *n*-C_6 and *n*-C_{10} paraffins increase with silica/alumina ratio and pass through maxima for ratios of from 17 to 19:1 (*32, 40*). Highly active mordenite cracking catalysts are of short life since they "coke-up" at several times the rate of silica-aluminas (*1*). Hydrogen–mordenite is more active than hydrogen–Y-zeolite for cracking of *n*-C_6 and *n*-C_7 paraffins (*31*).

Hydrogen–mordenites are active at 200°–260°C for the isomerization of *n*-butane, cyclohexane (*22*), and *o*-xylene (*23*). They also extensively crack cycloparaffins (*31*) and alkyl aromatics (*32*) under more severe conditions.

Under hydroprocessing conditions, precious-metal hydrogen mordenites are, like zeolite-Y-based catalysts, active for carbonium-ion type reactions involving dual function catalysts. With paraffinic fractions, particularly C_5 and C_6, the principal reaction is isomerization, for which a precious metal (Pt or Pd) exchanged decationized-mordenite is more active than similar catalysts based on Y-type zeolite or silica–alumina. Thus, Pt and Pd–H-mordenites will isomerize pentane and hexane at temperatures in the range 260°–330°C (*37, 39*), whereas we find Pd–H-Y-type only becomes active at temperatures in the range 340°–360°C. The mordenite catalyst, however, is not as active for paraffin isomerization as catalysts currently in use in commercially operating units for C_5/C_6 isomerization which operate at temperatures below 150°C.

Pd–H-mordenite also is reported to be more active than Pd–H-Y-type zeolite for liquid-phase isomerization of cyclohexane to methylcyclopentane (*21*).

Under more severe conditions, Pd or Pt–H-mordenite hydrocracks both paraffins and cycloparaffins, again being more active than the corresponding Y-type zeolite catalyst. We have found that n-C_6 and n-C_7 are converted essentially to C_3 and i-C_4, whereas n-C_{10} gives i-C_4 and i-C_5 primarily. Although n-C_6 and n-C_{10} paraffins are hydrocracked more rapidly than the corresponding cycloparaffins (cyclohexane and decalin), with binary mixtures of paraffin and the corresponding cycloparaffin, the latter are removed preferentially (*4, 24, 25*).

With alkyl aromatics, precious-metal H-mordenite catalysts are active for hydrogenation at low temperatures and hydrocrack at higher temperatures. Certain metal exchanged mordenites are effective for hydrogenation (*30*), dealkylation (*7*), transalkylation, disproportionation (*31, 38*), and isomerization reactions (*23*).

With light hydrocarbon mixtures, mordenite-based catalysts are capable of giving a variety of results, depending on the severity of the treatment. At relatively low temperature and with a moderate partial pressure of hydrogen, the catalyst will hydrogenate olefins and aromatics, but with little isomerization and cracking of the hydrogenated products. As the temperature is raised, the catalyst begins to isomerize the paraffins in addition to hydrogenating the aromatics and then, when isomerization has reached an equilibrium level, hydrocracking reactions set in. At first, cycloparaffins tend to be attacked preferentially but as the severity is further increased, all hydrocarbon types are broken down. Any sulfur present in the fraction is converted to hydrogen sulfide, excessive amounts, however, causing a lowering in the other reaction rates.

Catalytic Reactions of Higher Molecular Weight Hydrocarbons

In contrast to its superior activity with low molecular weight hydrocarbons, Pt–H-mordenite is less active than precious metal Y-types (*e.g.*, Pd–H-Y) for hydrogenating higher (polynuclear) aromatics or hydrocracking higher *n*-paraffins—*e.g.*, n-C_{16}, n-C_{18}—giving more isomerization. However, when complex mixtures of these higher *n*-paraffins with heavier hydrocarbons are hydrotreated, Pd and Pt–H-mordenites, unlike zeolite-Y-catalysts, hydrocrack selectively the *n*-paraffinic (and near n-paraffinic) molecules (*10, 11*). Since such molecules constitute a substantial part of the wax in heavy distillates from crude oils ("wax distillates"), catalytic hydrotreatment over a mordenite-based catalyst could be used to dewax such fractions with a view to improving their flow properties at low temperature. Possible applications of mordenite dewaxing therefore include preparation of low-pour-point fuel and diesel oils, lubricating oils, extenders, plasticizers, and process oils, and flexible microcrystalline wax.

The hydrotreatments are carried out at pressures in the range 250 to 2500 psig and temperatures from 230° to 510°C. The exact conditions used depend on the feedstock and extent of dewaxing required. The products from the hydrocracking of the wax in heavy distillate feedstocks boil below 177°C with C_3, C_4, and C_5 hydrocarbons predominating.

Fuel and Diesel Oils. As legislative pressure to reduce the sulfur content of fuel oils increases, new techniques are being sought to produce low-sulfur fuel oils. Unfortunately, many low-sulfur crude oils have high wax contents (*e.g.*, Libyan, Nigerian). Catalytic dewaxing of these low-sulfur, high-pour-point materials offers an alternative to desulfurization of high-sulfur, low-pour-point materials of, say, Middle East origin for the preparation of fuel oils.

While it is possible to treat an atmospheric residue from a Libyan crude direct, trace contaminants, particularly sodium, tend to poison the catalyst. However, this atmospheric residue can be vacuum-distilled to give a *ca.* 50 wt % yield of a 350°–550°C distillate containing 34 wt % wax, 72% of which is normal-paraffinic, and which is essentially free of catalyst poisons. This material can be catalytically dewaxed with Pt–H-mordenite in a *ca.* 70 wt % yield, resulting in a reduction of its pour-point from 115°F to values in the range 30° to 60°F. This low-pour-point material can be blended back with the vacuum residue, which contains the catalyst poisons and only *ca.* 2 wt % *n*-paraffinic wax, to form a low-sulfur, low-pour-point fuel oil blending component (*6*). With this type of operation, catalyst activity can be maintained for long periods.

Low-pour-point diesel oil may be produced similarly. If this is required to be of low sulfur content, it is feasible to combine the Pt–H-mordenite dewaxing with a conventional CoMo–alumina catalytic desulfurization step using a double catalyst bed (*9*).

Lubricating Oils. In conventional lubricating oils production, waxy distillates are refined in a series of steps which are solvent refining, solvent dewaxing, and finishing treatments (catalytic or adsorbent) for color and oxidation stability improvement. If in this sequence solvent dewaxing is replaced by mordenite catalytic dewaxing, a lower yield of oil is obtained as well as loss in viscosity index (VI), probably owing to catalytic removal of nonwaxy near-normal paraffins of high VI. However, on mordenite dewaxing the wax distillate prior to solvent refining, the selectivity for *n*-paraffins is improved and, after solvent refining, the oils obtained are similar in yield and properties to those from the "all-solvent" route. These are of satisfactory color and color stability, so no finishing treatment is required. This scheme has proved adequate for the lighter grades of lubricating oil (*12*) where the wax is essentially *n*-paraffinic. With the heavier grades, the proportion of *n*-paraffinic material in the wax decreases, *i.e.*, wax becomes more "microcrystalline," and catalytic dewaxing is inadequate in itself. Catalytic treatment may, however, still be of use even with these grades if it is followed by a solvent dewaxing step since the latter may be accomplished more easily and there is less oil entrained in the wax removed during the solvent dewaxing step.

Extenders, Plasticizers, and Process Oils. Materials in this class can be prepared from the extracts obtained when narrow wax distillate cuts are solvent-refined—with furfural, for example—in the production of lubricating oils. These materials must have a low pour-point. Hydrocatalytic treatment over a mordenite-based catalyst removes residual *n*-paraffinic wax with consequent reduction in pour-point of the product, which is obtained in good over-all yield.

Flexible Microcrystalline Wax. Microcrystalline wax is an essentially isoparaffinic wax obtained during the preparation of the heaviest grades of lubricating oil. Unlike the *n*-paraffinic waxes which are brittle, isoparaffinic microcrystalline wax is flexible, although its flexibility can be impaired by small amounts of *n*-paraffins.

The unique selectivity of mordenite can be used to remove these residual *n*-paraffins by hydrocracking, and the resultant microcrystalline wax has much improved flexibility (*8*).

Parameters Determining Selectivity. We believe that the peculiar selectivity of Pt–H-mordenite for hydrocracking normal and near-normal paraffins in high-boiling feedstocks could not have been predicted from the known adsorption and diffusion properties of mordenites (*loc. cit.*). However, extensive catalytic studies on the preparation of low-pour-point petroleum fractions have suggested to us that catalyst selectivity depends upon certain catalyst compositional parameters which have not been published yet in detail.

Optimum selective wax removal for diesel, fuel, and lubricating oil grades appears to occur for mordenite silica/alumina ratios in range *ca.* 16 to 20:1 (*11*), in reasonable agreement with the optimum noted previously for adsorption and *n*-hexane cracking studies. Catalyst selectivity for *n*-paraffins also appears to depend on certain feedstock compositional parameters. This is indicated by the improvement in selectivity observed on reversing solvent refining and catalytic dewaxing in the scheme for lubricating oil preparation mentioned previously.

Summary

In this paper we have tried to survey the catalytic properties of mordenite, making comparison where possible with faujasite-type zeolites. Mordenite shows a remarkable selectivity for *n*-paraffinic material in heavy distillates, and we have shown how this property might be used in a number of petroleum-refining applications.

Acknowledgment

Permission to publish this paper has been given by The British Petroleum Co., Ltd.

Literature Cited

(1) Adams, C. E., Kimberlin, J. R., Shoemaker, D. P., *Proc. Intern. Congr. Catalysis, 3rd, Amsterdam* **1964,** 2, 1310.
(2) Barrer, R. M., *Trans. Faraday Soc.* **1949,** 45, 358.
(3) Barrer, R. M., Peterson, D. L., *Proc. Roy. Soc. (A)* **1964,** 280, 466.
(4) Beecher, R., Voorhies, A., Jr., Eberly, P., Jr., *Ind. Eng. Chem. Prod. Res. Develop.* **1968,** 7, 203.
(5) Benesi, H. A., *J. Catalysis* **1967,** 8, 368.
(6) BP Co., Ltd., Belgium Patent **726,746** (1969).
(7) *Ibid.,* **734,099** (1969).
(8) BP Co., Ltd., French Patent **1,542,316** (1967).
(9) BP Co., Ltd., South African Patent **69/6671** (1970).
(10) BP Co., Ltd., U.K. Patent **1,088,933** (1967).
(11) *Ibid.,* **1,134,014** (1968).
(12) *Ibid.,* **1,134,015** (1968).
(13) Breck, D. W., Flanigen, E. M., "Molecular Sieves," p. 47, Society of the Chemical Industry (London), 1968.
(14) Cannings, F. R., *J. Phys. Chem.* **1968,** 72, 4691.
(15) Cannings, F. R., unpublished data.
(16) Curran, R., *et al., Tetrahedron Letters* **1968,** 14, 1669.
(17) Dubinin, M. M., Federova, G. M., *et al., Trans. Acad. Sci. USSR, Phys. Chem.* **1968,** 11, 2429.
(18) Eberly, P. E., Jr., *Ind. Eng. Chem. Fundamentals* **1969,** 8, 25.
(19) Eberly, P. E., Jr., *Ind. Eng. Chem. Prod. Res. Develop.* **1969,** 8, 140.

(20) Esso Research, Dutch Published Application, **6,900,715** (1969).
(21) Esso Research, U.K. Patent **1,143,140** (1969).
(22) Frilett, V. J., Rubin, M. K., *J. Catalysis* **1965,** 4, 310.
(23) Hansford, C. R., Ward, J. W., *J. Catalysis* **1969,** 13, 316.
(24) Hatcher, W. J., Jr., *Dissertation Abstr.* **1968,** B29, 589B.
(25) Hatcher, W. J., Voorhies, A., Jr., AIChE, 61st Ann. Meeting, **1968,** Preprint 20A.
(26) Mays, R. L., Pickert, P. E., "Molecular Sieves," p. 112, Society of the Chemical Industry (London), 1968.
(27) Meier, W. M., "Molecular Sieves," p. 20, Society of the Chemical Industry (London), 1968.
(28) Meier, W. M., *Z. Krist.* **1961,** 115, 439.
(29) Miale, J. N., Chen, N. Y., Weisz, P. B., *J. Catalysis* **1966,** 6, 278.
(30) Minachev, Kh. M., *et al., Izv. Akad. Nauk SSSR, Ser. Khim.* **1969,** 2, 481.
(31) Mobil Oil Corp., U.K. Patent **1,022,687** (1966).
(32) Norton Int. Corp., U.K. Patent **1,008,596** (1965).
(33) Pignzova, L. I., *et al., Kinetics Catalysis* **1969,** 10, 252.
(34) Sand, L. B., "Molecular Sieves," p. 71, Society of the Chemical Industry (London), 1968.
(35) Satterfield, C. N., *et al., AIChE J.* **1967,** 13, 731.
(36) Satterfield, C. N., Margetts, W. G., *Symp. Recent Advances Kinetics, Part II, Catalysis, 62nd Ann. Meeting AIChE,* **1969,** Reprint 56d.
(37) Shell International, Dutch Published Application, **6,807,837** (1968).
(38) Shell International, U.K. Patent **1,039,246** (1966).
(39) Shell International, U. S. Patent **3,190,939** (1965).
(40) Weller, S. W., Brauer, J. M., *Symp. Recent Advances Kinetics, Part II, Catalysis, 62nd Ann. Meeting, AIChE,* **1969,** Reprint 56c.

RECEIVED February 4, 1970.

Discussion

P. B. Venuto (Mobil Oil Corp., Paulsboro, N. J. 08066): Did the specific mordenite samples that were used to generate the hydrotreating-type data sorb 1,3,5-triethylbenzene ($\sim$ 8.5 Å)?

I. M. Keen: We did not specifically test these mordenite samples for sorption of 1,3,5-triethylbenzene. However, the catalyst silicon:aluminum atomic ratios used were in the range where both cumene (critical diameter $\sim$7.6 Å) and 1,3,5-triethylbenzene would be expected to be sorbed (Piguzova, L. I. *et al., Kinetics Catalysis* **1969,** 10, 252).

A. E. Hirschler (Sun Oil Co., Marcus Hook, Pa. 19061): You report on one of your slides that at 300°C the extent of isomerization of *n*-pentane was 65% *vs.* 15% for *n*-hexane. Usually, *n*-hexane is more readily isomerized, or more extensively isomerized, than *n*-pentane. How do you account for the reverse situation which you have reported and the magnitude of the difference between C_5 and C_6 paraffins? (The subject slide is not included in the paper.)

I. M. Keen: The slide data were given to show that mordenite-based catalysts are more active than their Y-type counterparts for isomerizing C_5 and C_6 paraffins. The feedstock used in this comparison was a commercial pentane/hexane blend, containing some isoparaffins. The C_5 and C_6 isomerization levels in this instance are defined as the percent weight of i-C_5 and 2,2-dimethylbutane (2,2-DMB) in the respective total C_5 and C_6 fractions. On this basis, the C_5 and C_6 isomerization conversion levels in the feedstock were *ca.* 35 and 2 wt %, respectively.

The data given for a reaction temperature of 300°C clearly showed the mordenite catalyst to be the more active for isomerization of both the C_5 and C_6 fractions. Conversions quoted were: precious-metal–H–mordenite: C_5 ~65 wt %, C_6 ~15 wt %; precious metal–H–Y: C_5 ~40 wt %, C_6 ~4 wt %. These data suggest that the pentane fraction may be slightly easier to isomerize over mordenite than the hexane. The equilibrium conversions to isopentane and 2,2-DMB at 300°C are in the vicinity of 65 and 18 wt %, respectively. A possible explanation is that impurities present—*e.g.*, cyclohexane and/or benzene—affect the rate of 2,2-DMB formation more than that of the isopentane.

H. A. Benesi (Shell Development Co., Oakland, Calif. 94608): With regard to the hydrogenation of olefins and aromatics over mordenite catalysts, wouldn't you agree that it is the noble metal rather than the mordenite support that is the active catalyst component? In this connection, what is the degree of dispersion of the noble metal? What proportion of the metal is on the outside surfaces of the mordenite crystals?

I. M. Keen: With regard to your first question, I would agree with you that the noble metal is the active catalyst component for hydrogenating olefins and aromatics, the zeolite acting as a high-surface-area support for the metal.

Secondly, the method of catalyst preparation *via* ion exchange ensures an initially high dispersion of noble metal. The catalyst is pre-calcined in air at 500°C and pre-reduced in hydrogen at *ca.* 400°C. In this form, some large metal crystallites have been observed by electron microscopy on the outside of the mordenite crystals. However, it is not possible at the moment to say what proportion of the metal is on the exterior surface of the crystals owing to difficulties in getting representative samples for examination. I would like to suggest that the fact that precious metal mordenites dewax raw (*i.e.*, not denitrogenated or desulfurized) vacuum distillates is secondary evidence for the dispersion of at least some of the metal on the interior surface of the mordenite. Apparently the noble metal is not poisoned by the sulfur- (or nitrogen-) containing hydrocarbons present.

J. W. Ward (Union Oil Co. of California, Brea, Calif. 92621): Could you detect platinum by x-ray diffraction on the Y faujasite catalysts?

I. M. Keen: We have detected little platinum in fresh Y faujasite catalysts by x-ray diffraction techniques. Preliminary investigations by other techniques—*e.g.*, electron microscopy, hydrogen chemisorption—have, however, suggested the presence of some largish crystallites of platinum, presumably on the surface/edges of the zeolite base crystals.

72

Cyclohexane Hydroisomerization Over Palladium–H–Mordenite

ALEXIS VOORHIES, JR., and J. R. HOPPER[1]

Louisiana State University, Baton Rouge, La.

A palladium–hydrogen–mordenite catalyst with a 10.8/1 silica/alumina mole ratio was evaluated for the hydroisomerization of cyclohexane. The rate of reaction followed a first-order, reversible reaction between cyclohexane and methylcyclopentane. The energy of activation for this reaction between 400° and 500°F was 35.5 ± 2.4 kcal/mole. Cyclohexane isomerization rates decreased with increasing hydrogen and cyclohexane-plus-methylcyclopentane partial pressure. These effects are compatible with a dual-site adsorption model. The change of the model constants with temperature was qualitatively in agreement with the expected physical behavior for the constants.

Catalytic reactions are the foundation of modern petroleum technology. These reactions are used for upgrading the octane numbers of gasoline, for converting gaseous material to high octane number liquid fuels, for cracking gas oils to more volatile hydrocarbons, for synthesizing petrochemicals, and many others. Isomerization reactions, singly or combined with other catalytic reactions, help achieve many of these desired conversions. Present-day isomerization catalysts generally have been of 2 types: the acid halide type (*2, 3*) and the "dual-function" type, which consists of a dispersed metal on a high-surface-area base. This base may be an amorphous solid (*1*), but more recently, crystalline zeolite bases have been employed (*6, 8, 9*).

The catalyst used for this study is of the more recent type of dual-function catalysts, wherein the base is a crystalline zeolite, mordenite. Palladium metal was dispersed on this support. A naphthenic isomeriza-

[1] Present address: Lamar State College of Technology, Chemical Engineering Department, Beaumont, Tex.

tion reaction was used to study the isomerizing properties of this mordenite, and a kinetic model for cyclohexane isomerization was developed.

Experimental

This study was performed in the Petroleum Processing Laboratories, Chemical Engineering Department, Louisiana State University. The project was sponsored by Esso Research and Engineering Co. The catalyst used was prepared by the Esso Research Laboratories, Humble Oil and Refining Co., Baton Rouge, La., from mordenite crystals obtained from the Norton Co. Ammonium mordenite was impregnated with 0.5% of palladium, pilled, crushed, sized, and heated to 1000°F in the presence of air to give Pd on H-mordenite catalyst. The properties of this catalyst are shown in Table I.

Table I. Pd–H–Mordenite Properties

SiO_2, wt. %	83.9
Al_2O_3, wt. %	13.3
SiO_2/Al_2O_3, mole ratio	10.8/1
Sodium, wt. %	0.05
Palladium, wt. %	0.55
Langmuir surface area, M^2/gram	532

The reactor system consisted of a 0.62-inch diameter fixed-bed reactor containing 15 cc of catalyst. The reactor was operated under isothermal and plug flow conditions. A fluidized sand bath was used to maintain isothermal conditions. Liquid cyclohexane (Phillips pure grade) was fed by a Ruska pump, and dry electrolytic hydrogen was fed through an orifice flowmeter. The product gas was measured with a wet test meter, and the product liquid was weighed after being separated. Both products were analyzed by a gas chromatograph.

Kinetic Model

Initial variable studies had shown that gas-to-particle mass transfer and intra-particle diffusion were not rate limiting. The reaction mechanism was assumed to follow a first-order reversible reaction. After confirming this assumption, the effect of temperature and pressure on this reaction was investigated by determining the effect on the rate constant.

The first order kinetic model was based on the following reaction, with k and k' the forward and reverse reaction rate constants, respectively:

$$\text{Cyclohexane (CH)} \underset{k'}{\overset{k}{\rightleftarrows}} \text{Methylcyclopentane (MCP)}$$

The rate of reaction, expressed as gram-moles of MCP produced per hour (N_m) per gram at catalyst (W_c), is represented by:

$$\text{rate} = \frac{dN_m}{dW_c} = k(C_n - C_m/K) \tag{1}$$

with C_n and C_m the concentration of CH and MCP, and K the thermodynamic equilibrium constant.

Integration of this equation results in a first order reaction expression in terms of the hydrogen-free mole fraction of MCP (Y_m) and the corresponding equilibrium value (Y_E), ρ_G the molar density of reactor gases, M the molecular weight of MCP, p the partial pressure of CH, P the total pressure, and W the weight of feed per hour per weight of catalyst:

$$\ln(1 - Y_m/Y_E) = -k\rho_G Mp/PWY_E \tag{2}$$

which can be used to determine a value for k. Equation 3 gives the relationship between the reactor composition and the contact time:

$$-\ln(1 - Y_m/Y_E) = \text{constant}/W \tag{3}$$

Figure 1 shows the assumed reaction order to be valid over the feed rate range of 1.6 to 8.1 grams of cyclohexane per hour per gram of catalyst.

Effect of Temperature on Over-All Rate

The effect of temperature on the over-all isomerization rate constant was determined using Equation 2. The temperature range was from 400° to 500°F; all other conditions were held constant. An activation energy of 35.5 ± 2.4 kcal/mole was determined from the Arrhenius-type plot shown in Figure 2. Values of 30 (*4*) and 35.4 (*5*) kcal/mole have been reported for temperature ranges of 600°–700° and 700°–800°F, respectively, for WS_2 and MoS_2 catalysts. The high activity of mordenite relative to these catalysts is apparent.

Effect of Pressure

In most gas–solid heterogeneous catalyst systems, the effect of pressure often is correlated with an adsorption model of the Langmuir-Hinshelwood type. The over-all rate constant for the first order reaction is related to the adsorption model constants by

$$k = k_oK_A/(1 + p_AK_A + p_RK_R + p_HK_H)^n \tag{4}$$

where k_o is the rate constant for the adsorption model, and p_AK_A, p_RK_R and p_HK_H are the partial pressure–adsorption constant products for CH, MCP, and hydrogen, respectively. The surface reaction for a reversible

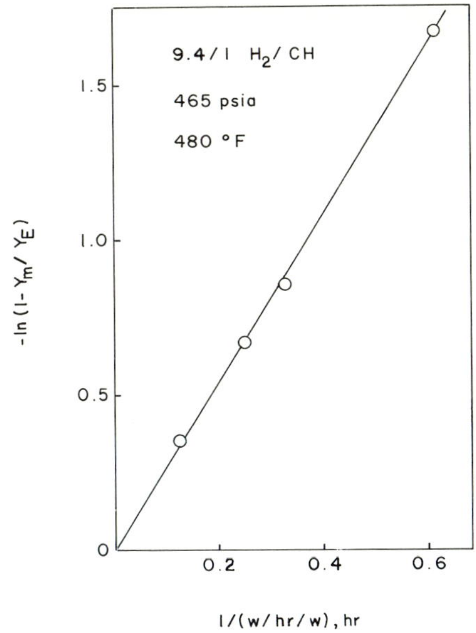

Figure 1. Reversible first order reaction

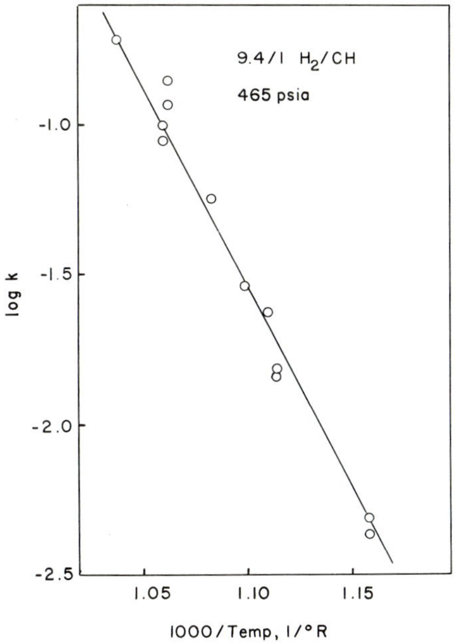

Figure 2. Arrhenius plot

unimolecular reaction, such as an isomerization reaction, generally proceeds by 2 possible mechanisms (7). These 2 surface reactions have been described as the "single-site" and the "dual site" surface reaction mechanisms and correspond to a value of 1 and 2 for the exponent "*n*" in Equation 4. A single-site mechanism describes a reaction involving only a single active site, whereas a dual-site mechanism describes a reaction of an adsorbed molecule only if it is adjacent to an unoccupied active site. In applying the adsorption model equations, a reasonable assumption that the adsorption constants for the hydrocarbons are essentially equal has been made—*i.e.*, $K_A = K_R = K_o$. The expressions in Equation 4 can be reduced then to:

single-site

$$\frac{1}{k} = \frac{1}{k_o K_o} + \left(\frac{1}{k_o}\right) p_o + \left(\frac{K_H}{k_o K_o}\right) p_H \tag{5a}$$

dual-site

$$\frac{1}{\sqrt{k}} = \frac{1}{\sqrt{k_o K_o}} + \left(\frac{\sqrt{K_o}}{\sqrt{k_o}}\right) p_o + \left(\frac{K_H}{\sqrt{k_o K_o}}\right) p_H \tag{5b}$$

Experiments were made at various partial pressures of cyclohexane and hydrogen, but at constant total pressure. Experiments were also made at constant partial pressure of hydrocarbon or hydrogen. The contact time was held constant in this latter case by holding the ratio of hydrocarbon partial pressure to feed rate constant. The rate decreased with an increase in total pressure and with an increase in either hydrocarbon or hydrogen partial pressure. An example of this data taken at 460° is shown in Tables II and III.

Table II. Effect of Total Pressure

Temperature, °F	—	460	—
Total pressure, psia	285	390	465
H_2/cyclohexane, mole ratio	—	9.40	—
k, cc/gram-sec	0.0759	0.0465	0.0375

Table III. Effect of Partial Pressure, 460°F

Total pressure, psia	465	485	665	685
H_2 partial pressure, psia	420	420	620	618
Hydrocarbon partial pressure, psia	45	65	45	67
k, cc/gram-sec	0.0375	0.0244	0.0272	0.0184

Equations 5 have been used with the variable pressure data to determine the constants. Variable pressure data were determined at 460°, 480°, and 499°F. Negative values of the adsorption constants were obtained

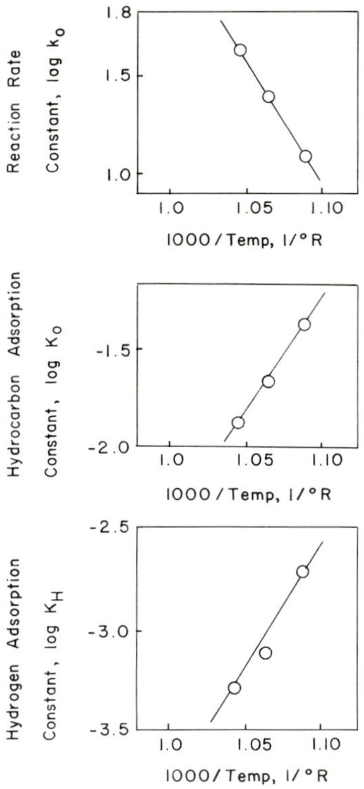

Figure 3. Arrhenius plot of adsorption model constants

at 2 of the 3 temperatures with the single-site model. All positive values were obtained with the dual-site model, which has been accepted as the more correct model because physically realistic values for the constants were obtained.

Effect of Temperature on Adsorption Model Constants

According to the adsorption-site theory, the model constants should follow an Arrhenius-type temperature relationship. An Arrhenius-type plot of the adsorption model constants is shown in Figure 3. The rate constant, k_o, increases with an increase in temperature, and the adsorption constants decrease with an increase in temperature. These opposing effects are in agreement with a physically realistic model. The activation energies found from these data are 29.3 kcal/mole for reaction, —28.9 kcal/mole for hydrocarbon adsorption, and —35.4 kcal/mole for hydrogen adsorption.

Literature Cited

(1) Ciapetta, F. G., Hunter, J. B., *Ind. Eng. Chem.* **1953,** 45, 147.
(2) Condon, F. E., "Catalysis," Vol. 6, p. 44, Reinhold, New York, 1958.
(3) Evering, B. L., "Advances in Catalysis," Vol. 6, p. 206, Academic Press, New York, 1954.
(4) Levitskii, I. J., Gonikberg, M. G., *Chem. Abstr.* **1962,** 56, 1097e.
(5) Maslyanskii, G. N., *J. Gen. Chem.* **1943,** 13, 540.
(6) Rabo, J. A., Pickert, P. E., Mays, R. L., *Ind. Eng. Chem.* **1961,** 53, 733.
(7) Thomas, J. M., Thomas, W. J., "Introduction to the Principles of Heterogeneous Catalysis," p. 458, Academic Press, New York, 1967.
(8) Voorhies, Alexis, Jr., Beecher, R. G., *Ind. Eng. Chem. Prod. Res. Develop.* **1969,** 8, 366.
(9) Voorhies, Alexis, Jr., Bryant, P. A., *Am. Inst. Chem. Engrs. J.* **1968,** 14, 852.

RECEIVED January 12, 1970.

Discussion

B. C. Gates (University of Delaware, Newark, Del. 19711): Did you observe catalyst aging, and can you make quantitative statements about it? Do you value the Langmuir-type rate equation as more than a curve-fitting equation?

J. R. Hopper: There was no aging of the catalysts as observed by a run at standard conditions made periodically during the testing. Although the model fits nicely into a physically realistic framework, it is considered to be primarily a mathematical fit of the data.

H. A. Benesi (Shell Development Co., Oakland, Calif. 94608): At the higher conversions of cyclohexane, were any products other than methylcyclopentane obtained?

J. R. Hopper: We did not observe any appreciable cracked products, usually less than 0.5%, and there was no benzene present in any of the products.

73

Synthetic Erionite and Selective Hydrocracking

H. E. ROBSON, G. P. HAMNER, and W. F. AREY, JR.

Esso Research Laboratories, Humble Oil & Refining Co.,
Baton Rouge, La. 70821

Erionite has been synthesized at 100°–150°C from a (Na,K) aluminosilicate gel with $SiO_2/Al_2O_3 = 10$. X-ray and electron diffraction results on the product show intergrowths of the related offretite structure, which is a large-pore zeolite. Adsorption capacity for n-hexane *is consistent with the density but adsorption rates are far slower than for zeolite A. Adsorption rates for* n-octane *are even slower but still better than for natural erionite. Hydrocracking tests on a C_5/C_6 naphtha show strong selectivity for converting normal paraffins to C_4^- gas, particularly propane. As temperature is increased, other components of the naphtha feed are cracked and selectivity decreases.*

Zeolite A, although an excellent adsorbent, has never seen wide use as a catalyst for petroleum processes. This is probably because of its low silica/alumina ratio which makes it unstable in the hydrogen form. Nevertheless, processes such as octane improvement and pour point reduction could benefit from a catalyst which would operate only on straight-chain hydrocarbons—*i.e.*, a catalyst based on a small-pore zeolite. Erionite appears to be the prime candiate for this service because it has both a three-dimensional 5-A pore system and high silica content ($SiO_2/Al_2O_3 \approx 7$).

Synthesis Experiments

The synthesis of erionite was reported by Zhdanov (*11*) in 1965. The medium was described as a mixed sodium–potassium aluminosilicate hydrogel at 90°–100°C but further details are not given. Breck and

Acara (*1*) had reported earlier the synthesis of zeolite T, which appears to be erionite as far as can be seen from the principal x-ray lines.

The results of our synthesis experiments are given in Figure 1, which plots 2 of the 4 composition variables required to describe the synthesis gel of 5 components ($Na_2O \cdot K_2O \cdot Al_2O_3 \cdot SiO_2 \cdot H_2O$). The system is comparatively less sensitive to the other 2 composition variables: $K_2O/(K_2O + Na_2O)$ and H_2O/SiO_2. Experiments in Figure 1 are at 0.25 $K_2O/(K_2O + Na_2O)$; the results would be nearly the same if this ratio were 0.20 or 0.30. Beyond this range, crystallization of other zeolite phases becomes predominant.

Silica sol was the SiO_2 source for all these experiments. An aluminate liquor was prepared by dissolving alumina trihydrate in hot NaOH–KOH solution. After cooling to room temperature, the $(Na,K)AlO_2$ liquor was blended with 40% silica sol using a high-speed mixer. Water content of the synthesis gel was held to a minimum; reduction of water has consistently given better erionite crystallinity. For high-silica formulas, H_2O/SiO_2 was 16; as silica was decreased, water content was reduced until at the low silica end, $H_2O/SiO_2 = 6$.

The gel was crystallized in closed containers to prevent evaporation loss. Crystallization time ranged from 1 to 5 days depending on gel composition and crystallization temperature. Figure 1 shows the ranges of gel composition where the product is substantially pure erionite. The boundaries are not sharp but rather represent areas of decreasing erionite crystallinity in the product.

Increasing the crystallization temperature from 100° to 150°C substantially increases the erionite crystallization area. At 150°C, good

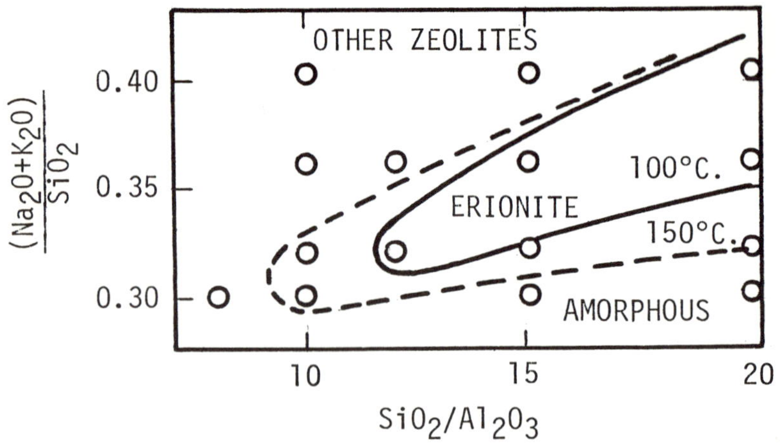

Figure 1. Erionite synthesis from silica sol, 0.25 $K_2O/(K_2O + Na_2O)$, minimum water

erionite products were obtained with 10 SiO_2/Al_2O_3 and 16 hours' crystallization time. The change in silica content of the erionite product is only from 7 to 6 SiO_2/Al_2O_3 as the gel changes from 20 to 10 SiO_2/Al_2O_3.

Comparison with Natural Erionite

Natural erionite, unlike most other natural zeolites, occurs in deposits large enough and pure enough to be used for commercial purposes (*3*). Several high-quality deposits are known in Nevada and Oregon. Natural erionite has some properties which are definite liabilities for catalytic purposes. Obviously, natural erionite contains a whole spectrum of impurities such as Fe, Ti, Ca, Mg, and Cu which may be objectionable. Further, the nature of the formation process practically guarantees variation in quality within a single deposit (*4*). Synthetic erionite, although more expensive, should be dependable in quality.

Table I. Erionite X-Ray Diffraction Patterns

		I/I_1	
hkl	*d*	*Natural*	*Synthetic*
100	11.47	100	100
101	9.37	8.5	—*
002	7.55	7.8	4.0
110	6.61	41	41
102	6.30	4.9	2.8
200	5.72	5.3	3.3
201	5.35	6.9	—*
103	4.61	4.1	—*
202	4.56	5.8	3.9
210	4.326	25	31
211	4.158	11	4.5*
300	3.814	14	12
212	3.754	40	36
104	3.587	21	15
302	3.401	0.8	0.6
220	3.304	17	11
213	3.278	6.2	1.3*
310	3.173	6.6	5.1
204	3.148	13	12
311	3.106	3.3	—*
312	2.923	6.5	4.4
400	2.860	27	30
214	2.842	21	25
401	2.811	22	5.7*
402	2.675	7.3	7.1
410	2.497	5.6	5.6
322	2.479	8.8	10

A second and more subtle area of difference is in the crystallography of the erionite phase itself (*10*). Table I compares x-ray diffraction intensities of low angle lines for a natural erionite (Jersey Valley, Nev.) and a synthetic erionite prepared at Esso Research Laboratories. The agreement is quite good except for those lines which have been marked by an astrisk indicating an intensity of less than half of that for natural erionite. Without exception, the designated lines (101, 201, 103, 211, 213, 311, and 401) have odd values for the "1" index. Further, their intensities are substantially less than the reference.

Such an effect is understandable in view of the distinction between erionite and offretite structures published by Bennett and Gard (*2, 9*). The designated lines are forbidden for the offretite structure. Gard has examined our synthetic erionite product by electron diffraction and found "disordered intergrowth with widely varying proportions of erionite and offretite structures" (*8*).

Adsorption Properties

Since offretite is a large-pore structure, intergrowth of offretite in the erionite phase would be expected to affect the adsorption properties. Table II compares adsorption capacities for natural and synthetic erionite with Zeolite A (Ca) and synthetic faujasite (Na) (4.8 SiO_2/Al_2O_3). As expected, the more dense erionite structure shows lower capacity (*5*). There is substantial agreement between natural and synthetic erionite capacity; the difference shows in adsorption rates (D/r^2). The low apparent diffusivity of *n*-paraffins in erionite is somewhat a mystery since there does not appear to be that much difference in pore dimensions between erionite and zeolite A as predicted from their structures (*6*). The difference cannot be attributed to crystallite size since the natural erionite sample (laths, 0.5 μ diameter or less) has finer crystallite size than any of the synthetic materials (1–5 μ).

Table II. Adsorption Capacities

	n-*Hexane*		n-*Octane*	
Sample	*Capacity*[a]	*Diffusivity*[b]	*Capacity*[a]	*Diffusivity*[b]
Faujasite (Na)	1.68	137[c]	1.06	46[c]
Zeolite A (Ca)	0.83	2.1	0.80	2.3
Natural Erionite	0.57	0.23	0.32	0.024
Synthetic Erionite (Na,K)	0.68	2.4	0.27	5.5
Synthetic Erionite (H)	0.50	10	0.35	5.1

[a] Millimoles adsorbate/gram zeolite (20- to 35-mesh granules).
[b] Apparent diffusivity (D/r^2) $sec^{-1} \times 10^3$ (*6*).
[c] Desorption.

The difference is more notable in *n*-octane adsorption which is shown in the last 2 columns of Table II. Zeolite A shows substantially the same capacity and adsorption rate for *n*-octane as for *n*-hexane. But for erionite, both natural and synthetic, *n*-octane capacities, and particularly the adsorption rates are substantially reduced. Here the difference between synthetic and natural erionite adsorption rate is quite large. It is possible that this is an effect of residual K^+ cations. However, simple exchange of Na^+ and K^+ for H^+ showed little change. We believe the more probable explanation is the intergrowth of offretite in the erionite crystal. The large offretite channels could give more rapid distribution of the sorbate molecule within the synthetic erionite crystal.

Hydrocracking Tests

These materials were tested as catalysts for the selective conversion of normal paraffins. The use of 5 A zeolites as shape-selective catalysts was first described by Eng (7); it has been developed subsequently by several authors. For this purpose, the exchanged zeolites were impregnated with 0.5% Pd, made into 14- to 35-mesh granules (self-bonded), and prereduced with hydrogen at 850°F. The feedstock was an Arabian C_5/C_6 naphtha stream, selected because of its high *n*-paraffin content (over 40%). The nitrogen content of the feed was less than 1 ppm, sulfur less than 10 ppm. Test conditions were 750°F (except where otherwise noted), 500 psig, 0.5 V/V/hr, and 2000 std. cu ft H_2/bbl. The catalyst was sulfided by adding 0.25% CS_2 to the feed. Liquid and gas products were evaluated by GC and MS analyses. Individual balance periods were 7 hours long; multiple tests on the same catalyst charge gave total exposures up to 200 hours.

Table III. Hydrocracking of C_5–C_6 Naphtha by Zn-Exchanged Zeolites

Catalyst Base	*Faujasite*	*Zeolite A*	*Natural Erionite*	*Synthetic Erionite*
$(Na_2O + K_2O)/Al_2O_3$	0.24	0.11	0.33	0.20
Conversion, Wt %				
n-Pentane	60	62	81	93
n-Hexane	79	86	98	99+
Other	52	15	25	36

Table III compares results from erionite catalysts with faujasite and zeolite A, all in the zinc-exchanged form. Zinc was used to obtain an active form of zeolite A which is stable to these conditions. Erionitc is less active than faujasite for conversion of the total naphtha feed to C_4

Table IV. Hydrocracking by Different Cationic Forms of Synthetic Erionite

Catalyst Base	*Zn-Erionite*	*H-Erionite*	*RE-Erionite*
$(Na_2O + K_2O)/Al_2O_3$	0.20	0.31	0.24
Conversion, Wt %			
n-Pentane	93	96	97
n-Hexane	99+	99+	99+
Other	36	41	27

Table V. Effect of Temperature on Hydrocracking with RE-Erionite

Temp., °F	*650*	*700*	*750*
Conversion, Wt %			
n-Pentane	52	91	97
n-Hexane	30	99+	99+
Other	8	15	27

Table VI. Feed and Product Composition, Wt %

Component	*Feed*	*Product*
C_1	0.0	0.7
C_2	0.0	3.2
C_3	0.0	36.6
i-C_4	0.0	1.7
n-C_4	0.2	6.6
i-C_5	9.8	7.1
n-C_5	16.2	1.4
Branched-C_6	38.7	34.4
n-C_6	25.0	0.2
Branched C_7	0.5	0.4
MCP	6.3	4.3
CH	1.0	0.6
MCH	0.0	0.2
Benzene	2.2	2.6

and lighter materials, but it is considerably more active for converting *n*-pentane and *n*-hexane. Zeolite A is still less active but is more selective for converting only the normal paraffin component. Compared with natural erionite, the synthetic erionite catalyst is more active for conversion of both total feed and the *n*-paraffin component. It is less selective for operating only on the normal paraffins.

Erionite can be used in other cationic forms including the hydrogen and rare earth forms. As expected, these give improved activity as shown in Table IV. The hydrogen form is the most active but also the least selective, possibly because the test conditions are too severe. The rare earth exchanged form (natural mixture less cerium) is just as active for converting normal paraffins but more selective. Again, the conditions are too severe, and better results can be achieved by accepting less than extinction of the *n*-paraffins. This is confirmed by Table V, which shows the temperature effect. At 650°F, *n*-hexane conversion is only 30%; this is consistent with lower adsorption rates observed for the higher *n*-paraffins. At 700°F and above, the greater stability of *n*-pentane becomes the controlling factor; *n*-hexane is substantially extinguished. Selectivity at 700°F is better than the earlier results with zeolite A at 750°F. Composition of feed and product from the rare-earth erionite catalyst at 700°F are given in Table VI.

The question remains why the other components, principally branched paraffins, are converted at all. Several explanations can be offered, none completely satisfactory. Not all the palladium is inside the zeolite cages but may be partially on external surfaces and nonzeolite components, amorphous material which is either the residue of incomplete crystallization or the product of zeolite decomposition in subsequent treatments. Since x-ray crystallinity is uniformly high, the amorphous component should be quite small. Branched paraffins can penetrate the zeolite surface far enough to be cracked. High temperature alters the selective adsorption properties of the zeolite, which were observed at low temperature. Offretite intergrowths provide enough surface in larger diameter pores partially to convert branched and cyclic molecules. There is some truth in all of these but we prefer the latter.

Literature Cited

(1) Breck, D. W., Acara, N. A., U. S. Patent **2,950,952** (1960).
(2) Bennett, J. M., Gard, J. A., *Nature* **1967,** 214, 1005–6.
(3) Deffeyes, K. S., "Molecular Sieves," Society of Chemical Industry, London, 1968.
(4) Deffeyes, K. S., *Am. Mineralogist* **1959**, 44, 501–9.
(5) Eberly, P. E., Jr., *Am. Mineralogist* **1964,** 49, 30–40.
(6) Eberly, P. E., Jr., *Ind. Eng. Chem. Prod. Res. Develop.* **1969,** 8, 140–4.
(7) Eng, Jackson, U. S. Patent **3,039,953** (1962).
(8) Gard, J. A., private communication.
(9) Sheppard, R. A., Gude, A. J., *Am. Mineralogist* **1969,** 54, 875–86.
(10) Staples, L. W., Gard, J. A., *Mineral. Mag.* **1959,** 32, 261–81.
(11) Zhdanov, S. P., *Izv. Akad. Nauk SSSR,* Ser. Khim. Nauk **1965** (6) 950–9.

RECEIVED February 10, 1970.

Discussion

F. W. Kirsch (Sun Oil Co., Marcus Hook, Pa. 19061): Did you infer any conclusions about reaction mechanism from the nature of the product distribution?

H. Robson: We were primarily interested in the hydrocracking of *n*-pentane and *n*-hexane to low molecular weight gases, principally propane. There were minor secondary effects such as isomerization.

I. M. Keen (British Petroleum Co., Ltd., Middlesex, England): First, how did you impregnate the palladium hydrogenation component onto your catalysts—*i.e.*, what salt did you use? Secondly, did you notice any differences in the hydrocracked product distribution from C_5–C_6 naphtha using your different ion-exchanged forms of the erionite catalyst?

H. Robson: Palladium was deposited on the zeolite powder by exchange with $Pd(NH_3)_4Cl_2$ solution. We did not observe significant differences in product distribution between cationic forms of erionite, but of course we were looking primarily for the disappearance of *n*-pentane and *n*-hexane.

R. C. Hansford (Union Oil Co. of California, Brea, Calif. 92621): To what extent might some amorphous silica–alumina be responsible for the poor selectivity—*i.e.*, are the synthetic products completely crystalline?

H. Robson: We know that the synthetic erionite product is highly crystalline from the intensity of x-ray diffraction peaks and the absence of the amorphous "halo." Unfortunately, this does not prove the sample is fully crystalline. If amorphous material is present, it should be at a very low level.

N. Y. Chen (Mobil Research & Development Corp., Princeton, N. J. 08540): It might be of interest to the audience, particularly to those who are not familiar with the application of zeolites in industrial catalytic processes, to mention that since the discovery of catalysis over shape-selective zeolite first published by Weisz and Frilette in 1960, a commercial process based on selective hydrocracking reactions similar to that reported in this paper has been in operation on a large scale in more than four of our refineries since 1967. A technical paper describing this process, known as the Selectoforming process, was published in 1968.

H. Robson: We have no particular comment except that priority of invention is determined by patents rather than publication in journals.

D. L. Peterson (California State College, Hayward, Calif. 94542): Did you examine the temperature dependence of selectivity and conversion of the Zn or H forms of either the synthetic or the natural erionites?

H. Robson: Reduction of reaction temperature improves selectivity of Zn and H forms of erionites, as well as rare earth exchanged forms.

D. A. Hickson (Chevron Research Co., Richmond, Calif. 94802): Can you comment on the stability of activity and product selectivity with time on stream?

H. Robson: Our tests were short-term screening tests; balance periods were six hours, total time on temperature up to two or three days. However, we were able to repeat the initial results after this much exposure. As far as can be determined from short-term tests, catalyst life appears satisfactory.

Question: Do you have any estimation of the internal *vs.* external surface area of the erionite?

H. Robson: Based on a particle size of about one micron observed in the electron micrographs, we estimate about 3 m^2/g external surface area, which is less than 1% of total surface area.

74

Hydrogen Generation by Steam Reformation of *n*-Hexane over Zeolite Catalysts

C. S. BROOKS

United Aircraft Research Laboratories, East Hartford, Conn. 06108

High-activity nickel and cobalt catalysts for hydrogen generation by the steam–hydrocarbon reforming reaction were prepared by ion exchange from synthetic zeolites. The activity of the catalysts for steam-reforming of hydrocarbons was evaluated with n*-hexane at 400° to 500°C. The specific reforming activity exceeded by factors of 2 to 30 that obtained for an active commercial nickel-on-alumina zeolite under comparable conditions. The superior activity was attributed to the high state of nickel dispersion in the catalysts prepared by ion exchange. The cobalt zeolites had comparable initial activity but rapidly lost their reforming activity owing to oxidation with reactant steam. Mordenite and faujasite zeolite catalysts with nickel contents less than 4 wt % demonstrated the highest specific reforming activity.*

The successful methane reforming catalysts (2) are primarily nickel-on-alumina with relatively heavy metal loadings ($\geqslant$ 15 wt %). The principal causes of catalyst degradation in the reforming of liquid hydrocarbons seem to be loss of catalytically active metal surface and heavy carbon deposition. It would appear fruitful to examine the difference between methane and higher hydrocarbon reforming and the properties desired in a steam-reforming catalyst. The obvious difference in the reforming of liquid hydrocarbons and of methane is that the desired catalyst should incorporate efficiency for C—C bond rupture with a minimum of coking.

The zeolites have several unique structural features which provide them with special interest as catalyst substrates. These features are extensive intralattice pore volume, a pore size of molecular dimensions, a large population of cation exchange sites, and the location of the exchange cation within the intralattice pore volume. These structural features are of special interest in various hydrocarbon reactions, notably hydrocarbon–water interactions, because of the possibilities for surface reactions within a lattice cage, providing an acid substrate to facilitate C—C bond rupture, and introducing catalytic metals in a highly disperse state on the crystal lattice surface and within the pore volume cages of the zeolites.

Experimental

Selection of Zeolites. The selected zeolites consisted of the 2 principal crystallographic types—*i.e.*, mordenite and faujasite—which provide an interesting range of intracrystalline cage volumes of 0.29 and 0.54 cm^3 per cm^3 zeolite, respectively; net charges per surface oxygen of 0.083 and 0.214 *e* units, respectively; and free cavity diameters of about 9 and 11.8 A, respectively (*3*).

A faujasite zeolite, such as Linde 13X (Union Carbide), has the empirical formula $Na_{86}(AlO_2)_{86}(SiO_2)_{106} \cdot 264\ H_2O$ which corresponds to a SiO_2/AlO_2 ratio of 1.23 and a cation exchange capacity (CEC) of 4.7 meq/gram. A mordenite zeolite, such as Norton sodium Zeolon (Norton Co.), has an empirical formula $Na_8(AlO_2)_8(SiO_2)_{40} \cdot 24\ H_2O$ which corresponds to a SiO_2/AlO_2 ratio of 5.0 and a CEC of 2.4 meq/gram. Clay binder to the extent of 17% was used with Linde 13X. The Norton Zeolon and the Linde 13X were in the form of 1/16-inch cylindrical pellets. The Linde Y zeolite (SK400) is also a faujasite type but with a greater Si:Al ratio (> 1.5) than the Si:Al ratio of about 1.23 characteristic of the X zeolite. The Y zeolite had a CEC of 4.04 meq/gram and was used in the form of 3/16-inch cylindrical pellets. Additional details on the properties of these zeolites are available in the published literature (*3, 4, 14*).

Catalyst Preparation. Catalyst preparation consisted of the exchange of nickel or cobalt nitrate for the sodium cation. Ratios of nickel or cobalt to sodium of 20:1 were used for maximum exchange and the ion exchanged zeolite pellets were leached with deionized water. Catalyst preparations were reduced for 16 hours at 400°C in a stream of hydrogen. Similar procedures have been reported (*5, 6*).

Catalyst Characterization. Chemical analyses, x-ray diffraction analyses, and gas adsorption procedures were used to characterize the composition, crystallographic character, and surface structure of the nickel and cobalt zeolite catalyst preparations. The chemical and x-ray procedures were standard methods with the latter described elsewhere (*11*). Carbon monoxide chemisorption measurements provide useful estimates of the surface covered by nickel atoms from the zeolite substrate (*10*).

Hydrocarbon Steam Reforming Tests in Microcatalytic Reactor. Catalyst performance for steam reforming of *n*-hexane to produce CH_4, CO, CO_2, and H_2 at 400° to 500°C was evaluated in a stainless steel reactor which has basically the same instrumentation used previously for oxidation catalyst studies (*9, 12*). The principal modifications consisted of the introduction of a dual injection metering pump, a vaporizer, a vapor mixer, and a watercooled condensate trap. Catalyst charges ranged from 1–12 grams and were interspersed through 1/8-inch alumina pellets (Norton Support SA5101) to form the catalyst bed (35 cm^3). The unconverted *n*-hexane was collected and weighed after the condensate was retrieved from the condensate trap. Effluent gases were analyzed by chromatography and mass spectrometry for all vapor components up to C_6 hydrocarbons.

Discussion

Catalyst Composition. Chemical compositions of typical nickel and cobalt zeolites are summarized in Table I. Based on the total CEC derived from the initial sodium composition, 23 to 37% of the Zeolon and 8.4% of the Linde SK400 exchange sites are occupied by nickel cations. In Zeolon, 55% of the exchange sites are occupied by cobalt cations. A ratio of 1.41:1 for cobalt to nickel on the Zeolon exchange sites resulted where nickel and cobalt were exchanged under comparable conditions.

Table I. Chemical Composition of Zeolites

Catalyst	*Cation*	*Wt %*	*CEC, Meq/Gram*
Na Zeolon	Na^+	5.33[a]	2.32
Ni-Co Zeolon	Ni^{2+}–Co^{2+}	1.52–2.15[b]	0.52 Ni^{2+}–0.73 Co^{2+}
Ni(Na) Zeolon	Ni^{2+}	3.24[b]	1.10
Co(Na) Zeolon	Co^{2+}	3.77[b]	1.28
Linde SK400	Na^+–Ni^{2+}	8.7–0.76[a]	3.78 Na^+–0.26 Ni^{2+}

[a] Atomic absorption.
[b] Wet analysis.

Crystallographic Character of Zeolites. It is essential for adequate characterization of the zeolite catalyst preparations that verification of structural integrity be made by x-ray diffraction analysis. Considerable information on the structure of zeolites, such as 13X, has been published (*5, 6*). X-ray diffraction analyses of several of the sodium, nickel, and cobalt zeolites established the principal interplanar *d*-spacings in Angstrom units and the relative intensity, 100I/Io, of these lines. The principal conclusions warranted by these x-ray diffraction data are that the zeolite structure was essentially intact after exchange of nickel for sodium in Zeolon but not in the case of 13X, a significant amount of nickel crystallites large enough for detection by x-ray were present in the hydrogen-reduced 13X nickel zeolite but not in the case of nickel Zeolon, and there was in no instance evidence of a significant amount of cobalt in crystallites large enough for detection by x-ray diffraction.

State of Dispersion of Metal. Chemisorption of carbon monoxide at 23°C and x-ray diffraction line broadening have been used to measure the state of dispersion of the nickel on 3 of these catalyst preparations. These results are given in detail in Ref. *10*. For example, the carbon monoxide chemisorption estimates of the nickel areas resulted in a specific nickel area of 30 m^2/gram of catalyst for the Zeolon support, compared with 10 m^2/gram of catalyst for the nickel-on-alumina in the case of G56. The x-ray diffraction line broadening measurements show that 58% of the total nickel surface area in the case of the G56 alumina supported catalyst is contributed by nickel crystallites greater than 100 A. On the other hand, only 0.4% of the nickel area for nickel Zeolon is contributed by nickel crystallites greater than 100 A. Furthermore, if the carbon monoxide chemisorption is expressed as molecules of carbon monoxide per nickel atom, a ratio of 1.03:1 is obtained for nickel Zeolon (Table III, Ref. *10*). This last result demonstrates that the nickel on the freshly prepared nickel Zeolon catalyst is present in an atomic state of dispersion.

Steam Reforming of *n*-Hexane Over Nickel and Cobalt Catalysts. The performance of several of the nickel and cobalt zeolite catalysts for steam reforming of *n*-hexane at 400°–500°C has been evaluated by short test runs with the reactor and the procedures described above (Table II). A Girdler reforming catalyst (G56) was tested under the same conditions as a comparative standard. All tests were conducted at a total pressure of 1 atm. Plateaus of sustained reforming activity were established within 1 hour. The cobalt catalysts lost essentially all reforming activity within 3 hours, presumably because of oxidation by steam. The space velocities reported are calculated in terms of theoretical hydrogen production based on the *n*-hexane injection rate and extent of conversion (Equation 2, Table II). The equation for the steam reforming of *n*-hexane with complete conversion to carbon dioxide is

$$C_6H_{14} + 12\,H_2O = 6\,CO_2 + 19\,H_2 \qquad (1)$$

Steam reforming was the primary reaction over these nickel catalysts. The presence of hydrocarbons (C_2 to C_5) which would indicate cracking reactions occurred to the extent of less than 10% in the reaction products. The presence of methane, which would indicate partial reforming, did not exceed 5% in the reaction products. There does not appear to be any significant difference in product selectivity for the *n*-hexane steam reforming reaction over nickel on the 2 quite different supports—zeolite *vs.* alumina. Carbonaceous residues accumulated in the case of all the nickel catalysts where reforming activity was sustained and the carbon deposition on the zeolite catalysts compared favorably with G56.

$$sv = \frac{\alpha FW\,(2.24 \times 10^4)\,(19)\,(3600)}{V} = \text{cm}^3/\text{ideal H}_2 \text{ per cm}^3 \text{ catalyst} \qquad (2)$$

Table II. Reforming of *n*-Hexane

Catalyst	*Temp., °C*	*Test Period, Min.*	α	SV Hr^{-1} *(Eq. 2)*	*K* $\frac{\mu\ moles}{G\ Cat\ Sec}$ *(Eq. 3)*
G56	500	40–455	0.36	1880	0.28
Ni Zeolon	500	73–475	0.54	3660	0.54
Co Zeolon	500	75	0.20	920	0.42
	500	200	0.006	71	0.015
G56	430	40–340	0.104	244	0.26
Ni Zeolon	430	60–385	0.114	556	0.60
Nickel Y	430	55–390	0.115	560	0.60

where F is *n*-hexane injection rate, moles per gram catalyst per sec, V is catalyst bed volume, cm^3, W is weight of catalyst/gram, and α is fraction *n*-hexane reformed to carbon monoxide, carbon dioxide, or methane. The reforming rate constant, K, defined as moles *n*-hexane reformed per gram catalyst per second, is given by the relation

$$K = F\ln\left[1/(1-\alpha)\right] \tag{3}$$

This assumes a first-order reaction with respect to the *n*-hexane. K' is the reforming rate constant defined as moles of *n*-hexane reformed per gram of nickel per second, and F represents moles of *n*-hexane injected per gram of metal per second.

The hydrogen production efficiency, H_p, which is the ratio of the actual to the ideal hydrogen production rate, is given by

$$H_p = H_2/19\ FW\alpha \tag{4}$$

where H_2 represents the hydrogen production rate as moles/second based on the effluent gas composition and FW represents the *n*-hexane injection rate as μmoles/second (Equation 1).

The reforming rate constants, K, referred to the total catalyst weight for all the nickel catalysts, fall within a comparatively narrow range of 0.26 to 0.60 μmoles per gram of catalyst per second. Cobalt Zeolon had an initial reforming rate constant, K, comparable to that of the nickel Zeolon but this declined rapidly to a much smaller value. The reforming rate constants, K', referred to the metal content, show that there is a significantly greater specific reforming activity by factors of 8 to 30 for the nickel zeolites, compared with the nickel-on-alumina catalyst (G56). This demonstrates that the high state of nickel dispersion on the zeolite catalysts does provide a higher specific activity nickel than obtained with the much higher nickel content of the nickel-on-alumina catalysts. This

Over Nickel and Cobalt Catalysts

K' $\frac{\mu\ moles}{G\ Met\ Sec}$ (*Eq. 3*)	H_p (*Eq. 4*)	*C, wt. %*	$\frac{Mol\ H_2O}{C\ atom}$
2.0	0.86	1.4	2.3
17.0	0.93	3.95	2.5
11.0	0.84	–	–
0.4	0.67	1.08	2.6
1.79	0.86	3.57	2.4
18.4	0.89	1.78	2.4
60.2	0.81	0.95	2.4

result is consistent with the greater nickel specific area determined by carbon monoxide chemisorption. This conclusion, however, appears to contradict earlier conclusions reached by Selwood (*13*) which were based on magnetic susceptibility measurements that a finite crystallite size large enough to form a ferromagnetic domain and significantly larger than a nickel atom is required for catalytic activity. The H_p values for all the nickel catalysts were high and fell within the relatively narrow range of 0.81 to 0.93, indicating that the reforming reaction proceeds well beyond methane to carbon dioxide, and that side reactions, such as cracking, occur to a minimal extent.

Conclusions

Both mordenite (nickel Zeolon) and faujasite (Linde Y SK400) zeolites with low nickel loadings, 3.24 and 0.76 wt %, respectively, have demonstrated activity for steam-reforming of *n*-hexane superior to a commercial nickel–alumina catalyst of much greater nickel loading. Furthermore, the reforming rates referred to the catalytic metal exceed that of the nickel–alumina catalyst by factors of 8 to 30. Tests of 4–8 hours' duration resulted in carbon accumulations comparing quite favorably with that of the nickel–alumina catalyst. However, the long-term stability and reforming activity of these catalysts must be established before they can be considered successful competitors with alkali-promoted nickel catalysts (*1, 2, 7, 8*) currently under consideration for steam reforming of liquid hydrocarbons. The structural integrity of catalysts exposed for extended periods to reforming process conditions should be established by x-ray diffraction determination of the zeolite crystallography and metal crystallite size, and by gas chemisorption measurements for determination of the persistence of catalytic metal surface area.

Acknowledgment

The author expresses appreciation to the United Aircraft Corp. for support of this work and permission to publish.

Literature Cited

(1) Andrew, S. P. S., *Ind. Eng. Chem. Prod. Res. Develop.* **1969,** 8, 321.
(2) Arnold, M. R., Atwood, K., Baugh, H. M., Smyser, H. D., *Ind. Eng. Chem.* **1952,** 44, 999.
(3) Barrer, R. M., *Ber. Bunsengesellschaft* **1965,** 69, 786.
(4) Breck, D. W., *J. Chem. Educ.* **1964,** 41, 678.
(5) Breck, D. W., Castor, C. R., Milton, R. M., U. S. Patent **3,013,990** (1961).
(6) Breck, D. W., Milton, R. M., U. S. Patent **3,013,982** (1961).
(7) Bridger, G. W., Wyrwar, W., *Chem. Process. Eng.* **1967,** 48, 101.
(8) Brogars, D. J., *Ind. Chemist* **1963,** 39, 177.
(9) Brooks, C. S., *J. Catalysis* **1965,** 4, 535.
(10) Brooks, C. S., Christopher, G. L. M., *J. Catalysis* **1968,** 10, 211.
(11) Klug, H. P., Alexander, L. E., "X-ray Diffraction Procedures," Wiley, New York, 1954.
(12) Kokes, R. J., Emmett, P. H., *J. Am. Chem. Soc.* **1960,** 82, 1037.
(13) Selwood, P. W., *Chem. Rev.* **1946,** 38, 41.
(14) Turkevich, J., *Catalysis Rev.* **1967,** 1, 1.

RECEIVED February 4, 1970.

Discussion

D. J. C. Yates (Esso Research Co., Linden, N. J. 07036): It is interesting that you find the same migration effect of nickel on NaX zeolites that I found some time ago on samples reduced under milder conditions than you used. Did you find less migration with nickel on Zeolon? I am also concerned with the use of CO for area measurement with nickel, as it is easy to make nickel carbonyl with reduced nickel even at quite low CO pressures. Did you check your results with hydrogen chemisorption?

C. S. Brooks: Regarding the question of the mobility of reduced nickel metal on oxide supports, I think there is no question that nickel does have a high mobility, and at relatively low temperatures. In order to have high dispersion, the most favorable conditions can be provided by starting with the highest possible state of dispersion, such as is possible by mounting the metal on a zeolite by cation exchange and by using a relatively low metal loading, preferably below 1 wt %. Of the zeolites examined here, the nickel mordenite with 3.2 wt % nickel and nickel Y faujasite with 0.76 wt % nickel demonstrated the highest state of metal dispersion and highest specific reforming activity. In the case of some of the nickel mordenite preparations, nickel metal crystallites were detected by x-ray diffraction analysis.

In regard to the second question raised about the possible mobility of nickel induced by the formation of nickel carbonyl during the CO chemisorption measurements, it is considered well established that the initial contact with CO after hydrogen reduction provides a reliable measure of the initial state of dispersion of the reduced nickel metal. The procedure used was evaluated in considerable detail (Ref. *10*). It was pointed out in discussion of this evaluation that nickel displacement by carbonyl formation might occur upon heating and desorption of the chemisorbed carbon monoxide upon completion of the initial CO chemisorption measurement.

In the account of this chemisorption work, hydrogen chemisorption was also described. Hydrogen chemisorption was consistently less than the CO (after correction for substrate contribution) and it is suggested that this difference is owing to the need for nickel crystallites large enough to provide adjacent sites for dissociative adsorption of hydrogen atoms, whereas the CO can adsorb as an undissociated molecule on individual surface nickel atoms.

Chemisorption measurements of either CO or hydrogen on transition metals mounted on zeolite substrates require considerable care for adequate interpretation. The zeolite lattices after high-temperature evacuation can adsorb large amounts of hydrogen, and divalent cations such as calcium can adsorb large amounts of CO.

75

Structural and Catalytic Properties of Nickel Modified Molecular Sieves

V. PENCHEV, H. MINCHEV, V. KANAZIREV, and I. TSOLOVSKI[1]

Institute of Organic Chemistry, Bulgarian Academy of Sciences, Sofia

The limit of stability of the crystal framework at different extents of Ni ion exchange of type A molecular sieve is shown by means of electron microscopy, differential thermal analysis, and x-ray diffraction. The data obtained from catalytic studies are in accord with the results of physical methods, showing preservation of the molecular sieve properties after reduction of the Ni ions. Metallic Ni aggregates on the external surface of the zeolite. In the dehydrogenation of cyclohexane and the hydrogenolysis of n-*hexane, type A molecular sieve shows the properties of metallic Ni on an inert support. When NiNaA is mixed mechanically with CaY, a typical bifunctional catalyst is obtained.*

Nickel catalysts on a support have been and continue to be a subject of multilateral studies, the results of which have helped the development of the theory of the polyfunctional catalyst. The catalytic properties of Ni molecular sieves in the presence of H_2 (*2, 7, 8*) have not been studied completely. Study of zeolite structural changes and the condition of the metal after thermal and hydrogen treatment have been comparatively poor. The same is true for the influence of zeolite type on the catalytic activity of the metal. This study aims to give additional information on the subject. As type A molecular sieve modified with Ni proves to be a suitable subject for clarifying some of the structural and catalytic peculiarities of zeolites, it is given main consideration.

[1] Present address: Institute of General and Inorganic Chemistry, Bulgarian Academy of Sciences, Sofia.

Experimental

For this study, Linde NaA zeolite was used. The ion exchange was carried out with 0.1*N* solution of $Ni(NO_3)_2$, according to our method (*6*). In Table I, some of the data obtained are shown. The acidity of the samples was determined by potentiometric titration with CH_3OK (*9*). The thermographic studies were carried out under conditions described in Ref. *11*. The reduction was checked on a sample previously reduced with H_2 under conditions similar to those in catalytic activation. The same procedure was carried out in N_2 flow for comparison. Figure 1 illustrates the results. X-ray studies of zeolite with Ni in ion form were carried out making use of Cr radiation (*10*). The x-ray patterns of the reduced samples were made with Cu radiation, the samples being placed in 0.5-mm diameter Lindeman capillary tubes. On platinum–carbon replica, electron microscope studies were carried out (*12*). In order to clarify the changes in the zeolite when heated, additional x-ray and electron microscope studies were carried out on Samples 3 and 4, heated at DTA speed up to 360°, 440°, 640°, 740°, 880°, and 1070°C. The catalytic properties were examined in the process of dehydrogenation of cyclohexane and hydrogenolysis and hydroisomerization of *n*-hexane. The tests under atmospheric pressure were carried out in a fixed-bed flow microreactor. The catalyst was reduced with H_2 in the catalytic reactor at 400°C during a 4-hour period. Dehydrogenation was carried out at 300°C, space velocity (F/W) $1.4h^{-1}$, mole ratio $H_2/CH = 5{:}1$, and dilution of the catalyst by an inert material. The conditions of hydrogenolysis and isomerization are as follows: temperature range: 300°–400°C, F/W $2h^{-1}$, H_2/CH 5:1 mole, and dilution with inert material. The products were analyzed by the gas chromatographic method.

Table I. Sample Data

No. of Sample	Ni, Wt %	Degree of Exchange, % Eqv	Acidity × 10^4, Mg Eqv H^+/Gram		DTA Data	
			in N_2	in H_2	Loss of Wt, %	Initial Temp. of Destruction, °C
1	0	0	0.50	0.45	22.4	840
2	3.6	18	1.56	3.45	–	–
3	4.0	23	–	–	24.0	780
4	5.9	34	0.60	1.16	25.0	750
5	10.6	55	0.71	1.12	26.3	640
4a	5.9	34	0.60	–	23.0	740
4b	5.9	34	–	1.16	22.0	755

Results and Discussion

The x-ray studies (*6*) show that after ion exchange, Ni zeolites retain their primary structure. More detailed information on the stability of the Ni zeolite was obtained by a thermographic method (Figure 1).

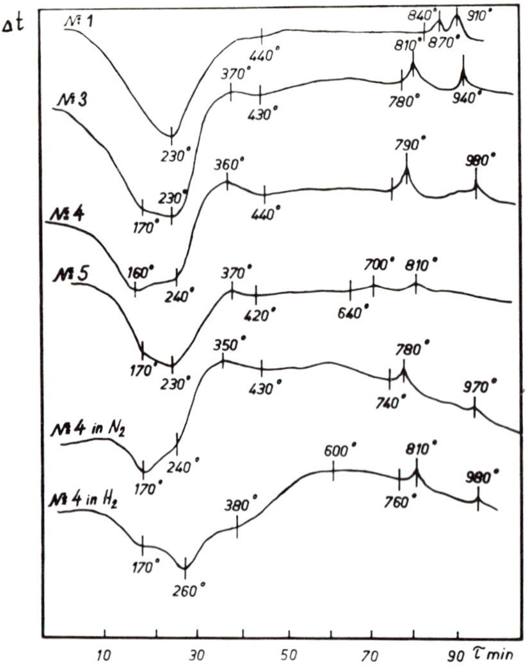

Figure 1. Differential thermograms of NaA and NiNaA zeolites

A pronounced endothermal effect (160°–240°C) and 2 high-temperature exothermal effects (700°–980°C) were observed in all samples. Ni forms are characterized by another exothermal effect (370°C), which is weaker than the other two. Comparison of DTA and DTG curves shows that the endothermal effect corresponds to the liberation of water from the molecular sieves. The water liberation is in stages, and complete dehydration occurs only after destruction of the molecular sieve framework. Even at temperatures higher than 600°C, about 1% well-bonded water remains in the zeolite. Although this quantity of water is small, it plays a definite part in the structural stability and the cation comparison (*4*). The observed character of dehydration is shown even better in the Ni sieves. The peak of the endothermal reaction is split in two, and its shift to lower temperatures has been observed. The total quantity of water adsorbed by the zeolite increases from 22 to 26% with increasing degree of ion exchange (Table I), which is in accord with the greater hydration ability of the Ni ions (*3*). When Ni is introduced, it decreases the relative dynamic thermal stability; the beginning of the first high-temperature exoeffect noted on the DTA plots, which marks the destruction of the framework (*1*), decreases from 840° to 640°C (Table I). The additional electron microscope (Figure 2c,d,e) and x-ray studies (Figure 3) show

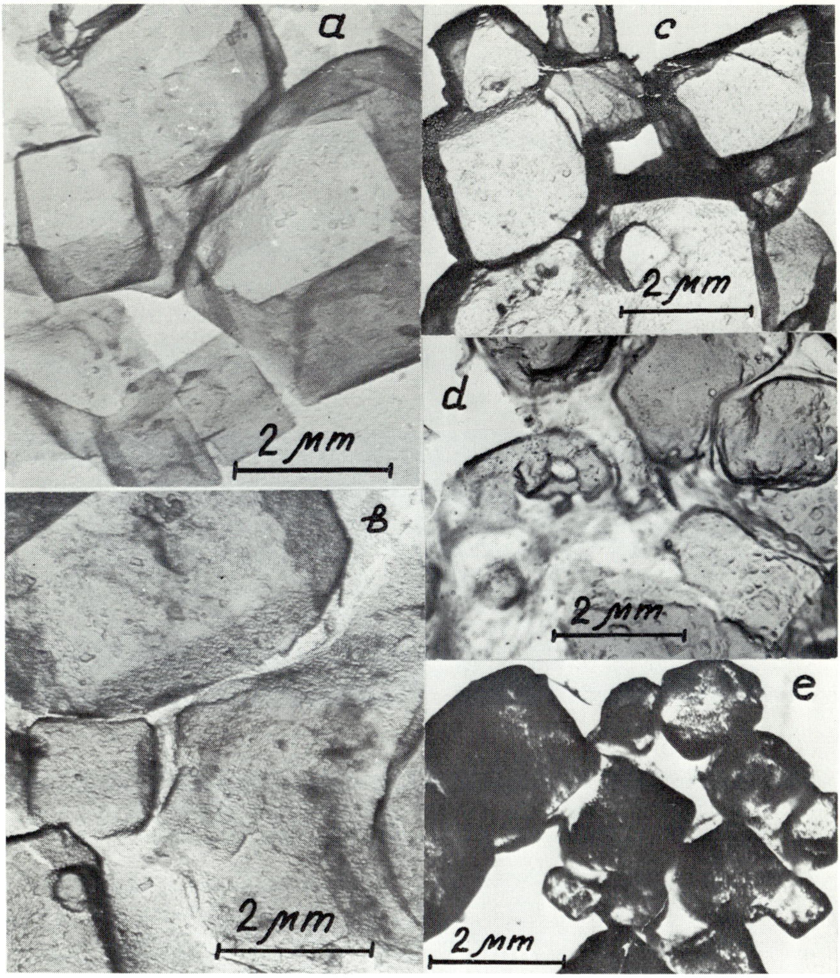

Figure 2. Electron microscope photos of zeolite samples

a. No. 1
b. No. 4
c. No. 4 heated to 640°
d. No. 4 heated to 870°
e. No. 4 heated to 1070°

that the zeolite structure remains unchanged until the beginning of the first high-temperature exoeffect and later transforms into a state amorphous to x-rays. At the end, a new crystal phase is formed with a nepheline structure. Thermographic studies for Sample 4 treated under conditions analogous to the catalytic activation show considerable changes in the endoeffect only characteristic of the reduced sample (Figure 1). Yet, after heating at 400°C, some nonreversible changes occur, marked by a decrease of the total adsorption capacity by 2–3% (Table I) and disappearance of some of the weaker x-ray lines (Figure 3).

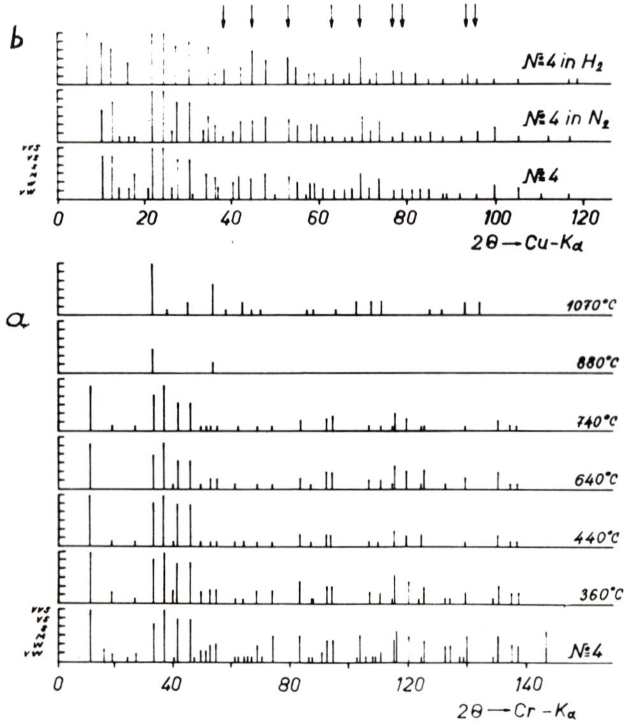

Figure 3. X-ray diffraction data on Sample 4; the arrows note lines characteristic of Ni and NiO

a. Heating in air
b. Heating to 400° in H_2 or N_2 flow

Although the cyclohexane molecule has a bigger critical diameter size (6.1A) than the entry openings of type A molecular sieve, dehydrogenation occurs to a considerable degree. This result confirms in a catalytic way for type A zeolites the fact, accepted for molecular sieves X and Y (*7, 13*), that after reduction with H_2, Ni aggregates on the external zeolite surface in crystals of considerable size. Table II shows that the specific activity of the metal decreases as its concentration increases on the catalyst. This effect might be considered an indication of an increase of Ni crystal size. X-ray studies (Figure 3) confirm the presence of metallic Ni. Determination of Ni crystal sizes from the line breadths is difficult because of the coincidence of all diffraction lines of Ni and NiO with the lines of the zeolite. The considerably lower specific activity of

Table II. Specific Activity and Concentration of Ni

Mol Sieve	*Quantity Wt % Ni*	*Dehydrogenation of Cyclohexane Mmole C_6H_{12}/ Gram Ni h*	*Hydrogenolysis of* n-*Hexane, Grams C_6H_{14}/Gram Ni h*		
			320°C	*350°C*	*400°C*
NiNaA	3.6	96.5	14.2	10.6	4.7
	5.9	87.4	9.7	8.2	3.9
	10.6	69.0	7.8	5.6	3.3
NiNaX	3.4	24.2			
	7.0	20.2			
	9.8	13.0			
NiNaY	from 3.0	unreactive			
	to 8.5	unreactive			

the Ni molecular sieves X and Y probably is related to a characteristic influence of the support.

The possibility of controlling the catalytic functions and properties of the Ni zeolites like typical bifunctional catalysts is followed in the hydrogenolysis and isomerization of *n*-hexane. In type A samples only hydrogenolysis occurs (Figure 4), as at higher temperatures intensive degradation to methane occurs. At this point, a decrease in activity takes place. In this reaction, as in the case of dehydrogenation, a decrease in the specific activity of Ni is observed when increasing its concentration in the catalyst. The fact that isomerization does not occur, although the acidity of the reduced zeolites is increased, shows that molecular sieve

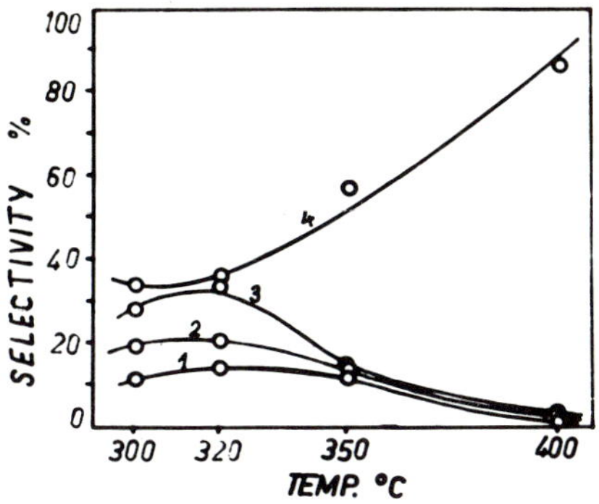

Figure 4. Temperature dependence of the selectivity at hydrogenolysis of n-*hexane*

1. *Propane*
2. n-*Butane*
3. n-*Pentane*
4. $C_1 + C_2$

Table III. Selectivity of Zeolite Samples

Sample	Yield, Wt %	Yield in % Converted Hexane, 350°C						
		$C_1 + C_2$	C_3	i-C_4	n-C_4	i-C_5	n-C_5	i-C_6
NiNaA	23.6	64.2	10.9	0	11.4	0	13.9	0
CaY	7.40	24.1[a]		8.5	4.1	7.0	4.5	51.8
NiNaA + CaY	28.5	18.3	8.3	6.7	5.4	7.7	7.1	46.5

[a] For $C_1 + C_2 + C_3$.

properties and crystal structure are preserved, which is in accord with the results obtained by thermographic and x-ray structural studies. The partial amorphousness observed in electron microscope pictures (Figure 2 a,b) probably concerns only the surface layers of the crystal. The catalyst NiNaA does not express its dual function mainly because of its molecular sieve properties. The potential possibility for this is shown when NiNaA (Sample 4) is mixed with a molecular sieve CaY, whose structure is more open and acidic properties are better expressed (5). Table III indicates that the selectivity of the reaction is shifted strongly to isomerization. This illustrates one of the possibilities for controlling the catalytic properties of dual functional catalysts—molecular sieves.

Literature Cited

(1) Berger, A. S., Yakovlev, L. K., *Zh. Prikl. Khim.* **1965,** 38, 1240.
(2) Borunova, N. V., *et al.*, *Materialy Vses. Soveshch. po Tseolitam, 2nd, Leningrad* **1964,** 380.
(3) Neimark, I. E., *Izv. Akad. Nauk USSR, Ser. Khim.* **1965,** 6, 959.
(4) Olson, D. H., *J. Phys. Chem.* **1968,** 72, 4366.
(5) Penchev, V., Kanazirev, V., Minchev, H., *Compt. Rend. Acad. Bulgare Sci.* **1968,** 22, 899.
(6) Penchev, V., Minchev, H., Bakurdjiev, I., Tsolovski, I., *Compt. Rend. Acad. Bulgare Sci.* **1968,** 21, 143.
(7) Rubinshtein, A. M., Minachev, Kh. M., *et al.*, *Izv. Akad. Nauk USSR, Ser. Khim.* **1968,** 4, 786.
(8) Selenina, M., Wencke, K., *Monatsber. Deut. Akad. Wiss. Berlin* **1966,** 8, 886.
(9) Strich, K. H., *et al.*, *Chem. Tech.* **1965,** 17, 489.
(10) Tsolovski, I., *Commun. Dept. Chem. Bulg. Acad. Sci.* **1970,** 2, in press.
(11) Tsolovski, I., Minchev, H., Penchev, V., *Commun. Dept. Chem. Bulg. Acad. Sci.* **1970,** 4, in press.
(12) Tsolovski, I., Minchev, H., Penchev, V., *Compt. Rend. Acad. Bulgare Sci.* **1970,** in press.
(13) Yates, D. J. C., *J. Phys. Chem.* **1965,** 69, 1676.

RECEIVED February 13, 1970.

76

Catalytic Properties of Synthetic Mordenite in Isomerization, Hydrogenation, and Hydroisomerization of Hydrocarbons

KH. MINACHEV, V. GARANIN, T. ISAKOVA, V. KHARLAMOV, and V. BOGOMOLOV

N. D. Zelinsky Institute of Organic Chemistry, Academy of Sciences, Moscow, U.S.S.R.

The catalytic properties of H-, Li-, Na-, K-, Mg-, Ca-, Zn-, Cd-, and Al-forms of synthetic mordenite in the reactions of cyclohexane and n-*pentane isomerization and benzene hydrogenation have been studied. The cation forms of mordenite that do not involve the metals of column VIII of the Mendeleyev Table show high activity in these reactions. To elucidate the mechanism of* n-*pentane isomerization, the kinetics of the reaction on H-mordenite have been studied. Carbonium ion is supposed to result from splitting off hydride ion from hydrocarbon molecule. Na-mordenite catalytic activity in benzene hydrogenation reaction decreases linearly with the increase of decationization. This indicates that cations are responsible for the catalytic activity of zeolite. The high activity of cations of nontransition metals in oxidation–reduction reactions seems to be quite unexpected and may provide evidence for some uncommon mechanism of benzene hydrogenation.*

The catalytic properties of synthetic faujasites (zeolite X and Y) have been thoroughly studied. Some of the catalysts prepared on the basis of faujasites have been widely used in industry. The activity of zeolite catalysts in many reactions (*10, 15*) depends to a considerable extent on their composition. In particular, the increase of silicon content (SiO_2/Al_2O_3 ratio) results in increasing catalytic activity. Synthetic mordenite with higher SiO_2/Al_2O_3 ratio than that of faujasites appears

Table I. Cyclohexane Isomerization on Synthetic Mordenite[a]

Catalyst	*HM*	*LiM*	*NaM*	*KM*	*MgM*	*CaM*	*ZnM*	*CdM*	*AlM*
Temperature, °C[b]	300	300–450			360	420	374	374	300
Methylcyclopentane yield, %[c]	39.2	No isomerization			9.9	6.2	27.1	34.6	18.7

[a] Pressure = 30 atm, H_2/C_6H_{12} (mol) = 3.2, V = 1 hour^{-1}.
[b] At higher temperatures, the process is not selective because of the increase of cracking product content.
[c] The amount of cracking products does not exceed 4 wt %.

to be more advantageous in this respect. In addition, this zeolite exhibits crystalline structure other than that of faujasites.

The information about synthetic mordenite properties was obtained in 1961 when Keough and Sand (*7*) found that H- and other forms of this crystalline aluminum silicate display high activity and selectivity in the reactions of hydrocarbon cracking and ethanol dehydration. Later this zeolite was shown (*1, 2, 5, 7, 8, 10–13, 15, 16*) an active catalyst in the reactions of isomerization, cracking, and alkylation of hydrocarbons and alcohol dehydration. However, the catalytic properties of mordenite have been studied insufficiently, compared with those of other zeolites.

The present paper reports the results of investigating the catalytic properties of a number of cation forms of synthetic mordenite in the reactions of isomerization, hydrogenation, and hydroisomerization of hydrocarbons.

Experimental

Catalysts. Li^+, Na^+, K^+, Mg^{2+}, Ca^{2+}, Zn^{2+}, Cd^{2+}, Al^{3+}, H-mordenite as well as hydrogen mordenite, with either 0.5% Pd or 5% Ni were used as catalysts. The mordenite cation modifications were prepared by treatment of NaM with the solutions of the corresponding salts to substitute the cations for Na^+. For each cation, the exchange degree was not less than 70% (equiv.). The hydrogen form of mordenite was obtained by treatment of Na-form with ammonium nitrate aqueous solution, followed by heating the NH_4-form thus obtained at 520°C in air flow. In this way, the samples with various Na^+–proton exchange degree were obtained. Palladium and nickel catalysts were obtained by soaking H-mordenite in H_2PdCl_4 and $Ni(NO_3)_2$ aqueous solutions. The catalysts obtained were dried at 110°C and pressed into pellets 4 × 4 mm without any binder.

Procedure. The catalytic properties of the samples were studied in a constant flow unit under hydrogen pressure. The reaction vessel was loaded with 10 cm^3 of catalyst in all runs. For experimental conditions refer to the corresponding section of this text.

Benzene and cyclohexane adsorption was studied on Na- and H-mordenite in a chromatographic thermodesorption unit (*3*) at various rates of programmed temperature changes.

Starting Substances. Benzene, cyclohexane, and *n*-pentane were used in the experiment. The chromatographic data showed no admixtures present in the substances.

Results and Discussion

Cyclohexane and *n*-Pentane Isomerization. The data on catalytic properties of cation forms of mordenite in cyclohexane isomerization are presented in Table I.

As seen from the table, the "hydrogen" form of mordenite is the most active isomerization catalyst. At temperatures as high as 300°C, the yield of methylcyclopentane on H-mordenite is 40%.

Al-mordenite exhibits a considerably smaller activity. The activity of two-valent forms is still lower (Zn- and Cd-forms are the most active of these, but they cause no cyclohexane isomerization at 300°C) while univalent modifications show no activity at all.

The "hydrogen" form of mordenite is an extremely active catalyst in paraffin hydrocarbon isomerization as well. For example, 50% of *n*-pentane at 280°C and 30 atm undergoes isomerization to produce isopentane. To attain the same degree of hydrocarbon isomerization with H-mordenite as compared with bifunctional catalysts involving Pt, Pd, and Ni per Al_2O_3 or zeolite, the temperature should be 70°–120° lower (*14*).

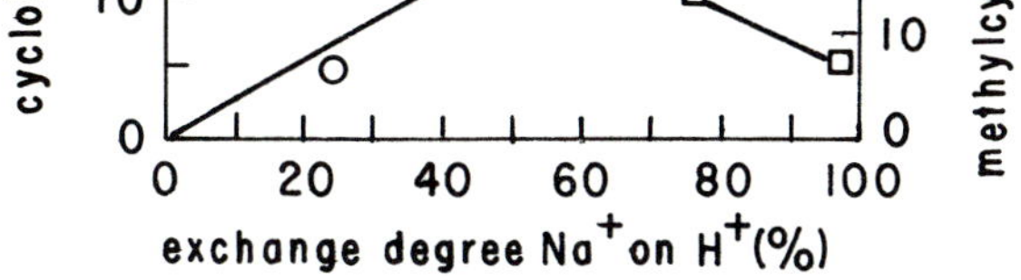

Figure 1. Dependence of isomerization and hydrogenation activity of mordenite on Na^+–H^+ exchange degree

Isomerization: T = *300°C;* P = *30 atm; LHSV = 1;* H_2/C_6H_{12} = *3.2*
Hydrogenation: T = *192°C;* P = *30 atm; LHSV = 5;* H_2/C_6H_6 = *5*

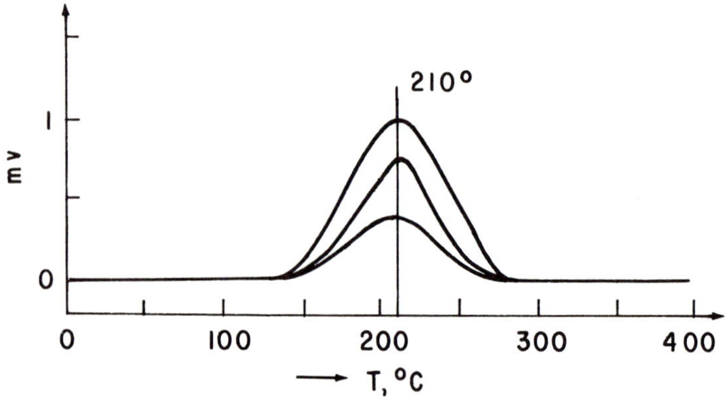

Figure 2. Dependence of desorption temperature on the amount of adsorbed cyclohexane

Thus, the activity of H-mordenite is thousands of times as high as that of the catalysts in question.

The consideration of mordenite isomerization activity as a function of Na^+–proton exchange (Figure 1) shows that the extent of isomerization of cyclohexane into methylcyclopentane increases linearly as the exchange degree increases. This dependence provides evidence for uniformity of surface catalytic activity, which has been confirmed by investigating cyclohexane thermodesorption from H-mordenite surface. Cyclohexane thermodesorption chromatograms produced at various programmed temperature changes display only 1 symmetric peak. The temperature of its maximum at the same rate of programmed temperature changes (Figure 2) is independent of the amount of cyclohexane preliminarily adsorbed—*i.e.*, the degree of surface coverage. According to (*3*), such regularities are characteristic of thermodesorption from the uniform surface.

Changes of activity with temperature in mordenite samples, with Na^+ ions replaced by H^+ to various extents, can be seen in Figure 3. With increase of exchange rate, the temperature decreases. Thus, the same degree of cyclohexane isomerization can be attained with samples of maximum and minimum rate of exchange at 270° and 420°C, respectively.

As indicated above (Table I), 2- and 3-valent cation forms of mordenite are considerably less active in cyclohexane isomerization than that of hydrogen. It is this property of mordenite-supported catalysts that distinguishes them sharply from zeolites of the faujasite type. Thus, in the case of faujasites, the activity of H-form (decationized form) and 2-valent modifications (such as Ca^{2+} and Mg^{2+}) in acid–base reactions (cracking, isomerization, alkylation) is the same, while H-mordenite is many times as active as any cation form under study.

The high catalytic activity of H-mordenite seems to be uncommon, since (*6, 9*) amorphous aluminum silicates and zeolites which do not involve the metals of column VIII are not active in isomerization of saturated hydrocarbons under similar conditions. In addition, as has been shown by this research, the introduction of metal into H-mordenite does not increase the catalyst activity in *n*-pentane isomerization. For example, the yield of isopentane on H-mordenite and 0.5% Pd/HM at $p = 30$ atm and $V = 1$ hour^{-1} is about 50 wt % (*12*).

The study of *n*-pentane isomerization kinetics on H-mordenite enabled us to bring out some peculiarities of this catalyst. In the absence of hydrogen, the reaction proceeded at a low rate. This probably results from hydrogen rearrangement. For example, in the presence of nitrogen at 280°C and 30 atm, the amounts of isopentane and cracking products were 5% and 5%.

Under the same conditions but in the presence of hydrogen, the yield of isopentane was 50%, with the same amount of cracking products. That hydrogen hinders the reaction was quite unexpected (Figure 4).

This is characteristic of bifunctional catalysts and can be accounted for by equilibrium shift at the first stage of the process (*9*):

$$n\text{-}C_5H_{12} \underset{+H_2}{\overset{-H_2}{\rightleftarrows}} n\text{-}C_5H_{10} \longrightarrow i\text{-}C_5H_{10} \underset{-H_2}{\overset{+H_2}{\rightleftarrows}} i\text{-}C_5H_{12}$$

I II III

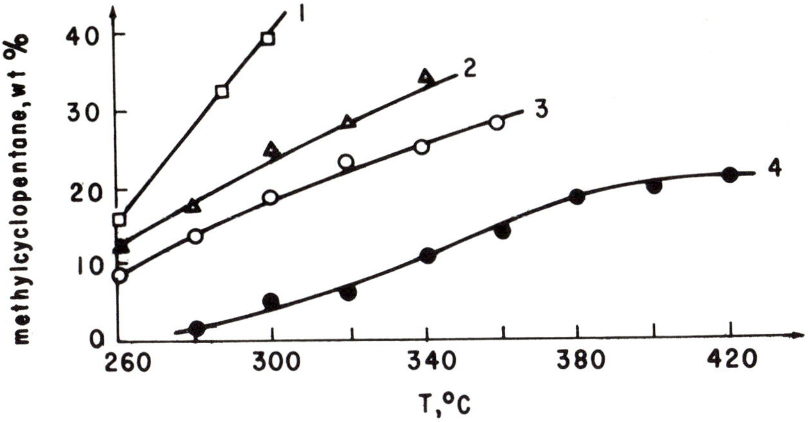

Figure 3. Dependence of isomerization activity on temperature for samples with different Na^+–H^+ exchange degree

Exchange degree:
1. 82.8% (equiv.)
2. 63.3% (equiv.)
3. 48.3% (equiv.)
4. 25.1% (equiv.)

However, H-mordenite cannot be considered as an actual bifunctional catalyst as it does not contain the metals of column VIII in its composition, which determine hydro–dehydrogenating properties of such catalysts. Yet, this argument does not seem sound enough for the reaction mechanism in question to be completely registered. It was necessary to find out whether H-mordenite itself might exhibit hydro–dehydrogenating properties. For this purpose, the possibility of the reactions of benzene hydrogenation and cyclohexane dehydrogenation has been verified.

H-mordenite shows certain activity in benzene hydrogenation (Figure 1) but it does not cause the reaction of cyclohexane dehydrogenation. By studying the reactions of benzene hydrogenation and cyclohexane isomerization on mordenite samples with various extent of replacement of Na^+ by H^+, it has been shown that hydrogenation and isomerization functions of the catalyst change antibatically (Figure 1). This is not in agreement with the above scheme of isomerization process.

We suggest that isomerization of saturated hydrocarbons on H-mordenite might proceed according to the following scheme:

$$n\text{-}C_5H_{12} + HM \rightleftarrows n\text{-}C_5H_{11}\text{–}M + H_2$$

$$n\text{-}C_5H_{11}\text{–}M \rightleftarrows iso\text{-}C_5H_{11}\text{–}M$$

$$iso\text{-}C_5H_{11}\text{–}M + H_2 \rightleftarrows iso\text{-}C_5H_{12} + HM$$

where HM is the hydrogen form of mordenite, $n\text{-}C_5H_{11}\text{–}M$, $i\text{-}C_5H_{11}\text{–}M$ are carbonium ions connected with the active centers of the catalyst.

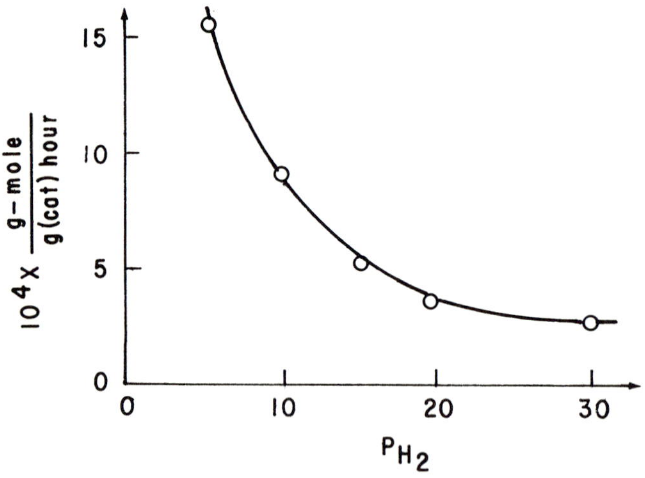

Figure 4. Dependence of isomerization rate on hydrogen pressure; T = *210°C;* P_{C_5} = *2 atm*

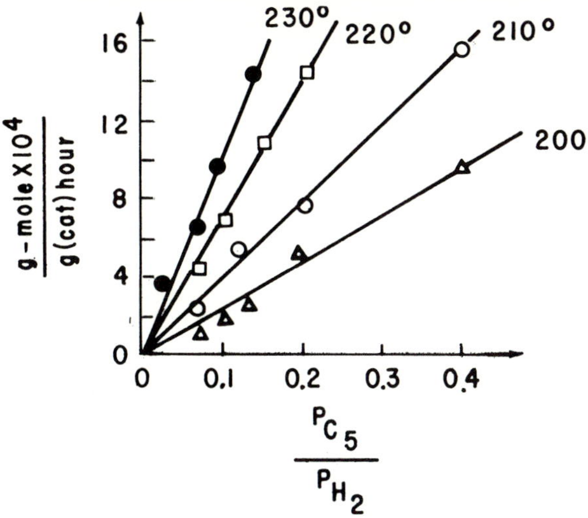

Figure 5. Dependence of isomerization rate on $\frac{P_{n-C_5}}{P_{H_2}}$

Assuming that carbonium ion isomerization is a slow stage (stage 2) while the other intermediate stages are fast and are in equilibrium, one may derive the following kinetic equation:

$$r = \frac{K_2\, a_1 \frac{P_{n\text{-}C_5}}{P_{H_2}} - \frac{K_{-2}}{a_3} \cdot \frac{P_{i\text{-}C_5}}{P_{H_2}}}{1 + a_1 \frac{P_{n\text{-}C_5}}{P_{H_2}} + \frac{1}{a_3} \frac{P_{i\text{-}C_5}}{P_{H_2}}} \qquad (1)$$

where r is the reaction rate, K is the rate constant of the corresponding stage, a is the equilibrium constant of the corresponding stage, and P is the partial pressures of the reactants.

From Equation 1, the rate of the forward reaction depends only on the ratio of n-pentane and hydrogen partial pressures. Figure 5 presents the experimental data in terms of $r \sim P_{A\text{-}C_5}/P_{H_2}$, showing that at all temperatures the experimental data fit well into the straight line. Thus, the rate of the forward reaction can be determined from the equation:

$$r = K \frac{P_{n\text{-}C_5}}{P_{H_2}} \qquad (2)$$

which is in agreement with Equation 1, provided the extent of the catalyst surface coverage with carbonium ions is small.

The dependence of the rate constant, K, in Equation 2 on temperature (Figure 6) shows the apparent activation energy in n-pentane isom-

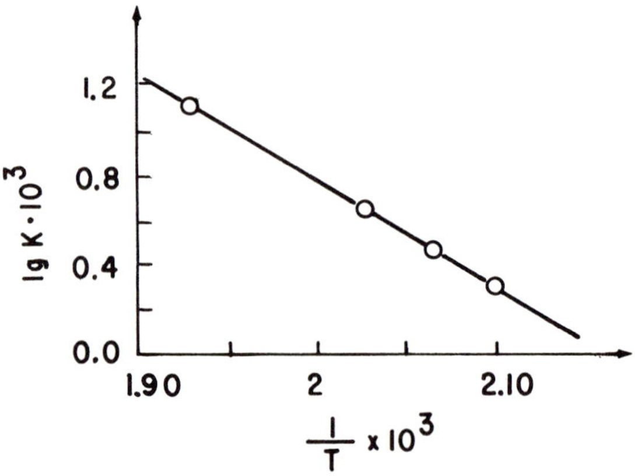

Figure 6. Dependence of rate constant, K, *on temperature*

erization reaction to be 31 kcal/mole. According to the suggested reaction mechanism, the apparent activation energy is the difference between the true activation energy of the adsorbed carbonium ion isomerization and the heat of its formation.

Thus, study of the kinetics of *n*-pentane isomerization on H-mordenite leads to the conclusion that the mechanism of the reaction in question is different from that of isomerization on bifunctional and metal–zeolite catalysts. This difference lies in the manner of carbonium ion formation. With bifunctional catalysts, carbonium ion originates with the attachment of a proton to the olefin molecule, while with H-mordenite it originates as a result of splitting off hydride ion from the saturated molecule of the starting hydrocarbon by mordenite proton, as has been suggested by the above reaction scheme.

Benzene Hydrogenation and Hydroisomerization. H-mordenite displays a marked hydrogenating activity at 250°C, $P = 30$ atm, volume rate 1 hour^{-1}. Under these conditions, the hydrogenation of benzene with Na-mordenite proceeds practically to completion, the reaction being selective. This seems to be quite unexpected since the catalyst does not involve the conventional hydrogenating components (Pt, Pd, Ni, etc.). The activity of mordenite in this reaction depends on Na^+ concentration. With the increase of the amount of sodium in the catalyst, its activity increases linearly. This provides evidence that Na^+ is involved in the hydrogenation reaction and the regions of catalytic activity are uniform. This is evidenced by thermodesorption data. The thermodesorption chromatograms exhibit only 1 benzene peak; its temperature maximum at a constant rate of temperature changes, 12°C/min, is 390°C, independent of the adsorption value.

The desorption activation energy calculated from the dependence of the temperature of the desorption peak maximum upon the rate of temperature changes is 27 kcal/mole. This value agrees with the isosteric heat of benzene adsorption on Na-mordenite (*4*), 25 kcal/mole. These values considerably exceed those of benzene adsorption heat on Na-faujasites. According to the data obtained, the latter values are 15–17 kcal/mole, depending on SiO_2/Al_2O_3 ratio. This difference accounts for the high activity of Na-mordenite in the reactions of benzene hydrogenation.

The participation of Na^+ in the hydrogenation process is confirmed also by the fact that the catalyst activity increases when Na-forms are obtained by the reverse exchange according to the scheme:

$$NH_4M \rightarrow HM \xrightarrow{\text{Na Cl}} NaM$$

$$NH_4M \xrightarrow{\text{Na Cl}} NaM$$

Because of the unusual behavior of Na-mordenite in the reactions of benzene hydrogenations, it seemed of interest to study other cation modifications of mordenite in this reaction. The data obtained with a number of cation forms are summed up in Table II.

Table II. Benzene Hydrogenation on the Cation Forms of Synthetic Mordenite at 250°C, $P = 30$ Atm, $V = 1$ Hour^{-1}

Catalyst	*HM*	*LiM*	*NaM*	*KM*	*MgM*	*CaM*	*ZnM*	*CdM*	*AlM*[a]
Cyclohexane yield, %	19.7	87.7	98.3	11.0	69.4	98.6	59.7	17.4	77.0

[a] At 200°C.

In addition to Na-form, Li-, Ca-, Al-, Mg- and Zn-mordenite also display high isomerization activity. The low activity of KM probably results from the large radius of the potassium cation.

Thus, the data presented provide evidence that the hydrogenating activity of mordenite catalysts is accounted for by the nature of the substituting cation.

To compare the hydrogenating activity of the cation forms of mordenite with that of H-form which contains the metals of column VIII, we have studied benzene hydrogenation on the catalysts 0.5% Pd/HM and 5% Ni/HM. Under the conditions indicated in Table II, the extent of benzene hydrogenation on these catalysts is 85 and 95%, respectively. Thus, the hydrogenating activity of certain cation forms of mordenite is not inferior to that of H-mordenite, which contains palladium and nickel. Benzene hydrogenation on these catalysts is accompanied by a considerable hydroisomerization to yield methylcyclopentane 30–40%.

Conclusion

The data presented in this paper suggest that mordenite in various cation modifications is a unique catalyst since it shows a high catalytic activity in the reactions of isomerization of saturated hydrocarbons and hydrogenation of aromatic compounds, with no metals of column VIII involved in its composition. In addition, the mechanism of hydrocarbon isomerization, though of carbonium-ion character, is different from those known for other catalysts. As for the mechanism of hydrogenation reactions on alkaline and alkaline-earth cation forms of mordenite, it is unlikely that the reaction would be the oxidation–reduction type since these cations lack *d*-electrons. More accurate conclusions concerning the mechanism of the hydrogenation reaction can be made after thorough investigation into the reaction kinetics, adsorption of the reactants on mordenite surface, and their IR-spectra study.

Literature Cited

(1) Adams, C. E., Kimberlin, C. N., Shoemaker, D. P., *Proc. Intern. Congr. Catalysis, 3rd,* **1965,** 2, 1310.
(2) Benesi, H., U. S. Patent **3,190,939** (1965).
(3) Cvetanovic, R. I., Amenomija, Y., *Advan. Catalysis* **1967,** 17, 103.
(4) Eberly, P. E., *J. Phys. Chem.* **1963,** 67, 2404.
(5) Frilette, V. I., Rubin, M. K., *J. Catalysis* **1965,** 4, 310.
(6) Garanin, V. I., Kurkchi, U. M., Minachev, Kh. M., *Kinetics Catalysis* **1968,** 9, 1080.
(7) Keough, A. H., Sand, L. B., *J. Am. Chem. Soc.* **1961,** 83, 3536.
(8) Miale, I. N., Chen, N. Y., Weisz, P. B., *J. Catalysis* **1966,** 6, 278.
(9) Mills, G. A., Heinemann, H., Milliken, T. H., Oblad, A. G., *Ind. Eng. Chem.* **1953,** 45, 134.
(10) Minachev, Kh. M., Garanin, V. I., Isakov, Ya. I., *Usp. Khim.* **1966,** 35, 2151.
(11) Minachev, Kh. M., Garanin, V. I., Isakova, T. A., Kharlamov, V. V., *Izv. Acad. Nauk SSSR Ser. Khim.* **1969,** 481.
(12) Minachev, Kh. M., Garanin, V. I., Kharlamov, V. V., Isakova, T. A., Senderov, E. E., *Izv. Acad. Nauk SSSR Ser. Khim.* **1969,** 1737.
(13) Pigusova, L. I., Prokof'eva, E. N., Dubinin, M. M., Bursian, N. P., Chavandin, Yu. A., *Kinetics Catalysis* **1969,** 10, 315.
(14) Topchieva, K. V., Minachev, Kh. M., Pigusova, L. I., Romanovsky, B. V., Isakov, Ya. I., Garanin, V. I., Thoang, Ho Si, Vitukhina, A. S., *Preprints, World Petrol. Congr., 3rd,* p. 379.
(15) Venuto, P. B., Landis, P. S., *Advan. Catalysis* **1968,** 18, 259.
(16) Voorhies, A. J., Bryant, Ph. A., *A.I.Ch.E. J.* **1968,** 14, 852.

RECEIVED February 27, 1970.

77

Present and Future Directions in Zeolite Research

W. F. AREY, JR.

Esso Research Laboratories, Baton Rouge, La. 70821

Molecular sieve zeolites have become established as an area of scientific research and as commercial materials for use as sorbents and catalysts. Continuing studies on their synthesis, structure, and sorption properties will, undoubtedly, lead to broader application. In addition, crystalline zeolites offer one of the best vehicles for studying the fundamentals of heterogeneous catalysis. Several discoveries reported at this conference point toward new fields of investigation and potential commercial utility. These include phosphorus substitution into the silicon–aluminum framework, the structural modifications leading to "ultrastable" faujasite, and the catalytic properties of sodium mordenite.

The task of summarizing the major results of the conference and suggesting the future direction of molecular sieve zeolites is not an easy one. In particular, it is difficult to summarize the major results because there were so many noteworthy contributions. Obvious and substantial progress is being made in all the major areas—*i.e.*, synthesis, structure and mineralogy, sorption and diffusion, and catalysis.

Breck made the following comment in the introductory lecture: "There seems to be a direct relation between the level of scientific interest and the area of major application." I believe that this meeting lends support to that statement. The high degree of representation of petroleum-oriented research organizations is evident.

One of the major applications of crystalline zeolites is as catalytic materials for petroleum refining processes. A prime use is in catalytic cracking, which is the major conversion process in today's refineries. Another growing use is in hydrocracking, such as the Unicracking–JHC process, which gives the refiner added flexibility in oils which can be processed and products which can be made.

To date, crystalline zeolite catalysts have been most effective in catalyzing carbonium ion reactions such as catalytic cracking and hydrocracking. Other carbonium ion reactions such as alkylation and isomerization also are catalyzed by certain forms of zeolites. I expect to see these applications expand—provided suitable catalyst compositions are developed to allow economically viable processes. Although X- and Y-type faujasite can be used in catalytic cracking and hydrocracking, the Y-type is preferred for paraffin–olefin alkylation. Y-type faujasite is suitable for use in hydroisomerization catalysts, but synthetic mordenite is also a promising material.

On the other hand, zeolite-based catalysts have not developed any significant interest in other important processes such as catalytic reforming, hydrogenation of olefins and aromatics, hydrodesulfurization and hydrodenitrogenation. This lack of utility is attributable, in part, to the high cracking activity exhibited by active zeolite-based catalysts.

At present, the major interest lies in catalysts based on synthetic X- and Y-type faujasite and synthetic mordenite.

With this background, I believe that the following results were highlights of the meeting:

- The substitution of materials other than silicon or aluminum into the zeolite framework—particularly the phosphorus substitution reported by Flanigen and Grose. It was, however, disappointing that this was successful with only one large-pore material, namely, type L zeolite.
- The elucidation of the structures involved in producing ultrastable faujasite reported by Maher, Hunter, and Scherzer. Although the discovery of the ultrastable faujasite was reported at the 1967 London meeting, these subsequent studies (and work by G. T. Kerr) are pointing to an unexpected method of altering the crystalline framework.
- The high activity of sodium mordenite for catalyzing benzene hydrogenation as reported by Minachev *et al.* This result is so unique that additional investigation is warranted.

It seems to me that the above constitute discoveries which could lead to new and different catalytic materials.

The continuing studies on the mechanisms involved in zeolite synthesis are leading to a better understanding of the controlling factors in this complicated reaction system. But more remains to be done. This also holds the promise of leading to the synthesis of new zeolitic structures.

The systematic investigations of the structure and sorption properties of natural and synthetic zeolites are certainly central to the body of knowledge that will aid in the future research and utilization of zeolites in various possible ways.

Nomenclature is influenced by structure. It would be desirable to have a better system of nomenclature. However, the question of just how much "fine structure" needs to be described may be difficult to answer.

Many natural zeolites have a variable composition. It may be that the degree of Si–Al order–disorder in synthetic zeolites may vary with preparation procedures. However, it would be of interest to determine what effect the order–disorder has on adsorptive and/or catalytic properties.

Several years ago, it was felt that catalytic studies using crystalline zeolites would dramatically aid in unraveling some of the mysteries of catalysis. This was because the crystalline zeolites had a more uniform surface than amorphous materials and the structure could be well defined. Some of the luster of this hope appears to be fading slightly. It seems that the "active sites" have a spectrum of intensities rather than being of a uniform character. Nevertheless, crystalline zeolites still offer one of the best vehicles for studying the fundamentals of heterogeneous catalysis.

Catalysts based on molecular sieve zeolites will certainly continue to be of much interest in petroleum refining. Improvements can be expected as time goes on. Radically different zeolite structures offer the most chance of major improvements. Several authors have indicated the likelihood of obtaining radical, carbanion, and other noncarbonium ion reactions over zeolite catalysts. Pursuit of this possibility may prove fruitful.

Separations were one of the first uses of molecular sieve zeolites. Although some iso–normal hydrocarbon separations plants are operating, growth in this area has not been large. Separations based on inherent adsorptive behavior (rather than sieve action) are becoming of more interest. This is true not only of the separation of dimethylnaphthalenes (as reported by Hedge at this meeting) but also of xylene isomers. Separations of this type to produce streams of interest in chemicals manufacture could grow.

As mentioned by Breck in the introductory lecture, molecular sieve zeolites should find use in various pollution control applications. Both their sorbent and catalytic properties make them technically feasible as agents for various aspects of pollution control and abatement.

In summary, crystalline zeolites offer not only useful articles of commerce but also an interesting research area involving many different scientific disciplines. Undoubtedly, the future will reveal more about these unusual materials, perhaps even in ways not clearly foreseen at present.

RECEIVED November 23, 1970.

INDEX

D

E